Principles of Electrical Measurement and Measuring Instruments

V. Jagadeesh Kumar

Principles of Electrical Measurement and Measuring Instruments

Springer

V. Jagadeesh Kumar
Emeritus Professor, Department
of Electrical Engineering
Indian Institute of Technology Madras
Chennai, Tamil Nadu, India

ISBN 978-981-96-5474-1 ISBN 978-981-96-5475-8 (eBook)
https://doi.org/10.1007/978-981-96-5475-8

This Springer imprint is published by the registered company Springer Nature Singapore Pte Ltd.
The registered company address is: 152 Beach Road, #21-01/04 Gateway East, Singapore 189721, Singapore

Preface

This book was in the making for more than a decade and I am happy to see that it has come to fruition. For over four decades, I was teaching courses at the undergraduate and postgraduate levels in this and allied subjects at the Indian Institute of Technology Madras, India. Along with the theory part there will be a laboratory part wherein students will practice what they have learnt in the theory classes.

Over these four decades, the topics I covered kept changing. The parts I taught for nearly half a semester in the earlier period are now covered within a week. Nearly 50% of the topics I was teaching earlier become obsolete or replaced with modern ones. I have kept some of these obsolete topics in the book because though these instruments are no longer manufactured, we still use the ones that were installed earlier and will continue to use them for at least a few years.

I could keep abreast in this field with help and advice from my peers and colleagues. In this aspect, I am grateful to all my teachers and would like to mention my Gurus Prof. Panchapakesan Sankaran and Prof. Vempati Gopala Krishna Murti. All the electronics I learnt are from Prof. Radhakrishna Rao. I am also grateful to all my students who kept me abreast of all the modern discoveries.

The supporting staff at the Fundamental and Measurements Laboratory, IIT Madras, helped me in augmenting the laboratory part as the topics changed. Again, special mention to: Mr. Vasudevan, Mr. Kalyanam, Mr. Adikesavalu, Mr. Jones Selvaraj, Mr. Umaithanu Pillai, Ms. Rekha, and Mr. Manigandan.

I place on record my sincere thanks to nonagenarian Prof. V. G. K. Murti for taking the efforts to read through the manuscript and pointing to me the "grave errors and unpardonable mistakes" I had made. I express sincere gratitude to the Indian Institute of Technology, Madras, Chennai 600036, India, for providing me with an "Emeritus Professor" position that helped me to a very large extant in bringing this book writing task to fruition.

Last, but not least, I am indebted to my family as without their help I would not have completed this book.

Chennai, India Dr. V. Jagadeesh Kumar

Competing Interests The author has no competing interests to declare that are relevant to the content of this manuscript.

Contents

About the Author

Dr. V. Jagadeesh Kumar (Life Member, IEEE) received the B.E. degree in ECE from the College of Engineering, Guindy, in 1978, and the M.Tech. and Ph.D. degrees from the Indian Institute of Technology (IIT) Madras, in 1980 and 1986, respectively. He was with King's College London in 1988, the Asian Institute of Technology, Bangkok, in 1996, the Braunschweig University of Technology in 1999, and RWTH Aachen University in 1999, 2007, 2011, 2013, 2015, 2017 and 2019. He is currently Professor Emeritus of Electrical Engineering at IIT Madras. He has published 60 journal articles (mostly in IEEE Journals) and presented 120 papers in international conferences. He had taught the UG course on Principles of Measurements and the PG course on Measurements and Instrumentation for four decades at IIT Madras.

Chapter 1
Units and Standards

There are two possible outcomes: if the result confirms the hypothesis, then you've made a measurement. If the result is contrary to the hypothesis, then you've made a discovery.
—Enrico Fermi

1.1 The Art of Measurement

What do we mean when we say, "We have made a measurement"? To understand this, let us look at some examples.

Example 1: Task: Measure the number of oranges in a box. We simply count the oranges in the box and say there are 65 oranges.
Example 2: Task: Measure the height of a person. The result is, say, 167 centimeters.
Example 3: Task: Measure the weight of the apples in the basket. The result is, say, 505 grams.

Now we do understand that through the process called "measurement", we quantify the extent or amount of what is being measured and express the result in terms of a

V. Jagadeesh Kumar, *Principles of Electrical Measurement and Measuring Instruments*,
https://doi.org/10.1007/978-981-96-5475-8_1

Table 1.1 Units of measurement that were in vogue

Quantity	Country		
	Britain	Greek	South India
Length	Inch, Foot, Furlong, Mile	Dactylos, Pous, Pechys [cubit], Plethron, Stadion	Nool, Angulam, Muzham, Kaadham
Weight	Ounce, Pound, Ton	Obolus, Mina, Talent	Tola, Palam, Veesai
Volume	Pint, Quart, Gallon	Cotyla, Chous, Metretes	Aazhakku, Padi, Marakkal

number and a unit as follows:

65 Oranges
167 centimeters
505 grams

> *If it can't be expressed in figures, it is not science; it is opinion.*
> —Robert Heinlein

In the past, both the numbers and units were distinctly different in different parts of the world. Table 1.1 illustrates the units of measurement used for different quantities in different parts of the world.

The earliest unit for measurement is concerned with length. In most parts of the world, the length measurements were based on parts of the body. The Egyptian pechys [cubit] and the muzham used in the southern part of India are the same and indicate the length between the tip of the middle finger and the elbow of the arm of a person. Though the cubit is not in use, muzham is still used in southern India, especially in Tamil Nadu, for the sale of a string of knit flowers.

1.2 International System of Units

After years of debating to have a uniform system of units all over the globe, in 1960, several countries came together and agreed to have a single uniform system of units for measurement. This system is now called the "International System of Units" (Le Systeme international d'Unites) or simply SI units.

Today the SI system has been officially recognized and adopted by nearly all countries. Only a very few countries (Liberia, Marshall Islands, Micronesia, Myanmar, Palau, Samona, and notably the United States of America) have not yet legally adopted the SI system. Though the usage of SI in scientific and engineering exchanges around the globe is 100%, the amount of usage in different countries varies with each nation. For example, in India (where only SI unit is legal), it is quite prevalent in measuring the extant of land the unit "feet" and "square feet" are used, even in official documents. The salient landmarks on the timeline of the evolution of SI units are given below.

1.2.1 Brief History of SI Units

1799: Introduction of the metric system and first prototype meter and kilogram.
1832: Gauss advocates adoption of "metric" system, later Weber joins.
1874: Metric system abandoned, and the centimeter-gram-second (cgs) system comes into vogue due to politics.
1875: The meter, kilogram, second (MKS) system was proposed and accepted.
1901: Giogi links m, kg and s to other units (A, V, J, H, F).
1936: mkgsA system proposed (popularly called MKS system or metric system) and approved in 1946.
1954: Officially kelvin and candela were taken as base units.
1960: Officially the mkgsAKcd becomes International System of Units (SI units).
1971: "mole" is added as the seventh base SI unit.

1.2.2 The Base Units of SI

The SI system of units consists of seven base units and several derived units. The seven base SI units are listed in Table 1.2.

Table 1.2 Base units of SI system

Serial no.	Quantity	Unit	Abbreviation/Symbol
1	Length	meter	m
2	Mass	kilogram	kg
3	Time	second	s
4	Electric current	ampere	A
5	Temperature	kelvin	K
6	Amount of substance	mole	mol
7	Luminous intensity	candela	cd

Note The American spelling for the unit of length is "meter". This unit is spelt in the original "Le Systeme international d'Unites" document as "metre". Today both spellings are accepted as correct. In addition to the above seven units which have dimensions, two non-dimensional units (just numbers), representing plane and solid angles, and called auxiliary units, are also taken as part of the basic SI system of units as shown in Table 1.3.

Table 1.3 Auxiliary units in SI

Serial no.	Quantity	Unit	Symbol	In terms of SI base units
1	Plane angle	radian	rad	m/m
2	Solid angle	steradian	sr	m^2/m^2

In order to fix how long a meter is or how much a kilogram is or what constitutes the duration of a second … SI lists a set of definitions for the base and derived units. The definitions of the seven SI base units are given below with the corresponding abbreviation.

meter [m]

One meter (spelt also as metre), the basic unit of length, is defined as the distance that light travels in vacuum in 1/299792458th of a second.

kilogram [kg]

One kilogram, the basic unit of mass, is defined as the mass of an international prototype in the form of a platinum-iridium (90% platinum, 10% iridium) cylinder kept at Sevres in France.
Note: The kilogram is the only base unit that is still defined in terms of a material prototype unit. Efforts are on to replace the prototype platinum-iridium cylinder by a sphere made from single crystal silicon containing a certain exact number of molecules. The kilogram is also the only SI base unit with a prefix (kilo) already in place.

second [s]

One second, the basic unit of time, is defined as the time taken for 9192631770 periods of radiation (oscillation) corresponding to the transition between the two hyperfine levels of the ground state of the cesium 133 atom to occur.

ampere [A]

One ampere, the basic unit of electric current, is defined as that current which produces a force equal to $2\ 10^{-7}$ newton (kg m/s^2) per meter of length of the conductors when passing through two straight parallel conductors of infinite length of negligible circular cross section, placed 1 meter apart in vacuum.
The unit of ampere is named to honor the French scientist Andre Ampere (1775–1836).
Note: The somewhat impractical definition of the ampere is expected to be replaced by a new definition by improving the accuracy of the known numerical value of *e*, the fundamental unit of charge and linking the ampere to it.

kelvin [K]

One kelvin, the basic unit of temperature, is defined as 1/273.15th of the thermodynamic temperature of the triple point of water.

The unit of kelvin is named after the British scientist William Thomson who became the first Lord Kelvin (1824–1907).

candela [cd]

One candela, the basic unit of luminous intensity, is defined as the intensity in a given direction, of a light source that emits monochromatic radiation of frequency of 540 tera hertz (10^{12} Hz or THz) and has a radiant intensity in that direction of 1/683 watt per steradian.

mole [mol]

One mole, the basic unit of substance, is defined as the amount of substance that contains as many elementary units as there are atoms in 0.012 kg of carbon 12. When the unit "mole" is used, the elementary entities must be specified (may be atoms, molecules, ions, electrons, other particles, or specified groups of such particles).

1.2.3 Derived Units in SI

Apart from the base units, SI defines several derived units. Derived units are those which can be re-written (expressed) using a combination of SI base units. The derived units in SI are grouped into two distinct categories: (i) Derived units with special names (listed in Table 1.4) and (ii) Derived units without special names (listed in Table 1.5).

Apart from the above units, SI also permits continued use of some existing units. For example, the unit of degree Celsius (also known as centigrade), in use for the measurement of temperature, is accepted as an alternative unit for kelvin. 0 °C is 273.15 K. Table 1.6 lists other "non-SI units" permitted by the Comité International des Poids et Measures (CIPM) for everyday use.

The definitions of some of the prominent derived SI units are given next.

becquerel

One becquerel (pronounced as "be ka rel"), the unit that describes the quantum of radioactivity, defines the radioactivity when 1 nucleus decays per second in a radioactive material.
The unit "becquerel" is named to honor the French Physicist Henri Becquerel (1852–1908).

coulomb

One coulomb, the unit of electric charge, is the amount of electric charge transported in 1 second by a direct current (dc) of 1 ampere.
The unit "coulomb" is named to honor the great French Physicist Charles-Augustin de Coulomb (1736–1806).

Table 1.4 Derived units in SI system with special names

Quantity	Name of the unit	Symbol	In terms of	
			Other SI units	SI base units
Absorbed dose	gray	Gy	J/kg	$m^2\ s^{-2}$
Activity (radionuclide)	becquerel	Bq	–	s^{-1}
Capacitance	farad	F	C/V	$A^2\ s^4\ kg^{-1}\ m^{-2}$
Catalytic activity	katal	kat	–	$mol\ s^{-1}$
Celsius temperature	degree Celsius	°C	–	K
Dose equivalent	sievert	Sv	J/kg	$m^2\ s^{-2}$
Electric charge	coulomb	C	–	A s
Electric conductance	siemens	S	A/V	$A^2\ s^3\ m^{-2}\ kg^{-1}$
Electric Resistance	ohm	Ω	V/A	$kg\ m^2\ s^{-3}\ A^{-2}$
Electromotive force	volt	V	W/A	$kg\ m^2\ s^{-3}\ A^{-1}$
Energy	joule	J	N m	$kg\ m^2\ s^{-2}$
Force	newton	N	–	$kg\ m\ s^{-2}$
Frequency	hertz	Hz	–	s^{-1}
Illuminance	lux	lx	lm/m^2	$cd\ sr\ m^{-2}$
Inductance	henry	H	Wb/A	$kg\ m^2\ s^{-2}\ A^{-2}$
Luminous flux	lumen	lm	Cd sr	$m^2\ m^{-2}\ cd$
Magnetic flux	weber	Wb	V s	$kg\ m^2\ s^{-2}\ A^{-1}$
Magnetic flux density	tesla	T	Wb/m^2	$kg\ s^{-2}\ A^{-1}$
Power	watt	W	J/s	$kg\ m^2\ s^{-3}$
Pressure	pascal	Pa	N/m^2	$kg\ m^{-1}\ s^{-2}$

Note When we write the temperature in Celsius, the symbol degree is prefixed and should be read as such. When kelvin is used to indicate the temperature, one should neither use the degree symbol nor include the "degree" when referring to the measurement.
Example Temperature of triple point of water at sea level is 273.15 K (should be read as two hundred and seventy-three point one five kelvin). Boiling point of water at sea level is 100 °C (should be read as hundred degree Celsius).

farad [F]

One farad, the unit of the capacitance indicating an electrical system's capacity to store charge (electricity) in electrostatic field(s), stores 1 coulomb of charge, and develops a potential difference of 1 volt. One farad is a rather large unit and the units microfarad $(\mu F = 10^{-6}F)$, nano farad $(nF = 10^{-9}F)$, and picofarad $(pF = 10^{-12}F)$ are more commonly used in practice.
The unit "farad" is named to honor the great English scientist Michael Faraday (1791–1867).

gray

Table 1.5 Derived units in SI system without special names

Quantity	Name	XXX
Area	square meter	m^2
Volume	cubic meter	m^3
Speed, velocity	meter per second	m/s
Acceleration	meter per second squared	m/s^2
Wave number	reciprocal meter	m^{-1}
Mass density	kilogram per cubic meter	kg/m^3
Specific volume	cubic meter per kilogram	m^3/kg
Current density	ampere per square meter	A/m^2
Magnetic field strength	ampere per meter	A/m
Luminance	candela per square meter	cd/m^2

Table 1.6 Non-SI units recognized by CPIM

Quantity	Unit
Angle	degree (°), minute (') and second ("){π rad = 180°}
Area	hectare (ha) = 10^4 m^2
Energy	electron volt (eV)$\{1\text{eV} = 1.6021\ 10^{-19}\text{J} = 160.21\ \text{ZJ}\}$
Mass	tonne (t)$\{1\text{t} = 10^3\text{kg}\}$
Pressure, stress	bar {1 bar = 100 kPa}
Time	minute (m), hour (h) and day (d)
Volume	litre or liter (l or L) = 10^{-3} m^3

One gray, the unit for the measurement of ionizing radiation, is the amount of radiation absorbed when 1 joule of energy is taken in by a mass (matter) of 1 kilogram.
The unit "gray" is named to honor Louis Harold Gray who proposed a similar concept (1905–1965).

henry

One henry, the unit of inductance (self or mutual), is the inductance that produces 1 weber of magnetic flux for a current of 1 ampere.
The unit "henry" is named to honor the American scientist Joseph Henry (1797–1867).

hertz [Hz]

One hertz, the unit that describes the frequency of a periodic phenomenon, indicates that one complete cycle of the phenomenon occurs every second.
The unit "hertz" is named to honor the German scientist Heinrich Rudolph Hertz (1857–1894).

joule [J]

One joule, the unit for energy, is the amount of work done when an applied force of 1 newton moves through a distance of 1 meter in the direction of the force.
The unit "joule" is named to honor the English physicist James Prescott Joule (1818–1889).

katal

One katal, the unit for catalytic activity (quantifies catalytic activity) of enzymes and other catalysts, is the amount of enzyme or catalyst required for breaking 1 mole of the substance being broken in 1 second.

lumen

One lumen, the unit for luminous flux, is the luminous flux emitted (uniformly) by a light source of intensity 1 candela across a solid angle of 1 steradian.

newton [N]

One newton, the unit of force, is the force required to give a mass of 1 kilogram an acceleration of 1 meter per second per second.
The unit "newton" is named to honor the great English scientist Sir Isaac Newton (1642–1727).

ohm [Ω]

One ohm, the unit of resistance of an electrical conductor, develops 1 volt across it when 1 ampere flows through it. The ohm has a special symbol Ω (the capital Greek letter "omega").
The unit "ohm" is named to honor the German physicist Georg Simon Ohm (1789–1854).

pascal [Pa]

One pascal, the unit of pressure, is the pressure generated by a force of 1 newton acting on an area of 1 square meter.
The unit "pascal" is named to honor the French mathematician and physicist Blaise Pascal (1623–1662).

siemens

One siemens, the unit of electric conductance, is that conductance which permits a current of 1 ampere to flow through it when 1 volt is applied.
The unit "siemens" is named to honor the German engineer Ernst Werner von Siemens (1816–1892).

sievert

One sievert, the unit of ionizing radiation dose, is 1 joule of ionizing radiation absorbed by 1 kilogram of biological mass.

Though gray and sievert have the same dimension (J/kg) they are different. Sievert points to the biological effects of ionizing radiation and gray records the absorbed dose of radiation, a physical quantity.
The unit "sievert" is named to honor the Swedish medical physicist Rolf Maximilian Sievert (1896–1966).

tesla

One tesla, the unit of magnetic flux density, is the flux density that occurs when 1 weber of flux is uniformly distributed over an area on 1 square meter.
The unit "tesla" is named to honor the Serbian-American physicist and electrical engineer Nikola Tesla (1856–1943).

volt [V]

One volt, the unit of electric potential, is the difference of potential between two points of an electrical conductor when a DC current of 1 ampere flowing between those points dissipates a power of 1 watt.
The unit "volt" is named to honor the Italian physicist Alessandro Giuseppe Anastasio Volta (1745–1827).

watt [W]

One watt, the unit of power, is the power when 1 joule of energy is consumed or delivered per second.
The unit "watt" is named to honor the Scottish engineer James Watt (1736–1819).

weber

One weber, the unit of magnetic flux, is the flux that produces 1 volt if the flux decays uniformly to zero in 1 second.
The unit "weber" is named to honor the German physicist Wilhelm Eduard Weber (1804–1891).

1.2.4 Standard Prefixes for SI Units

Some SI units are either too small or too large for practical use. Hence, their use tends to be cumbersome in some applications. For example, to represent one thousandth of a meter, we need to write 0.001 m. Similarly, a hundred thousand meters is to be written as 100 000 m. It would be nice if we are able to represent small and large values in a concise manner. In order to facilitate this, these values are expressed in terms of powers of ten and a prefix is added to the unit to indicate that a multiplication factor has been added. The well-known prefix kilo indicates that the number presented must be multiplied by one thousand. For example, one kilometer is one thousand meters. The full range of SI approved prefixes with their symbols and abbreviations along with other popular names in practical use (if any) are given in Table 1.7.

Table 1.7 Standard prefixes in SI

Multiplicative factor	Prefix	Symbol	Value	Popular name of the prefix
10^{24}	yotta	Y	1 000 000 000 000 000 000 000 000	
10^{21}	zetta	Z	1 000 000 000 000 000 000 000	
10^{18}	exa	E	1 000 000 000 000 000 000	
10^{15}	peta	P	1 000 000 000 000 000	
10^{12}	tera	T	1 000 000 000 000	trillion
10^{9}	giga	G	1 000 000 000	billion
10^{6}	mega	M	1 000 000	million
10^{3}	kilo	k	1 000	thousand
10^{2}	hecto	h	100	hundred
10^{1}	deca	da	10	ten
10^{0}			1	one
10^{-1}	deci	d	0.1	one tenth
10^{-2}	centi	c	0.01	one hundredth
10^{-3}	milli	m	0.001	one thousandth
10^{-6}	micro	μ	0.000 001	one millionth
10^{-9}	nano	n	0.000 000 001	one billionth
10^{-12}	pico	p	0.000 000 000 001	one trillionth
10^{-15}	femto	f	0.000 000 000 000 001	
10^{-18}	atto	a	0.000 000 000 000 000 001	
10^{-21}	zepto	z	0.000 000 000 000 000 000 001	
10^{-24}	yocto	y	0.000 000 000 000 000 000 000 001	

Note The symbol used for micro is the Greek letter μ and is pronounced as "mu".

Some of the multiplication factors listed in Table 1.7 have special names that were in vogue even before SI units came into existence. For example, the multiplication factor 1 000 000 (Mega in SI) is also popularly known as one million. Historically, in olden days British usage (Britain and Australia), the term "billion" was used to refer to a million million [$(10^6)^2 = 10^{12}$, now known as the "billion" in the long scale]. Similarly, in the old British and Australian usage, a "trillion" [the "trillion" in the long scale] referred to a million million million $(10^6)^3$]. Current British English, however, has generally adopted the American equivalents, which are as specified in Table 1.7.

Table 1.8 Preferred SI multiples and submultiples in engineering

Sl. No	Quantity	SI unit	Preferred multiples and submultiples
1	Length	m	nm, μm, mm, km
2	Area	m^2	mm^2, km^2
3	Volume	m^3	mm^3
4	Time	s	ps, ns, μs, ms
5	Velocity, speed	m/s	km/h *or* kmh^{-1}
6	Frequency	Hz	μHz, mHz, kHz, MHz, GHz, THz
7	Power	W	pW, nW, μW, mW, kW, MW, GW, TW
8	Current	A	fA, pA, nA, μA, mA, kA
9	Charge	C	pC, nC, μC, kC
10	Electric potential difference	V	nV, μV, mV, kV, MV
11	Capacitance	F	pF, nF, μF, mF
12	Resistance	Ω	μΩ, mΩ, kΩ, MΩ, GΩ
13	Conductance	S	μS, kS, mS
14	Inductance	H	pH, nH, μH, mH

Using the prefixes, the value 0.001 m can now be written as 1 10^{-3} m or more concisely 1 mm (read as one millimeter). Similarly, 1 000 000 Ω can be written as 1 MΩ (read as one mega ohm). The unit of pressure, pascal, is not a big unit even in terms of ordinary use. So pressure is generally expressed in terms of 1 000 Pa (kPa) or 1 000 000 Pa (MPa). The atmospheric pressure at sea level is 101 325 Pa and we denote it as 101.325 kPa.

In engineering, the usage of only a few selected SI prefixes is preferred to enable additional standardization. For example, we never write 0.010 A as 1.0 cA (one centiampere) Instead, we prefer to write it as 10 mA. The units with the preferred prefixes are listed in Table 1.8.

1.2.5 Rules for Writing the Value of a Measurement

1.2.5.1 Rules for Writing the Units

Now we know that when we record the result of a measurement, we must write the value in terms of a number and a unit. SI stipulates various rules for writing the number and the SI unit. These rules are made to avoid ambiguities in the interpretation of the result and to convey the essential information about its limitations to a third person who has no knowledge of the measurement process, or the details of instrument used. They are listed below.

- The unit symbol must be placed after the numerical value with a single space between the numerical value and the unit symbol. This rule is also applicable for the symbol "%" (percentage) and ºC. The only exception to this rule is when writing the value of an angle θ in degrees, minutes, and seconds.

  Example: $\theta = 33°22'44''$ (*Note*: No space is to be given between the number and the unit symbols). Some more *examples* are tabulated below.

Correct representation	Wrong representation
Room temperature = 30.1 °C	30.1°C; 30.1 ° C
22 kΩ resistor	22kΩ resistor; 22 k Ω resistor
100 mL	100mL; 100 m L; 100×10^{-3} L
33.5 %	33.5%

- Only one unit must be written for indicating the result.

For example, 1.234 m is correct and should not be written as 1 m, 23 cm and 4 mm. The only exception to this rule is when an angle is expressed in degrees, minutes, and seconds.

Example $\theta = 12°34'56''$ (to be read as Twelve degrees, thirty four minutes, and fifty six seconds).

- The symbol for the unit must always be written in normal typeface (SI does not permit italics or bold face for writing the symbol for units).

  Proper The total weight of the apples in the crate is 123.4 kg.
  Improper The total weight of the apples in the crate is *123.4 kg*.
  The total weight of the apples in the crate is **123.4 kg**.

- The unit cannot have any annex or appendix.

Examples V_{1max} = 123 V is correct and should not be written as V_1 = 123 $\mathrm{V_{max}}$.
Resistance per tray is 100 Ω and not Resistance is 100 Ω/tray.

- Though it is not part of SI rules, it is better to stick to writing variables in italics as is done above. Over the years, symbols to represent different variables have also become sort of standardized. For example, the symbols θ or ϕ are invariably used to represent angles (sin θ, cos ϕ, and so on). Table 1.9 lists standard symbols that are normally reserved for certain types of quantities.

Table 1.9 Standard symbols in vogue

No.	Quantity	Quantity symbol	Unit symbol
1	Admittance	Y	S
2	Angle (plane angle)	$\theta, \phi, \alpha, \beta, \gamma...$	rad, also ..°, ..', .."
3	Angular frequency	$\omega = 2\pi f$	rad/s
4	Angular velocity	ω	rad/s
5	Apparent power	S	VA
6	Area, surface area	A, S	m^2
7	Average value of a current	$\bar{I}, \bar{\imath}, I_{av}$ *Note 1*	A
8	Capacitance	C	F
9	Conductance	G	S
10	Conductivity	σ, γ	S/m
11	Coupling coefficient	k, k_{ab}	–
12	Current density	J	A/m^2
13	Current phasor	$I, I, \bar{I}, \underline{I}$ *Note 1*	A
14	DC component of the emitter current containing AC and DC components	I_E	A
15	Density	ρ	kg/m^3
16	Dissipation factor (= tan δ)	d	–
17	Effective value of the AC component of the emitter current	I_e	A
18	Efficiency	η	Ratio or percent
19	Electric charge	Q	C (also A s)
20	Electric current	I	A
21	Electric dipole moment	p	C m (also A s m)
22	Electric field strength	E	V/m
23	Electric flux	ψ	C
24	Electric flux density, displacement	D	C/m^2
25	Electric polarization	P	C/m^2
26	Electric potential	V, φ	V
27	Energy	E, W	J, also kW h, eV
28	Force	F	N
29	Frequency	f	Hz
30	Impedance	Z	Ω
31	Incremental value of a current	ΔI	A
32	Instantaneous value of the AC component of the emitter current.	i_e	A
33	Instantaneous value of a varying current	i	A
34	Instantaneous value of the emitter current carrying AC and DC	i_E	A
35	Length	L	m
36	Linear speed, velocity	v, u	m/s
37	Loss angle	δ	rad
38	Magnetic field strength	H	A/m

(continued)

Table 1.9 (continued)

No.	Quantity	Quantity symbol	Unit symbol
39	Magnetic flux	Φ	Wb
40	Magnetic flux density (induction)	B	T
41	Magnetic moment	m	A m^2
42	Magnetic potential difference	u, U	A
43	Magnetic vector potential	A	Wb/m
44	Magnetomotive force (mmf)	F, F	A (also AT)
45	Mass	m	kg
46	Mass moment of inertia	J	kg m^2
47	Mutual inductance	M, L_{ab}	H
48	Peak value of a current	$\hat{I}$, I_m, I_p	A
49	Permeability	μ (μ_0 *for free space)*	H/m
50	Permeance = (Reluctance $)^{-1}$	Λ	H
51	Permittivity (Capacitivity)	ε (ε_0 for free space)	F/m
52	Phase difference	Φ, φ, θ	rad (also in °)
53	Potential difference, voltage	V, U	V
54	Power factor	λ, PF	–
55	Power, active power	P	W
56	Poynting vector	S, P, $\vec{P}$	W/m^2
57	Pressure	p	Pa
58	Quality factor	Q	–
59	R.M.S. value of a current	I, I_{rms}	A
60.	Reactance	X	Ω
61	Reactive power	Q	VAr also var
62	Relative permeability	μ_r	Number, unitless
63	Relative permittivity	ε_r	Number, unitless
64	Reluctance	R, R_m, $\Re$	H^{-1}
65	Resistance	R	Ω
66	Resistivity	ρ	Ω m
67	Rotational speed	n	s^{-1}
68	Self-inductance	L, L_{aa}	H
69	Solid angle	Ω	sr
70	Surface density of charge	σ	C/m^2
71	Time	t	s
72	Time constant	τ	s
73	Time of one cycle, period	T	s
74	Torque	T	N m
75	Velocity of electromagnetic waves	c	m/s
76	Volume	V, v	m^3
77	Volume density of charge	ρ	C/m^3

(continued)

Table 1.9 (continued)

No.	Quantity	Quantity symbol	Unit symbol
78	Wave length	Λ	m

Notes

- If the symbol $\bar{I}$ is used for the phasor of a current, then the same symbol should not be used to denote the average value of a current. IEEE standards committee recommends the use of $\underline{I}$ for a phasor and $\bar{I}$ for indicating the average value of a current. This rule is applicable for voltages as well.
- It is desirable that the letter symbols for a particular quantity and its associated unit are different.

For example, if we denote a potential on a node in a circuit as V, we may have to write its value as $V = 100$ V (the value for V is assumed to be 100 V) in print and, worse still, as V = 100 V in handwritten form, where italics are not used. To circumvent this awkwardness, we generally associate a suitable subscript with the symbol. For example, a particular voltage in a system can be represented as $V_P, V_1, V_2, V_a, V_b, V_{ab}\ldots$

In such a case, when we write $V_P = 100$ V, the meaning becomes very clear. The use of the symbol U for potential difference as a quantity is a better practice. In fact, Indian Standard specification IS: 3722 and Continental standards recommend the use of the letter symbol U for voltage, but in practice most continue to use the traditional symbol V for indicating the electric potential difference as is common practice in the U.K. and the U.S.A.

- A unit in symbolic form cannot stand alone without a number symbol. "There are 10^{12} μm in 1 km" is correct. "There are trillion μm in 1 km" is wrong and should be written as "There are trillion micrometers in a kilometer".

"Volume is expressed in cubic meters" is correct and "volume expressed in m^3" is wrong.

- Units must be written contiguously and cannot be split.

Correct way of writing

The concentration of oxygen is 5 g/L
Sensitivity of O_2 molecules is 5000/m^3
The ion emission rate is 10^{10}/s

Wrong way of writing

The concentration is 5 g of oxygen/L
Sensitivity is 5.0 10^3 O_2 molecules/m^3
The emission rate is 10^{10} ions/s

- Any unit may take only one prefix. For example, "micro micro meter" (μμm) is incorrect and should be written as "picometer" (pm).
- Most prefixes which make a unit bigger are written in capital letters (mega M, giga G, tera T, etc.), but when a prefix converts the unit to a smaller value then the lower case (m n p, etc.) is used. Exceptions to the former are the kilo (maybe to avoid confusion with kelvin K), hecto (may be to avoid confusion with henry H), and deca.
- From Table 1.4, it is seen that many units are eponymous, that is, they are named after persons who had carried out pioneering work in the field in which the unit is used. When named in full, such a unit must be written with all letters in lower case (newton, volt, pascal, etc.), but written with a capital letter (N V Pa, etc.) when abbreviated. All other generic units (kilogram, meter, and second) must be written with lower case letters when spelt in full as well as when abbreviated (kg, m, and s). An exception to this rule is the liter which can be abbreviated with a capital letter L. May be due to the fact that if liter is abbreviated as a lower case "l" it could be mistaken for a "1" (one).
- Units written in abbreviated form should never be pluralized with an "s" appended at the end. For example, "m" can indicate either "meter" or "meters". "Kilograms" should be abbreviated as "kg" and not "kgs" because "kgs" would represent "kilogram second".
- Other than permitted units and prefixes, other forms of abbreviations such as ppm (parts per million) or ppt (parts per thousand) are not recognized in SI.
- An abbreviation (such as K, A, J, N, g, Pa ⋯, etc.) is never followed by a full-stop unless it is written at the end of a sentence.
- When writing multiple measurements, units must accompany each number as indicated below.

Correct representation

The dimensions (length by width) chosen were 1.2 m by 3 m
The temperature varied in the range 273 K—330 K
The voltage was varied in the 0 V—5 V range.
100 m ±1 m OR (100±1) m

Wrong representation

The dimensions (length by width) chosen were 1.2 x 3 m
The temperature varied in the range 273–330 K
The voltage was varied in the 0–5V range.
100 ± 1 m *or* 100 m ±1

1.2.5.2 Rules for Writing the Numbers

We have so far discussed the rules to be observed for indicating the units in the result of a measurement. It is common practice to indicate the numerical part of the result with upright numerals (not italics). Additionally, SI stipulates the following rules for writing the numeral part in a standardized fashion.

- The SI preferred way of showing a decimal fraction is to use a comma (123,456) to separate the whole number (123) from its fractional part (456/1000). The practice of using a "decimal point", as is common in many countries, is acceptable in SI provided that the point is placed only on the bottom edge of the line in which the numbers are written.

Example 123.456. The point should never be placed in the middle as in "123·456".

- When writing large numbers, the numerals may be divided into groups of 3 separated by spaces (or half-spaces) so as to make the numbers easier to read. The numbers should never be separated by commas.
- As far as possible, prefixes must be used. For example, writing the value as 1.23 10^{10} Hz should be avoided. Instead, the value should be written as 12.3 GHz. Similarly, 0.001 234 g should be written as 1.234 mg.
- Like the units, the prefixes also cannot be used alone. For example, it is correct to write: "*The change in the refractive index was 1.2 10^{-6}*". It is wrong to write: "*The change in the refractive index is 1.2 μ*".
- Only the significant digits must be written, to clearly indicate the reliable extent of the result obtained in the measurement.

1.2.5.3 Significant Digits

An instrument needs to be employed in making any measurement. The smallest value that can be unambiguously read with an instrument is called the *resolution* of that instrument. The number part of the result of a measurement, along with the proper unit, should clearly reflect this fact (that is the resolution of the measurement).

Let us look at some *examples:*

- Figure 1.1 shows an ammeter wherein a pointer moves on a scale of 150 divisions. Obviously, the minimum value that can be unambiguously read is 1 division out of 150 and hence the resolution is 1/150 (0.0066...) of the full scale. If we take this scale to represent a 150 mA ammeter, then the resolution is 1/150th of 150 mA and hence is 1 mA. The reading from Fig. 1.1 can be then written as $10\underline{0}$ mA. The underscore here indicates the significant digit; also called the least count. When we write the result as $10\underline{0}$ mA, it shows (to others who had not witnessed the reading being taken) that the reading obtained is closer to $10\underline{0}$ mA than to either 99 mA or 101 mA (±1 variation on the significant digit). It is customary that the underlining is done on the lowest significant digit instead of on all the significant digits.

- Figure 1.2 shows another scenario. Here we know for sure that the reading lies between 89 mA and 90 mA and we may write the reading as $8\underline{9}.5$ mA.

  Note: The underscore is still on the significant digit. Hence a reading should always portray the true resolution of the instrument employed for obtaining that measurement.

- Figure 1.3 shows yet another measurement, this time a resistance is measured with an ohmmeter. The correct way to represent the result of this measurement is:

$$1.\underline{2}\ \mathrm{k}\Omega.$$

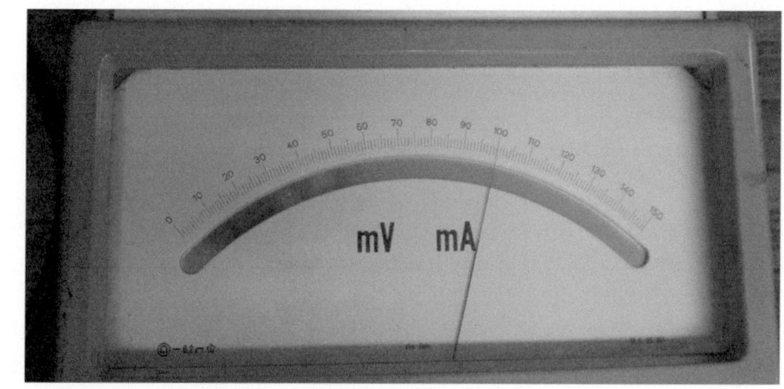

Fig. 1.1 Reading with a PMMC meter (courtesy Siemens, Germany)

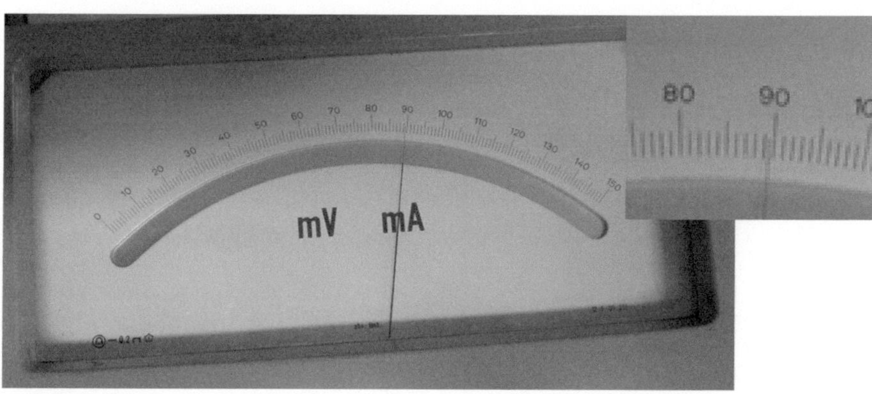

Fig. 1.2 Reading with a PMMC meter (courtesy Siemens, Germany). Inset: Enlarged view

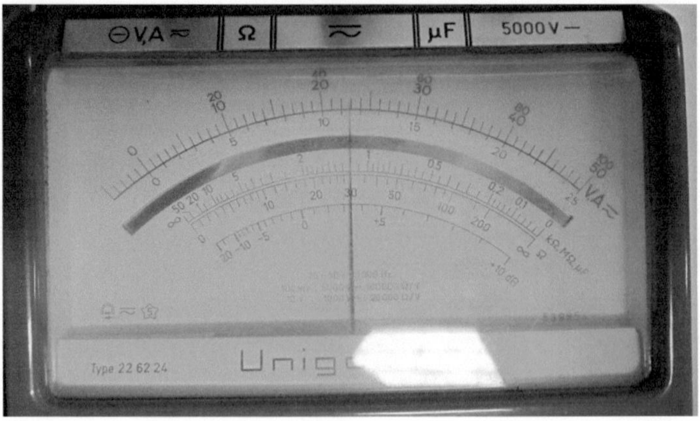

Fig. 1.3 Measurement of a resistance using an analog multimeter

Though 1.2 kΩ and 1200 Ω are mathematically equivalent they are not the same when results of measurement are indicated. The resolution in the measurement as portrayed in Fig. 1.3 is 0.1 kΩ or 100 Ω. If the result is indicated as 1200 Ω, then it implies that the measurement is made with a resolution of 1 Ω.

Hence when results of a measurement are indicated, $1.\underline{2}\ \text{k}\Omega \neq 120\underline{0}\ \Omega$!

- The underlining of the lowest significant digit is often omitted with the understanding that the resolution is the value of one unit of the smallest digital position displayed in the result.

  Thus, the resolution is 1 V if the reading is indicated as 90 V. Accordingly, the resolution is 0.1 V if the reading is written as 90.0 V and the resolution is 0.01 V when the reading is indicated as 90.00 V and so on. Thus, $90 \neq 90.0 \neq 90.00$ when you are indicating the result arising out of a measurement.

1.2.5.4 Propagation of Significant Digit

- In many instances, we may use the results of several measurements to compute the value of a required parameter. It may turn out that the instruments utilized for these measurements have different resolution levels and thus the measurements have significant digits in different decimal positions. In such a case, the end result must be written to reflect the influence of that measurement which has the lowest number of significant digits.

For example, we may measure the voltage V_R across and current I_R through a resistor R and compute its resistance as $R = V_R/I_R$.

Say, in a particular measurement, the value of V_R is $100.0\underline{0}$ V (number of significant digits is 5) and the value of I_R is $1\underline{0}$ mA (number of significant digits is 2).

Then $R = V_R/I_R = 100.00/1\underline{0} = 1\underline{0}\ \Omega$.

Here the result is expressed correctly with the smaller number of significant digits (2 digits) of the two measurements.

Now if the current is measured as 10.0 mA (three significant digits) then R will be

$R = V_R/I_R = 100.0\underline{0}/10.\underline{0} = 10.\underline{0}\ \Omega$ (number of significant digits is 3)

Rounding off of significant digit

When we compute a result using several measured values, its numerical value may contain a greater number of digits than the least number of significant digits. For example, in a DC circuit, if V_{dc} is measured as 1.251 V and I_{dc} is measured as 10 mA then power P_{dc} is computed as

$$P_{dc} = V_{dc} I_{dc} = 12.51 \text{ mW}.$$

However, this is not the proper way of expressing the computed value as the result can have only two significant digits. We can write the result as

$$P_{dc} = V_{dc} I_{dc} = 1\underline{2}.51 \text{ mW}.$$

In the above case, we had shown the position of the significant digit by an underscore. SI also allows that we can drop the excess digits. The dropping of excess digits is governed by the "rounding off rule" of SI. If the excess digits to be dropped are > 5000…, then a one is added to the lowest significant digit and the excess digits are dropped. In the above example 51(digits to be dropped) > 50, hence the result is indicated as

$$P_{dc} = 1\underline{3} \text{ mW}.$$

Here the underscore indicates the position of the significant digit. On the other hand, if the digits to be dropped are say, < 5000.., then the digits are dropped and the lowest significant digit is maintained as it was. If for example the voltage is 1.24499 V and current is 10.0 mA in the above DC circuit then power is

$$P_{dc}(\text{as computed}) = 12.\underline{4}499 \text{ mW};$$

$$\text{Value of digits to be dropped} = 499 < 500. \text{ Hence after rounding off,}$$

$$P_{dc} = 12.4 \text{ mW}.$$

It may so happen that the digits to be dropped are exactly 5000. Then a one is added to the significant digit if it is odd and retained as it is if it is even. For *example,* if the computed result is 12.$\underline{4}$500 then the rounded off value is 12.4. On the other

Table 1.10 Examples of "rounding off"

Number before rounding-off	Fineness of rounding-off	Number after rounding-off	Significant figures
39.2949	0.01	34.29	4
34.187	0.1	34.2	3
0.3465	0.001	0.346	3
0.3435	0.001	0.344	3
2210.4	10	2.21×10^3	3
2210.4	1	2.210×10^3	4
0.0496	0.1	0	–
0.0496	0.01	0.05	1
0.0496	0.001	0.050	2

hand, if the number to be rounded off is 6$\underline{3}$.500 then the rounded off value is 64. Table 1.10 lists some more *examples*.

1.2.6 Quick Recapture of Do's and Don'ts for Writing SI Units

Writing a unit in a sentence: When the unit name is written *in full*, use the lower case letter as its beginning letter, even if it is a unit named after a person, such as ampere, joule, kelvin, watt, ohm, etc. Similarly, use the lowercase letter at the start of all prefixes written in full, such as mega, micro, etc.

Examples A few newtons, several amperes, 1.5 megavolts, 2 milliohms.
However, if a sentence begins with the unit's name, its first letter should be in capitals as per normal practice.

Example Ampere is the name of SI unit of current.
Symbols of units and prefixes: In writing the symbols, use the capital letter if the unit has been named after a person.

Examples 2 N, 50 Hz, 2 A.
For units except those named after persons, and of prefixes exa (E), peta (P), tetra (T), giga (G), and mega (M), use lowercase letters.

Examples 2 s, 5 ms, 2 km, 100 MVAr.
Plural number: Symbols of units are identical both in singular and plural. For example, write 5 V to indicate 5 volts and do not write it as 5 Vs.

Spacing before the unit symbol: Allow a space between the number and the unit symbol.

Examples 2.1 mA not 2.1mA, 4 kΩ not 4kΩ.

Units with prefixes: There should be space between the prefix and the unit symbol.

Example 3.5 mA and not 3.5mA.

But a space must be provided between the individual symbols of a compound unit symbol if they are not separated by a solidus (/).

Example Write N s/m^2 for newton second per metre squared, and not Ns/m^2.

Periods: Periods or full stops should not be used at the end of symbols except when required for normal punctuation, e.g., at the end of a sentence.

Examples A voltage of 10.1 V appears across the resistor R_1.

The voltage across the resistance R_1 is 10.1 V.

Number of prefixes for a unit: Do not use compound prefixes by juxtaposition of several SI prefixes.

Examples use pF not μμF, use nm not mμm.

Powers of units: When a multiple of a particular unit is raised to a power, the power applies to the whole number, and not to the particular unit alone.

Examples The symbol kV^2 means (kV)^2 or 10^6 V^2 and not k(V^2) or 10^3 V^2.

The symbol mm^2 means 10^{-6} m^2 and not 10^{-3} m^2.

Solidus or negative index: Do not apply more than one solidus (/) in any combination.

Examples Use 5 m/s^2 and not 5 m/s/s.

Parentheses or negative powers may be employed as alternatives.

Examples 5 m/s^2 or 5 m s^{-2}, 50 V/s or 50 V s^{-1}, 2 J/(m s K) or 2 J m^{-1} s^{-1} K^{-1}.

Large numbers: In scientific literature, large numbers are written in groups of three with a space in between, as in 101 325 N/m^2, and not with a comma in place of space as in 101,325 N/m^2. The grouping should be done on both sides from the decimal, as in

23 456.789 01

But a sequence of four figures is permitted under certain circumstances, such as in the case of a date (3 October 1933), designation of a standard (IS: 3722–1983), etc. The use of punctuation marks like commas for breaking up numerals is not recommended by IS: 3722–1983.

Bankers and accountants may have a different view on this business of leaving gaps in numerals. In any event, this rule is eminently forgettable when you make out your next cheque for twenty two thousand rupees!

Numerical values: Appropriate units and multiplying factors should be used to adjust the presented numerical data to a value between 0.1 and 1000. This minimizes risks of error through misreading.

Examples

(a) $R = 101.451$ kΩ instead of $R = 101\ 451$ Ω
(b) $L = 1.234$ mH instead of $L = 0.001\ 234$H

While computing the final result using measured values, one should not be carried away with the large number of digits that the calculator or computer throws up. Care must be taken to preserve and show the significant digits. For *example*, the voltage V_R across a resistor is measured as 48.5 V (number of significant digits is 3) and the current through that resistor, I_R, is measured as 2.1 A (2 significant figures). If you substitute these values in your calculator, you may get a result $R = V_R/I_R =$ 23.0952381 Ω. Do not give the result displayed by the calculator as the final result but round it off as $R = 23$ Ω. The value of current itself is in doubt in the second digit and therefore all the digits beyond the second in the computed value of R do not have any validity or significance. Do not let your calculator lure you into claiming higher levels of significance than is warranted by the initial data.

1.3 Standards

The word "standards" has two different connotations when referred to, in the fields of measurement and testing. Physical objects that provide "*standard devices*" for making proper measurements as per SI specifications correspond to the "standards" in the field of measurement. For example, a precise 1.000000 Ω, four terminal standard resistance corresponds to this category. On the other hand, "standards" can also refer to standardized methods and procedures for tests to be conducted for determining the characteristics of instruments or several other devices. Examples are:

(i) International Electro Technical Committee Standards—acronym IEC standard (visit http://www.iec.ch/),
(ii) The Bureau of Indian Standards, acronym IS standard (visit http://www.bis.gov.in/)
(iii) Deutsche Industrial Normen, acronym DIN standard (visit https://www.din.de/en)
(iv) American National Standards Institute, acronym ANSI standard (visit https://www.ansi.org/)

Apart from these "standards" organizations, we also have the "International organization for standardization (ISO)" that quantitatively and qualitatively assess the practices of organizations. If an organization conforms to a particular set of rules then they are certified to be as per, say ISO9001:2015 standard.

1.3.1 *The Physical Standards*

A physical standard, as the name indicates, is a physical device that provides an exact value of a quantity in terms of the pertinent SI defined unit. For example, the "standard cell" that was proposed by Edward Weston in 1893 provided a constant voltage of 1.018638 V across its terminals and was in use for a long time as the *voltage standard* for measurement and comparison. Figure 1.4 illustrates a typical "Weston standard cell". Today, the 1.000 000 000 0 V, Josephson voltage standard (number of significant digits 11 and resolution 1 in 10^{-10}) is employed as the standard for voltage measurement and comparison. Apart from the "voltage standard", other physical devices such as "standard resistance", "standard capacitance", "standard weights", and standard gage blocks" are also physical entities (reference devices) conforming to the SI unit definitions. Photographs of a typical *standard resistor* and a *standard capacitor* are shown in Fig. 1.5.

The physical standards are classified mainly into three categories:

(i) Primary standards
(ii) Secondary standards and
(iii) Working or Laboratory standards

1.3.1.1 Primary Reference Standards

An example of the primary standard is the reference platinum-iridium kilogram mass maintained by the Bureau International des Poids et Mesures in Sevres, France. Apart from BIPM, other national organizations such as:

(i) National Physical Laboratory (NPL), India

Fig. 1.4 Weston standard cell

Fig. 1.5 Standard resistor (top) and standard capacitor (bottom)

(ii) National Physical Laboratory (NPL), United Kingdom, and
(iii) National Institute of Standards and Technology (NIST), United States of America also maintain primary and secondary reference standards.

Besides the "standard kilogram", these institutions also maintain "standard" devices for other SI units, both fundamental and derived. Another example of a primary reference standard is the "atomic clock" built using the cesium 133 element. For a long time, the reference standard for a meter was a "reference scale (ruler)" maintained by the BIPM and many national "standards" organizations. Today, the reference standard for the meter is no longer defined by a physical object. In 1983, the standard meter was redefined as the distance light travels in vacuum during 1/ 299,792,458 of a second.

1.3.1.2 Secondary Reference Standards

Since the primary reference standards are very valuable and cannot/should not be used on a day-to-day basis, secondary reference standards are created and utilized. For example, many of the national laboratories house stainless steel balls of one kilogram which are used to set standards for manufacturing ordinary weighing machines. These secondary reference standards are very close approximations of primary reference standards.

1.3.1.3 Working (Laboratory) Standards

Working standards (also popularly known as laboratory standards) are used in testing at regional laboratories within a country to provide calibration facilities for industry/ common users (in proximity to their workplace) to have their measurement setup to have a traceable relationship to the secondary and thus to the primary standards. Depending on the application, a working reference standard may be a close approximation to the secondary reference standard or can deviate a little from the secondary reference standard. For example, the 1-kilogram platinum-iridium primary reference standard is expected to be exactly 1.000 000 kg. On the other hand, a mass of 1.000 kg may be accepted as a working standard. The "weight box" (A wooden box housing different weights such as 100 g, 50 g, 20 g...... 10 mg) that we have used for measurement during scientific experimentation in our school/college is *calibrated* by a local testing laboratory and certified. When the local testing laboratory "calibrates" the weights in the weight box, they compare these weights with their "laboratory standard" weights and determine that the difference is within acceptable limits set by the "OIML R 111-1: 2004(E) standard set by Organization International de Metrrologie Legale" for classification of laboratory weights. Periodically, the testing laboratory calibrates (compares and corrects) its weights at the apex national standards laboratory, which in turn maintains the level of "calibration" by comparing its weights with the BIPM. Figure 1.6 illustrates the traceability that occurs as an outcome of this chain of calibrations. Figure 1.6 also indicates the resolution at each stage and captures the "traceability" for measurements made using a "weight box" and a "screw gage".

1.3.2 Standards for Standardized Methods and Procedures

On the other hand, "standards" when referred to in the realm of testing, details standardized methods and procedures for tests to be conducted for determining the characteristics of not only instruments but several other devices and even practices. Tests conducted as per standards help compare various "products" and "services". In this context, "standards" may also spell out properties and characteristics of devices so that these devices conform to a particular class or genre of devices. The standards such as:

"ASTM A385: Standard Practice for Providing High-Quality Zinc Coatings (Hot Dip)", "DIN 33958: Bicycles-Lighting equipment and dynamos", "IEC 17025:2005: General requirements for the competence of testing and calibration laboratories", "IS 1132: Method for partial discharge measurement in instrument

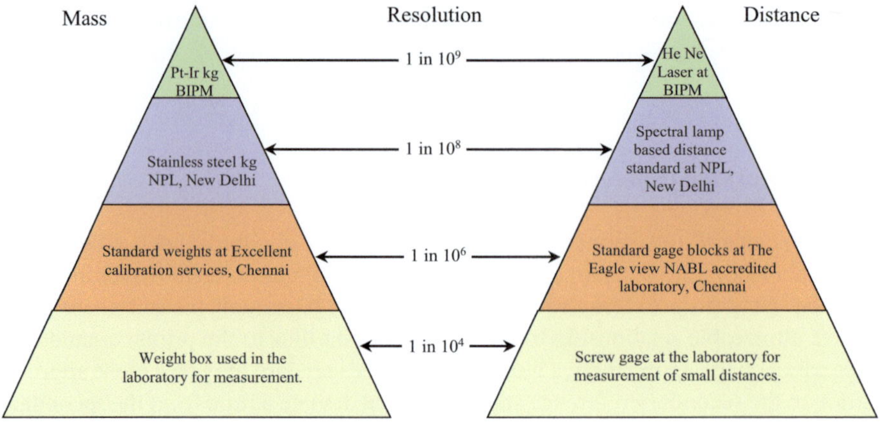

Fig. 1.6 Traceability of a measurement through calibration

transformers", and ISO 9001:2008 Quality Management Systems, all belong to the category of "Standardized methods and procedures".

Of these,

The standard ASTM A385: Standard Practice for Providing High-Quality Zinc Coatings (Hot Dip)" the standard prepared by the **A**merican **S**ociety for **T**esting and **M**aterials organization, details the step-by-step procedure to coat zinc on a target using the hot dip method.

The standard DIN 33958: Bicycles-Lighting equipment and dynamos", prepared by Deutsches Institut für Normung (German Institute for Standardization) lists all important characteristics (size, terminals, material to be used for construction, etc.) for a manufacturer to follow so that the manufacturer can claim that their bicycle dynamo or light is "as per DIN 33958" standard.

The standard IS 1132: Method for partial discharge measurement in instrument transformers, prepared by the Bureau of Indian Standards (BIS), lists the method to be employed for testing the health of the insulation of an instrument (current or voltage) transformer. It also lists how the test report is to be prepared.

The ISO 9001:2008 Quality Management Systems, the standard from the International Standards Organization (ISO), portrays the different procedures and norms to be adopted by an organization so that the organization can then claim (after due inspection by a certifying authority of ISO) that their products and services are as per ISO9001 certification. You may come across markings such as "IIT Madras is the first educational institution in India to be certified as per ISO9000:2001".

1.4 Calibration of Measuring Instruments

We measure using an instrument. It may be a simple ruler (meter scale) for the measurement of a distance, or it could be a weighing machine to know our weight. How can we be sure that the measurement undertaken with such instruments gives results in accord with the appropriate definition in SI?

To assure the correctness of our measurement, each instrument manufacturer "*calibrates*" the instrument being sold and certifies that the instrument will give measurements correctly. The term "correctly" can give different meanings depending on the situation. For example, a voltmeter, say meter 1, gives a reading of 1.0 V (2 significant figures) when measuring:

1.000 000 V(videsignificantdigitinsection 1.2)

and another, say meter 2, can provide a reading of 1.000 V (four significant digits) for the same voltage. We can then say that the readings of both meters are correct.

Now, if the voltage being measured is 1.006 023 V, the first meter will still read 1.0 V and the second meter may read 1.006 V. We may, once again in general, say meter 2 reads more correctly than meter 1. Now another meter, meter 3, may give a reading of 1.010 V when measuring the input of 1.006 023 V. Then it is prudent to say that reading obtained from meter 3 is incorrect. From the foregone example, it is understood that it is nearly impossible for any manufacturer to produce instruments that read correctly all the time. Even if an instrument reads "correctly" (i.e., reads what it should under the given situation), the measured value may not reflect the actual value of the input due to the limited number of significant digits the instrument may possess.

Manufacturers of measuring instruments have a bounden duty to inform the buyer about the possible error the user may get when using their instrument. To this end, they specify the maximum possible difference between the "actual value" and the "value measured" by the instrument (known as the error) under specified conditions. This difference is always expressed as a percentage as given below:

$$\text{Error} = \frac{\textit{Measured value} - \textit{Actual value}}{\textit{Actual value}} \; 100\ \%$$

To provide this information on the errors that can arise, the manufacturer must test the instrument. Testing of an instrument with a view to quantifying the errors in its readings is termed as "*Calibration*" of the instrument. For example, to calibrate a 100 V full-scale voltmeter, the manufacturer will measure "standard voltages" in the range 0 V–100 V (say in steps of 5 V) and record the errors in the readings obtained. Here, standard voltages refer to voltages that are correctly known with the number of significant digits higher than (at least one digit more than) that provided by the instrument. Calibration can be based on either the "absolute method" or on the "comparison method".

In the absolute method of calibration, shown in block schematic form in Fig. 1.7, the instrument is given an input which is as defined by the SI unit definition and

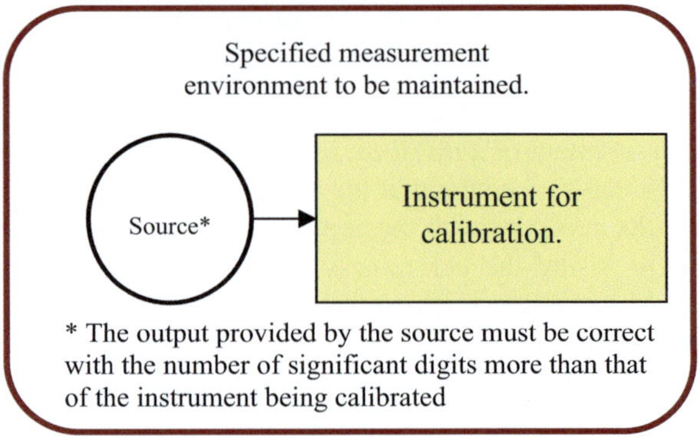

Fig. 1.7 Schematic of absolute method of calibration

its reading adjusted to read exactly the value of the input or error at that point of measurement.

For example, if we want to calibrate a thermometer at 0 °C, then the thermometer can be placed in a reference temperature bath (which contains both ice and distilled water in equal proportion and the bath maintained at an ambient pressure of one atmosphere) and adjustments made on the thermometer so that it indicates 0 °C.

In the comparison method of calibration (shown in block schematic form in Fig. 1.8), as the name indicates, the instrument to be calibrated is compared with a standard instrument of the same type and the errors are *noted* or *correction* is applied to the instrument being calibrated. The standard instrument can be a primary standard or secondary or working standard depending on the type of instrument being calibrated. Without loss of generality, the standard meter/instrument being used for calibration must have an accuracy specification an order of magnitude higher than the instrument being calibrated and possess "significant digits" more (at least +1) than the meter/instrument being tested. For example, to calibrate an ammeter possessing an error of 1.0% and three significant figures, the standard ammeter should have an error specification of 0.1% or better and the number of significant digits four or higher.

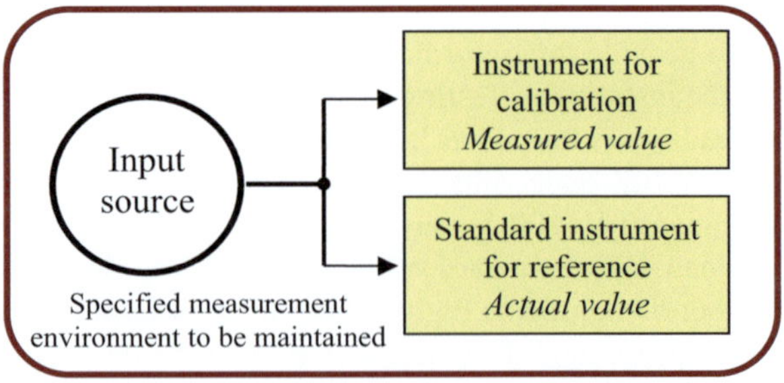

$$Error = \frac{Measured\ value - Actual\ value}{Actual\ value} 100\ \%$$

Fig. 1.8 Schematic of comparison method of calibration

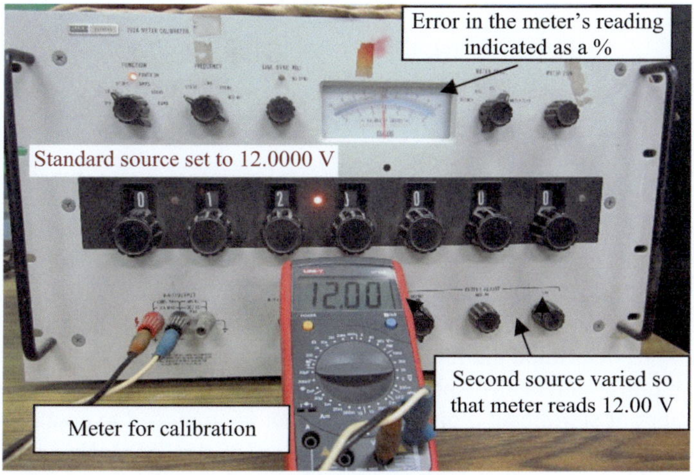

Fig. 1.9 A typical calibrator (Courtesy: Fluke corporation)

Many manufacturers provide "*calibrators*" that use the comparison method of calibration. Such a calibrator would house the source and the standard instrument in a single unit and thus facilitate ease of calibration (either for adjustment on the instrument being calibrated or for the measurement of its errors). Figure 1.9 shows a calibrator for calibrating voltmeters, ammeters, and resistances. As already brought out, the process of calibration brings *traceability.*

1.5 Problems and Exercises

E1.1. Write the units given below in terms of the seven fundamental units of SI.

(a) Joule	(b) Volt	(c) Coulomb	(d) Watt
(e) Newton	(f) Pascal	(g) Tesla	(h) Henry

E1.2. How many significant digits does each of the following numbers have?

(a) 0.00305	(b) $3.05\ 10^{-3}$	(c) 30500	(d) $3.05\ 10^{5}$
(e) 305.0	(f) 305	(g) −0.1234	(h) 0.1E1.3

E1.3. Round off the following numbers to three significant digits each.

(a) 4.076	(b) 4.6743	(c) 4.7550	(d) 4.7650
(e) 4.7499	(f) 4.7501	(g) 4.4501	(h) 4.4499

E1.4. Round off the numbers in exercise E1.3 to two significant digits each.

E1.5. A digital frequency meter displays the frequency of the AC supply mains on a 3-digit display. Determine the minimum number and maximum number of significant digits that can be obtained with this meter. If this meter displays 50.0 Hz, determine the resolution in this measurement.

E1.6. A digital counter can display a maximum value of 199999999. Determine the number of significant digits. If the counter is employed to count the number of periods of a signal whose frequency is 123.456789 MHz in 1.000000 s, what value it will display. Determine the resolution for this measurement.

E1.7. Write the answer for the current I restricting it to the correct significant digits of the computation:

$$I = 1.1334\ \text{A} + 0.3344\ \text{A} + 123.45\ \text{mA}$$

E1.8. A 0–100 mA PMMC-type ammeter has a full-scale deflection of 100° with marking at every degree. Determine the resolution of the meter. Determine the minimum and maximum number of significant digits that can be obtained using this meter.

E1.9. The measured values of four resistors are 5.321 kΩ. Find the total resistance and express the result to the correct number significant digits, when the four resistances are connected in series.

E1.10. The synchronous speed of an induction motor is 1500 rpm. The speed of the rotor is read as 1.48 on a tachometer with a scale x1000 rpm. Calculate the slip speed to the correct number of significant digits.

E1.11. Express the voltage drop across a 5.321 kΩ resistance to the correct number of significant figures if the current through the resistors is measured to be 2.1 mA.

E1.12. The voltage across a resistor R is measured as 43.41 V and the current flowing through it is measured as 2.5 A. Find the value of R to the correct number of significant digits.

E1.13. An 18 pole AC motor is running on a supply main whose frequency is measured as 49.66 Hz. Express the synchronous speed of the motor to the correct number of significant figures.

E1.14. The following are the measured values of the currents and voltages in the circuit of the figure shown in Fig. 1.10. Given: I_1, the current through voltage source E_1 is 0.53 A, I_2, the current supplied by voltage source E_2 is 1.27 A. The voltage of E_1 is 10.1 V, and that of E_2 is 15.7 V. R_o is separately measured and found to be 5.102 Ω. Evaluate R_1 and R_2 and express the results to correct significant figures.

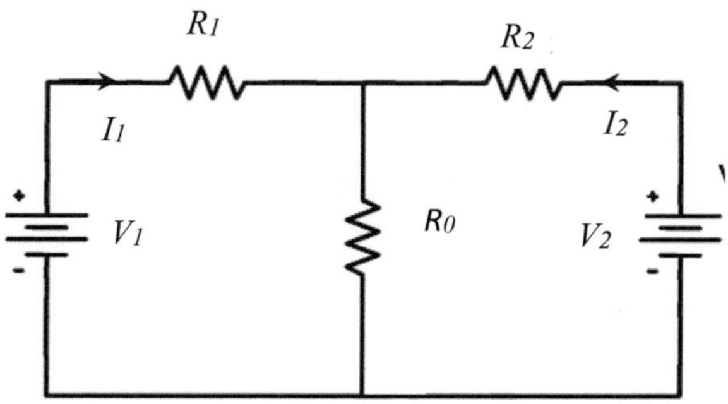

Fig. 1.10 Resistive Circuit

E1.15. If in exercise E1.13, a less refined instrument was used to measure R_o as 5.1 Ω, evaluate R_1 and R_2.

E1.16. Correct the errors in the following:

(a) Length of a rod = 1 m 27.45 cm
(b) Current is expressed in kAs
(c) T = 33°K
(d) A current of one million amperes was recorded
(e) $C = 12$ μμ F
(f) The values of resistances are in kΩ.
(g) $P = 10$ pa
(h) Area = 10 × 10 cm
(i) Pressure is 106 Pascal.
(j) The figure of merit of a voltmeter is expressed in kΩ per V.

Further Readings

R1.1 http://www.bipm.org/en/si/
R1.2 http://en.wikipedia.org/wiki/International_System_of_Units
R1.3 http://physics.nist.gov/cuu/Units
R1.4 IEEE Std 100-1996 IEEE Standard Dictionary of Electrical and Electronic Terms, Sixth Edition
R1.5 IEEE Std 260.1-1993 American National Standard Letter Symbols for Units of Measurement (SI Units, Customary Inch-Pound Units, and Certain Other Units)

R1.6 IEEE Std 280-1985 American National Standard for Mathematmatical Signs and Symbols for Use in Physical Sciences and Technology IEEE Std 280-1985 IEEE Standard Letter Symbols for (R1997) Quantities Used in Electrical Science and Electrical Engineering
R1.7 IEEE Std 315-1975 IEEE Graphic Symbols foe Electrical and (R1993) Electronics Diagrams (Including Reference Designation Letters) (Includes supplement 315A-1986, R1993)
R1.8 ISO Standard: http://www.iso.org/iso/home.htm
R1.9 SI 10-1997 (IEEE/ASTM) Standard for Use of the International System of Units (SI)—The Modern Metric System.
R1.10 IS: 10005-1994: SI units and recommendations for the use of their multiples and of certain other units.
R1.11 IS: 2-1960: Rules for rounding off numerical values
R1.12 http://www.bis.org.in/
R1.13 http://www.astm.org/

Chapter 2
Analog Indicating Instruments

In physical science the first essential step in the direction of learning any subject is to find principles of numerical reckoning and ***practicable methods for measuring*** *some quality connected with it.*
—Lord Kelvin

2.1 Analog Indicating Instruments

When humans started using electricity, instruments to measure various electrical quantities such as electric potential (voltage, V), current (ampere, A), power (watt, W), and usage of energy in terms of ampere hour (Ah) for DC systems or watt-hour (Wh) for AC systems were needed. Though the SI unit for energy is joules, today we measure electrical energy in terms of a select unit called kilowatt-hour (which equals 3.6 MJ of energy) and is written in short form as kWh. One kilowatt-hour (1 kWh) is called one "*unit*" of electrical energy. By the time commercial usage of electricity started and meters were needed to measure all these electrical quantities, people were very familiar with taking measurements of distances with a ruler (scale)

V. Jagadeesh Kumar, *Principles of Electrical Measurement and Measuring Instruments*,
https://doi.org/10.1007/978-981-96-5475-8_2

or graduated tape as well as measurement of angles with a protractor. Hence, the instruments developed to measure electrical quantities also mimicked the ruler or protractor. In these electrical measuring instruments (also called meters), a pointer moved on a graduated scale to indicate the value of the electrical current or voltage or power that is being measured as shown in Fig. 1.1 of Chap. 1.

Since the value being measured is indicated on a scale by a pointer and the movement of the pointer is continuous (smooth), these instruments are called "***analog indicating instruments***", in contrast to "***digital indicating instruments***", which displays the value being measured by a finite array of digits. The permanent magnet moving coil meter, popularly known as the PMMC meter, (also known as the DC ammeter based on the **Jacques-Arsène d'Arsonval**'s galvanometer principle) is most probably the very first electrical "*direct indicating meter*" or "*direct reading meter*" to be invented. It should be noted that the d'Arsonval galvanometer though was invented earlier to the PMMC meter, is not a meter that directly measures the value of any electrical quantity. The PMMC meter, as we know it now, was invented by Edward Weston in 1886 and a patent was granted to him by the United Kingdom patent office in 1888.

2.2 The Permanent Magnet Moving Coil (PMMC) Meter

The operation of a permanent magnet moving coil meter (PMMC meter) relies on the fact that a force is generated when a conductor (wire) carrying an electric current interacts with a magnetic field. If we place a conductor carrying a current I_c amperes inside a uniform magnetic field of flux density B_m tesla as shown in Fig. 2.1, the conductor (wire) will experience a force F_L in a direction perpendicular to both the field and current directions. This force F_L is called the Lorentz force, and the direction of the force is determined by the right-hand rule. Open the palm of your right hand and place a pointed stick vertically on your palm, pointing away from it. Now if you position your hand such that your thumb points to the direction of the current in the conductor and the rest of the fingers align with the direction of the flux as shown in Fig. 2.2, then force F_L will be in the direction of the pointed end of the stick. The magnitude of the force can be easily derived as

$$F_L = L_w B_m I_c Sin(\varphi). \tag{2.1}$$

In Eq. (2.1), L_w is the length of interaction between the conductor and the magnetic field and φ is the angle between the conductor (assumed to have a straight-line shape as in Fig. 2.1) and the flux vector (B_m).

To convert this force into a movement of a pointer on a scale, the meter is designed as shown in Fig. 2.3. A rectangular shaped coil, wound with a thin copper or aluminum wire, is attached to two stems, one at the top and the other at the bottom as shown in Fig. 2.3. The coil has N turns and physical dimensions of length l_c, width W_c, and thickness T_c. Both the top and bottom stems are pivoted using a pin and socket

Fig. 2.1 The Lorentz force

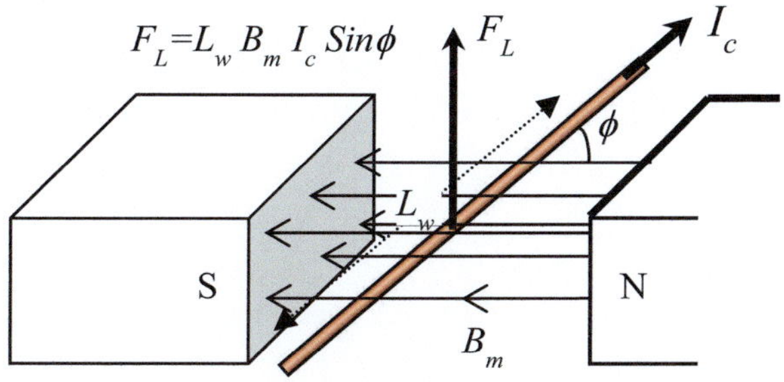

Fig. 2.2 The right-hand rule

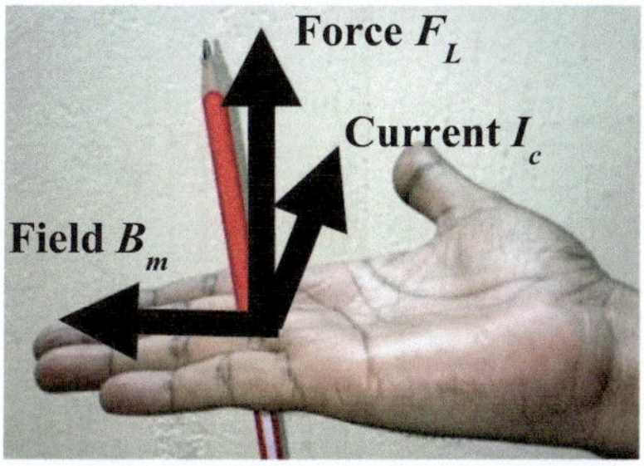

or bearing arrangement and is placed such that the vertical sides of the coil occupy and can freely rotate in an annular space created between two semi-circular shaped pole pieces as shown in Fig. 2.3, encompassing a stationery central cylinder made of ferromagnetic material of high permeability. The coil is free to rotate but the rotation is restrained by two helical springs (popularly called hair springs) attached to the stems (one spring to the top stem and the other spring to the bottom stem). A pointer attached to the top stem moves on a graduated scale as indicated in Fig. 2.3. The magnetic flux required in the annular space is provided by a permanent magnet. The cross section of the PMMC meter depicted in Fig. 2.3 along the vertical plane marked as a—a' is shown in Fig. 2.4a. From Fig. 2.4a, we see that the magnetic flux in the annular space, say with a density of B tesla, will be in the radial direction and hence the angle φ subtended by the sides of the coil to the magnetic field will always be 90° provided the coil stays within the annular space created by the pole pieces and the central cylinder. When a current I_x ampere flows through the coil, a couple of Lorentz forces with values as specified by Eq. (2.1) with $Sin(\varphi) = 1$ (as $\varphi = 90^{\nu}$) are generated as shown in Fig. 2.4b. These two forces combine (form a couple) and generate a rotating torque. The torque τ_{I1} generated by one turn of the coil is obtained by multiplying the width W_c of the coil and the force F_L generated by the turn and hence can be derived as

$$\tau_{I1} = W_c l_p B I_x. \tag{2.2}$$

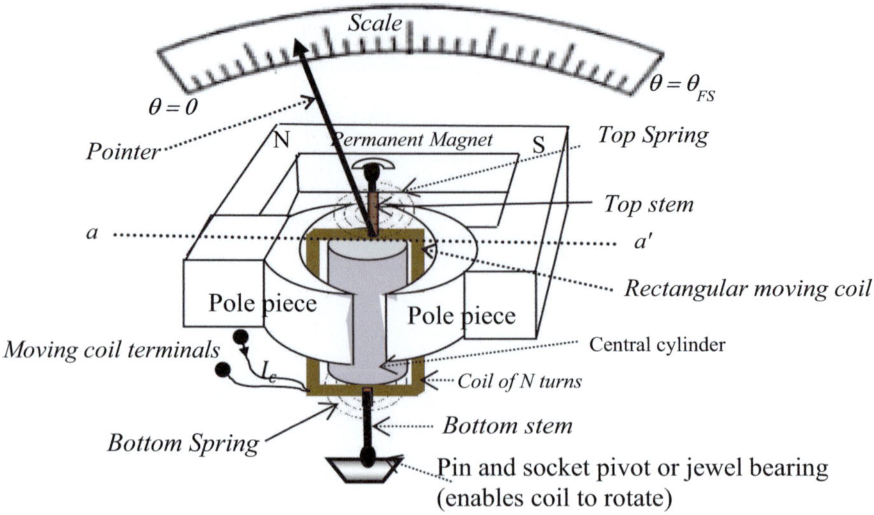

Fig. 2.3 The PMMC meter—with a rectangular magnet

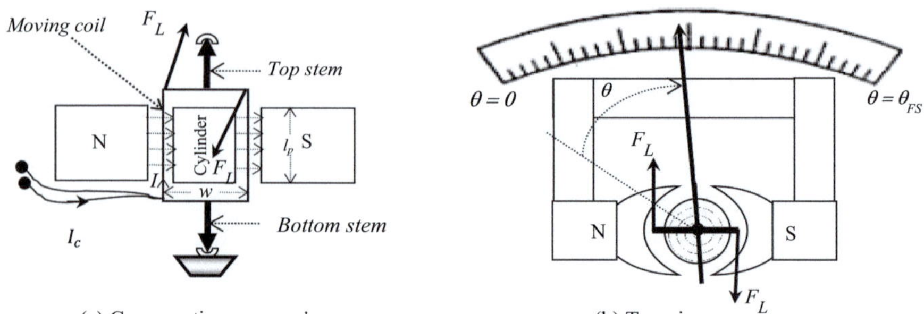

(a) Cross section on *a* - *a'*

(b) Top view

Fig. 2.4 The PMMC meter: Cross section and top views

Here, l_p is the length of the sides of the coil that interacts with the magnetic flux (that is the height of the magnetic pole pieces and hence $l_\mathrm{p} < l_\mathrm{c}$). The total torque τ_I developed by N turns of the coil (if all turns are in one layer) carrying a current I_x is

$$\tau_{I1} = NW_c l_p B I_x. \tag{2.3}$$

In practice, the coil is made of few layers. Then the turn in the innermost layer will have width w and the outer most layer will have the distance between the sides as $(w + 2b)$, where b is the thickness of the coil. Hence, the average distance will be $(w + b)$. Then the torque generated by the current is

$$\tau_I = N(W_c + b)l_p B I_{x.} \tag{2.4}$$

If we substitute $(W_c + b)l_\text{p}$ as the area A, Eq. (2.4) can be rewritten as

$$\tau_I = BANI_{x.} \tag{2.5}$$

Here, $A = \left[(W_c + b)l_\text{p}\right]$ is the *area of interaction* between the coil and the magnetic flux and **not** the area enclosed by the coil. This torque will rotate the coil (in clockwise direction for the condition shown in Fig. 2.3) and hence wind or unwind the springs, depending on the way the springs are mounted. The springs with a spring constant K_s (also called *torsion coefficient* or *torsion elastic modulus*) will apply a restoration torque τ_R when twisted by an angle θ, where

$$\tau_R = K_s\theta. \tag{2.6}$$

The coil and along with it, the pointer, will move through an angle θ till the torque τ_I generated by the current I_x equal the restoration torque τ_R generated by the springs. Under such a condition,

$$\tau_R = K_s\theta = \tau_I = BANI_x. \tag{2.7}$$

Rearranging Eq. (2.7), we get

$$\theta = \tfrac{BAN}{K_s}\, I_x = K_M\, I_x. \tag{2.8}$$

In Eq. (2.8),

$$K_M = \left(\tfrac{BAN}{K_s}\right), \tag{2.9}$$

and is known as the meter constant (with units °/μA). Here, we see that a current I_x through the coil of a PMMC meter will deflect its pointer by an angle θ. Thus, the PMMC meter is basically an ammeter that translates a current into a proportional angle of movement of a pointer that is easily measured with a graduated scale kept underneath the pointer. In Fig. 2.3, the permanent magnet shown in Fig. 2.3 is a rectangular magnet (popularly called the bar-magnet). In the initial days, the permanent magnets used (not only in PMMC meters but also in general use) were of the U-shaped (popularly known as the horseshoe) magnet. Figure 2.5 depicts a PMMC meter with a U-shaped (horseshoe) magnet.

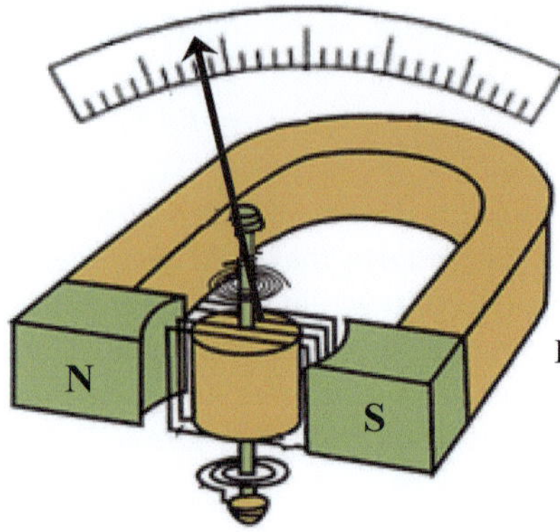

Fig. 2.5 The PMMC meter with horse-shoe magnet

2.2.1 Full Scale, Resolution, and Sensitivity of a PMMC Meter

Once a PMMC ammeter is designed and built, the values of $B,\ A,\ N$, and K_s are set and the value of K_M is also determined. Again, for a given design, the maximum angle θ_{FS} that the coil can traverse on the scale (vide Fig. 2.3) is set. The value θ_{FS} is called the full-scale deflection angle and the current, say, I_{FS}, that deflects the pointer by an angle θ_{FS} is called the full-scale current of the PMMC ammeter. For example, if a PMMC meter is designed with a full-scale deflection angle θ_F of 100° and K_M is adjusted such that 1 μA of current through the coil deflects the pointer by one degree, then a current (we call this the full-scale current I_{FS} of the meter) of 100 μA will deflect the pointer by 100°. A current of 51 μA will deflect the pointer by an angle of 51° and so on. This will be true if and only if the springs used have a linear characteristic and possess no hysteresis. If the scale is marked in steps of 1°, then each marking (division) will correspond to 1 μA and the instrument will become a 100 μA full-scale *PMMC ammeter* of resolution 1 μA. As mentioned in Chap. 1, the term "resolution" points to the ability of the meter to resolve the smallest quantity being measured without any ambiguity and is always indicated either as the smallest step size in the input that can be measured without any doubt, thus for the 100 μA full-scale PPMC meter referred above, the resolution is 1 μA. The resolution is also indicated as a non-dimensional quantity by dividing the smallest step that can be measured without ambiguity by the full-scale value. For the 100 μA full-scale PMMC ammeter, the resolution is 1 μA in 100 μA or 1/100 or simply 1 in 100. The ratio of the output to its corresponding input at the resolution level, applicable for any measuring instrument such as the PMMC meter, is called the sensitivity of the meter. For the 100 μA PMMC ammeter, the sensitivity is

$$Sensitivity = \left.\frac{\Delta Output}{\Delta Input}\right|_{At\ the\ resolution\ level} = \frac{1^\circ}{1\mu A} = 1^\circ/\mu A. \tag{2.10}$$

2.2.2 Static and Dynamic Characteristics of a PMMC Ammeter

In the preceding sections, we viewed the PMMC meter as an electromagnetic apparatus as it contains electrical as well as magnetic elements. It is also an *electromechanical apparatus* as it is made of mechanical parts: the springs, the stem, and the moving (rotating) coil. Hence, let us look at the electromechanical aspects of the meter. The moving part of the PMMC meter is a rotational system having moment of inertia J_c of the coil, damping D_c (mostly viscous damping in addition to a negligibly small frictional damping due to the pivots) and spring(s) having a constant K_s. The response of the electromechanical system can be studied for

(i) Static input. That is the current through the PMMC meter is of constant value and does not change with time (a DC current) and
(ii) Dynamic input. That is the current through the PMMC meter varies with time (AC current).

2.2.2.1 Static Characteristic of a PMMC Ammeter

First, let us understand the behavior of a typical PMMC meter for static inputs. Even when we apply a static input (say 50 μA for the 100 μA full-scale ammeter, discussed earlier), when the current is first applied or simply the system in which the meter is connected is switched on, the meter gets a current that goes from zero (before switching on) to 50 μA just after switching on the current (or the system). In such a case, the pointer of the meter should move from zero and stop at the 50 μA marking. Once the current starts flowing through the coil of the meter, the coil and hence the pointer will start from its zero position, rotate, and settle at the final position. Once the pointer stops and indicates a value on the scale, we say the meter has attained the *steady state*. The period between the start of the movement of the pointer and the time at which steady state occurs is called the *transient period*. Thus, the response of the PMMC meter for a static input has two distinct phases:

(a) The transient period and
(b) The steady state.

The response of the meter (movement of the pointer) during the transient period is called the transient response of the meter. The final value indicated by the meter after it has reached steady state is called the *steady-state value*.

(A) Transient response of a PMMC meter:

The transient response of the movement of the pointer is decided by the *dynamics* of the electromechanical system described by the differential equation:

$$J_c \frac{d^2\theta(t)}{dt^2} + D_c \frac{d\theta(t)}{dt} + K_s\, \theta(t) = BAN\; i(t). \tag{2.11}$$

In Eq. (2.11), J_c is the moment of inertia, D_c is the damping (viscous and frictional) constant, and K_s is the spring constant of the rotating part of the PMMC meter. Equation (2.11) reflects the fact that the deflection $\theta(t)$ of the pointer at a general time instant t in the transient period is accompanied by an angular acceleration $\left(\frac{d^2\theta(t)}{dt^2}\right)$ and an angular velocity $\left(\frac{d\theta(t)}{dt}\right)$. Once the steady-state condition is attained, both the acceleration and velocity become zero. A look at Eqs. (2.8) and (2.11) tells us that Eq. (2.8) is the steady-state solution of Eq. (2.11), wherein both angular acceleration and angular velocity are zero.

Applying Laplace transform to Eq. (2.11), we get (assuming zero initial conditions)

$$J_c s^2\theta(s) + D_c s\theta(s) + K_s\theta(s) = BAN\ I(s). \tag{2.12}$$

Rearrangement of Eq. (2.12) results in

$$\theta(s) = \frac{BAN}{(J_c s^2 + D_c s + K_s)}\ I(s). \tag{2.13}$$

This can further be put in the form

$$\theta(s) = \frac{\left(\frac{K_S}{J_C}\right)}{\left(s^2 + \frac{D_C}{J_C}s + \frac{K_S}{J_C}\right)}\frac{BAN}{K_s}I(s) = \frac{\left(\frac{K_S}{J_C}\right)}{\left(s^2 + \frac{D_C}{J_C}s + \frac{K_S}{J_C}\right)}K_M\ I(s). \tag{2.14}$$

Substituting $\omega_n = \sqrt{\frac{K_s}{J_c}}$ and $\xi = \frac{D_c}{2\sqrt{J_c K_s}}$ in Eq. (2.14), we get

$$\theta(s) = \left(\frac{\omega_n^2}{s^2 + 2\xi\omega_n s + \omega_n^2}\right)K_M I(s). \tag{2.15}$$

Equation (2.15) is the standard way of representing a second-order low-pass system in the Laplace domain. Here, ω_n is the undamped natural frequency of the system and ξ is called the damping constant. Hence, a PMMC meter is inherently a second-order low-pass response system.

When we apply a static input, say a DC current of magnitude I_m amperes at $t = 0$ to a PMMC meter, the current ideally must rise from zero to the value I_m at $t = 0$ and thereafter remain constant as shown in Fig. 2.6. Such an input is called a *step input* of magnitude I_M. This can be mathematically defined as

$$i(t) = \begin{matrix} 0 & t < 0 \\ I_m & t \geq 0 \end{matrix} \quad \Rightarrow \quad i(t) = I_m\ u(t). \tag{2.16}$$

In Eq. (2.16), $u(t)$ is called the unit step function and is defined as

$$u(t) = \begin{matrix} 0 & t < 0 \\ 1 & t \geq 0 \end{matrix}. \tag{2.17}$$

Fig. 2.6 Step input (current) given to a PMMC meter

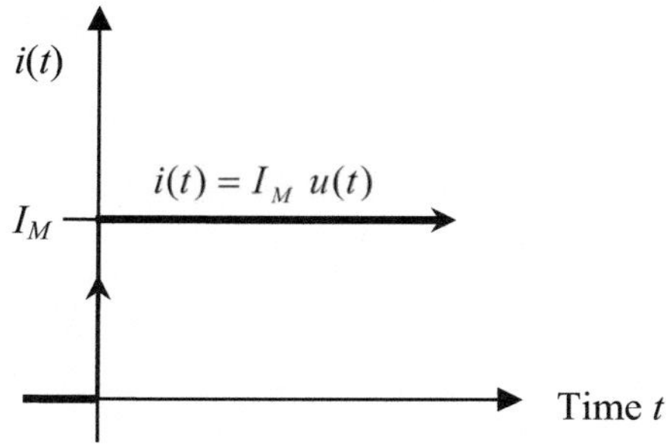

For the step input current as described by Eq. (2.17), the Laplace transform $I(s)$ is

$$I(s) = L\{i(t)\} = L\{I_m u(t)\} = \tfrac{1}{s} I_m. \tag{2.18}$$

Substituting Eq. (2.18) in Eq. (2.15) and applying inverse Laplace transform to the result yields

$$\theta(t) = L^{-1}\theta(s) = L^{-1}\left[\tfrac{1}{s}\left(\tfrac{\omega_n^2}{s^2+2\xi\omega_n s+\omega_n^2}\right)K_M I_m\right], \tag{2.19}$$

where L^{-1} indicates the inverse Laplace transform operation. Using partial fraction, the inverse Laplace transform in Eq. (2.19) can be found once the roots of the quadratic equation $s^2 + 2\xi\omega_n s + \omega_n^2 = 0$, say α and β are determined. Depending on the value of the damping constant ξ, the nature of the roots α and β vary and so does the solution to Eq. (2.19). The three distinctly different conditions that would arise depending on the value of the damping coefficient ξ can be summarized as follows:

(i) Overdamped condition with $\xi > 1$,
(ii) Critically damped condition with $\xi = 1$, and
(iii) Underdamped condition with $\xi < 1$.

(i) If $\xi > 1$, then roots α and β will be real and distinct (different). The solution to Eq. (2.19) can now be obtained using partial fraction (vide Annexure 1 for details) as

$$\alpha = \left(-\xi + \sqrt{\xi^2 - 1}\right)\omega_n \textit{ and } \beta = \left(-\xi - \sqrt{\xi^2 - 1}\right)\omega_n, \tag{2.20}$$

resulting in

$$\theta(t) = \theta_M\left(1 + \frac{\left(-\xi - \sqrt{\xi^2-1}\right)e^{\omega_n\left(\xi+\sqrt{\xi^2-1}\right)t} - \left(-\xi + \sqrt{\xi^2-1}\right)e^{\omega_n\left(\xi+\sqrt{\xi^2-1}\right)t}}{2\sqrt{\xi^2-1}}\right)u(t). \tag{2.21}$$

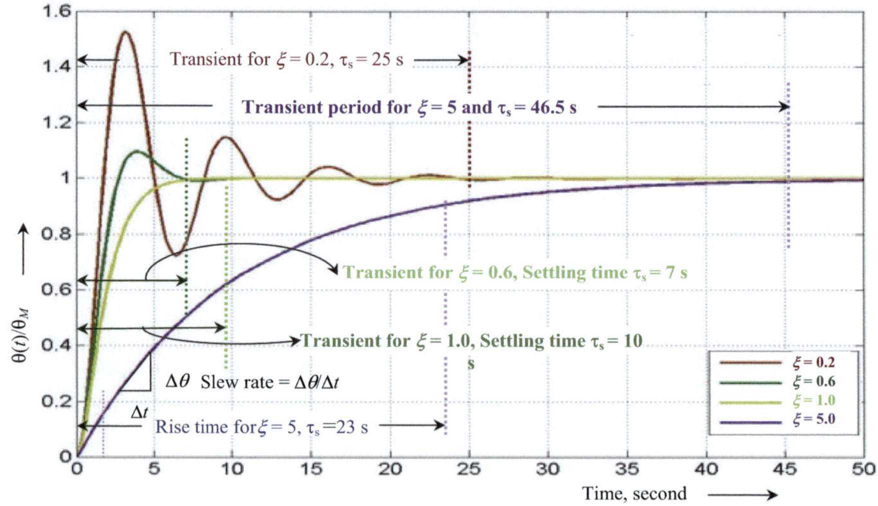

Fig. 2.7 Step response of a PMMC meter

The condition $\xi > 1$ is popularly called the ***overdamped condition***. The normalized time response of an overdamped second-order system for a unit step input is shown in Fig. 2.7.

(ii) If $\xi = 1$, then roots α and β will be real and equal with $\alpha = \beta = \omega_n$ resulting in

$$\theta(s) = \theta_M \left(\frac{\omega_n^2}{(s+\omega_n)^2} \right) \frac{1}{s} \tag{2.22}$$

In this situation, we get the solution, once gain using partial fractions (details in Annexure 1) as

$$\theta(t) = \theta_M \left[1 - (1 + t\omega_n)\, e^{-\omega_n t} \right] u(t). \tag{2.23}$$

The response of a critically damped PMMC meter is also plotted in Fig. 2.7.

(iii) If $\xi < 1$ (popularly known as *underdamped* system), then roots α a β will be complex conjugates (vide annexure 1 for detailed solution) as

$$\alpha = \left(-\xi + j\sqrt{1-\xi^2} \right) \omega_n \; and \; \beta = \left(-\xi - j\sqrt{1-\xi^2} \right) \omega_n, \tag{2.24}$$

$$\theta(t) = \theta_M \left[1 - \frac{e^{-\xi \omega_n t}}{\sqrt{1-\xi^2}} Cos(\omega_d t + \phi) \right] u(t). \tag{2.25}$$

In Eq. (2.25),

$$\phi = -Tan^{-1}\left(\frac{\xi}{\sqrt{1-\xi^2}}\right) \text{ and } \omega_d = \omega_n\sqrt{1-\xi^2}. \tag{2.26}$$

Here, ω_d is called the "*damped natural frequency*" of an underdamped PMMC meter (system). The dynamic response of a typical second-order system as described by Eq. (2.15) for four different values of damping coefficient ξ (one overdamped condition, one critically damped condition and two underdamped conditions) are given in Fig. 2.7. When the damping constant is equal to one (critically damped condition), the pointer (along with the coil, stem, and spring) rotates smoothly and after a brief transient period settles to indicate the final value. When the damping constant ξ is more than one (overdamped condition), the pointer deflects, once again smoothly but takes much longer time to reach steady state compared to the situation when $\xi = 1$. When the damping constant is less than one (underdamped condition), we see that the pointer oscillates for a while (overshoots and undershoots the expected value for brief periods of time) before settling down to the steady-state value. It is also clear from Fig. 2.7 that the settling time is optimal when the damping coefficient is slightly less than one. The peak amplitude of the first overshoot and the number of noticeable over/under shoots are dependent on how low the value of the damping constant is compared to 1. As an approximation, the first peak of the overshoot can be calculated if $\xi < 1$ as

$$PercentageOvershoot = 100e^{-\frac{\pi\xi}{\sqrt{1-\xi^2}}} \text{ for } \xi < 1. \tag{2.27}$$

Dynamic response of the meter: We have already observed that dynamic inputs are signals that vary in time. The response of a PMMC meter (or any system) to such signals is called the *dynamic response* of the meter (or the system). These signals can further be classified into:

(i) Repetitive and
(ii) Non-repetitive or random signals.

Measurements concerning a non-repetitive quantity are basically outside the scope of an analog indicating instrument. Hence, we will confine the following discussion to repetitive or periodic inputs. To ascertain the response of a PMMC meter (or any system for that matter) for any type of input signals, it would be easier if the input is expressed as a set of (combination of) standard signals. The response of the meter (or system) for standard signals can be determined a priori. If the meter (or system) can respond well to all the standard signals that make up a particular type of dynamic signal, then we can say that the meter (or system) will respond well to that particular dynamic signal. One such standard signal is the sinusoidal signal represented by

$$i(t) = \sqrt{2}\, I\ Cos\ (\omega t)\text{A}. \tag{2.28}$$

In Eq. (2.28), ω is the angular frequency (rad/s) of the sinusoidal signal (current i) and $\omega = 2\pi f$, f being the frequency of the sinusoid, in hertz (Hz). When we apply the sinusoidal current, the PMMC meter will first go through a transient phase before reaching steady state. The steady-state responses (deflection angle θ_ω) of the meter for a range of ω, is known as the frequency response of the meter. This is obtained from Eq. (2.15) by simply replacing the Laplace variable s by $j\omega$. In this manner, the steady-state response of the meter deflection for an input current $i(t)$ specified in the above para can be obtained from (2.15) as

$$\theta_\omega(t) = \frac{\omega_n^2}{\sqrt{[2\xi\ \omega\ \omega_n]^2 + [\omega_n^2 - \omega^2]^2}} K_m \sqrt{2}\, I\, Cos\,(\omega t + \partial), \tag{2.29}$$

with

$$\partial = -tan^{-1}\left(\frac{2\xi\ \omega\ \omega_n}{\omega_n^2 - \omega^2}\right). \tag{2.30}$$

As the function $Cos(\omega t)$ can take both positive and negative values, the coil of the meter generates positive (say clockwise rotation) and negative (anticlockwise rotation) torques, following the cosine function at the frequency of the current. From Eq. (2.29), it is seen that as the frequency of the current through the PMMC meter increases, the peak excursions of the angle of deflection of the meter decreases for a given *I*. The magnitude response of the meter as a function of angular frequency derived from Eq. (2.29) is plotted, for the three damping conditions, in Fig. 2.8. The character of all the three responses is known as the low-pass type as only low-frequency variations of the input (coil current, in our case) are adequately passed on to produce proportional variations of the output (coil deflection $\theta_\omega(t)$, in our case). It is easily seen that if we send a sinusoidal current having a frequency much higher than the natural frequency of the meter, the meter will not respond, the pointer showing no deflection at all. Thus, in the low-frequency region, the pointer of a PMMC meter oscillates and shows zero deflection (not respond) in the high-frequency region. Since a reliable reading cannot be obtained in either case, the PMMC meter is unsuited for the measurement of an AC current (a repetitive dynamic input). Moreover, in the interest of better sensitivity, the value of K_s is chosen to be as small as practically possible to build a meter with the lowest feasible full-scale current I_{FS}.In turn, a small value of K_s results in a small value for ω_n. Consequently, the value ω_n for a practical PMMC meter is less than 20 rad/s. **Thus, for all practical purposes a basic PMMC meter is a *DC ammeter***, or at best an ammeter that can measure the *DC* component embedded in a current having a recurrent (periodic) non-sinusoidal waveform. In the latter case, it is the average value of the driving torque generated by the moving coil that corresponds to the steady-state deflection of the movement.

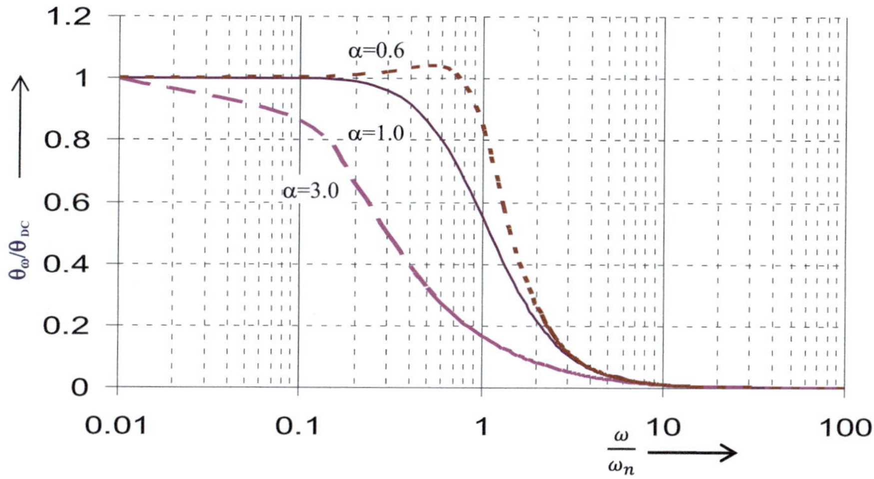

Fig. 2.8 Frequency response of a PMMC meter

2.2.3 Settling Time, Rise Time, and Slew Rate of a PMMC Meter

It is seen from Fig. 2.7, that after the application of the current (or switching on the DC circuit in which the meter is connected), the meter takes quite some time to reach the final value. We see from the transient analysis that the response contains exponential terms and hence theoretically the meter will take an infinite amount of time to reach the final value! However, as can be seen from Fig. 2.7, we cannot sensibly detect any variation in the pointer deflection after a *finite* period and can therefore assume that the pointer has reached the final value or close enough to it, to assume that steady state is reached. The term *settling time* gives us a measure of the speed with which steady-state point is reached. Whenever that term is used, it is essential to specify the *closeness* to the final value that is assumed. For example, a typical specification for indicating settling time would be

> Settling time = 1.2 s for the reading to settle within ±3% of the final value (not to reach the final value!)

If the settling time is specified without expressly stating the limits on the *closeness,* then it is presumed that the meter reading is within ±1% of the steady-state value. We can derive the settling time of an underdamped second-order system for the transients to be within ±1% to be

$$t_S \approx \frac{4.6}{\xi\omega_n}. \tag{2.31}$$

It is seen from Fig. 2.7 that by making the damping constant slightly less than unity, the settling time can be made optimal. Hence, the damping constant of PMMC meters (not only for PMMC meters but also for almost all types of analog-indicating instruments which have a moving system attached to a pointer that moves on a scale) is designed to possess a damping factor slightly less than one ($\approx$0.8) to give just one overshoot (half period of oscillation at the damped natural frequency) and settle fast. Please note that Eq. (2.31) is not valid for critical or overdamped systems and hence should not be used for such systems.

Another important measure that can be obtained from the response curves of Fig. 2.7 is the *rise time* t_R. Rise time is defined as the time taken for the pointer to rise from 10 to 90% of the final reading as indicated in Fig. 2.7. It is seen that the t_R for an underdamped system is less than the rise time of the critically damped system which in turn is less than that of the overdamped system. In the field of dynamic measurements, the *slew rate* is more important than the settling time or the rise time. The slew rate is defined as the maximum rate at which the output can rise (slope of the output characteristics). In the case of a PMMC meter, the output is the angle of deflection of the pointer, hence the slew rate can be mathematically expressed as

$$slew\ rate = \left.\frac{d\theta}{dt}\right|_{max}. \tag{2.32}$$

Like the rise time, for a simple instrument such as the PMMC meter which is not expected to work in the dynamic environment, slew rate is not important and hence is never specified.

2.2.4 Constructional Features of a PMMC Meter

Design considerations limit the full-scale deflection θ_{FS} in a PMMC meter to about 100°. To manufacture a PMMC meter to measure as small a current as possible, the full-scale current (the current needed to deflect the pointer by an angle θ_{FS}) must be small. To this end, we should select the product of *BAN* as high as possible and K_s as small as possible, as Eq. (2.8) suggests.

2.2.4.1 Selection of the Permanent Magnet

The most important parameters of a permanent magnet material for use in a PMMC meter are the remnant flux density B_R and the coercive field intensity H_C indicated in the *B*-*H* loop shown in Fig. 2.9. B_R should be as high as possible, as this value determines the actual magnetic strength *B* obtainable in the airgap of the PMMC meter. If H_C is small, then the probability of the permanent magnet getting demagnetized is high and hence H_C should also be high for a good permanent magnet. Thus, for a good permanent magnet material, both B_R and H_C should be as high as possible. Of the available materials, the alnico-type material was the most popular for use in a PMMC meter. This is an alloy made with aluminum, nickel, cobalt, copper, and iron. Some alnico magnets developed later are made with some amount of titanium

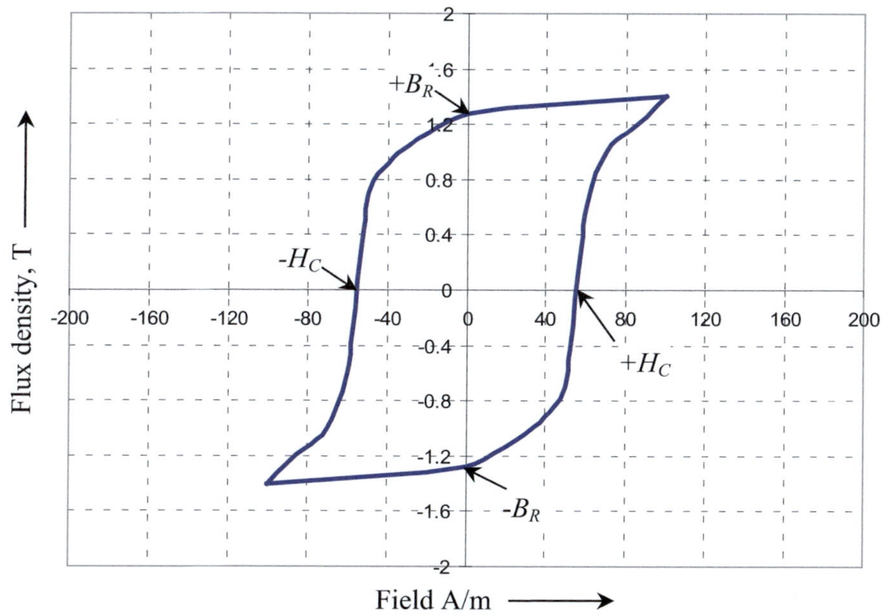

Fig. 2.9 Typical B-H loop of an alnico permanent magnet

also. Table 2.1 lists the properties of typical alnico magnets. After the invention of ceramic-type permanent magnets, such as the Ferroxdure ($BaFe_{12}O_{19}$), the alnico magnets were replaced with these modern types. With a given permanent magnet, the flux density developed in the annular air gap decreases as the air gap increases. Hence, it is prudent to design the air gap to be optimal for a particular meter design. The value of B depends mainly on the magnetic strength (flux density) offered by the permanent magnet PM and the width of the air gap (between the pole pieces and the central cylinder).

2.2.4.2 Criteria for the Selection of the Coil (Area A and Number of Turns N)

The selection of area of interaction A is constrained by the physical size of the permanent magnet and the overall size of the meter itself. For obvious reasons, A cannot be more than a few square centimeters. Selection of N is also constrained by the physical size. The larger the value of N, the thickness b of the coil will be large resulting in larger value of air gap thus reducing B available in the air gap. This opposing criterion necessitates the selection of an optimum combination of the air gap width and the number of turns in the coil. To keep the weight of the moving coil system minimum, the coils are wound with aluminum wire instead of copper wire as is normally preferred in electrical devices.

Table 2.1 Properties of different Alnico magnets

Alloy name	Al%	Ni%	Co%	Cu%	Ti%	Fe%	$(BH)_{max}$ kAT/m	B_R T	H_C kA/m
Alnico 1	12	21	5	3	–	59	11.14	0.72	37.4
Alnico 2	10	19	13	3	–	55	13.53	0.75	44.6
Alnico 3	12	25	–	3	–	60	10.74	0.70	38.2
Alnico 4	12	28	5	–	–	65	10.74	0.56	57.3
Alnico 5	8	14	24	3	–	51	43.77	1.28	50.9
Alnico5 DG	8	14	24	3	–	51	51.73	1.33	53.3
Alnico 5.7	8	14	24	3	–	51	59.68	1.35	58.9
Alnico 6	8	16	24	3	1	58	31.04	1.05	62.1
Alnico 8	7	15	35	4	5	34	42.18	0.82	131.3
Alnico 8HC	8	14	38	3	8	29	39.79	0.72	151.2
Alnico 9	7	15	35	4	5	34	71.62	1.06	119.4

2.2.4.3 Criteria for the Selection of the Stem and Pivots (Bearings)

The stem and the pivot of a normal PMMC meter design must take care of only two properties, namely,

(i) ability to handle the weight of, and at the same time.
(ii) offer very low friction for the rotational movement of the moving coil system.

Initially, the pivot joint was of the pin and socket type as indicated in Fig. 2.10a and later changed to the jewel bearing type as shown in Fig. 2.10b. The name jewel bearing was given since the bearing was made of hard stones, normally semiprecious stones, used in making jewels. Jewel bearings were invented by Nicolas Fatio de Duillier and Pierre Jacob Debaufre in 1702 to reduce friction in the gear assembly and balance wheel of watches and later adapted to meters. The hard and slick surface of a jewel exhibits a lower coefficient of friction with metal. For example, the static coefficient of friction of steel-on-steel is 0.58, while that of sapphire-on-steel is 0.15. Until the twentieth century, the jewel bearings were made of natural gems such as sapphire, ruby, or diamond. In 1902, a process to grow artificial sapphire crystals was invented, making jewel bearings much cheaper. Modern jewel bearings are all made of synthetic sapphire or ruby called corundum which is one of the hardest substances known. The only difference between synthetic sapphire (normally blue in color) and synthetic ruby (normally red in color) is that different impurities are added to get two different colors. There is no difference in their properties as jewel bearings.

2.2.4.4 Criteria for the Selection of the Springs

The lower the value of the spring constant K_S the higher will be the value of the meter constant K_M and hence the better will be the sensitivity for the meter. Once again,

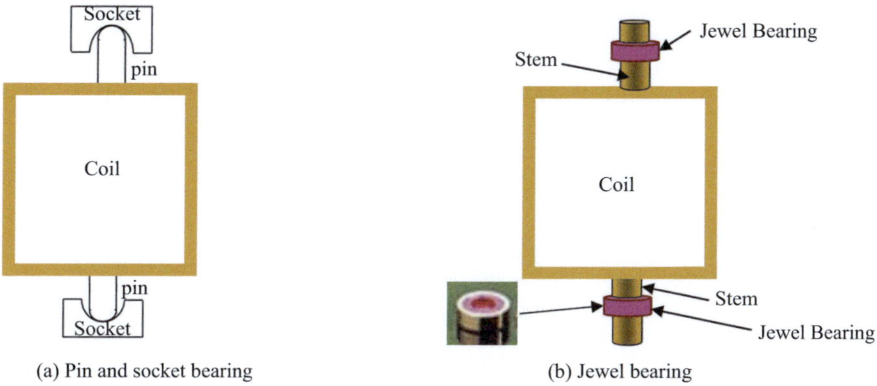

Fig. 2.10 Suspension of coil in a PMMC Meter

the physical size constraints decide the spring size and hence K_S. Any spring will have characteristics as indicated in Fig. 2.11. The initial zone where the deflection of the spring is zero till a threshold in the applied torque is reached is called the dead zone. To obtain a linear torque versus deflection angle characteristics, the two springs in the ammeter are operated away from zero. To accomplish this, the two springs are initially rotated to a small angle and are attached to the stem such that they counterbalance each other. The top spring is normally wound clockwise and balanced by the bottom spring (so that it exerts equal but opposite tension created by the top spring) mounted in the anticlockwise direction. Since the springs are already twisted and are operating away from the initial (zero) point, the springs will never come to operate in the dead zone region and hence will always operate in the linear region of the torque versus angle characteristic. The initial balancing torque will be just adequate to avoid the dead zone not only in meters with a zero to full-scale markings but also for meters designed to have a center zero. Such center zero meters will respond in the clockwise direction for positive current and respond in the anticlockwise rotation for negative current. In either direction, the springs operate in the linear region of the torque versus deflection angle characteristic.

Initially, hardened steel was the material used for manufacturing the spring and later phosphor bronze was used. With optimal design, manufacturers were able to manufacture PMMC meters with a full scale in the range of few hundred microamperes down to few microamperes. If one wants to make an overly sensitive meter with a full-scale current in the nanoampere range, then the spring dimensions (thickness and width) must be so small that it would be impractical to manufacture them. Even if we did, the springs would start to sag due to their self-weight. To overcome this problem, instead of a spring, manufacturers used taut elastic bands of flat spring leaves to provide the restoration torque as shown in Fig. 2.12. The taut band suspension systems resulted in PMMC meters with very low full-scale currents in the range of nanoamperes, but these meters became very delicate. Since the weight of the coil

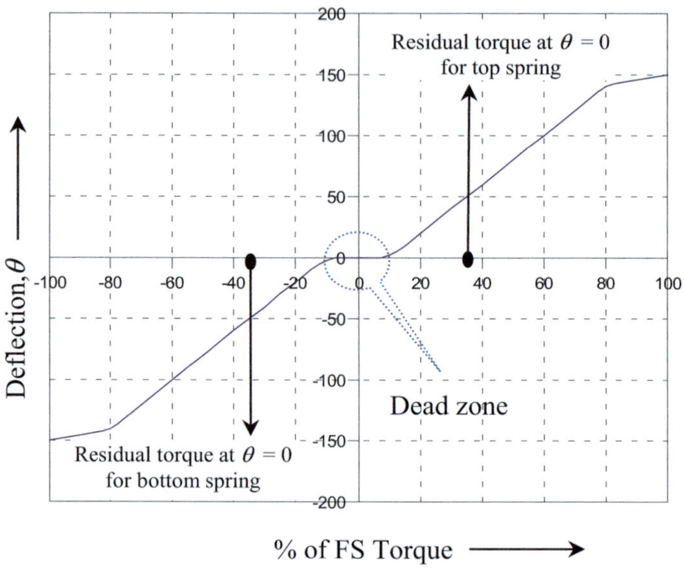

Fig. 2.11 Characteristics of spiral springs

was no longer carried by the pivoted stem but are handled by the taut bands themselves, any shock or vibration impinged on the meter would put a large stress on the taut bands resulting in the breaking of the bands. Hence, taut band meters must be handled with care.

Fig. 2.12 Taut band suspension

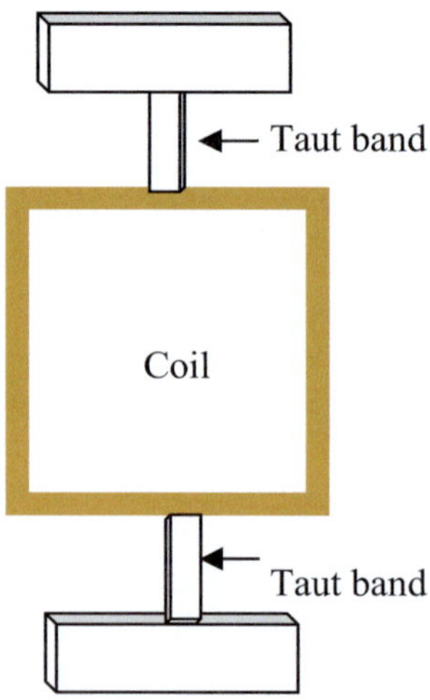

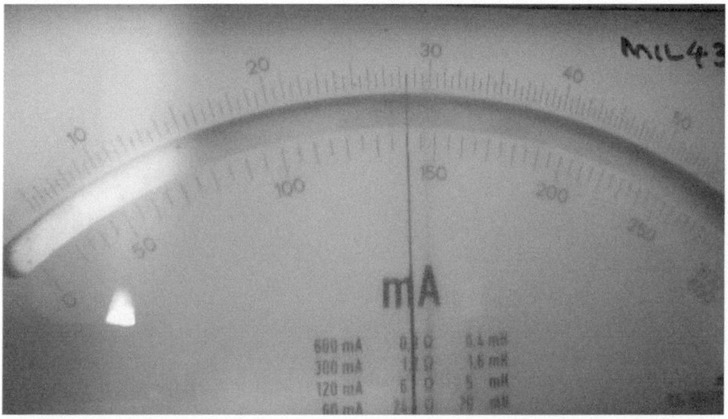

Fig. 2.13 Use of a mirror to avoid parallax error

2.2.4.5 Criteria for the Selection of the Pointer

The pointer of the PMMC meter is invariably made with aluminum (to keep the weight of the moving system as small as possible). The pointing side is fabricated with a thin long structure and is balanced by a thick short tail structure. The position of the pointer on the scale should be ascertained with the eyes of the viewer exactly on top of the pointer. If the pointer is viewed at an angle, then the pointer may appear to be coinciding with a wrong marking on the scale. The error that would be made in reading the meter with an erroneous viewing angle is called the parallax error. To avoid parallax error and help a reader to read correctly with the right viewing position manufacturers resorted to:

(i) Making the pointer with a blade having sufficient dimension in the z direction. In such a case if the viewing angle is not right, the pointer will appear thick and if viewed at the right angle (right on top of the pointer) the pointer will become a slim line.
(ii) The most popular method for avoiding parallax error is to place a mirror underneath the pointer as shown in Fig. 2.13. To take a proper reading, the reader must keep the direction of his/her sight to align the pointer and its image on the mirror. Some manufacturers replaced the pointer altogether with an optical system (vide Fig. 2.14) that moved a light spot with a marker on the scale of the meter to avoid parallax error. Such meters are popularly called the *light spot meters*. However, to get a reading from such meters, an auxiliary power for creating the light spot is required.

2.2.4.6 Controlling the Damping Constant of PMMC Meters

As already mentioned, the damping constant is adjusted to have a value slightly less than unity. In the initial days, the additional damping required was provided by

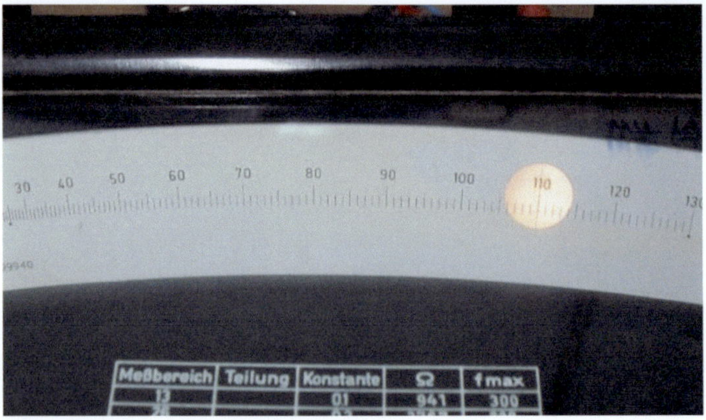

Fig. 2.14 Moving spot meter that avoids parallax error

attaching a vane to the spindle and encompassing the vane in a box like structure thus creating the viscous damping (called *air-damping* in this case). Air-damping is not popular in PMMC meters. The additional damping in PMMC meters, if required is provided using electrical damping (also called electrodynamic damping). Additional electrical damping is achieved by positioning another coil, namely the damping coil of N_D turns, along with the main moving coil and connecting a resistor across or simply short circuiting the damping coil. When the coils are in transient, they cut across the magnetic field lines and hence a voltage $v_D = N_D AB \frac{d\theta}{dt}$ will be induced in the damping coil. This voltage will force a current $i_D = \frac{v_D}{R_D} = \frac{BAN_D}{R_D}\frac{d\theta}{dt}$ in the damping coil, where R_D is the total resistance in the damping coil circuit. This current i_D will react with the flux and provide an opposing force to the movement of the coil and hence Eq. (2.11) will have to be modified as

$$J_C \frac{d^2\theta}{dt^2} + D_C \frac{d\theta}{dt} + \frac{(BAN_D)^2}{R_D}\frac{d\theta}{dt} + K_S\,\theta = BAN\; i(t). \tag{2.33}$$

Rearranging Eq. (2.33), we get

$$J_C \frac{d^2\theta}{dt^2} + (D_C + D_E)\frac{d\theta}{dt} + K_S\,\theta = BAN\; i(t). \tag{2.34}$$

In Eq. (2.34),

$$D_E = \frac{(BAN_D)^2}{R_D} \tag{2.35}$$

is the damping coefficient due to electrical damping. A comparison of Eq. (2.34) with that of Eq. (2.11) indicates that the damping is partly provided by the mechanical (viscous) damping D_C and additional damping is provided by the electrical damping D_E. In most PMMC meter, the damping coil is simply realized by winding the moving coil on a closed aluminum former. The aluminum former works as a shorted one turn damping coil.

2.2.4.7 Zero Adjustment of PMMC Meters

When a meter is manufactured, it may happen that the pointer may not exactly sit right on the starting (zero) value. Hence, a suitable mechanical design is made to move the pointer of the meter by few degrees on the scale in either direction. Moving the binding post of one the springs (invariably the top spring as it is the closest to the viewing side of the meter) the initial tension applied to that spring is varied resulting in the movement of the pointer. This adjustment is called the zero adjustment. This situation, namely, the pointer not resting on zero when no input is applied (called drift of the zero point or simply zero drift) can arise in usage as well. Hence, the zero-adjustment mechanism (normally a turn-able screw head on the meter) is made available to the user.

2.2.4.8 Full-Scale Adjustment (Calibration) of PMMC Meters

Even with a careful design, it may not be possible for a manufacturer to make all meters conform to the designed full-scale current. For example, whatever be the care taken, magnets manufactured even in one lot will not produce identical flux density B_R values and at best the manufacturing process can be controlled to make the B_R values lie only in a small range of values, say, B_{Rmin} and B_{Rmax}. A meter manufactured with a magnet having flux density B_{Rmin} will possess a K_M less than the meter manufactured with a magnet possessing B_{Rmax}. And neither meter will provide the required K_M. This kind of variation is also applicable for the springs and the air gap.

Thus, it is essential that a mechanism is available for a manufacturer to make small adjustments after the meters are manufactured and assembled to compensate for small deviations from the expected values that can occur in the parameters of the different parts and bring in the same full-scale current for all meters of a particular design. For the PMMC meter, this aim is achieved by making the design assuming the flux density of the permanent magnet to be 90% of B_{Rmin}. If the magnet fitted on the PMMC meter is B_{Rmax}, then the excess 10% must not enter the air gap. The excess 10% flux is diverted away from the air gap through an adjustable shunt path as shown in Fig. 2.15. By simply varying the position of the blade (that provides a variable reluctance shunt path to the air gap flux) shown in Fig. 2.15, the air gap flux density B is varied and thus desired K_M can be achieved. The variation in the shunt path flux can, not only compensate for variations in B_{mag} of the permanent magnet but also can be used to compensate for small variations in the spring constants and the air gap itself from meter to meter. This adjustment process is called final adjustment or single point calibration of the meter. For example, if the PMMC meter is a 100 μA full-scale ammeter, after zero adjustment, exactly 100 μA is passed through the

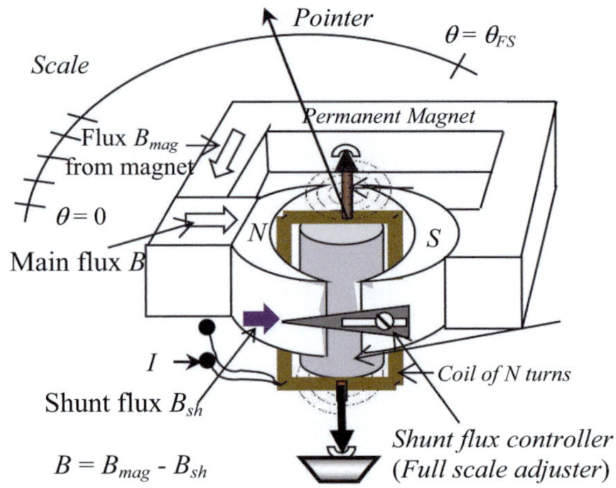

Fig. 2.15 Shunt magnetic path for calibration

meter and the position of the shunt path blade is varied till the pointer comes to the 100 μA mark and this position is sealed with glue. This calibration process requires sending the correct value of full-scale current through the meter and hence is not made available to the user.

2.2.5 *PMMC Ammeters*

The PMMC meter, as already brought out, is basically an ammeter suitable for reading only DC currents of value less than or equal to the full-scale current of the meter. Since the electrical current is made to flow through the coil of the PMMC meter, the practical meter, looked at from the electrical circuit point of view, will possess a resistance R_M and an inductance L_M in series with an ideal ammeter A as indicated in Fig. 2.16a. For DC operation (The PMMC meters can work only for DC) the inductance has no influence in steady-state response and hence the reduced equivalent circuit of Fig. 2.16b would suffice. In these figures, the circle marked with **A** inside it represents an ideal ammeter with zero resistance and hence zero voltage across it.

2.2.5.1 Scale Changing in PMMC Ammeters

Once a basic PMMC ammeter of given full-scale value is available, it is easy to obtain an ammeter of higher full-scale value by using a shunt as indicated in Fig. 2.17a. To understand this aspect easily, we consider an example. Let us take a basic PMMC (DC) ammeter of full-scale current $I_F = 50\ \mu$ A possessing an internal resistance R_M (resistance of the moving coil) of 19.9 Ω and we need a DC ammeter (new full

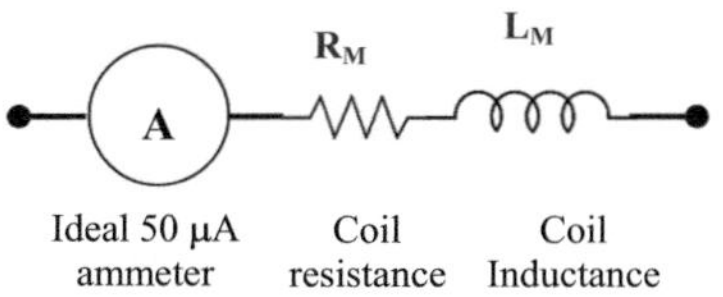

(a) Equivalent circuit of a 50 μA PMMC ammeter

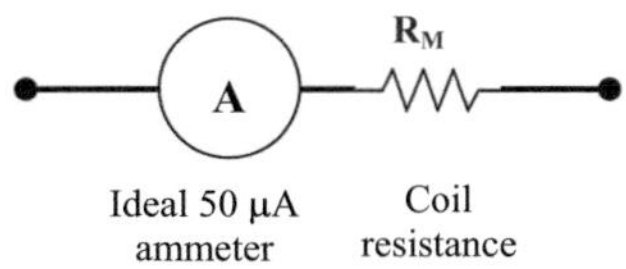

(b) Simplified Equivalent circuit of (a) applicable to steady state dc operation

Fig. 2.16 Equivalent circuit of a PMMC meter calibration

scale) with $I_{NF} = 10$ mA. In the latter, when 10 mA flows through the parallel circuit shown in Fig. 2.17a, the current through the basic meter must be 50 μA to show full-scale reading. In this condition, the voltage developed across the parallel circuit is $I_F R_M (= 19.9\Omega\ x\ 50\ \mu\text{A} = 995\ \mu\text{V})$. The current through the shunt resistance at this operating point is $(I_{NF} - I_F) = (10\ \text{mA} - 50\ \mu\text{A})$. Then the value of the shunt resistance can be easily calculated as

$$R_{SH} = \frac{R_M I_F}{I_{NF} - I_F} = \frac{19.9 * 50 * 10^{-6}}{10 * 10^{-3} - 50 * 10^{-6}} = 0.1\ \Omega. \tag{2.36}$$

Thus, by connecting a shunt resistance of value 100 mΩ in parallel to the 50 μA meter, the 50 μA meter can be converted to a 10 mA full-scale ammeter. It should, of course, be noted here that the numerical value of the reading on the meter scale must be divided by 5 to get the value of the current in milliamperes, read by converted meter. From this example at first, it looks as though it is possible to convert the basic meter to any full-scale value, which is not true as practical limitations come into picture. For example, if the above 50 μA meter is to be converted to a 10 A full-scale ammeter then the shunt resistance required is 99.50 μΩ, which is not practical. We can slightly alter this situation and construct a 10 A meter using the scheme given in Fig. 2.17b and obtain a reasonable value for the resistors. We now add a series resistance R_{SE} of 9980.05 Ω and now the drop across the meter and the series resistance is $99999.95\ \Omega\ x\ 50\ \mu\text{A} = 0.49999975\ \text{V}$. The shunt resistance $0.4999975/9999.95 = 50$ mΩ. This scheme though at first glance looks ingenious,

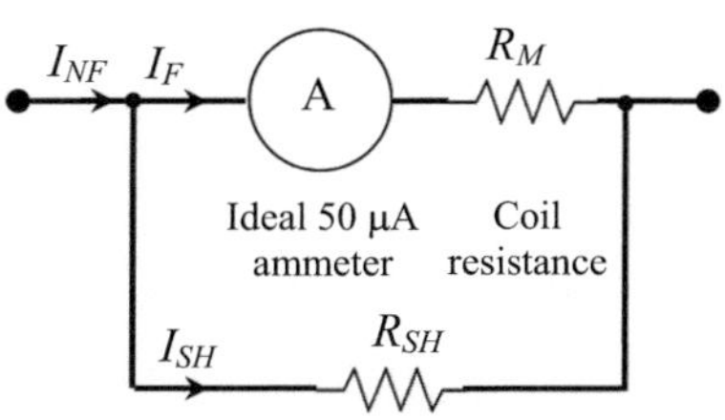

(a) Single Shunt

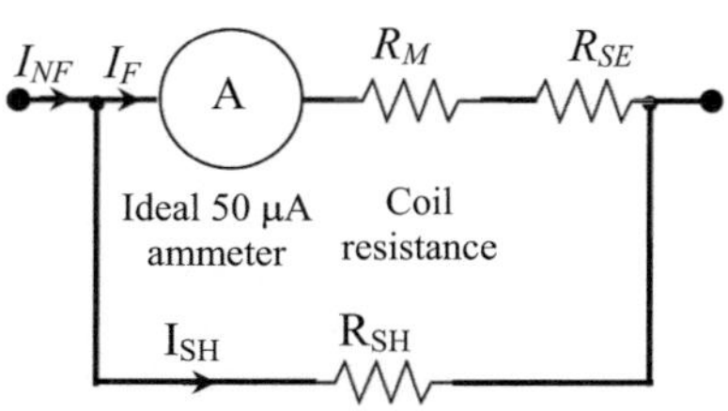

(b) Alternate: Series and Shunt

Fig. 2.17 Ammeter scale changing

is again not very practical as the 50 mΩ shunt resistance must dissipate 5 W of power when 10 A is being measured!

2.2.5.2 Multi-range Ammeters

It is possible to have several shunts on the meter and connect one or more than one in parallel to get several full scales. Fig. 2.18 shows a basic PMMC ammeter converted into three different ammeters of full-scale values, FS_1, FS_2 and FS_3 using three different shunts SH_1, SH_2 and SH_3 of appropriate values and a single-pole three-way switch. Such meters are called multi-range ammeters and are quite common.

2.2.6 *PMMC Voltmeters*

It is also possible to convert the basic PMMC ammeter into a DC voltmeter using the circuit scheme shown in Fig. 2.19. The series resistance R_S in Fig. 2.19 is chosen depending on the full-scale voltage required for the voltmeter. Let us say that we want to convert the basic 50 μA meter into a 5 V full-scale voltmeter. In such a case, when 5 V is applied to the circuit in Fig. 2.2.19, 50 μA should flow through the PMMC meter resulting in full-scale deflection. Hence, the total resistance of the circuit $R_M + R_S$ must be

$$R_M + R_S = \frac{V_F}{I_F} = \frac{5\ V}{50\ \mu A} = 100\ k\Omega.$$

Hence, $R_S = 100 - R_M = 99.9801$ kΩ. Hence, by connecting a resistance of value 99.9801 kΩ in series to the 50 μA basic PMMC ammeter, we can realize a 5

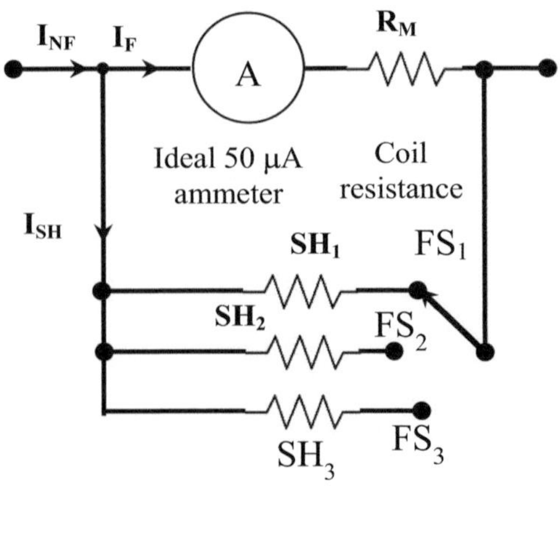

Fig. 2.18 Multi-range ammeter

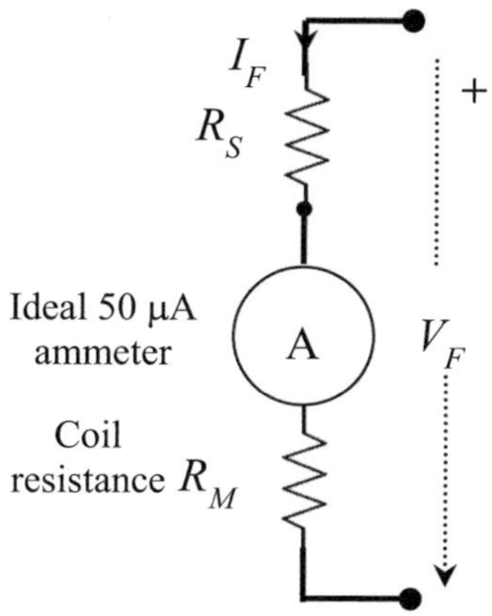

Fig. 2.19 Converting an ammeter into a voltmeter

V full-scale PMMC DC voltmeter. In this manner, by connecting a series resistance of appropriate value, a basic PMMC ammeter can be converted to a DC voltmeter with a full-scale voltage of our choice.

2.2.6.1 Sensitivity (Figure of Merit) of Voltmeters

Ideally, a voltmeter should indicate the voltage without drawing any current from the circuit, but a practical voltmeter will have to draw a current, however small it is. The quality of a voltmeter is reflected in the current drawn by it. The smaller the value of the current drawn, the better is the voltmeter. A quantitative measure of this quality is the *sensitivity* of a voltmeter, defined by the reciprocal of the current drawn by it at full scale in mA and expressed in kΩ/V. If the basic 50 μA meter we have been considering is converted into a voltmeter, the sensitivity of the voltmeter would be

$$\text{Sensitivity} = 1/50\,\mu\text{A} = 20\ \text{k}\Omega/\text{V}.$$

2.2.6.2 Multi-range Voltmeter

Like the multi-range ammeter, it is also possible to design a multi-range voltmeter as shown in Fig. 2.20. It is possible to combine the functions of a multi-range ammeter and a multi-range voltmeter, add facility for the measurement of resistance too in a single instrument called a multi-meter. The resistance measurement can be accomplished using either a series-type ohm meter or the shunt-type ohm meter. Since such types of multi-meters have become extinct, we shall not discuss them here.

Fig. 2.20 Multi-range PMMC voltmeter

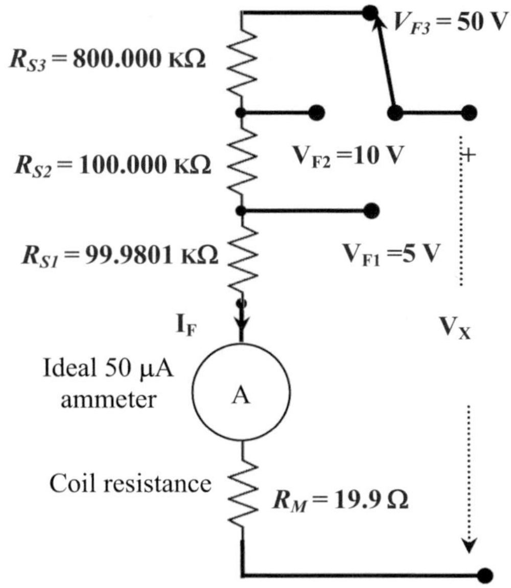

2.3 The Electrodynamometer-Type Instruments

Though the electrodynamometer-type meter became popular for the measurement of power and all the present-day analog indicating wattmeters are of this type, the principle of electrodynamometer was originally developed by Wilhelm Eduard Weber in 1845 for the accurate measurement of DC current! The PMMC meter was not even invented at that time. Even after the PMMC meters came into vogue, the electrodynamometer-type ammeters were used as standard meter for calibrating the PMMC meters. It took almost half a century for the electrodynamic principle developed for the measurement of current to be applied for the measurement of power. The reason is simple; all those working in the field of electricity at that time were more interested in measuring electric current, voltage, and energy (watt-hour). Hence, the race was on for nearly half a century in developing ampere-hour meters, which automatically provide watt-hours in the then prevailing DC systems rather than developing a true wattmeter.

2.3.1 *Electrodynamic-Type Ammeter*

The operating principles of an electrodynamic type of ammeter and a PMMC-type meter are similar, the difference being that the magnetic field in the electrodynamic type of meter is produced by an electromagnet instead of a permanent magnet as is in the PMMC-type meter. Figure 2.21 shows the constructional features of an electrodynamometer (also called electrodynamic type) ammeter. The moving coil arrangement of the PMMC meter is retained as such in the electrodynamic type of meter, namely, a moving coil of N_M turns suspended with two pivots (top and bottom), restrained by two springs, attached with a pointer that moves on a graduated scale and freely rotates between the zero and full-scale angles. The moving coil is positioned to interact with the magnetic flux produced by the fixed coil which is fabricated in two identical sections of $N_F/2$ turns and placed on either side of the moving coil. The fixed and moving coils are magnetically coupled. When they carry currents i_{F} and i_{M}, the energy $\Im_M$ stored in the magnetic field is

$$\Im_M = \tfrac{1}{2}L_F i_F^2 + \tfrac{1}{2}L_M i_M^2 + M i_F i_M . \tag{2.37}$$

Here, L_F is the self-inductance of the fixed coil, L_M is the self-inductance of the moving coil, and M is the mutual inductance between the coils. While the self-inductances L_{F} and L_{M} remain fixed, M varies as the moving coil rotates and is a function of the deflection angle θ. We will now develop an expression for the instantaneous driving torque developed by the electrodynamometer movement with coil currents i_F and i_M and at a deflection angle θ. We consider a dynamic situation when the moving coil is still in motion. In this condition, the electrical power input p_e into the coil system goes partly to increase the magnetic stored energy w_{mag} and partly to supply the mechanical power p_{mech} needed to drive the moving part of the meter. The following analysis leads to expressions for these three components.

Terminal voltages v_F and v_M of the fixed and moving coils are

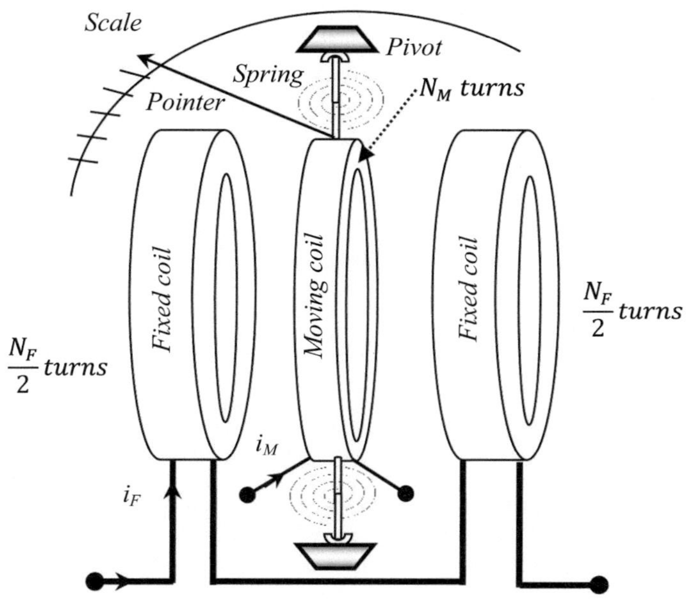

Fig. 2.21 Electrodynamometer-type meter: Fixed and Moving Coils

$$v_F = \frac{d}{dt}(L_F i_F + M i_M) = L_F \frac{di_F}{dt} + M \frac{di_M}{dt} + i_M \frac{dM}{dt}, \tag{2.38}$$

$$v_M = \frac{d}{dt}(L_M i_M + M i_F) = L_M \frac{di_M}{dt} + M \frac{di_F}{dt} + i_F \frac{dM}{dt}. \tag{2.39}$$

Power input p_e into the coils is

$$p_e = v_F i_F + v_M i_M = i_F L_F \frac{di_F}{dt} + i_F M \frac{di_M}{dt} + i_M L_M \frac{di_M}{dt} + i_M M \frac{di_F}{dt} + 2 i_M i_F \frac{dM}{dt}. \tag{2.40}$$

Rate of increase of the magnetic stored energy, p_{mag} is

$$\begin{aligned} p_{mag} &= \frac{d}{dt}\left(\frac{1}{2}L_F i_F^2 + \frac{1}{2}L_M i_M^2 + M i_F i_M\right) \\ &= i_F L_F \frac{di_F}{dt} + i_M L_M \frac{di_M}{dt} + i_F M \frac{di_M}{dt} + i_M M \frac{di_F}{dt} + i_M i_F \frac{dM}{dt}. \end{aligned} \tag{2.41}$$

Mechanical power output p_{mech}:

$$p_{mech} = \tau_\theta \frac{d\theta}{dt} = p_e - p_{mag} = i_M i_F \frac{dM}{dt} = i_M i_F \frac{dM}{dt} \frac{dt}{d\theta}. \tag{2.42}$$

Equation (2.42) simplifies to

$$\tau_\theta = i_M i_F \frac{dM}{d\theta}. \tag{2.43}$$

The torque that drives the system is thus the product of the two coil currents at that time and the rate of change of the mutual inductance M with respect to the deflection angle at that position. The expression does not depend on whether the *moving coil* is in motion or at rest. A point of interest to note is that, when the *moving coil* is in motion, the coil system draws an extra power equal to $2 i_M i_F \frac{dM}{d\theta}$ from the electrical supply than when the former is at rest. One half of this extra power serves to affect the movement against the resisting mechanical agencies (inertial, damping, and control) and the other half goes to increase the magnetic stored energy.

2.3.1.1 Steady-State Deflection θ

The analysis of the response of an electrodynamic-type ammeter varies depending on the current being measured is (i) DC or (ii) AC as detailed below.

(i) **Currents I_F and I_M are DC currents**: If the coils carry DC currents I_F and I_M, then a steady deflection θ would occur when the steady driving torque $\tau_\theta = I_M I_F \frac{dM}{d\theta}$ equals the restraining torque $K_s\theta$ of the control springs. Therefore, we get

$$\theta = \left(\frac{1}{K_S}\right)\frac{dM}{d\theta}I_F I_M. \tag{2.44}$$

(ii) **Currents i_F and i_M are AC currents**: We now develop an expression for the steady-state deflection θ of the meter when i_F and i_M are two periodic currents of the same frequency. As in the case of a PMMC meter, the electrodynamometer instrument also has a rotating part with a small natural frequency. The pointer can therefore show swings only in response to the low-frequency components in the driving torque (See Sect. 2.1.1). For applications involving currents at power frequencies and above, these swings are negligibly minute, and the pointer deflection is steady. It turns out that this steady deflection θ corresponds to the average value of the varying τ_θ over one period. This in turn depends on the average value of $(i_F i_M)$. Thus, θ for AC periodic currents is given by

$$\theta = \frac{1}{K_S}\frac{dM}{d\theta}\left(\frac{1}{T}\int_0^T i_F i_M\, dt\right). \tag{2.45}$$

Here, the rate of change of M refers to its value at the deflection θ and $\frac{1}{T}\int_0^T i_F i_M\, dt$ is the average over one period T. We will now consider the various forms in which the electrodynamometer instrument is used for the measurement of current and voltage.

2.3.1.2 Use as a Milliammeter

In this class of ammeters, the fixed and moving coils are connected in series and carry the line current i as shown in the Fig. 2.22. Here, R_M and R_F are the coil winding resistances and L_M and L_F are the self-inductances of the moving and fixed coils. Now, $i_F = i_M = i$, then $i_F i_M = i^2$. We know that the average value of i^2 for a general periodic current i of period T is defined as the square of its effective value I_{rms}. Thus, the meter deflection for a general periodic current i is proportional to the square of its effective value and hence:

$$\theta = \frac{1}{K_S}\frac{dM}{d\theta}I_{rms}^2. \tag{2.46}$$

For the practically important special cases of DC and sinusoidal currents, $\theta = \frac{1}{K_S}\frac{dM}{d\theta}I^2$, where the symbol I represents either the strength of the DC current or the effective value (rms value) of the sinusoidal current, as the case may be. Unlike a PMMC instrument, the electrodynamic type of ammeter is a dual-purpose AC/DC ammeter, and the same scale gives either the AC rms or the DC value of the

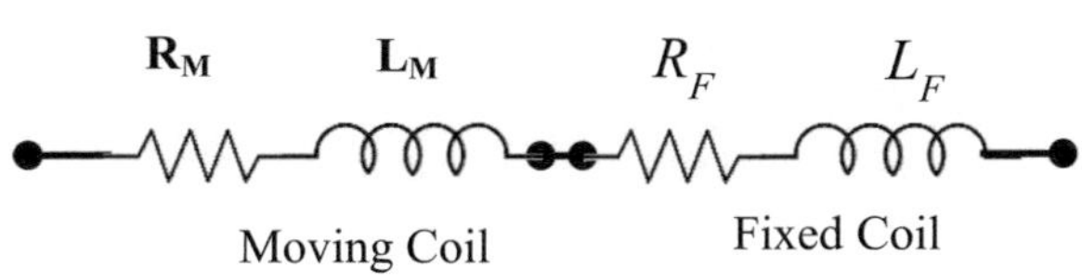

Fig. 2.22 Electrodynamic milliammeter—Connecting the fixed and Moving coils in series

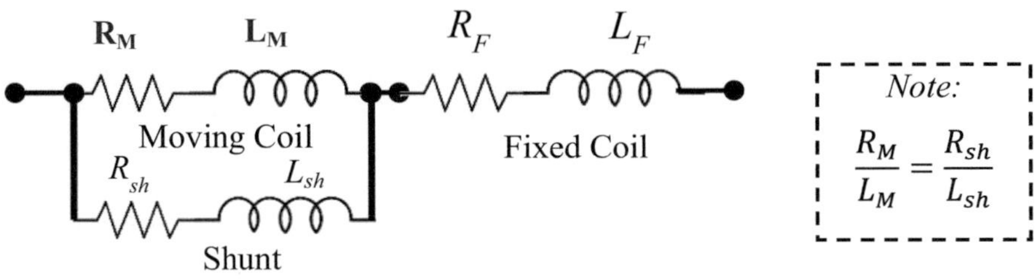

Fig. 2.23 Electrodynamic ammeter—Shunt in parallel with Moving coil

current that is fed to the meter. This property enables the meter to be calibrated on DC with the available DC standards and subsequently used on AC without any further calibration. That the same calibration holds for both is facilitated by the fact that the magnetic circuits are free of iron and its attendant hysteresis, eddy- current and nonlinear effects, which would modify the scale factor on AC use. Due to the absence of iron, the working flux density (typically 5 mT) in this type of instrument is much lower than the density of about 500 mT obtainable in a PMMC instrument. Consequently, the electromagnetic forces are that much lower. Hence, the lowest full-scale value achievable in these meters is in the milliampere range and not in the microampere range.

2.3.1.3 Use as an Ammeter for Higher Currents

The moving coil in an electrodynamometer instrument is wound with very thin wire and hence its current-carrying capacity is limited to the milliampere range. The use of a shunt in parallel with the moving coil, as shown in the Fig. 2.23, enables a self-contained ammeter with full-scale value in the range of amperes. The shunt should have an inductance/resistance ratio equal to that of the moving coil branch. This is necessary for the moving coil current i_M, in AC usage, to have the correct fraction of and the same phase as the line current i.

2.3.1.4 Use of Electrodynamometer-Type Ammeter as a Voltmeter

As in the case of a PMMC meter, a high resistance in series with the fixed and moving coils allows an electrodynamometer voltmeter to be built. However, only sensitivities of the order of 100 Ω/V can be achieved in this meter as compared with a typical figure of 20 kΩ/V achievable with a PMMC voltmeter. Hence, electrodynamic-type voltmeters are normally built only for AC use at a specified frequency (say 50 Hz) and the impedance is suitably adjusted. In such AC voltmeters, the meter calibration will not be valid at other frequencies and on DC.

2.3.1.5 Additional Comments

Shielding: Due to its very weak operating magnetic field, an air-cored dynamometer instrument needs to be protected from the influences of external magnetic fields. To this end, the entire coil system is enclosed in a casing of high permeability to bypass the external fields.

Damping: A standard provision in all pointer-type electrical instruments is a damping mechanism for bringing the pointer quickly to rest after any transition. *Air-damping* is the mechanism normally adopted in the electrodynamometer and the moving-iron instruments to be covered later in Sect. 2.3. The device used is a light aluminum vane attached to the rotating spindle and free to move in a stationary chamber closed at one end. The forces of compression and suction during the pointer movements provide the required damping.

Scale shape: The mutual inductance M between the fixed and moving coils is clearly dependent on the deflection angle θ. Its rate of change $\frac{dM}{d\theta}$, which influences the value of the deflecting torque, has its largest value when the moving coil is in a plane perpendicular to the planes of the fixed-coil sections. The reduction in this value is not significant when the coil deviates from this central position by less than about 45° on either side.

Accordingly, the scale of an electrodynamometer instruments is arranged to have a length of about 90°, with the maximum $\left(\frac{dM}{d\theta}\right)$ position at its centre. Assuming that $\left(\frac{dM}{d\theta}\right)$ is constant over this working range, we can approximate the formula for steady-state deflection in these meters as KI^2 or KV^2 (as the case may be), where K is a constant.

This implies, for example, that if the current is doubled, the deflection is increased fourfold. The nonlinear scale of an ammeter of this type, shown in Fig. 2.24, illustrates this stretching effect, particularly in the midrange.

2.3.2 Electrodynamometer-Type Wattmeter

Though the electrodynamometer principle was invented to measure DC currents and found to work equally well for AC currents, it was later applied (more than half a century after its invention) to measure power (in both DC as well as AC circuits). Soon it became popular for use in wattmeters rather than ammeters and voltmeters. In an electrodynamometer-type wattmeter, the fixed coil called the current coil carries the circuit current i_L. The moving coil, called the voltage coil (in the olden times it was called the potential coil), is made to carry a current proportional to the voltage of the circuit through the series connection of an appropriate resistance Fig. 2.25 shows the connection of an electrodynamometer-type wattmeter to measure power fed to a load from an alternating voltage source of period T. The series resistance R_V is taken to include the voltage coil resistance as well. We further assume that the

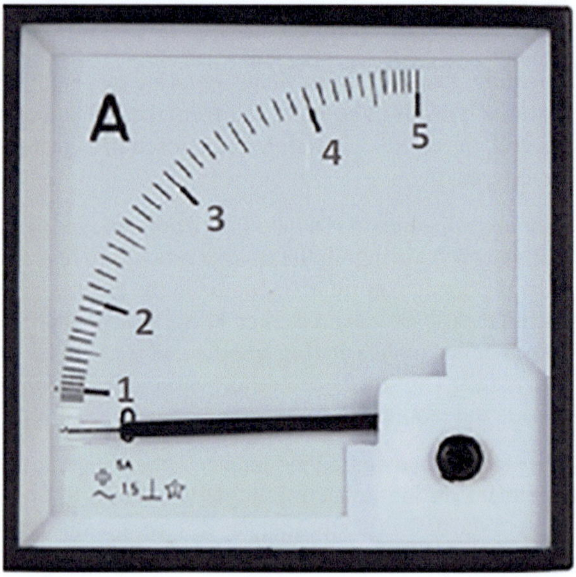

Fig. 2.24 Nonlinear scale of electrodynamic-type ammeter

inductance and resistance of the fixed coil and the inductance of the moving coil are negligible, two simplifications which are valid in practice. From the earlier analysis of an electrodynamometer instrument in Sect. 2.2.1, we deduce the deflection angle θ as

$$\theta = \frac{dM}{d\theta}\frac{1}{K_S}\left[\frac{1}{T}\int_0^T \frac{v_L}{R_V} i_L \, dt\right]. \tag{2.47}$$

We note here that the integration (averaging function) arises due to the fact the moving coil of the electrodynamometer-type wattmeter can respond and deflect by the angle θ only due to the DC value of the torque generated by the two currents (currents through the fixed and moving coils). We also note that the simplifying assumptions made earlier enable us to express:

$$i_M = \frac{v_s}{R_V} = \frac{v_L}{R_V}.$$

Taking $\frac{dM}{d\theta}$ to be a constant, we now get

$$\theta = K_W\left[\frac{1}{T}\int_0^T v_L i_L \, dt\right] = K_W \, P_L. \tag{2.48}$$

Here, $K_W = \frac{dM}{d\theta}\frac{1}{K_S R_V}$ is the meter constant of the wattmeter and P_L is the active or average power consumed by the load.

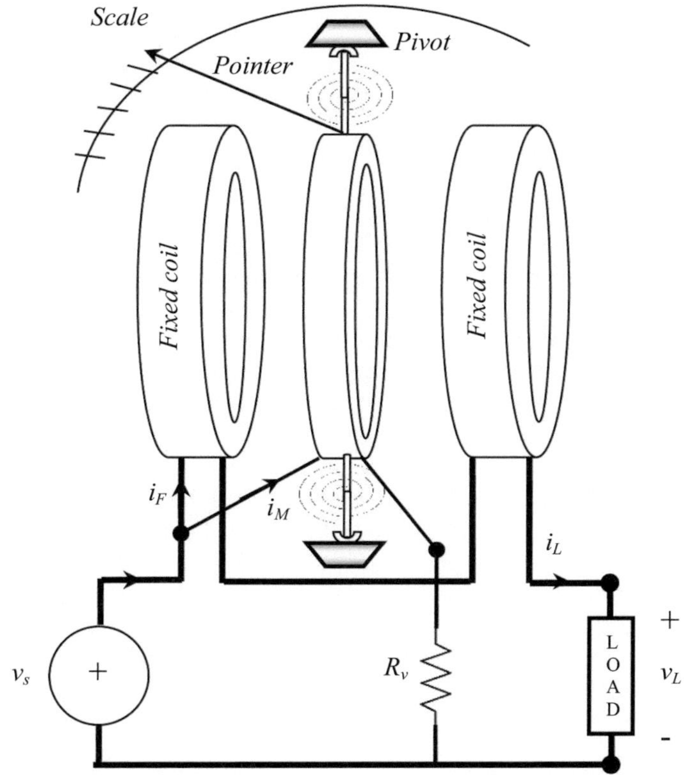

Fig. 2.25 Electrodynamometer wattmeter

2.3.2.1 Constructional Features of Electrodynamometer-Type Wattmeters

The wire size (diameter) of the fixed coils is determined on the basis of the maximum load current for which the wattmeter is to be used. This current is specified as the rated current or the full-scale current, I_F. The maximum current the moving coil can safely carry multiplied by the resistance R_V (sum of the resistance of the moving coil and the value of the resistance connected in series with the moving coil) determines the maximum (load) voltage that can be applied to the voltage terminals, normally marked as rated or full-scale voltage V_F. If the wattmeter is used in a DC circuit, then I_F and V_F are the full-scale DC current and voltage specification of the wattmeter. On the other hand, if the wattmeter is used in an AC circuit, then I_F and V_F are the rms values of the full-scale current and full-scale voltage, respectively. Four terminals are provided on the wattmeter to facilitate its connection in a circuit to measure the power to a load. Two of these terminals marked M and L have the current coil connected internally between them. Externally M and L are connected to the mains (power source) and the load on one side of the supply. The voltage coil along with total resistance R_V is internally connected between the other two terminals. One of these, labelled COM is connected to either M or L. The label of the fourth terminal indicating the rated voltage of the meter, for example, 250 V is connected to the other side of the supply. If the wattmeter has multiple voltage ratings, say 150 V/ 300 V/600 V, then three corresponding terminals are provided instead of one and the appropriate terminal is to be used.

2.3.2.2 Specification of the Current and Voltage in a Wattmeter

The fixed coils of electrodynamometer-type wattmeters, as seen from Fig. 2.25, carry the load current. Hence, depending on the maximum load current for which the wattmeter is to be used, its wire size (diameter) is determined. The maximum current that the fixed coil can safely carry is specified as the full-scale current, I_F. The maximum current the moving coil can safely carry multiplied by the total series resistance (sum of the resistance of the moving coil and the value of the resistance connected in series with the moving coil) determines the maximum (load) voltage that can be applied to the voltage terminals, normally marked as full-scale voltage V_F. If the wattmeter is used in a DC circuit, then I_F and V_F are the full-scale DC current and voltage specification of the wattmeter. On the other hand, if the wattmeter is used in an AC circuit, then I_F and V_F are the rms values of the full-scale current and full-scale voltage, respectively.

2.3.2.3 Wattmeter Ratings and Variations

To prevent damage to its moving mechanism, a wattmeter cannot obviously be used where the power to be read is above its full-scale value P_F. The proper usage of a wattmeter therefore demands that none of the three quantities, viz. the load voltage V_L, current I_L, and power P_L does not exceed their respective safe value of the wattmeter in use, namely V_F, I_F, or P_F. In AC circuits, where these wattmeters are normally used, the load power factor also comes into play. As a result, compliance with some two of the above safe limits will not necessarily make certain that the third condition is also met. Furthermore, there exist wattmeters, where the full-scale mark P_F differs from the value of the product $V_F I_F$. Therefore, unlike in a voltmeter or ammeter, where the knowledge of one rated value (voltage or current) is sufficient for its usage, the data of a wattmeter is complete only if the three ratings V_F, I_F, and P_F are known.

It is also common to give the wattmeter data by an equivalent set of specifications, viz. V_F, I_F and λ, where λ is a reference value of power factor. The understanding is that the full-scale reading P_F is obtained on the wattmeter if the load voltage (potential coil voltage) is V_F and the load current (current coil current) is I_F and the load power factor is λ. In other words, $P_F = \lambda\ V_F\ I_F$. Most wattmeters have $\lambda =$ 1. These are called unity power factor (UPF) type of wattmeters. The second type of wattmeters are designed for a low value of λ, typically 0.2. These are called low power factor (LPF) wattmeters. The power factor specification λ of the wattmeter is not to be confused with the actual load power factor λ_L in any application.

We now proceed to clarify further the observations in the two foregoing paras using two specific examples of wattmeter specifications:

(i) **300 V, 1 A, UPF**: This is a UPF wattmeter with a full-scale reading of 300 W, occurring for example with $V_L = 300$ V, $I_L = 1$ A, and $\lambda_L = 1$.

Now consider the following three sets of loads and the corresponding wattmeter readings P_L that would ensue.

$$\begin{aligned} &V_L = 300\ \text{V};\ I_L = 1.25\ \text{A};\ \lambda_L = 0.8;\ P_L = 300\ \text{W} \\ .\ &V_L = 300\ \text{V};\ I_L = 2\ \text{A};\quad\ \ \lambda_L = 0.2;\ P_L = 120\ \text{W} \\ &V_L = 400\ \text{V};\ I_L = 1\ \text{A};\quad\ \ \lambda_L = 0.6;\ _L = 240\ \text{W} \end{aligned}$$

The wattmeter is overloaded in all the three cases, as either the current coil current is more than its rated value of 1 A or more than the rated voltage is applied to the voltage coil. However, the dial reading on the wattmeter has not exceeded the full-scale mark in all the cases. Clearly, this manner of overloading cannot be deduced from observing the wattmeter dial alone. This risk is present whenever the load power factor λ_L is less than the reference power factor λ. It is therefore advisable in practice to simultaneously monitor the load current and voltage, using an ammeter and voltmeter or by any other appropriate means.

(ii) **250 V, 2 A, 0.2 PF**: This is an LPF wattmeter with a full-scale reading of 100 W, occurring for example with $V_L = 250$ V; $I_L = 2$ A; $\lambda_L = 0.2$.

As in UPF wattmeters, where also current or voltage overloading can occur without any specific indication showing up from the meter dial. This contingency can arise with a load power factor less than the reference value of 0.2. As an example, consider the case of:

$V_L = 250$ V ; $I_L = 3$ A; $\lambda_L = 0.1$, where the current coil is overloaded but the meter reading of 75 W is well below the full-scale mark.

LPF wattmeters have another limitation. For load power factors more than the reference power factor, the full ranges of rated voltage V_F and of rated current I_F would not be available for use. Consider, for instance, a load power factor of 0.4. With $V_L = V_F = 250$V and $I_L = I_F = 2$A, the load power would be 200 W and the pointer would be forced to go beyond the scale causing possible damage to the moving system. Only If $V_L I_L$ does not exceed 250 VA, can the wattmeter operate safely with this value of load power factor. Under this condition (current coil of an LPF wattmeter carrying full-scale current), full-scale DC voltage cannot and should not be applied. If this is the case, what is the purpose of an LPF wattmeter? LPF wattmeters are especially useful for power measurement in an AC circuit operating at a low power factor. For example, a transformer under its no-load condition draws a small amount of power from the source at a very low power factor. Let us take the example of a 230 V/230 V, 1 kVA isolation transformer with estimated values

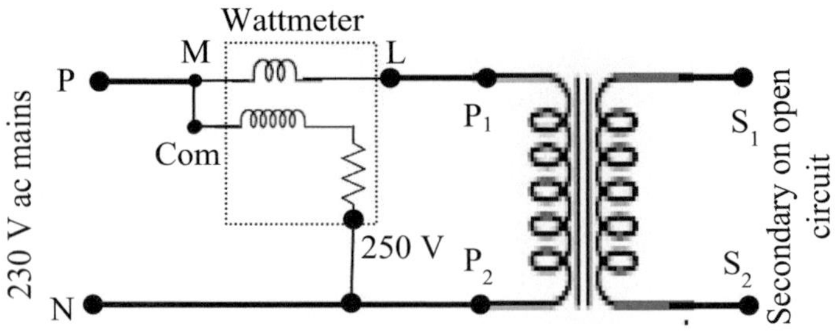

Fig. 2.26 No load loss measurement on a transformer

of 0.4 A and 20 W for its no-load current and no-load power input. The wattmeter connection to measure this power is shown in Fig. 2.26. If a 250 V, 0.5 A, UPF wattmeter is used in the circuit, the power reading (of about 20 W) should be taken below the 20% mark of its full-scale of 125 W. On the other hand, if a 250 V, 0.5 A, 0.2 PF wattmeter is used, the reading of about 20 W is taken near the 80% mark of its full-scale of 25 W. In the next chapter on errors in measurement we will learn why errors are minimized by taking readings as close to the full scale as possible in indicating instruments. This is the justification for the need of LPF wattmeters for use in low power factor AC circuits.

2.4 The Moving Iron-Type Instruments

Though the utility value of alternating current was demonstrated by Guillaume Duchenne in 1855, the actual use of alternating current for domestic and industrial applications started nearly four decades later. Till then for the measurement of DC voltage and currents, the PMMC and electrodynamic-type meters were sufficient. Once the AC system became popular, new types of measuring instruments were needed. To fulfill this requirement, several types of AC ammeters and voltmeters were invented. Of these, the moving iron-type meter is by far the most popular type. Though the original idea was used in the ampere gauge developed by Kelvin, the moving iron meter in its present form was invented by Professor W.E. Ayrton? and his co-worker John Perry in 1879. In a moving iron-type meter, the stem, the springs, and the pointer moving on a scale are retained as in the earlier types of meters. Instead of the moving coil (as in PMMC or electrodynamic-type meters) a curved piece of iron with high permeability is attached to the stem. Another iron piece of similar shape is fixed to the inner wall of a coil having N turns as indicated in Fig. 2.27. When a current i_C (in the direction indicated in Fig. 2.27) is sent through the coil, the coil becomes an electromagnet (say with the flux lines emanating at the top of the coil, thus creating a north pole at the top). This flux magnetizes both the movable iron plate attached to the stem and the fixed iron plate, with north poles at the top. Now these plates possessing similar poles at the top and bottom repel each other and

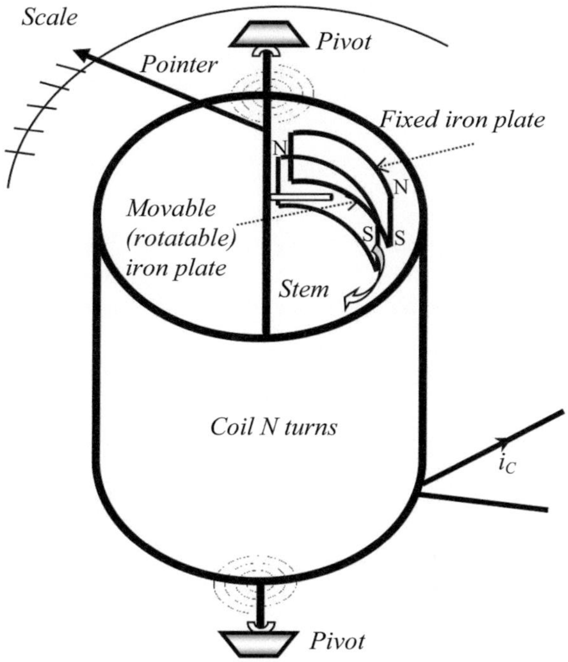

Fig. 2.27 Moving iron-type ammeter—repulsion type

thus generate a force as indicated in Fig. 2.27. Since the fixed plate cannot move, the force moves (rotates) the movable plate and along with it the stem and the pointer are rotated.

Now if we reverse the current through the coil then the flux will be such that the bottom part of the coil becomes the north pole. The magnetization of the fixed and movable iron plates also will reverse, and now both will have north poles at the bottom and south poles at the top. Once again, the plates experience a repulsive force in the same direction as in the previous case. Thus, the moving plate along with the stem and the pointer will rotate clockwise as before. Hence, like the electrodynamometer-type ammeter, the deflection is always positive whether we send a positive current or negative current through the coil. As the moving plate rotates away from its rest position, the overlap area between the two parallel magnetized plates diminishes, thereby reducing their mutually demagnetizing effects. This results in an increase of the effective magnetic flux linkages of the coil for the same current. The coil inductance L therefore turns out to be an increasing function of the deflection angle α. We now derive an expression for T_d which is the driving torque generated in the instrument. The analysis is quite similar to what we used in Sect. 2.2.1 for an electrodynamometer instrument, where we equated, on instantaneous value basis, the electrical power inflow to the system to the sum of the rate of increase of the stored magnetic field energy and the power needed to rotate the moving system against the opposing mechanical forces. We examine this balance of power flow at an arbitrary instant t during the transition of the moving system from its initial position to its final steady-state position.

2.4.1 Electrical Power Input P_E into the Coil

Let the deflection of the moving system at this instant be α and let the instantaneous values of the coil current and its terminal voltage be i and v. Note that as α varies with time, L, which is a function of α also varies with time. Now we have

$$v = \frac{d}{dt}(Li) = L\frac{di}{dt} + i\frac{dL}{dt}. \tag{2.49}$$

Instantaneous power p_e drawn by the coil.

$$p_e = vi = Li\frac{di}{dt} + i^2\frac{dL}{dt}. \tag{2.50}$$

The rate of increase of energy in the magnetic field, p_{mag}, is

$$p_{mag} = \frac{di}{dt}\left(\frac{1}{2}Li^2\right) = Li\frac{di}{dt} + \frac{1}{2}i^2\frac{dL}{dt}. \tag{2.51}$$

The mechanical power output p_{mech} is

$$p_{mech} = \tau_d\frac{d\alpha}{dt} = p_e - p_{mag} = \frac{1}{2}i^2\frac{dL}{dt} = \frac{1}{2}i^2\frac{dL}{dt}\frac{dt}{d\alpha}. \tag{2.52}$$

Simplifying Eq. (2.52), we get

$$\tau_d = \frac{1}{2}i^2\frac{dL}{d\alpha}. \tag{2.53}$$

This expression makes it clear that the driving torque at any given values of t and α is dependent only on the coil current at that time and the rate of variation of coil inductance with respect to the deflection angle α at that position. It is valid whether the moving plate is in motion or is at rest. Now, as we had observed on earlier occasions, for periodic currents in the frequency band of practical relevance, it is the average value of τ_d (over its own period) that governs the steady-state deflection θ of the pointer. It is the value of the deflecting torque to be reached by the opposing torque of the stretched control springs to establish the static equilibrium. If K_s is the spring constant and θ is the steady-state deflection, we then have

$$(\tau_d)_{avg} = K_s\theta = \frac{1}{2}\frac{dL}{d\theta}\left(i^2\right)_{avg} = \frac{1}{2}\frac{dL}{d\theta}\left(\frac{1}{T}\int_0^T i^2 dt\right) = \frac{1}{2}\frac{dL}{d\theta}I_{eff}^2. \tag{2.54}$$

Equation (2.54) reduces to

$$\theta = \frac{1}{2K_s}\frac{dL}{d\theta}I_{eff}^2. \tag{2.55}$$

Here, $\left(\frac{dL}{d\theta}\right)$ is $\left(\frac{dL}{d\alpha}\right)$ evaluated at $\alpha = \theta$. From Eq. (2.55), it is obvious that this ammeter works equally well for DC and AC currents as I_{eff} for a DC current I_{dc} being I_{dc} itself. Once again, like the electrodynamic type of ammeter, the moving iron ammeter will also have a square-law scale, if $\left(\frac{dL}{d\alpha}\right)$ is a constant. However, a square law implies a severe crowding of divisions (if marked for equal increments of current) at the lower end of the scale, making the latter unusable. Therefore, manufacturers design the shape of the iron plates suitably and tweak the $\left(\frac{dL}{d\alpha}\right)$ characteristic to obtain a scale with a more uniform spacing of divisions as shown in Fig. 2.28. Moving iron instruments are simple in construction and are inexpensive. As there is no need to feed current into the moving system, they are more robust than PMMC and dynamometer instruments. Even though they can be used both on DC and at different AC frequencies, the calibration will be different because of variations in the coil reactance and hysteresis and eddy current effects in metal parts.

Like the PMMC and electrodynamometer-type ammeters, the moving iron-type meters also can be converted to read higher currents by employing a shunt and as a voltmeter by employing a series resistance. The fact that for a given range, a moving iron ammeter possesses higher inductance than the electrodynamometer-type ammeter should be kept in mind when extending the range of a moving iron-type AC ammeter or voltmeter.

The moving iron mechanism illustrated in Fig. 2.27 utilizes the repulsion between two soft iron pieces that are magnetized by the same external magnetic field. It is also possible to make a moving iron deflection mechanism using the attraction force that gets developed between magnetic poles of opposite polarity between the fixed plate

Fig. 2.28 The near uniform scale of a moving iron-type ammeter

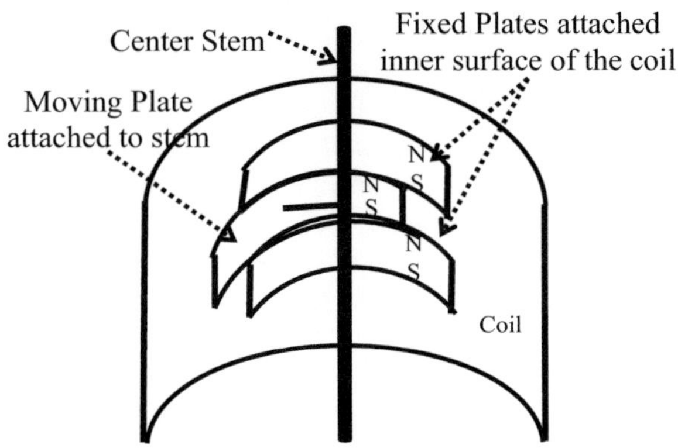

Fig. 2.29 Fixed and moving plates of attraction-type moving iron ammeter

and the moving iron plate as illustrated in Fig. 2.29. Here, it is seen that the fixed plate is split into two and the movable plate sandwiched between the two halves. The poles induced in the nearby edges of the fixed and movable iron plates are always of opposite polarity to each other irrespective of the direction of the current in the coil. Hence, an "attraction" force is always developed between them. Whether a moving iron-type meter exploits repulsion or attraction force, the meter characteristics are always governed by Eqs. (2.53) and (2.55). Moving iron instruments are simple in construction and are inexpensive. As there is no need to feed current into the moving system, they are more robust than PMMC and dynamometer instruments. Even though they can be used both on DC and at different AC frequencies, the calibration will be different because of variations in the coil reactance and hysteresis and eddy current effects in metal parts. Thus, moving iron instruments (ammeters or voltmeters) are popular only as meters operating at power supply frequency (50 Hz of 60 Hz).

2.5 The Rectifier-Type Instrument

Following the development of electronics and availability of the diode rectifier, the PMMC meter was suitably augmented to read both DC and AC utilizing the rectifiers. A typical rectifier-type PMMC ammeter is shown in Fig. 2.30a. If we take the current i in Fig. 2.30 to be a sinusoid of rms value I as given in Fig. 2.30b, the meter current i_M will be as indicated in Fig. 2.30b. Since the PMMC meter responds to the average value of i_M, the meter deflection angle θ will be

$$\begin{aligned}\theta &= K_M \tfrac{1}{T}\int_0^T i_M\,dt = \tfrac{\sqrt{2}}{T}\left(\int_0^{\frac{T}{2}} I\ Sin(\omega t)\ dt - \int_{\frac{T}{2}}^{T} I\ Sin(\omega t)\ dt\right)\theta \\ &= K_M \tfrac{2\sqrt{2}}{\pi} I.\end{aligned} \tag{2.56}$$

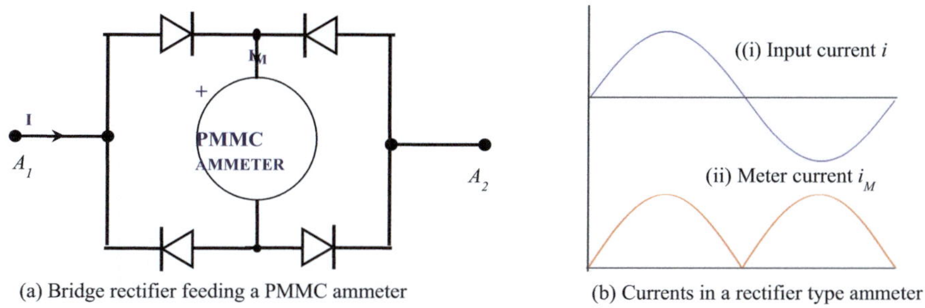

(a) Bridge rectifier feeding a PMMC ammeter

(b) Currents in a rectifier type ammeter

Fig. 2.30 Rectifier-type (PMMC) ammeter

Thus, the rms value I of the current i can be obtained by multiplying the deflection angle θ by a factor $\frac{\pi}{2\sqrt{2}} \approx 1.110$ and noting the current value marked there or in other words the markings on the scale can be suitably altered to incorporate the multiplication factor of 1.110. Table 2.2 illustrates this concept for a 100 mA full-scale PMMC meter having a full-scale deflection angle of 100°. Of course, when we use his meter for DC current measurement, every 1° corresponds to 1 mA as indicated in row 2 of Table 2.2. However, for AC current the meter will have to be marked as indicated in row three of Table 2.2.

It is also possible to have the same scale (as in Table 2.3) for both AC and DC by incorporating a shunt for DC measurement as indicated in Fig. 2.31. Please note that the basic PMMC meter employed in Fig. 2.2 is the same as in Fig. 2.30 and possess a sensitivity of 1°/mA. Here, for measuring AC current, the switch S is kept open and for measuring DC current switch S is closed. The required value of R_{SH} is 9.031731 R_M, where R_M is the PMMC meter's internal resistance.

It should be noted that a rectifier-type AC ammeter is always marked to read 1.110 times the average value of the full wave rectified average value of the current and hence would indicate the rms value only for a sinusoidal current. If an AC current

Table 2.2 Scales for AC and DC measurement in a rectifier-type instrument

Current in mA	0	10	20	30	40	50	60	70	80	90	100
Marking angle θ for DC	0°	10°	20°	30°	40°	50°	60°	70°	80°	90°	100°
Marking angle θ for AC	0°	9°	18°	27°	36°	45°	54°	63°	72.1°	81.1°	90.1°

Table 2.3 Same scale for measurement of DC/AC (rms) currents in a rectifier-type ammeter

i (DC[@] or rms AC*) mA	0	10	20	30	40	50	60	70	80	90	100
Marking angle θ	0°	9°	18°	27°	36°	45°	54°	63°	72.1°	81.1°	90.1°

@ Switch S in Fig. 2.31 is closed and R_{SH} comes in parallel to R_M. * sinusoidal current

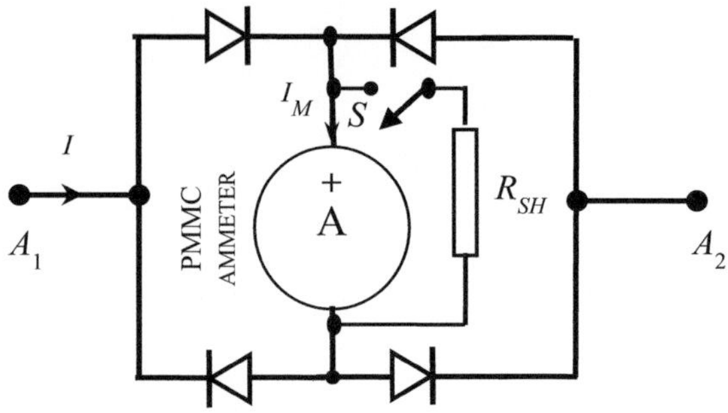

Fig. 2.31 Rectifier-type (PMMC) meter with same scale for DC and AC

possessing any other waveform (examples: pulse, triangular, square) is measured with a rectifier-type ammeter, the ammeter will not indicate the rms value. The value indicated will always be 1.110 times the average of the rectified current. Another important point to be noted is that the voltage drop across the rectifier-type ammeter terminals A_1, A_2 will be $\{I(R_M + 2R_{ON}) + 2V_\gamma\}$ volts, where I is the value of the DC current or rms value of the AC current, R_M is the PMMC meter's internal resistance, R_{ON} is the forward (ON) resistance of the diodes employed and V_γ is the forward drop (cut-in voltage) of the diode for the scheme shown in Fig. 2.30. For the scheme of Fig. 2.31, the drop will be $I_{dc}\ \frac{R_{SH}R_M}{R_{SH}+R_M} + 2R_{ON} + 2V_\gamma$ for DC current and $I_{ac}\ (R_M + 2R_{ON}) + 2V_\gamma$ for AC current. While the inclusion of the diodes for AC current measurement is essential, for DC current measurement, we do not require the diodes. Hence, it will be prudent to use the PMMC meter without the diodes for DC and with diodes for AC. Like a PMMC or electrodynamometer-type ammeter or moving iron ammeter, the full-scale reading of a rectifier-type ammeter can also be extended by simply connecting a shunt resistor of suitable value as shown in Fig. 2.17. Again, a rectifier-type PMMC-type ammeter can be converted into an AC voltmeter to read AC voltages by simply connecting a series resistor as in Fig. 2.19.

2.6 The Induction-Type Watt-Hour (Energy) Meter

Another important parameter to be measured in an electrical network is the energy consumed by an implement or apparatus or all electrical apparatuses in a house. This meter is aptly called an *energy meter* but is also known as a *watt-hour meter*, since electrical energy is historically measured using the term "*watt-hour*" instead of a joule, which is really a watt-second.

2.6.1 *The* Unit *of Electricity*

If an electrical apparatus consumes a power of 1.0 W and is in operation for 1 hour, then the energy consumed by that apparatus1 watt-hour or simply 1 Wh. Obviously 1 Wh = 1 W × 3600 s = 3600 watt-seconds or joules. Since 1 Wh is too small a quantity in the context of electrical power 1 kWh (that is 1000 Wh) is taken as the basic unit of energy consumption in electrical apparatus. Thus, *one unit of electrical energy* is 1 kWh or 3.6 MJ.

2.6.2 *The Single-Phase Induction-Type Energy Meter*

For a long time, the energy meter utilized in electrical power utilities was invariably the induction-type energy meter. Though today these meters are no longer manufactured, many of them are still in use and will continue to be used, especially in households, for quite some time. Figure 2.32a shows a simple block schematic of a typical single-phase induction-type energy meter. Figure 2.32b shows the actual components of a single-phase energy meter assembly. The meter consists of a central circular disc made of thin aluminum sheet (≈0.5 mm) attached at its center to a pivoted spindle so that the disc can freely rotate. A mechanism is attached to the spindle to count the number of revolutions made by the disc and indicate a proportional number of the units of energy consumed. Like the basic PMMC or electrodynamic instruments, the mounting of the rotating disc is accomplished using either pin and socket arrangements or jewel bearings. A current coil having N_c turns and a voltage coil having N_v turns are placed on the disc as shown in Fig. 2.32. As the name indicates, the current coil, connected in series with the load, carries the current i_L drawn by the latter. The voltage coil connected across the load is fed with the load voltage v_L.

Let $v_L = \sqrt{2}V_L Cos(\omega t)$ and $i_L = \sqrt{2}I_L Cos(\omega t - \theta)$. The alternating magnetic flux $\emptyset_v$ set up by the voltage coil produces eddy currents i_{ev} in the disc. Similarly, the flux $\emptyset_i$ set up by the current coil generates eddy currents i_{ei} in the disc. The phasor diagram shown in Fig. 2.33 indicates the relative phases of these sinusoidal quantities. Their rms values have the following proportionality relations, where the constants K are independent of the magnetic and electrical variables involved.

$$\emptyset_v = \frac{K_1}{\omega} V_L; I_{ev} = K_2\omega\emptyset_v; \quad \emptyset_i = K_3 I_L \, and \, I_{ei} = K_4\omega\emptyset_i. \tag{2.57}$$

The eddy currents i_{ev} and i_{ei}, flow over wide paths in the disc surrounding the projections of the respective poles. A significant portion of each current also crosses the flux of the other pole. The spread of these currents is not specifically shown in Fig. 2.34 to avoid clutter. It is the interactions between the eddy current i_{ev} with $\emptyset_i$ on one hand and that between i_{ei} with $\emptyset_v$ on the other, which generate the driving torque responsible for the rotation of the disc. To understand this, let us imagine that the total eddy current i_{ei} flows in an equivalent filamentary path crossing $\emptyset_v$ as shown in

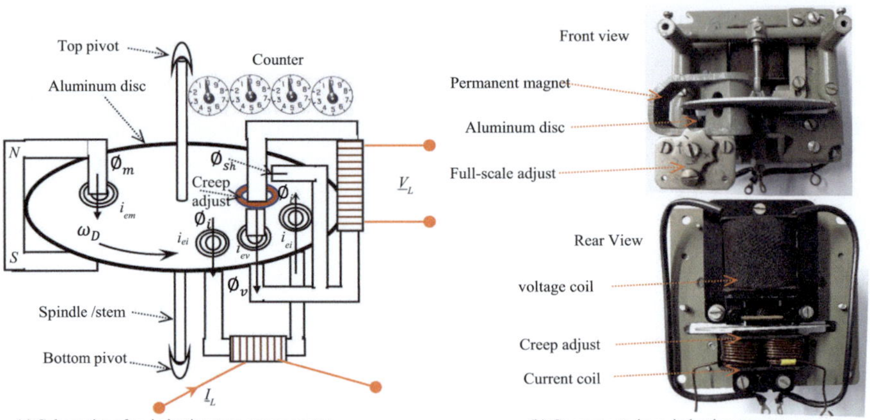

(a) Schematics of an induction type energy meter (b) Components in an induction type energy meter

Fig. 2.32 Induction-type watt-hour (energy) meter

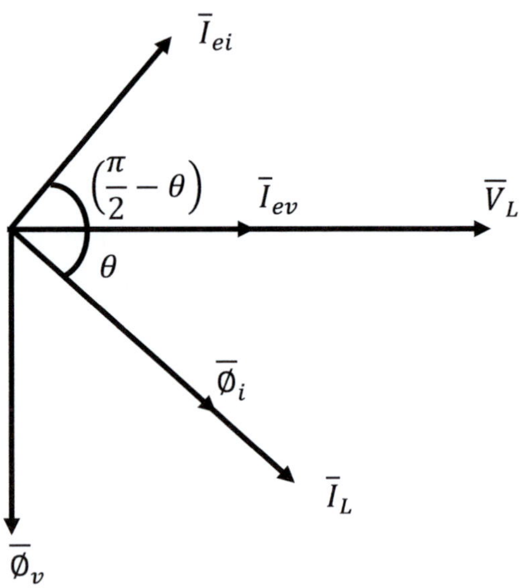

Fig. 2.33 Phasors in Induction-type watt-hour (energy) meter

Fig. 2.34. The instantaneous force generated by their interaction is proportional to the product of i_{ei} and B_v, the flux density in the region (which itself is proportional to the flux $\emptyset_v$). As a result, the average torque τ_1 acting on the disc by the eddy current produced by the voltage coil reacting to the flux produced by the current coil is given by

$$\tau_1 = K_5(i_{ev}\emptyset_i)_{average} = K_5 I_{ev}\emptyset_i Cos\ \theta = K_5(K_2\omega\emptyset_v)(K_3 I_L)Cos\ \theta. \tag{2.58}$$

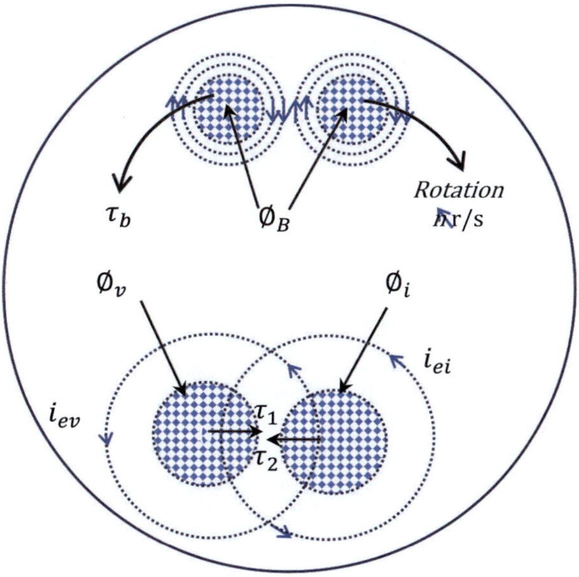

Fig. 2.34 Fluxes and Eddy currents and torques on the disc

Equation (2.58) further reduces to

$$\tau_1 = K_5(K_2V_L)(K_3I_L)Cos\ \theta = K_1K_2K_3K_5(V_LI_LCos\ \theta). \tag{2.59}$$

Similarly, the torque τ_2 generated by the interaction i_{ei} with φ_v can be similarly worked out as follows:

$$\tau_2 = K_6(i_{ei}\emptyset_v)_{average} = K_6I_{ei}\emptyset_vCos(\pi - \theta) = -K_6(K_4\omega\emptyset_i)\left(\frac{K_1V_L}{\omega}\right)Cos\ \theta. \tag{2.60}$$

Equation (2.60) further reduces to

$$\tau_2 = -K_6(K_4K_3I_L)(K_1V_L)Cos\ \theta = -K_1K_3K_4K_6(V_LI_LCos\ \theta). \tag{2.61}$$

The total driving torque T_d on the disc is therefore

$$\tau_d = (\tau_1 - \tau_2) = K_1K_3(K_2K_5 + K_4K_6)P_L = K_dP_L, \tag{2.62}$$

where

(i) $P_L = V_LI_L\ \cos\ \theta$, the load power and
(ii) K_d is a constant, which depends on the number of turns in the two coils, the permeability of the soft ferromagnetic cores employed, the resistivity of the aluminum disc and the geometry of the entire assembly.

While the energized voltage and current coils generate the driving torque τ_d needed for the disc rotation, it is the permanent magnet near the edge of the disc which restrains the speed of rotation and is consequently called the *brake magnet*. As the disc rotates, a part of it cuts the flux in the air gap of this magnet. The currents thereby induced in this region (*according to the familiar Blv rule*), interact with the same flux to produce a force opposing the rotation (*Bli rule*). The braking torque τ_b so generated can easily be shown to be

$$\tau_b = K_b r_d \O_B^2 n, \tag{2.63}$$

where:

(i) The constant K_b depends on the pole-face geometry and the resistivity of the disc material,
(ii) r_d is the radial distance from the midpoint of the disc to the effective center of force generation,
(iii) $\O_B$ the flux of the brake magnet, and
(iv) n the speed of rotation in revolutions/second.

When the braking torque equals the driving torque, the disc attains a steady speed given by

$$n = \left[\frac{K_b}{K_b} r_d \O_B^2\right] P_L = K_P P_L r/s. \tag{2.64}$$

The above equation establishes that the speed of rotation of the aluminum disc is proportional to the load power P_L. Therefore, count N of the revolutions of the aluminum disc in a particular interval of time is a measure of the energy W (integral of power P_L) consumed in that interval. In this context, a number K_M called the meter constant, specifying the number of revolutions per unit of electricity, namely, one kWh, is useful in the testing of meters. Example meter constant $K_M = 200$ rev/kWh. In normal use, the units of electrical energy consumed are directly read on the meter dial which is a component of a counting/registering system employing gear trains and registers. Initially, the display was accomplished by cascaded divide-by-ten gear train as shown in Fig. 2.32, but later replaced with a digital counter.

To understand the working of the meter, let us assume that the load is purely resistive carrying a current $i_L = \sqrt{2}\ I_L \text{Cos}\omega t$ ampere in phase with its voltage $v_L = \sqrt{2}\ V_L \text{Cos}\omega t$ volts. Let the flux produced by the current coil be $\O_i$, the flux produced by the voltage coil is $\O_v$ and the flux produced by the permanent magnet is $\O_m$. The flux $\O_i$ will produce eddy current i_{ei}, the flux $\O_v$ will produce eddy current i_{ev} and the flux by the permanent magnet will produce an eddy current i_{em} once the disc starts rotating. Figure 2.35a illustrates the phasor diagram of the load voltage, load current, flux produced by the current and voltage coils and the eddy currents on the aluminum disc. The flux $\O_i$ produced by the current coil will be in phase with the current phasor. If the voltage coil is ideal, then flux $\O_v$ will be in quadrature with it as shown in Fig. 2.35a. The Lorentz force will be created only when the spatial angle between the current and the flux directions is not 0°. Hence, the eddy current i_{ei} will not produce a Lorentz force with the flux $\O_i$ and will interact only with the flux $\O_v$ and similarly the eddy current i_{ev} will only interact only with $\emptyset_i$. Both interactions will result in the Lorentz force on the aluminum disc, in the direction shown in Fig. 2.34. The value of the average force F_D due to these flux-eddy current interactions will be

$$F_D = K_D \frac{1}{T} \int_0^T v_L i_L dt. \tag{2.65}$$

In Eq. (2.65), the constant K_D depends on:

(i) Number of turns in the current coil,
(ii) Number of turns in the voltage coil,
(iii) The permeability of the soft ferromagnetic cores employed, and
(iv) The resistivity of the aluminum disc.

This average force F_D will act on the disc and create a rotating torque τ_D in the direction of the force with the value:

$$\tau_D = r_D F_D = r_D K_D \frac{1}{T} \int_0^T v_L i_L dt. \tag{2.66}$$

In Eq. (2.66), r_D is the radial distance from the center of the disc to the point on the disc the force is generated. If nothing is attached to the disc this torque will accelerate the disc and the disc speed will increase without a limit. However, once the disc starts rotating with an angular velocity ω_D rad/s, the flux $\emptyset_M$ produced by the permanent magnet will generate a torque in the opposite direction of rotation and thus produce a breaking torque τ_M given by

$$\tau_M = r_M F_M = r_M \emptyset_M \frac{d\emptyset_M}{dt} = K_M r_M \emptyset_M^2 \omega_D. \tag{2.67}$$

In Eq. (2.67), r_M is the distance from the center of the disc to the point of application of flux $\emptyset_M$ and the constant K_M depends on the area of the pole pieces of the permanent magnet and the resistivity of the aluminum disc. With the two torques (that is propelling torque τ_D and breaking torque τ_M), the disc will attain a steady sate speed ω_D till the propelling torque τ_D equals the breaking torque τ_M. Thus, at steady state, we have

$$\tau_D = \tau_M = K_M r_M \emptyset_M^2 \omega_D = r_D K_D \frac{1}{T} \int_0^T v_L i_L dt. \tag{2.68}$$

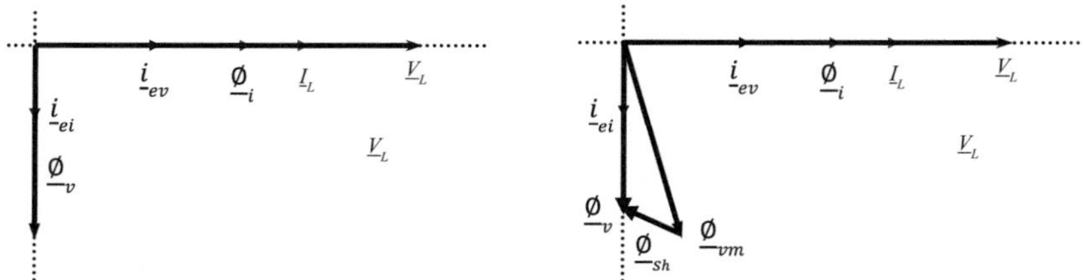

Ideal voltage coil

Practical voltage coil – Flux $\emptyset_{sh}$ is adjusted till $\emptyset_v$ subtends 90°

Fig. 2.35 Phasor diagram of induction-type energy meter for UPF Load

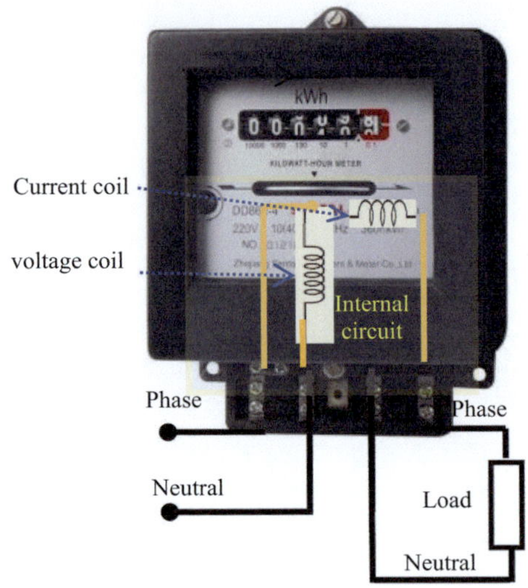

Fig. 2.36 Single-phase energy meter connection

Rearranging Eq. (2.68), we get

$$\omega_D = K_E \frac{1}{T} \int_0^T v_L i_L dt. \tag{2.69}$$

In Eq. (2.69), we have $K_E = \frac{r_D K_D}{K_M r_M \varnothing_M^2}$. With the voltage $v_L = \sqrt{2}\, V_L \text{Cos}\omega t$ V and current $i_L = \sqrt{2}\, I_L \text{Cos}\,(\omega t + \theta)$ A, Eq. (2.69) reduces to

$$\omega_D = K_E V_L I_L Cos(\theta). \tag{2.70}$$

Equation (2.70) indicates that the speed ω_D of the aluminum disc is proportional to the power $P_L = V_L I_L \text{Cos}(\theta)$ drawn be the load. All we need to do to obtain energy is to integrate the disc speed ω_D which is proportional to the load power P_L. We can integrate the disc speed ω_D by simply counting the number of rotations N_D of the disc. Hence, the energy consumed is

$$N_D = K_E \int P_L dt = Energy\, cnsumed. \tag{2.71}$$

Initially, the counting was accomplished by cascaded divide-by-ten gear train as shown in Fig. 2.32a, but later replaced with a digital counter as shown in Fig. 2.36 which is a single-phase energy meter.

2.6.2.1 Full-Scale Adjustment

Once an induction-type energy meter is designed, the number of rotations per kilowatt-hour is set. For example, an energy meter may have 200 rev/kWh as the meter specification. To enable the manufactured unit to fulfill this specification, the load current i_L and voltage v_L are set at full-scale values, with power factor set to unity. Then the position of the permanent magnet is adjusted in the radial direction to obtain 200 rev/kWh. Such an adjustment is called the full-scale unity power factor adjustment or simply *full-scale adjustment.*

2.6.2.2 Power Factor Adjustment

In the foregone discussion, we assumed that the voltage coil flux will lag the voltage by 90° as shown in Fig. 2.35a. In a practical case we know that flux $\varnothing_{vm}$ produced by voltage v_L lags $\underline{V}_L$ by an angle less than 90° due to the winding resistance of the voltage coil as shown in Fig. 2.35b. This would lead to errors in the measurement of energy, especially for low power factor loads. To correct this, the voltage coil flux $\varnothing_{vm}$ is split into two parts with one part $\varnothing_v$ going through the aluminum disc and the other part $\varnothing_{sh}$ shunts back to the main core as indicated in Fig. 2.32a and the phasor diagram of Fig. 2.35b. By adjusting $\varnothing_{sh}$, we ensure that $\varnothing_v$ phasor is exactly at 90° to the applied voltage phasor $\underline{V}_{L.}$ This adjustment is termed as "*power factor adjustment*" and is normally done with the values of v_L and i_L at full-scale and power factor adjusted to be 0.5.

2.6.2.3 Creep Adjustment

The disc and the associated counting mechanism will possess mechanical friction. If the torque to overcome this friction is extracted from the generated torque τ_D, once again we will have errors in the measurement of energy. Therefore, we generate an additional torque using the "*shaded pole*" principle. The position of a short, circuited sleeve is adjusted to generate a torque to compensate the frictional torque. It is also possible to split the flux $\varnothing_v$ into two parts, namely, $\varnothing_v'$ and $\varnothing_c$ and generate a torque using $\varnothing_c$ alone. The value of $\varnothing_c$ is adjusted so that the torque generated by $\varnothing_c$ overcomes the friction of the moving disc. The friction of any rotating mechanism will change with the usage. Thus, in induction-type energy meters the friction may change in due course and thus the frictional torque will differ from the compensating torque. If the friction reduces in due course, this difference will rotate the disc even when the current and hence power is zero. Such slow rotation of disc is called "*creeping*" and hence this adjustment is called *creep adjustment*. The creep adjustment is done as given below. With the current set to zero, the rated voltage is applied, and the creep adjustment is made (normally by turning a screw that moves the relevant part) till the disc starts rotating. Then the adjustment is reversed till the disc stops moving.

2.6.2.4 Calibration of Induction-Type Watt-Hour Meter

From the foregone discussion, we understand that there are three parameters to be adjusted, namely,

1. Full-scale adjustment,
2. Power factor adjustment, and
3. No-load (creep) adjustment.

The above are intertwined and one adjustment disturbs the other two, calling for readjustment of the latter. Hence, the need for repetitive adjustments makes the calibration of an induction-type energy meter quite tedious and time consuming. The following procedure is normally utilized for calibration, with the connections as shown in Fig. 2.36.

Step 1: **Creep adjustment**: The current is made zero, the creep adjustment is varied till the disc starts moving and brought back to the point where the disc just stops rotating.

Step 2: **Full-scale adjustment**: Once creep level is adjusted, then the full-scale adjustment is accomplished as given next. Voltage and current values are adjusted full-scale, UPF conditions. For example, if the energy meter specification is 5A/230 V, then voltage to the potential coil is set as 230 V, the current through the current coil is set at 5 A with zero phase angle. The position of the permanent magnet is adjusted till the speed of the disc exactly matches with the designed full-scale speed. The disc speed is normally given in terms of number of rotations per kilo-watthour. Example: 200 rev/kWh.

Step 3: **Power factor correction**: Now the voltage and current are kept at full-scale values and the power factor is set as 0.5. The power factor adjustment is now accomplished to ensure that there is zero error in the energy meter's disc speed.

The above three steps are repeated till no change is required in any of the above three steps.

2.6.3 Single-Phase and Three-Phase Energy Measurement

The energy meter shown in Fig. 2.36 is a single-phase energy meter and its connection is as shown in the figure. If we need to measure energy in a three-phase connection, then we have the following options:

(i) For a three-phase four-wire system, three independent single-phase energy meters with one meter for each phase may be used as shown in Fig. 2.37a. For a three-phase three-wire system, two independent single-phase energy meters would suffice with the meters connected (normally in R and B lines) as shown in Fig. 2.37a.

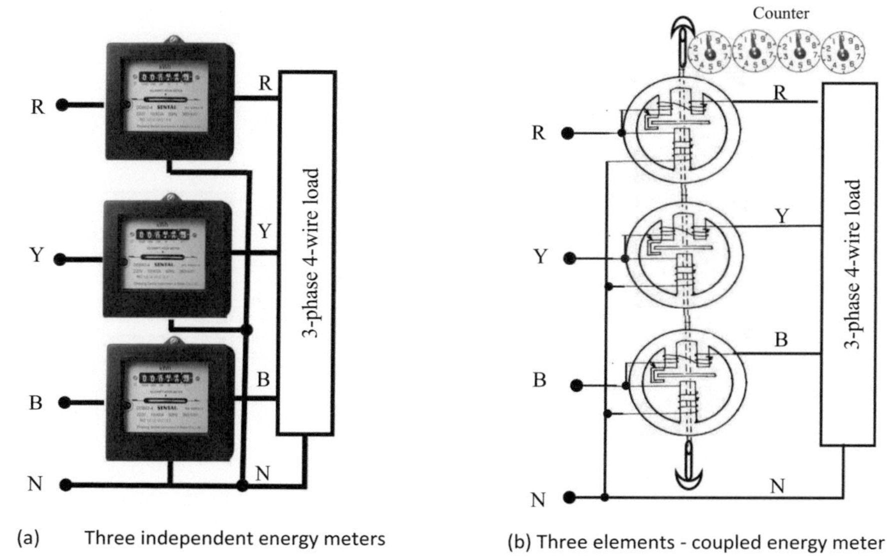

Fig. 2.37 Three-phase four-wire energy measurement

(ii) It is also possible to employ three discs combined in one spindle and build an integrated single energy meter for a three-phase four-wire system as shown in Fig. 2.37b. A meter with such an arrangement is known as "3-*element energy meter*". Similarly for energy measurement on a three-phase three-wire load, we need to use only two discs combined in a central shaft as shown in Fig. 2.37b. Such meters are called *three-phase* two-*element energy meters.*

It may be noted while the three independent or three-element energy meters will have their voltage specification as the phase to neutral voltages, the voltage specification for two energy meters or two-element meter solutions must be for the line-to-line voltage. It may also be noted that the diagram of the energy meter in Fig. 2.32a shows the cores of current and voltage coils separately, only to explain the principle of operation. However, in a practical induction-type energy meter the core is an integrated one as illustrated in Figs. 2.37b and 2.38b. Also, the current coil is wound around two poles for better space utilization.

2.7 Cryptic Indication of Meter Parameters

If we look at the indicating instruments that we learnt about so far, irrespective of the type of instrument (PMMC or electrodynamic or moving iron), the face of the meter looks same with a pointer moving on a scale. Hence, it would be hard to identify the meter just by looking at the face of the meter. Even if the manufacturer wishes to

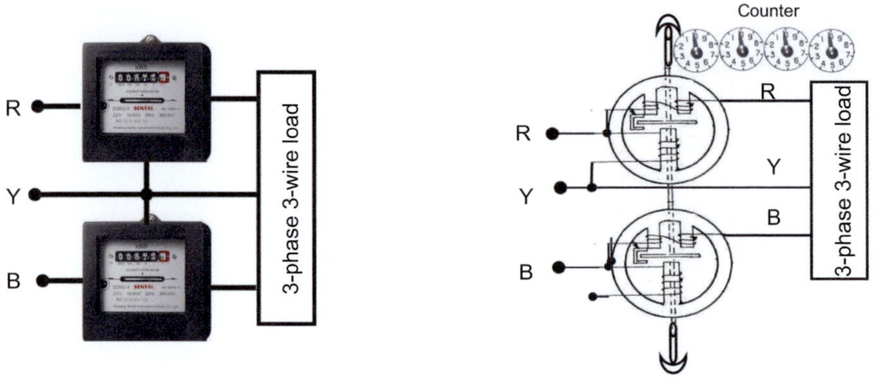

(a) Two independent energy meters (b) Two elements (discs) - coupled energy meter

Fig. 2.38 Three-phase three-wire energy measurement

present the relevant meter details in words on the dial, they would be hard to read and will only clutter the face of the meter. Hence, to identify the details of the meter (type, accuracy, AC or DC, insulation level, and mounting method), meter manufacturers print cryptic information on the face of the meter.

Figure 2.39 indicates front faces of three different meters. The cryptic information starts with a symbol and is followed by numbers/symbols. In Fig. 2.39a, the first symbol ∩ indicates the meter is of PMMC type. The second symbol ⊥ indicates that the meter must be mounted vertical (perpendicular to ground plane) and the third number (1.5) indicates the meter is of accuracy class 1.5 (Worst case error is ± 1.5% of full-scale of 15 A).

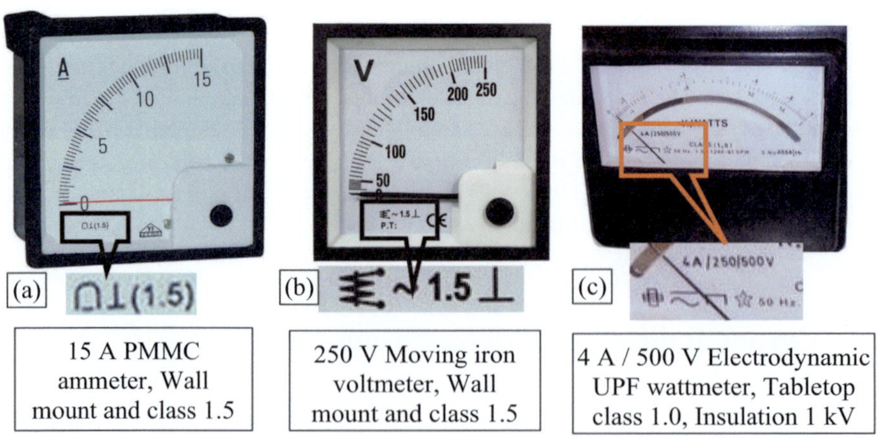

Fig. 2.39 Face plate engravings of Three different meters

In Fig. 2.39b, the first symbol indicates that the 250 V full-scale voltmeter is of moving iron type, the second symbol ~ indicates that the meter is calibrated only for AC and the third number identifies the meter to be of accuracy class 1.5 and the last symbol instructs that the meter must be vertically(wall) mounted.

Figure 2.39c shows that the wattmeter is of electrodynamic type (symbol ⊕), full-scale current is 4 A, full-scale voltage can be 250 V or 500 V (can be selected), can be used for DC or AC (symbol ≂), to be placed horizontally (symbol ⌐¬), insulation level is 1 kV (Symbol ☆) and calibrated at 50 Hz. The complete standardized cryptic code is as given in Table 2.4.

2.8 Problems and Exercises

E2.1 A PMMC meter is manufactured with the following specifications:

$\theta_{FS} = 100^0$; $I_{FS} = 1$ mA; $B = 1$ mT $\pm$ 200μT; $N = 100$; Area of the interaction of the coil with $B = 7.5$ cm^2.

(a) Calculate the specification (spring constant) of the springs required.
(b) Assuming the spring constant of manufactured springs will be in the range of ($K_S \pm 20\%$), determine the ratio with which the shunt flux and the main flux to be split so that every meter that is manufactured can be adjusted to obtain full-scale deflection angle for the specified full-scale current.

E2.2 Two DC ammeters of full scale of 10 A are required. Two PMMC ammeters A_1 and A_2 are available with the specifications 1 mA/100 mV and 1 mA/150 mV, respectively.

(a) Calculate the values of shunts required to convert these ammeters into 10 A full-scale ammeters.
(b) If by mistake, the shunts are interchanged (shunt of A_1 is put on A_2 and vice versa), what would be the full-scale currents of the new ammeters A_1 and A_2?

E2.3 Show how you will convert a 100 μA full-scale PMMC DC ammeter having a coil resistance of 10 Ω into a 10 V full-scale DC voltmeter. What is the sensitivity of this 10 V full-scale DC voltmeter?

E2.4 A PMMC voltmeter has a PMMC basic ammeter having a meter resistance R_m and the series resistance R_s. The coil of the PMMC ammeter (R_m) is made of copper. Temperature coefficient of resistance of copper is 0.00427 K^{-1}. Determine the ratio R_s/R_m that would ensure that a change in temperature of 10 K produces an error of less than $\pm 1\%$ in a reading obtained with this voltmeter. Assume R_s to be temperature invariant.

E2.5 A critically damped PMMC movement serves as a recording meter (Instead of the pointer, a writing stylus is attached to the moving coil).

Table 2.4 Symbols that can be inscribed on a meter's face

Sl. no	Cryptic symbol/ information	Meaning	Remarks
1		Meter is of PMMC type	Only for DC
2		Meter is PMMC and rectifier type	For both AC and DC
3		Meter is electro-dynamometer type	For both AC and DC
4		Meter is moving iron type	For both AC and DC
5	⊓	Meter is to be placed (mounted) horizontally	Tabletop meter
6	∠45°	Meter is to be placed (mounted) at an angle 45°	Bench mounted meter
7	⊥	Meter is to be placed (mounted) vertically (90°)	Wall mounted meter
8	=	Meter can measure only DC quantities	DC meter
9	~	Meter is suitable for measurement of AC only	AC meter
10	≃	Meter can measure both AC and DC quantities	AC/DC meter
11	1.0	Accuracy class is 1.0	FS error ± 1%
12		Insulation level 1 kV	
13	$\underline{0}$ .. $\underline{50}$ .. 200	Calibrated at DC and 50 Hz but can be used till 200 Hz.	
14	!	Special instructions. Read the manual before using the instrument	
15	$2\,J_N$	200% over the rated current is permitted	Only for LPF wattmeters
16	IPXY	Enclosure is as per ingress protection class XY.	X: 1 to 6 Y: 1 to 8

Note Please see Table 2.5 for the complete listing of the Ingress Protection class abbreviations

Table 2.5 Ingress protection (IP) class

Contact with solids (Number X)		Number	Contact with water (Number Y)	
	Protected against a solid object > 50 mm coming into contact (example hand)	⇔ 1		Protected against encountering water droplets
	Protected against a solid object > 12.5 mm coming into contact (example finger)	⇔ 2		Protected against encountering water droplets at 15°
	Protected against a solid object > 2 mm coming into contact (example screwdriver)	⇔ 3		Protected against encountering water spray at 60°
	Protected against a solid object > 2 mm coming into contact (example thin wire)	⇔ 4		Protected against encountering water spray at any angle
	Protected against limited dust ingress	⇔ 5	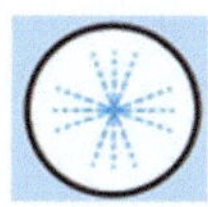	Protected against encountering light water jets at any angle (Example light train)
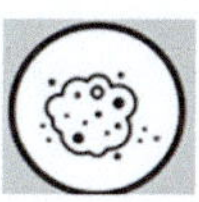	Protected against dust ingress (airtight)	⇔ 6		Protected against encountering heavy water jets at any angle (Example Heavy rain)
Example IP66 Can be used in a very dusty environment and the meter will not go bad even if the meter is exposed to rain.		7 ⇒		Protected against immersion in water up to 1 m for brief period (< 30 minutes)
		8 ⇒		Protected against immersion in water up to 4 m.

(a) What should be the natural frequency of the PMMC movement so that the 5th harmonic of power supply frequency of 50 Hz is reproduced with an error $< \pm 5\%$ in its magnitude?

(b) Determine the phase shift introduced in recording at (1) 50 Hz and (2) 5th harmonic of 50 Hz with this recorder.

E2.6 A potential divider consists of a 100 kΩ in series with 400 kΩ and is shunted across a 25 V DC source. A 5 V FS DC voltmeter with figure of merit of 20 kΩ/V is used to measure the voltage across the 100 kΩ resistor. What value will the voltmeter read?

E2.7 A dynamometer-type wattmeter possessing a voltage coil of resistance of 2000 Ω is used to measure the power in a certain load. The voltage coil is connected across the load side and the wattmeter reads 250 W. If the load voltage is 250 V, what is the power being taken by the load?

E2.8 A voltage $[100\ sin(\omega t)+40\ cos(3\omega t-\pi/6)+50\ sin(5\omega t+\pi/4)]$ is applied to the voltage coil of a wattmeter. A current $[8\ sin(\omega t)+6\ cos(5\omega t-2\pi/3)]$ A is passed through the current coil. What will be the reading of the wattmeter? What percentage of this power is due to the fundamental component? Neglect instrument errors.

E2.9 The voltage coil of an electrodynamic-type wattmeter has an inductance of 8 mH and resistance of 2000 Ω. What is the deviation in reading of the instrument, when measuring the power in an inductive load having phase angle of 89° at 50 Hz. Neglect power consumed by the meter. (Assumed that the meter reads correctly in the absence of 8 mH inductance)

E2.10 The resistance of the coils of a wattmeter are 0.01 Ω and 1000 Ω, respectively, and both are non-inductive. The load current is 20 A and the voltage applied to the load is 30 V. Show the two ways in which the voltage coil can be connected. Find the error in the reading in each case.

E2.11 A wattmeter reads 5.54 kW, when its current coil is connected in the R phase and its voltage coil is connected between the R phase and the neutral of a symmetrical 3 phase system, supplying a balanced inductive load of 30 A at 400 V. What will be the reading on the meter, if the current coil connection remains unchanged, but the voltage coil is now connected between the Y and B phases instead of between R phase and neutral phase sequence is RYB.

E2.12 Show that power in a three-phase three wire balanced/unbalanced load can be measured using only two wattmeters. Derive expressions for the readings of the two wattmeters for (a) General case (b) Balanced load.

E2.13 Power in a balanced load supplied by a 400 V, RYB sequence, three-phase three-wire source is measured using the two-wattmeter method. The wattmeter, whose current coil is connected in R phase reads 3000 W and the wattmeter, whose current coil carries the current in B-phase reads zero. Determine the line current and the load impedance assuming a (a) star connected load and (b) delta connected load.

E2.14 A 5 A full-scale moving iron ammeter is built with springs having spring constant $K_s = 3$ nNm/rad. Determine the $(dL/d\theta)$ in H/rad of the meter, if the full-scale deflection is 60^0.

E2.15 A moving iron voltmeter is made by connecting a series resistance of 1980 Ω to a moving iron ammeter possessing a resistance of 20 Ω and an inductance of 3 mH. Show how to make the voltmeter read the same in DC as well as at 50 Hz sinusoidal AC. Determine the figure of merit for the voltmeter in DC and 50 Hz AC.

E2.16 Show how you will convert a 1 mA/100 mV PMMC meter into (1) 300 V AC Voltmeter and (2) 1 A AC Ammeter (i) with different scales (ii)with the same DC scale for both DC and AC. Assume availability of ideal diodes.

E2.17 If the diodes used in problem E2.17have an ON resistance of 0.5 Ω, cut-in voltage of 0.6 V and a reverse resistance of 10 MΩ. What will be the error introduced in the (i) 1 A AC ammeter and (ii) the 300 V AC voltmeter.

E2.18 A 100 V rectifier-type voltmeter reads 60 V. If the waveform of the voltage being read is symmetrical square wave, determine the correct rms value of the voltage. What is the percentage error in this reading?

E2.19 If the waveform in problem 1 above is symmetrical triangular wave, determine the correct the rms value of the voltage. What is the percentage error in this reading?

E2.20 A 240 V, 5 A single-phase energy meter has a specification of 300 rev/kWh. At full-scale unity power factor, the disc makes 19 revolutions in 100 seconds. Determine the error in the reading.

Chapter 3
Errors in Measurement

When a man tells you that he knows the exact truth about anything, you are safe in inferring that he is an inexact man. Every careful measurement in science is always given with the probable error ... every observer admits that he is likely wrong and knows about how much wrong he is likely to be.
—Bertrand Russel
The Scientific Outlook 1931

3.1 Errors in Measurement

Whenever a measurement is made some amount of error is invariably introduced in the measured value. It is practically impossible to make an error-free measurement. The error which can happen due to the imperfections of the measurement system is called "*Systematic Error*" in the measurement. Such errors remain the same in repeated measurements with the same system under the same conditions. An error can also be introduced by variable internal or external factors such as electromagnetic and other interferences as well as noise. These errors are called "*Random Errors*".

V. Jagadeesh Kumar, *Principles of Electrical Measurement and Measuring Instruments*,
https://doi.org/10.1007/978-981-96-5475-8_3

Table 3.1 Different ways of representing an error in a measurement

Sl. no	True value x_o	Measured value x_m	Absolute error δx	Relative error ε_x	Relative error ε_x %
1	2.0 V	2.2 V	+0.2 V	0.10	+10%
2	2.0 V	1.8 V	−0.2 V	−0.10	−10%
3	250 V	255 V	+5.0 V	0.02	+2%
4	250 V	245 V	−5.0 V	−0.02	−2%

These are likely to vary in repeated measurements even with the same measurement system. While the systematic errors govern the "*Accuracy*" in a measurement, the errors that arise due to random influences affect the "*Precision*" in a measurement. We will learn more about these terms later.

Quantitatively, the error in a measurement is defined as the difference δx between the measured value x_m and the true value x_o. That is $\delta x = (x_m - x_o)$. However, the term relative error ε_x, defined as $\left(\frac{\delta x}{x_o}\right)$ is commonly used in error analysis because it gives a better perception of the stringency of an error than the absolute error δx. It is obvious that an error of 0.2 V made in measuring 2 V is worse than an error of 5 V made while measuring 250 V. The relative error is often expressed as a percentage as

$$\varepsilon_x = \left[\left(\frac{\delta x}{x_o}\right)100\right]\%.$$

Table 3.1 illustrates these terms.

3.2 Systematic Errors

Systematic errors may arise due to the following causes:

(i) The measurement process itself may load or disturb the existing situation and cause a change in the value of the measured variable. The altered value, even if measured accurately, becomes erroneous. Normally this type of error is referred to as a "*Loading Error*". The *loading error* is also termed a "*Systemic Error*" as it is due to the system of measurement.

(ii) The instruments employed in making the measurement have their own errors. Termed as "*Instrument Errors*", they clearly affect the result.

(iii) Other reasons are as follows:

   (a) Mathematical approximations: Example—using a linearized model of a sensor characteristic,

(b) Assumptions made: Example: Treating an AC standard resistor as inductance-free, and
(c) Environmental changes like temperature being different from an assumed standard value.

Systematic errors can happen when a measurement is made, due to the:

(i) Measurement process itself and is called "*Systemic Error*" or "*Loading Error*".
(ii) Measuring instrument employed for making that measurement. Such an error is called the *Instrument Error*.

3.2.1 Loading (Systemic) Errors

Let us look at the circuit shown in Fig. 3.1a. From the given values, we can easily calculate the current as 1.0 A. However, if we want to measure this current, then we need to insert an ammeter as shown in Fig. 3.1b. Once we insert the ammeter, its internal resistance (say 100 mΩ) comes in series with the 10 Ω and the current will now be

$$\frac{10}{10.1} = 0.99009900\ldots.\ \mathrm{A}$$

Thus, even if the ammeter reads correctly (has zero *instrument error*), the very act of introducing the ammeter had changed the current to be measured from 1.0 A to 0.99009900… A. We should have measured 1.0 A, but instead ended up measuring it as 0.99009900… A. Thus, our measurement has an error δI, which is defined as

$$\mathit{Measured\ Value} - \mathit{True\ value} = \delta I = -9.900990099\ \mathrm{mA} \tag{3.1}$$

Normally systematic error is expressed as a percentage as

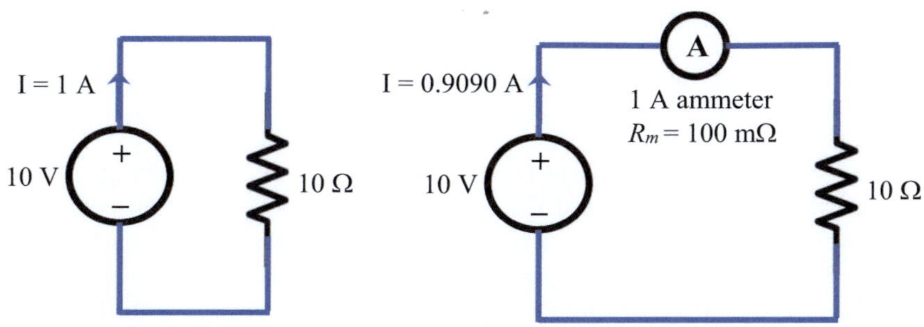

(a) The dc circuit. (b) Ammeter added to measure the current.

Fig. 3.1 Use of an ammeter to measure current in a dc circuit

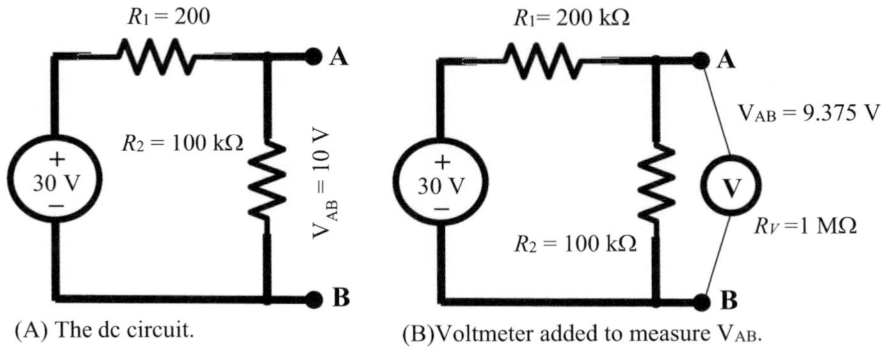

Fig. 3.2 Use of a voltmeter and resulting systemic error in a dc circuit

$$\varepsilon_I\% = \left(\frac{Mearsure\ Value - True\ Value}{True Value}\right)100\% = -0.990099\% \tag{3.2}$$

Please note here that the ammeter in Fig. 3.1 is assumed to have *zero instrument errors*. If not, the instrument error will also be added to this error. Similarly, in Fig. 3.2a, we can calculate the voltage $V_{AB} = 10$ V. If we use a voltmeter having an internal resistance R_V of 1.0 MΩ as shown in Fig. 3.2b to measure the voltage V_{AB}, the internal resistance 1.0 MΩ of the voltmeter will come in parallel to the 100 kΩ resistor, resulting in the resistance across terminals A–B to become.

$$R_{AB}\frac{R_1 R_V}{R_1 + R_V} = \frac{(100)10^9}{(1.1)10^6} = 90.909\ \text{k}\Omega.$$

The voltage V_{AB} measured by the voltmeter will be

$$V_{AB} = 30\frac{(99.909)10^3}{(299.909)10^3} = 9.375\ \text{V} \tag{3.3}$$

Hence, the percentage error ε_V % due to the introduction of the voltmeter is

$$\varepsilon_V\% = \frac{(9.375 - 10)}{10}100 = -6.25\% \tag{3.4}$$

Thus, we see that whenever we introduce a measuring instrument, the quantity we want to measure gets altered and that leads to an error. Since the error is due to the measurement system itself, it is termed a *systemic error*. A good measurement engineer must be aware of this aspect and should ensure that the *systemic error* is as small as possible by choosing the right measuring instruments within the given constraints.

3.2.2 *Instrument Errors in a Measurement*

We need a measuring instrument to make a measurement. Thus, any measurement we make will be affected by the characteristics of the instrument employed for making the measurement.

3.2.2.1 Errors in a Measurement due to the Resolution of the Measuring Instrument

When we make a measurement, we express the result in terms of a number and a unit. Most quantities we measure are analog in nature and can vary continuously. Hence, the number part would need many digits for proper delineation in a general case. For example, the voltage that we need to measure may be 1.2345678901234…. V. However, we do not have an instrument that can determine its value correct to so many digits. If the voltmeter that we employ can measure only in steps of 0.1 V, then this value may be read as 1.2 V. Hence, we incur an error $\delta V = -0.0345678901234\ldots.$ V and

$$\varepsilon_v\% = \tfrac{-0.0345678901234}{1.2345678901234}100 = -2.7999991\% \tag{3.5}$$

Thus, due to the limited number of *significant digits,* that a measuring instrument can provide, we will not be measuring the exact value. Let us take an example of a 150 mA full-scale ammeter with a scale marked at every milliampere as shown in Fig. 1.1. Hence, any value of current that we can read without any ambiguity or use of interpolation is an integral number of milliamps. The smallest value or the smallest step change that we can distinctly read using an instrument is called the "*resolution*" of the instrument. The resolution in the present case of a 150 mA full-scale meter is 1 mA. Let the current measure be100 mA. If the meter is perfect, then the pointer will move on to the 100th milliampere marking, and we will read the current as 100 mA.

Now if the current is, say, 100.01 mA, the pointer will move just (1/100)th of a division above the marking labeled as 100. We may not even notice that the pointer has moved beyond the 100th mark and hence would still read the current as 100 mA. Similarly, if the current is 99.99 mA, then also the movement of the pointer may not be perceptible, and we will still read the value as 100 mA. Even if we observe the position to be distinctly away from a scale mark, say for a current of 100.24 or 99.75 mA, we have no way of unambiguously judging the amount of deviation from the nearest scale mark (100 mA in our example.).

We therefore acknowledge our inability to determine the correct value of the current from our observation and record its value of 100 mA. In short, when we measure a current with this meter, we can read the current only in steps of 1 mA. Thus, when we read a current as 100 mA, then we are sure that the current is closer to 100 mA than to 99 or 101 mA. In other words, we are stating that the value of current can be anywhere between 99.5 and 100.5 mA. Many times, it may so happen that the pointer may be in between two markings making it difficult to judge to which of the two is closer. This may happen for, say, a current of 99.45 or 99.55 mA. We may then be tempted to record the value as 99.5 mA. In such cases, we need to indicate the *correct significant digit* by placing an underscore as $9\underline{9}.5$ mA. By doing so, we clearly indicate that the reading was obtained using a meter whose resolution is 1 mA but we have interpreted the reading as 99.5 mA.

From the above discussion, we conclude that the largest error caused by the limitations of the scale-marking is one-half of the smallest step size on the scale (resolution). Thus, the worst-case relative error introduced by the finite resolution of a measuring instrument is

$$Error(due\ to\ resolution) = \pm\frac{0.5\ Resolution}{True\ value}100\% \tag{3.6}$$

In Eq. (3.6), we use $\pm$ to indicate that the actual reading may be less than or more than the read value. In this section, we assumed that the instrument itself had no errors, i.e., the pointer of ammeter considered deflects exactly to the 100 mA scale mark for a current of 100.0000 mA. It is possible that the meter itself has some errors and may give a reading of 99 mA or 101 mA, for an actual current of 100 mA. Such errors are called *instrument errors*, which we will consider in the next section. After the digital meters were invented, the resolution itself is called "*quantization*" and the error due to the limited resolution of the digital measuring instrument is called the "*quantization error*".

3.2.2.2 The Instrument Error

In Chap. 2, we derived the expression for the deflection angle of a PMMC meter as

$$\theta = \left(\frac{BA_cN_c}{K_s}\right)I \tag{3.7}$$

When a manufacturer makes such a meter, A_c and N_c can be kept exactly the same across different units produced. However, the value of flux density B and the spring constant K_s will vary from one unit to the other, as it is practically not possible to manufacture identical magnets or springs. As was brought out earlier in Chap. 2, this variation is handled by the manufacturer by doing "*calibration*", wherein the value of B is adjusted so that (B/K_s) attains a value that ensures a full-scale deflection at the designated full-scale current. However, over a period B and K_s will vary due to natural causes. For example, for a meter, let us assume that (B/K_s) is set during calibration to yield 1°/mA (so that 100 mA gives a deflection of 100°). Let initially the flux density of the permanent magnet is B. After a year let us assume that the flux density of the permanent magnet has reduced to 0.99 B. Now (B/K_s) will cause a deflection of 0.99°/mA and the meter will deflect only 99° for 100 mA. Thus we will read the current wrongly as 99 mA with this 1-year-old meter, resulting in an error of −1.0 %. Every manufacturer of an instrument estimates this variation and indicates the effect of such variations as an error that may result if that instrument is used for a measurement. Since this error crops up due to the measuring instrument, we term this error as "*instrument error*". In a practical situation, the error may vary, depending on the reading obtained as indicated in Fig. 3.4 Consolidating all the available data, the manufacturer will take the maximum error (*worst-case error*) and express it as a percentage of full-scale value and call it as a *full-scale error* as. (Fig. 3.3)

$$Full\ Scale\ Error = \left(\frac{(Measured\ Value - True\ Value)_{max}}{Full\ Scale\ Value}100\right)\% \tag{3.8}$$

In Eq. (3.8), the $(MeasuredValue - TrueValue)_{\max}$ is the maximum error of the meter, which can occur at any reading and not necessarily at full scale. However, this worst-case error is expressed as a percentage of the full-scale value. If we take a reading in such a meter, say at the half full scale, then the percentage worst-case error in the value read will be double. It may be four times if the reading is obtained is one fourth of the full scale and so on. This situation is illustrated in Fig. 3.4a, where the same width of error band stretches from zero to full scale. The word "*maybe*" is used to indicate that this is only a worst-case scenario, and the actual error may be even zero! Fig. 3.4b illustrates the effect of "Error in reading". The full-scale error is also expressed as *accuracy class* of an instrument. Many "standards" such as IEC 60051, EN 60051 and IS1248 define the "*accuracy class*". Table 3.1 describes the class of instrument and the corresponding full-scale accuracy of direct reading analog indicating instruments as per IS1248:2003 specifications. These standards,

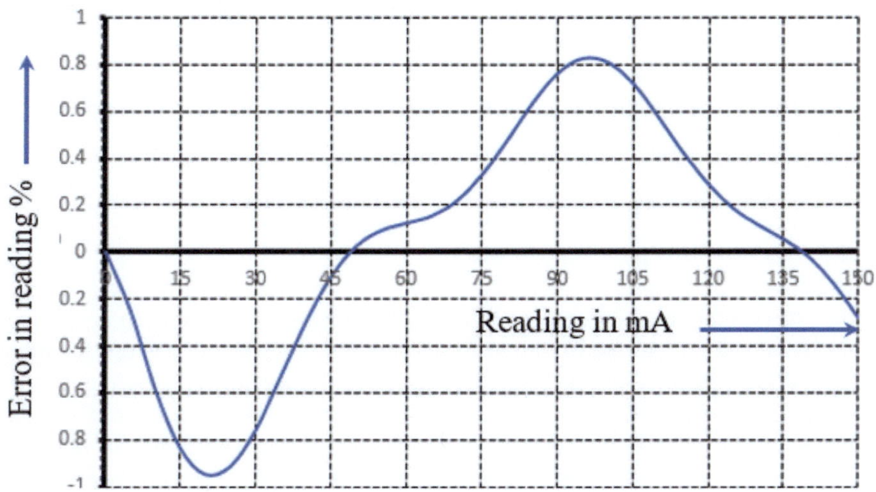

Fig. 3.3 Error as a function of reading in a150 mA full-scale ammeter

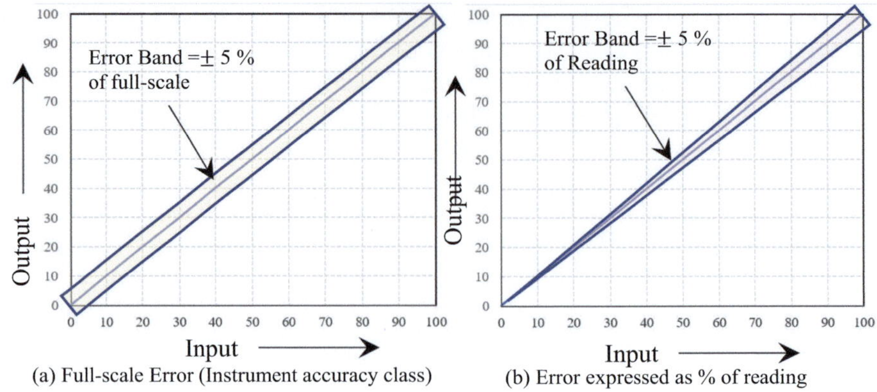

Fig. 3.4 Instrument error

apart from specifying the full-scale accuracy, also provide additional specifications such as limits on overload, mounting instructions, and labeling methods. Whenever a measuring instrument has full-scale accuracy specification, care must be taken to get a reading as close to the full scale as possible so that the *worst-case instrument error* is minimal as can be seen from the examples given below.

Example 1 A 100 mA full-scale, class 1 ammeter shows a reading of 95 mA. What is the worst-case error in the reading?

Class 1 instrument $\Rightarrow$ worst-case error is $\pm 1\%$ of full scale $\Rightarrow \pm 1$ mA. Therefore, error in the reading of 95 mA is ± 1 mA or $\left(\frac{1}{95}100\right) = \pm 1.05\%$. Hence, worst-case instrument error at this reading is $\pm 1.05\%$.

Example 2 A 100 V full-scale, class 2 voltmeter shows a reading of 50 V. What is the worst-case error in the reading?

Class 2 instrument $\Rightarrow$ worst-case error is ± 2 % of full scale $\Rightarrow \pm$ 2 V (Table 3.2).

Therefore, worst-case error in the measurement of 50 V is ± 2 V or $\pm\left(\frac{2}{50}100\right) = \pm 4.0$ %

3.2.2.3 Errors in Digital Instruments

Modern day digital instruments come with a few varied ways of reporting instrument errors. Some manufacturers of digital measuring instruments still use the accuracy class specifications as given in Table 3.1. Once the number of digits in the display (OR number of bits, if the digital output is a binary number) is known, the resolution of the meter is known. For example, if a digital meter has a 3 ½ digits display, then the value displayed value can only be in the range: 0000–1999. Typically, we call this a 2000-counts digital instrument. Hence, the resolution of a 3½ digit digital meter is 1 in 2000. Practically this is 1 in 1999, but the zero state is also counted as a valid

Table 3.2 Accuracy class as per IS1248

Sl. No	Accuracy class	Full-scale error %
1	Class 0.05	0.05
2	Class 0.1	0.1
3	Class 0.2	0.2
4	Class 0.3	0.3
5	Class 0.5	0.5
6	Class 1	1.0
7	Class 1.5	1.5
8	Class 2	2.0
9	Class 2.5	2.5
10	Class 3	3.0
11	Class 5	5.0

level and hence we take the resolution as 1 in 2000! Some manufacturers of digital measuring instruments specify the errors as a percentage of full-scale + a few counts. The following examples illustrate this.

Example 3 Error Specification: 200 mV full-scale, 3½ digit voltmeter, ± 1 % of full scale + 10 counts.

If we measure any value on this meter, then the first part of the error specification is 1 % of full scale, that is error is 2 mV. That is straightforward. The plus 10 counts means that any value (0001 to 1999) can have an error of ±10 counts.

Which means $\pm\left(\frac{10}{2000}200\right)\text{mV} = \pm 1$ mV and hence the total worst-case error is (2 mV + 1 mV) = ± 3 mV at all readings.

Some manufacturers of digital meters specify the accuracy as percentage of reading plus counts. In such a case, the error must be calculated as given in the example given below.

Example 4 Error Specification: 2 V full-scale, 4½ digit voltmeter, ±0.5 % of reading +10 counts. 10 counts correspond $\pm\left(2\frac{10}{20000}\right) = \pm 1$ mV. Now, if the reading is 1 V, then the error is (0.5 % of 1 V) = 1 mV.

Hence, total error in the reading = ± (5 mV + 1 mV) = ± 6 mV, that works out to ± 0.6 % of the reading.

Note: Systematic error is quantifiable, and the errors connected with accuracy can be calculated and if enough information is available, even be corrected. Hence, systematic error is "*calculable and correctable*".

3.2.3 Propagation of Systematic Error

Oftentimes it is more convenient and practical to calculate the value of a quantity from the measured values of some associated quantities rather than measure the required quantity directly.

Example 5 For example, if we need to measure power into a DC circuit, we can simply measure the terminal voltage V_c and current I_c fed to the circuit and compute the power in the circuit as $P_C = V_c\, I_C$. If v % and i % are the respective percentage errors in measured values of V_c and I_c, then what would be the percentage error p % in the computed P_c? To calculate this, let us put in what we know:

$$\Delta v = |(V_c)_{Measured}| - |(V_c)_{True}| \text{ and } \Delta i = |(I_c)_{Measured}| - |(I_c)_{True}|$$

$$v\,\% = \tfrac{\Delta v}{V_c} 100 and i\,\% = \tfrac{\Delta i}{I_c} 100 \tag{3.9}$$

While discussing percentage errors in a *general* case, we are interested only in their worst-case magnitudes and hence we are not considering the signs of the absolute errors in the above expressions. However, we recognize that each of these errors can be either positive or negative and give appropriate consideration to this fact in the analysis below. Let ΔP be the magnitude of the error in the value of power arrived at, after multiplying the measured values of the voltage and current. Accordingly, the worst-case *positive* error in the calculated power occurs when the measured values of both voltage and current have positive errors. This would be

$$+\Delta P = (V_c + \Delta v)(I_c + \Delta i) - P_c = V_c \Delta i + I_c \Delta v + \Delta v \Delta i\,. \tag{3.10}$$

The worst-case *negative* error in the calculated power occurs when both the measured values have negative errors. This would be

$$-\Delta P = (V_c - \Delta v)(I_c - \Delta i) - P_c = -V_c \Delta i - I_c \Delta v + \Delta v \Delta i \tag{3.11}$$

It is obvious that the positive error, worked out in (3.10), has a larger magnitude and we take that to be ΔP, the worst-case error in the computed P_c. Dividing the left-hand side of Eq. (3.10) by P_c and the right-hand side by $V_c I_c (= P_c)$ and multiplying all terms by 100, we get

$$\begin{gathered} p\,\% = \tfrac{\Delta P}{P_c} 100 = \tfrac{V_c \Delta i + I_c \Delta v + \Delta v \Delta i}{V_c I_c} 100 = \\ \tfrac{\Delta i}{I_c} 100 + \tfrac{\Delta v}{V_c} 100 + \left(\tfrac{\Delta i}{I_c} 100\right)\left(\tfrac{\Delta v}{V_c} 100\right) \tfrac{1}{100} = v\,\% + i\,\% + \tfrac{v\,\% i\,\%}{100} \end{gathered} \tag{3.12}$$

Since $v\,\%$ and $i\,\%$ are small, we can neglect their product i*n the last term of* Eq. (3.12) and condense the equation as

$$p\,\% \approx v\,\% + i\,\% \tag{3.13}$$

To sum up, when we multiply two measured values to determine a dependent third quantity the percentage errors in the measured values simply add-up to establish the worst-case percentage error in the result.

Example 6 With the same voltage and current, we can also compute the resistance R_c of the network as $R_c = \frac{V_c}{I_c}$. If that is the case, then percentage error r % in R_c can be obtained following a similar analysis as before. The upper limit of error in the calculated resistance (obtained by the division of the larger value of numerator by the smaller value of denominator) is.

$$\Delta R = \frac{(V_c+\Delta v)}{(I_c-\Delta i)} - R_c = \frac{V_c}{I_c}\left(\frac{1+\frac{\Delta v}{V_c}}{1-\frac{\Delta i}{I_c}}\right) - R_c = R_c\left[\left(\frac{1+\frac{\Delta v}{V_c}}{1-\frac{\Delta i}{I_c}}\right) - 1\right] \tag{3.14}$$

Equation (3.14) simplifies to

$$\frac{\Delta R}{R_c} = \frac{\Delta v}{V_c} + \frac{\Delta i}{I_c} + \frac{\Delta v}{V_c}\frac{\Delta i}{I_c} \tag{3.15}$$

Similarly, the lower limit of error (obtained by the division of the smaller value of numerator by the larger value of the denominator) can be derived as

$$\Delta R = \frac{(V_c-\Delta v)}{(I_c+\Delta i)} - R_c = R_c\left[\left(\frac{1-\frac{\Delta v}{V_c}}{1+\frac{\Delta i}{I_c}}\right) - 1\right] \tag{3.16}$$

Simplifying Eq. (3.16), we get

$$\frac{\Delta R}{R_c} = \frac{\Delta v}{V_c} + \frac{\Delta i}{I_c} - \frac{\Delta v}{V_c}\frac{\Delta i}{I_c} \tag{3.17}$$

The positive error in (3.15) has a larger magnitude than the negative error in (3.17) and we take it to be the worst-case error ΔR. Then the percentage error $r\,\%$ in R_c will be

$$r\,\% = v\,\% + i\,\% + \frac{v\,\% i\,\%}{100} \approx v\,\% + i\,\% \tag{3.18}$$

Equation (3.18) indicates that if you divide one measured quantity by another measured quantity to derive a third quantity, the relative errors add-up in the result.

Example 7 Let us take a node in a DC circuit as in Fig. 3.5, where three elements are connected at a node and the currents through those elements are I_1, I_2, and I_3. If we measure, say, I_1 and I_2, I_3 can be computed using Kirchhoff current law, namely, the algebraic sum of all currents at a node is zero. If the directions of I_1, I_2, and I_3 are as in Fig. 3.5, then we have.

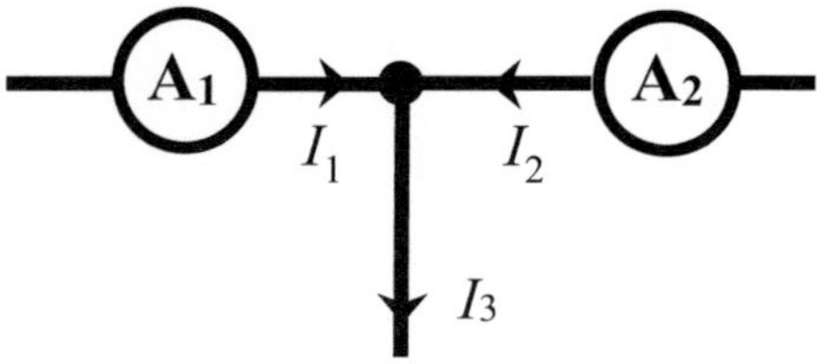

Fig. 3.5 Currents at a node

$$I_3 = I_1 + I_2$$

If we have systematic errors of ΔI_1 and ΔI_2 (corresponding to systematic percentage errors of ε_1 % and ε_2 %) in I_1 and I_2, then we can write I_3 as

$$I_3 \pm \Delta I_3 = I_1 \pm \Delta I_1 + I_2 \pm \Delta I_2 \pm \Delta I_3 = \pm \Delta I_1 \pm \Delta I_2 \quad (3.19)$$

In this case, it is clear that the positive and negative errors that can occur in I_3 will have the same maximum magnitude and thus,

$$\Delta I_3 = \Delta I_1 + \Delta I_2 \quad (3.20)$$

Dividing Eq. (3.20) by I_3 and multiplying with 100, we get

$$\frac{\Delta I_3}{I_3} 100 = i_3\% = \frac{I_1}{I_3}\frac{\Delta I_1}{I_1} 100 + \frac{I_2}{I_3}\frac{\Delta I_2}{I_2} 100 \quad (3.21)$$

Equation (3.19) can be simplified as

$$\varepsilon_3\% = \frac{I_1}{I_3}\varepsilon_1\% + \frac{I_2}{I_3}\varepsilon_2\% \quad (3.22)$$

In Eq. (3.22), $\varepsilon_1\%$, $\varepsilon_2\%$ and $\varepsilon_3\%$ are the percentage errors in currents I_1, I_2, and I_3. It should be noted here that while I_1 and I_2 are measured, I_3 is computed from I_1 and I_2.

Example 8 In Fig. 3.5, ammeter A_1 is 100 mA full-scale, class 2 instrument, reading 100 mA. Ammeter A_2 is 100 mA full-scale, class 1 instrument reading 50 mA. Then $I_3 = I_1 + I_2 = 100 + 50 = 150$ mA. Error ε_1 is 1% and error ε_2 is.

$$\frac{100}{50} 1 = 2\,\%$$

$$\varepsilon_3\,\% = \frac{I_1}{I_3}\varepsilon_1\,\% + \frac{I_2}{I_3}\varepsilon_2\,\% = \frac{100}{150} 2\,\% + \frac{50}{150} 2\,\% = 2\,\% \quad (3.23)$$

Example 9 If the currents I_A, I_B and I_C are as indicated in Fig. 3.6, then we have.

Fig. 3.6 Currents at a node

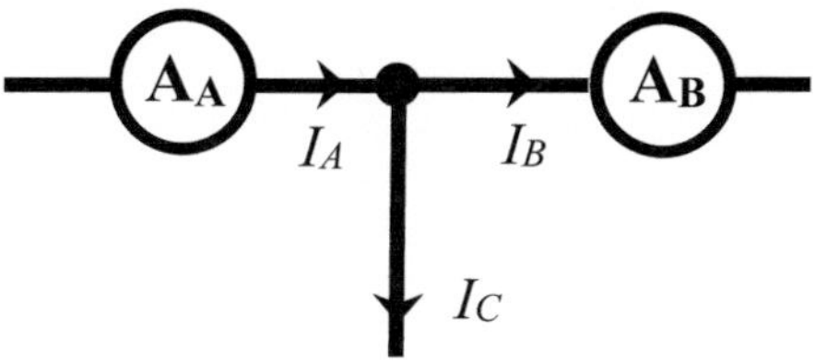

$$I_C = (I_A - I_B)$$

Let the systematic error in I_A and I_B be ΔI_A and ΔI_B, respectively (corresponding to systematic percentage errors of $\varepsilon_A\%$ and $\varepsilon_B\%$). Since I_C is calculated as the difference between two measured values, it will have its largest positive value when the error in the measured value of I_A is positive and the error in the measured value of I_B is negative. Accordingly,

$$I_C + \Delta I_C = (I_A + \Delta I_A) - (I_B - \Delta I_B) \tag{3.24}$$

Similarly, the negative limit of the calculated value of I_C, we have

$$I_C + \Delta I_C = (I_A - \Delta I_A) - (I_B + \Delta I_B) \tag{3.25}$$

After substituting $I_C = (I_A - I_B)$, both the above equations reduce to

$$\Delta I_C = \Delta I_A + \Delta I_B \tag{3.26}$$

Dividing all terms in Eq. (3.23) by I_C and multiplying with 100, we get

$$\tfrac{\Delta I_C}{I_C}100 = \tfrac{I_A}{I_C}\tfrac{\Delta I_A}{I_A}100 + \tfrac{I_B}{I_B}\tfrac{\Delta I_B}{I_B}100. \tag{3.27}$$

Equation (3.27) can be simplified as

$$\varepsilon_C\,\% = \left(\tfrac{I_A}{I_C}\varepsilon_A\,\% + \tfrac{I_2}{I_3}\varepsilon_2\,\%\right) \tag{3.28}$$

Example 10 In Fig. 3.6, ammeter A_A is a 100 mA full-scale, class 2 instrument, reading 100 mA. Ammeter A_B is a 150 mA full-scale, class 1 instrument reading 75 mA. Then.

$$I_C = I_A - I_B = 100 - 75 = 25\ \text{mA}.$$

Here, error ε_A in I_A is 2% and error ε_B in I_B is $\frac{150}{75}1 = 2\,\%$ and hence error ε_C in I_C is

$$\varepsilon_C\,\% = \tfrac{I_A}{I_C}\varepsilon_A\,\% + \tfrac{I_B}{I_C}\varepsilon_B\,\% = \tfrac{100}{25}2\,\% + \tfrac{75}{25}2\,\% = 14\,\% \tag{3.29}$$

From Eq. (3.29), we see that though we are using ammeters of class 1 and class 2, the error in the computed value of (I_A–I_B) is quite high. If we had used even a class 5, 25 mA full-scale ammeter to directly measure I_C, we would have incurred only 5% error. Thus, whenever it is necessary to find the difference of two comparable quantities then it is better to find the difference and measure rather than measure and then find the difference!

In the foregone examples, we saw that the errors propagate in computations involving measured quantities possessing systematic errors. In multiplication and division processes, the percentage errors simply add-up. However, when addition or subtraction is involved, the errors add-up but each multiplied by a weighting factor, namely, the ratio of the individual value to the final value. Though we can compute the propagation of errors as above for any simple situation, it would be nice if we can derive a general formula.

3.2.3.1 Formula for Determining the Propagation of Systematic Error

Let Y be computed as $Y = f(X_1, X_2, \ldots X_N)$ with $X_1, X_2, \ldots X_N$ being the measured values with systematic errors ε_1 %, ε_2 % ... ε_N % and $f(X)$ being any function. Then we can write

$$\Delta Y = \tfrac{\partial f}{\partial X_1}\Delta X_1 + \tfrac{\partial f}{\partial X_2}\Delta X_2 + \ldots + \tfrac{\partial f}{\partial X_N}\Delta X_N. \tag{3.30}$$

In Eq. (3.30), the Δs are actual error in respective quantities, namely, the computed value Y and the measured values $X_1, X_2, \ldots X_N$. Dividing all the terms in Eq. (3.30) by Y and multiplying by 100, we get

$$\tfrac{\Delta Y}{Y}100 = \tfrac{\partial f}{\partial X_1}\tfrac{X_1}{Y}\tfrac{\Delta X_1}{X_1}100 + \tfrac{\partial f}{\partial X_2}\tfrac{X_2}{Y}\tfrac{\Delta X_1}{X_2}100 + \ldots + \tfrac{\partial f}{\partial X_N}\tfrac{X_N}{Y}\tfrac{\Delta X_N}{X_N}100. \tag{3.31}$$

Simplifying Eq. (3.31), we get

$$\varepsilon_Y\% = \tfrac{X_1}{Y}\tfrac{\partial f}{\partial X_1}\varepsilon_{X_1}\% + \tfrac{X_2}{Y}\tfrac{\partial f}{\partial X_2}\varepsilon_{X_2}\% + \ldots + \tfrac{X_N}{Y}\tfrac{\partial f}{\partial X_N}\varepsilon_{X_N}\%. \tag{3.32}$$

Let us check this formula for the power in a DC circuit, $P_c = V_c I_c$ with v % and i % as percentage errors in measuring the voltage V_c and the current I_c.

$$p\,\% = \left(\tfrac{V_c}{P_c}\tfrac{\partial(V_c I_c)}{\partial V_c}v\,\% + \tfrac{i_c}{P_c}\tfrac{\partial(V_c I_c)}{\partial I_c}i\,\%\right). \tag{3.33}$$

Evaluating Eq. (3.33), we get

$$p\,\% = \left(\tfrac{V_c I_c}{P_c} v\,\% + \tfrac{I_c V_c}{P_c} i\,\%\right) = v\,\% + i\,\%. \tag{3.34}$$

Comparison of Eq. (3.34) with Eq. (3.12) shows that Eq. (3.34) is only approximate (as in Eq. (3.13)) as it has failed to capture the small part, namely,

$$\frac{v\%i\%}{100}$$

in the error. This comes about because Eq. (3.34) is a first-order approximation and ignores higher order derivatives, which should be included for a more accurate estimation. Hence, it should be used only when errors are small.

3.3 Random Errors

The voltage across the 470 kΩ resistor in the circuit of Fig. 3.7 must be a DC voltage as a DC current of 40 nA is passing through it. However, if we look at the voltage across the 470 kΩ resistor, its waveform will be as shown in the oscilloscope screen of Fig. 3.7. The reason for this is the presence of thermal noise voltage, inherent in any resistor. Even for a resistor R not connected in a circuit, as shown in Fig. 3.8, the voltage v_R across its terminals will be as given in the same figure. However, the average value of such v_R will be zero. In 1926, John Johnson of Bell laboratory discovered this and found that noise increases as temperature increases. Later, Harry Nyquist found that this noise voltage is due to the temperature driven Brownian motion of charge carriers within the resistor and derived Eq. (3.35) to represent the average "noise power".

$$\overline{v_n^2} = 4K\theta R\Delta f \tag{3.35}$$

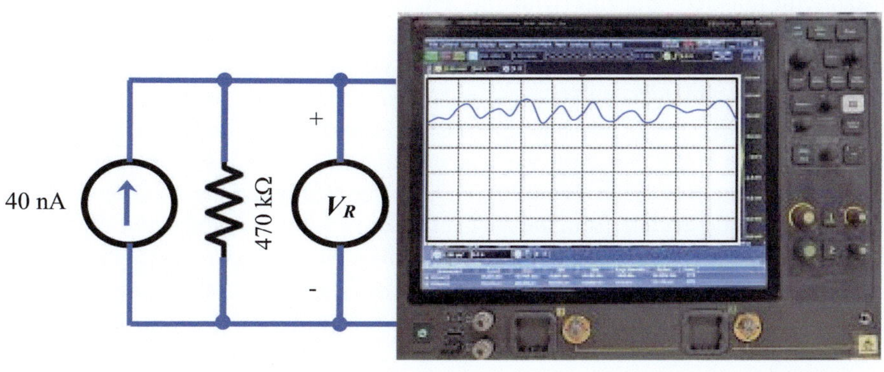

Fig. 3.7 Voltage affected by random (thermal) noise

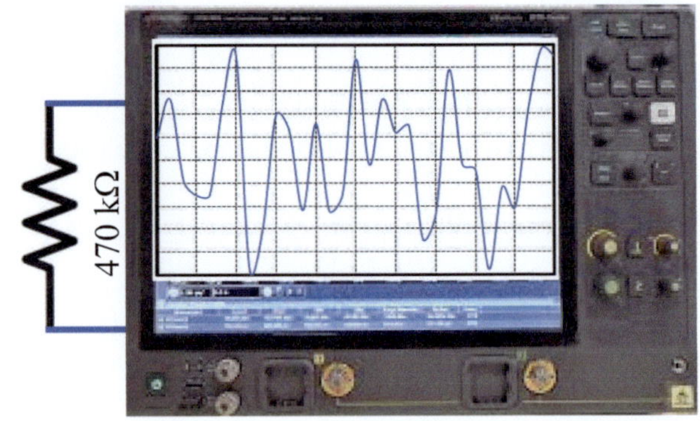

Fig. 3.8 Thermal noise

In Eq. (3.35), v_n is the noise voltage, K is Boltzmann constant ($\approx$1.38 10^{-23} J/K), θ is the temperature of the resistor in kelvin, R is the value of resistance in ohms and Δf is the frequency bandwidth in hertz over which the noise power is considered. Since Brownian motion is random in nature, the voltage across the resistor also will have a magnitude that randomly fluctuates. Named after its discoverer, it is called Johnson's noise. Since the noise is due to temperature, it is also popularly called thermal noise. Being random in nature and unbiased, the voltage contains all the frequencies and has a flat frequency spectrum. Because of this the thermal noise is also referred to as white noise (white color contains all frequencies in the visible spectrum). Not only a resistor, but also many other electrical and electronic components generate not just thermal noise, but also different types of noise.

Due to the presence of the thermal noise in the 470 kΩ resistor of Fig. 3.8, the voltage across it is not a pure DC voltage of 18.8 mV but has an irregular waveform as shown on the oscilloscope screen. Interference from external factors such as electric or electromagnetic fields can also bring in a similar situation. How do we measure the DC voltage corrupted by noise under these circumstances? The fact that most noise voltages—white noise, for example—have zero average gives us a clue. Theoretically, if an infinite number of voltage readings are taken at all possible instants, the noise components (with different signs and magnitudes) of the individual readings cancel out in the sum of the readings and the average of the readings is the true DC value. This however is not practical. We can take only a finite number of readings, say N and deduce the *best possible value* from these readings. We make N measurements of v_R at N possible time instances and obtain the values V_1, V_2, V_3, $\ldots V_N$. as shown in Fig. 3.9. Now we need to find the *best possible value*, say $\overline{V}_R$ that approximates the true value. From Fig. 3.9, we see that individual measured values V_1, V_2, V_3, $\ldots V_N$ deviate from the as—yet-unknown *best possible value* $\overline{V}_R$ by d_1, d_2, d_3, … d_N, which are called deviations. The *best possible value* $\overline{V}_R$ can be chosen such that it has no bias to any of the deviations d_1, d_2, d_3, $\ldots d_N$. Mathematically expressing this condition, the sum of all the deviations d_1, d_2, d_3, $\ldots d_N$ is zero, i.e.,

$$\sum_{j=1}^{N} d_j = 0. \tag{3.36}$$

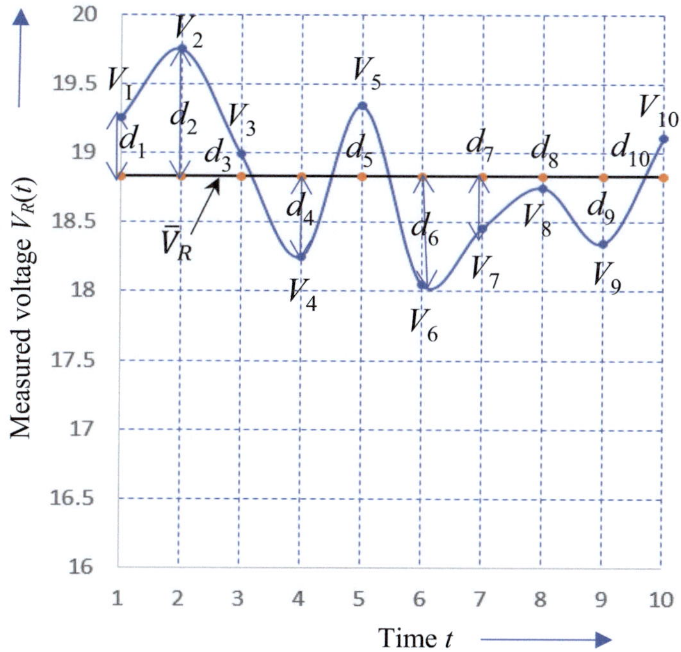

Fig. 3.9 Measured voltage v_R

Substituting $d_j = \left(V_j - \overline{V}_R\right)$ in Eq. (3.36), we get

$$\sum_{j=1}^{N}\left(V_j - \overline{V}_R\right) = 0. \tag{3.37}$$

Evaluating the summation in Eq. (3.37), we get

$$\sum_{j=1}^{N} V_j - N\overline{V}_R = 0 \rightarrow \overline{V}_R = \frac{1}{N}\sum_{j=1}^{N}\left(V_j\right). \tag{3.38}$$

Equation (3.38) indicates that when we have random variations in our measurements (maybe due to noise or interference), best possible result can be obtained by making a large number of measurements and obtaining the average value of those measurements. Thus, the condition that the sum of all deviations must be zero provides the *best possible value* $\overline{V}_R$ so that it has no bias to any of the deviations d_1, d_2, d_3, $\ldots d_N$. There is another approach to the determination of the *best possible value*. This is: "The principle of least squares", a statistical approach often used in data fitting. Applied to our case, it simply states that the most probable value of an unknown quantity on which observations have been made is that value that makes the sum of squares of errors (deviations d_j in our case) a minimum. This requires us to ensure that the sum of the squares of the deviations, viz.

$$\sum_{j=1}^{N}\left(d_j^2\right)$$

is a minimum. Once again if we substitute $d_j = \left(\left(V_j - \overline{V}_R\right)\right.$, we must minimize $\sum_{j=1}^{N}\left(V_j - \overline{V}_R\right)^2$. This leads to

$$\frac{d}{d\overline{V}_R}\left(\sum_{j=1}^{N}\left(V_j - \overline{V}_R\right)^2\right) = 0.$$

Expanding, we get

$$\frac{d}{d\overline{V}_R}\left(\sum_{j=1}^{N}\left(V_j^2 + \overline{V}_R^2 - 2V_j\overline{V}_R\right)\right) = 0. \tag{3.39}$$

Evaluating Eq. (3.39) by reversing the summation and differentiation, we get

$$\sum_{j=1}^{N}\left(\frac{d}{d\overline{V}_R}\left(V_j^2 + \overline{V}_R^2 - 2V_j\overline{V}_R\right)\right) = 0\sum_{j=1}^{N}\left(2\overline{V}_R - 2V_j\right) = 0. \tag{3.40}$$

Here we should note that the differentiation is with respect to $\overline{V}_R$ and hence $\frac{d}{d\overline{V}_R}\left(V_j^2\right) = 0$. Evaluating the summation, in Eq. (3.40), we get

$$\overline{V}_R = \frac{1}{N}\sum_{j=1}^{N}\left(V_j\right). \tag{3.41}$$

This is exactly the same result that we had for the best value using the criterion of "zero sum" of all deviations. Hence, whenever we have to make measurements when random errors are present, we must make a large number of measurements of the same quantity and take the average of all the measured values as the *best possible value.* Once we have obtained the average value, then a quantitative measure of the deviations needs to be included in the result of the measurement. There are three possible ways we can do this, as given below.

(i) Bipolar Peak Deviation
(ii) Unipolar Peak Deviation
(iii) Standard deviation.

3.3.1 Bipolar Peak Deviation (Tolerance)

One way we can express the deviations is to obtain the maximum value of the positive deviation d_P and the negative deviation d_N and express the result as

$$\textit{Readings obtained is expressed as} \quad \overline{V}_R{}^{\substack{+d_P \\ -d_N}}. \tag{3.42}$$

Here it should be noted that deviations can be positive ($+d_P$) or negative ($-d_N$) and hence there will be a maximum in the set of positive deviations and maximum in the set of negative deviations (minimum measured value). When we express the deviations in this fashion, we get what is known as *bipolar* (both positive and negative) deviations. Many times, the positive and negative deviations are expressed with just one value as $\overline{V}_R^{\pm d_m}$. Deviations can occur not only when random errors are present in a measurement. Deviations may also occur in manufacturing products. Let us say that we need to manufacture cylinders (or rods) of diameter $D = 1$ cm and length $L = 10$ cm. Obviously we will not be able to manufacture several cylinders with exact dimensions of 1.0000 and 10.000 cm. Depending on our ability to control these dimensions during manufacturing, we will get deviations in the length and diameter of the manufactured cylinders. In such a case we can say that the cylinders are manufactured within the tolerance limits as

$$D^{\substack{+d_{pd} \\ -d_{Nd}}} \textit{ and } L^{\substack{+d_{pL} \\ -d_{NL}}}. \tag{3.43}$$

In Eq. (3.43) d_{pd} and d_{Nd} are the worst-case positive and negative deviations in the diameter of manufactured rods and similarly d_{pL} and d_{NL} are the worst-case positive and negative deviations in the length of the manufactured rods. Though the positive and negative deviations can differ, many times, we simply put one value for both positive and negative limits of deviations as

$$D^{\pm d_d} \text{and} L^{\pm d_L}.$$

3.3.2 Unipolar Peak Deviation (Tolerance)

We may need to fit components to manufacture a product. For example, when we manufacture a fan, the central hanging rod and the rotating part of the fan need to be mated with the help of a couple of bearings. If the rod diameter is more than the inner diameter of the bearing, it will not be possible to fit the bearing onto the rod.

Hence, the diameter of the rod is specified as

$$D_{rod}{}^{+0}_{-d_{Nr}}. \tag{3.44}$$

And the inner diameter of the bearing as

$$D_{bearing}{}^{+d_{Prb}}_{-0} \tag{3.45}$$

Equation (3.44) indicates that rod's diameter can never exceed D_{rod} and only can be $<D_{rod}$ to the extent of d_{Nr}. Similarly, Eq. (3.45) indicates that the bearing's inner diameter can never go below $D_{bearing}$ and only can be $>D_{bearing}$ to the extent of $+d_{Prb}$. Hence, in such cases, we can tolerate deviations in only one direction and thus the deviations become "Unipolar". When we deal with deviations in manufactured product dimensions, it is customary to use the word "tolerance" rather than deviation. It may be noted here that the specification of the length of the rod would have both the positive and negative tolerances. Hence, in the manufacturing vocabulary, we deal with bipolar and unipolar tolerances as appropriate. Like the specification of accuracy in terms of percentage error, the deviation is also expressed as a percentage for the same reasons, as

$$d\,\% = \frac{d}{\overline{X}}\,100, \tag{3.46}$$

where $\overline{X}$, the average value of X is the relevant reference. One should not confuse the percentage deviation with the percentage error. To differentiate between the two, we always specify the random error as "tolerance" in certain situations. In the full specification of a resistor, we need to include the tolerance also. For example, a typical specification of a resistor may be

Sl. No	Parameter	Value
1	Value	100 Ω
2	Power rating	0.5 W
3	Type	Metal Film, Axial leads
4	Tolerance	±0.5 %
5	Packaging	Reel

Here, the ±0.5 % tolerance means that the value of any of the 100 Ω resistor from that reel of package will be within the values 99.5 Ω and 100.5 Ω. It should be clearly understood that error indicates how close a measurement is to the correct (true) value. On the other hand, deviation simply tells us how repeatable the measurements were and thus does not indicate how close the average of the measured value is to the true value. We can visualize this using the well-known "*dart experiment*", wherein competitors throw darts (small arrows) from a distance on a board with concentric circles as shown in Fig. 3.10. The aim of the game is to throw three darts accurately and consistently. If the darts are thrown to be within a certain area, the points earned are high. The central (the smallest) circle is called the bullseye, and a player is normally expected to hit the bullseye consistently. Figure 3.10 shows the throw results of four players. We see from Fig. 3.10 that player 1 is both accurate and precise. Player 2 is not accurate but precise. Player 3 is accurate compared to player 2, but not precise as compared to player 1 or 2, while player 4 is neither accurate nor precise.

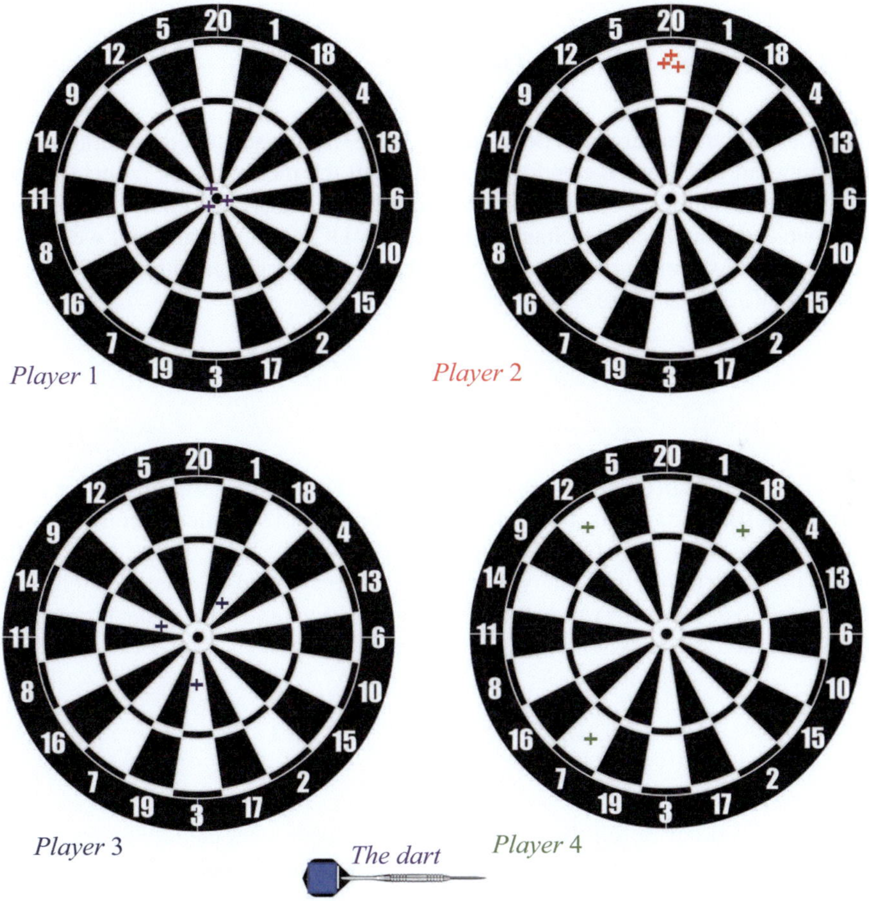

Fig. 3.10 Dart Game to illustrate accuracy and precision

3.3.3 Average Deviation

We can also represent the deviations in a measurement involving random errors as $\overline{V}_R \pm \overline{d}$, where $\overline{d}$ is the average deviation, and is computed as

$$\overline{d} = \frac{1}{N}\sum_{j=1}^{j=N}\left|d_j\right| \tag{3.47}$$

In Eq. (3.44), we take the average of *the magnitude* of the deviations as $\sum_{j=1}^{N} d_j = 0$. The average deviation is of academic interest and is rarely used in practice.

3.3.4 RMS and Standard Deviations

If we look at the measurement portrayed in Fig. 3.10, we see that the average value is nothing but a DC voltage and if we connect all the measurement points then we see that the deviations look like a time varying AC voltage riding on the DC voltage. Thus, we can visualize the voltage in Fig. 3.7 as DC + AC. In general, we always represent the AC voltage by its rms value. Hence, the readings in Fig. 3.10 can be written as $\overline{V}_R \pm d_{rms}$. Here,

$$d_{rms} = \sigma_{rms} = \sqrt{\frac{1}{N}\sum_{1}^{N} d_j^2} \tag{3.48}$$

The symbol σ_{rms} is in common use instead of d_{rms}. Though the term rms makes sense to us as electrical engineers, the above formulation is not statistically correct, as out of the N deviations, only (N–1) are independent, the N_{th} being the negative of the sum of the other (N–1) deviations. Hence, the correct way to calculate the reference deviation is to use only the (N–1) independent deviations and arrive what is popularly called the standard deviation d_{std} (also written as σ_{std}) as

$$d_{std} = \sigma_{std} = \sqrt{\frac{1}{N-1}\sum_{1}^{N-1} d_j^2} \tag{3.49}$$

To sum up, in order to foil the random errors in our measurement, we take a large number of independent measurements of the same quantity, calculate their average value and standard deviation and express the results as

$$\overline{V}^{\pm\sigma} or\left(\overline{V} \pm \sigma\right).$$

To differentiate between *the systematic error* (due to the measurement system and or due to the instrument) from *the deviation* due to random errors, it is customary to indicate the effect of random errors (deviation) as *precision* in the measurement and the standard deviation as a *precision index.*

3.3.5 The Histogram

The measured data can also be plotted in a different way with the x-axis depicting the measured value and the y-axis providing the number of readings obtained within a given range of measured values. Figure 3.11 plots the data, as described above, for a total of 100 readings obtained by measuring the voltage across a resistor. Such a diagram as depicted is called the *histogram.* The heights of the blocks as portrayed here is dependent on the total number of readings N_T. Hence, for similar distributions of measured values, the histogram may look different for different values *of* N_T.

To compare different sets of measurements, it is better to normalize the data and plot the histogram accordingly. Normalization is accomplished by simply dividing the number of readings N_i within a range by the total number of readings N_T. For the histogram shown in Fig. 3.11, the normalized histogram (obtained by dividing each column by the total number of measurements made, namely, 100) is illustrated in Fig. 3.12. From Fig. 3.12, we see that the heights of the column bars are now fractions. If we add all the column bar heights, we get.

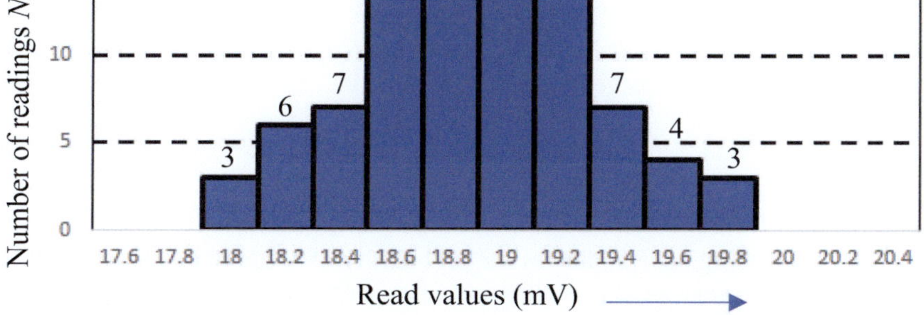

Fig. 3.11 Histogram of measurement affected by thermal noise

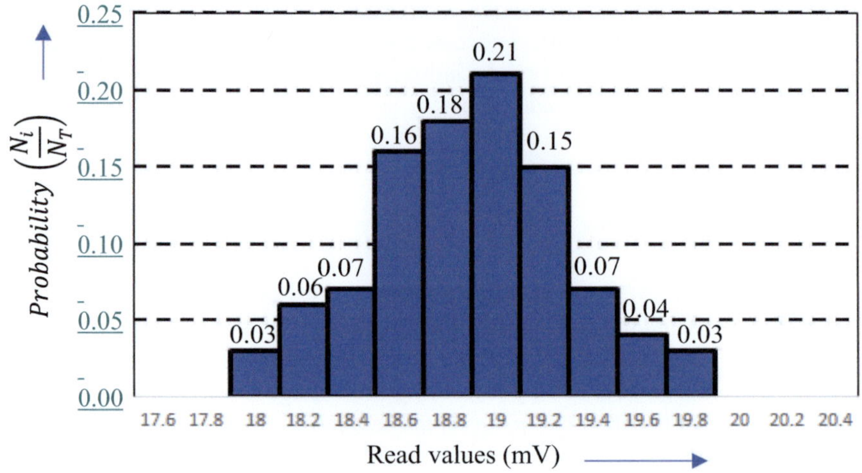

Fig. 3.12 Normalized histogram of measurement affected by thermal noise

$$\sum_{i=1}^{N} \frac{N_i}{N_T} = \frac{N_T}{N_T} = 1. \tag{3.50}$$

3.3.6 *The Probability Density*

The height of the column bar, say for values around 19.0 mV is

$$\frac{26}{130} = 0.20$$

and indicates that out of the total 130 readings, the number obtained in the range from 18.95 to 19.05 mV is 26. In other words, we can say that the probability of obtaining a reading in the range 18.95 to 19.05 mV is

$$\frac{26}{130} = 0.20.$$

Thus, the *y*-axis of Fig. 3.12 indicates the probability of obtaining a reading within the corresponding range of values. We note that while the column heights add-up to 1, the total area under the stepped envelope equals *w*, where *w* is the width of each segment on the *x*-axis. If we reduce *w* in order to get a better resolution for the selection of ranges, the number of segments increase, the column heights go down and the total area reduces to the new value of *w*. This gives rise to difficulties when we wish to extend the concept of the histogram to a similar chart which enables us to find the probability of readings falling within any arbitrary range and not within pre-defined intervals w as in the histogram. For this purpose, we need a continuous curve, which can be imagined to be a stepped figure with w tending to zero. If w tends to zero, so will the total area of all columns. Needless to say, this approach fails. The way to circumvent the problem is to make the y-axis represent

$$\frac{1}{w}\left(\frac{N_i}{N_T}\right) and\ NOT \left(\frac{N_i}{N_T}\right).$$

Now it would be the area of the ith column which gives the probability of a reading in the ith range. The sum of the areas of all columns would be unity, irrespective of the value of ω. The y-axis now represents what is known as the probability density $p(x)$ (probability per unit width). The density in each block being constant, the density multiplied by the width of the block gives the probability of a reading within its range, as already observed. With a change in the y-axis scale factor, the layout of Fig. 3.12 itself can be viewed as conforming to this revised convention, as seen from Fig. 3.13. The way is now clear for transition to the continuous case. As w tends to zero, the stepped pattern in the histogram yields a smooth *probability distribution* curve as in Fig. 3.14. Here, the y-axis represents the probability density $p(x)$ at a point x. It means that the probability of a reading in a small range Δx around x is $(p(x)\,\Delta x)$ and the probability of a reading within any arbitrary range (x_1, x_2) is.

$$\int_{x_1}^{x_2} p(x)dx.$$

The probability of obtaining one or the other of the entire set of readings is 1 and so the area under the curve will also be 1. It would be nice if we can represent the continuous curve of Fig. 3.14 as a mathematical (analytical) function. If we look at the history of measurements, many functions were proposed to represent the probability distribution of a set of measured data that possess random errors. Of all the functions that were proposed, the function proposed by the famous mathematician Carl Friedrich Gauss withstood the test of time and today we use the Gaussian function as the standard function to represent the probability function of a set of measured data, mainly due to the reason that it is found to be optimal in representing the normalized histogram of the measured data. The Gaussian function as given by Gauss is

$$y = he^{-x^2h^2} \tag{3.51}$$

Figure 3.15 shows the graph of Eq. (3.48), with $h = 1$. If we wish to construct a Gaussian function to match the curve of Fig. 3.15, the first thing we need to do is

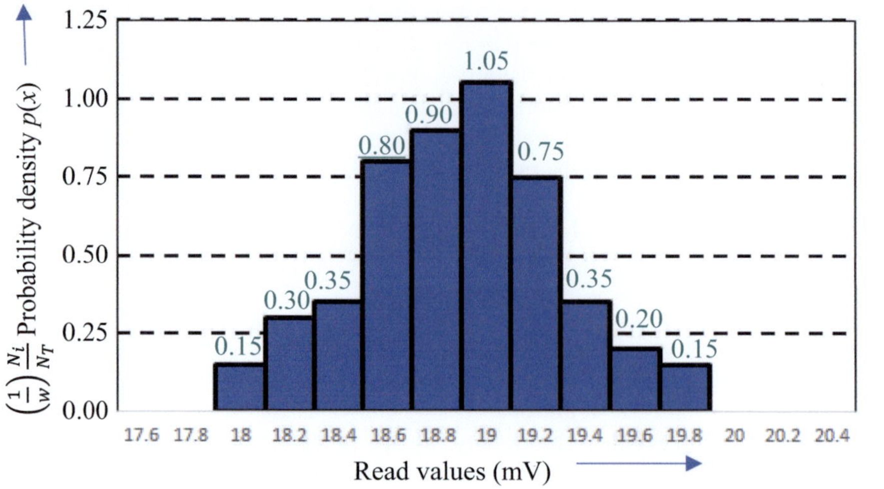

Fig. 3.13 Probability density histogram of the data of Fig. 3.12

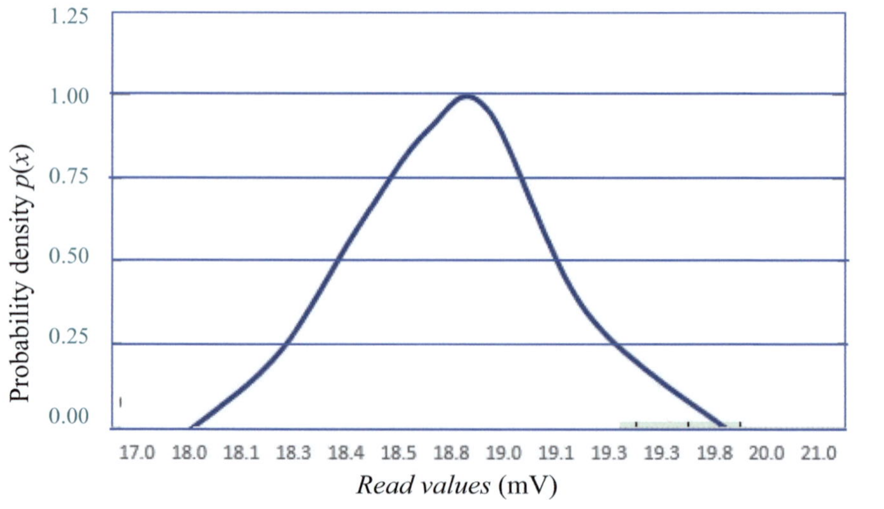

Fig. 3.14 Probability density distribution of data presented in Fig. 3.12

to shift the curve of Fig. 3.15 to the right side by the value of the average value of the readings. The second thing that we should do is to ensure that the area under the curve is 1. Since $\int_{-\infty}^{\infty} y = \int_{-}^{\infty} e^{-x^2} = \sqrt{\pi}$, we need to modify Eq. (3.51) as

$$y = \frac{h}{\sqrt{\pi}} e^{-(x-\bar{x})^2 h^2} \tag{3.52}$$

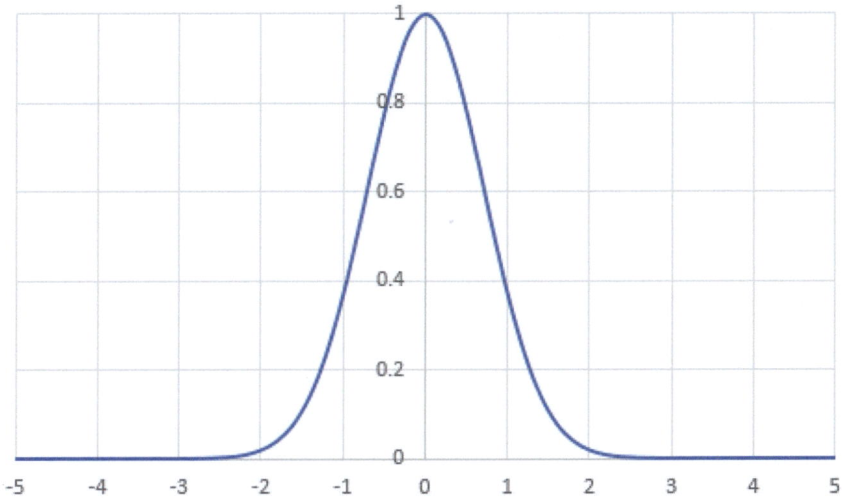

Fig. 3.15 The gaussian function

Now if we compute the standard deviation with the Gaussian function as given in Eq. (3.49), we get

$$\sigma = \frac{1}{h\sqrt{2}} \quad h = \frac{1}{\sigma\sqrt{2}} \tag{3.53}$$

Substituting the value of h in terms of σ as given in Eq. (3.53) in Eq. (3.52), we get

$$y = \frac{1}{\sigma\sqrt{2\pi}} e^{-\frac{1}{2}\left(\frac{x-\bar{x}}{\sigma}\right)^2}. \tag{3.54}$$

A typical Gaussian curve representing a measured data set (with average value 2.5) is shown in Fig. 3.16. Thus, once we calculate the average value and the standard deviation from the measured data set, we can represent that dada set using Eq. (3.50). We use the Gaussian function wherever natural distribution occurs and hence Gaussian distribution is also popularly called as "*The Standard distribution*" or "*Normal distribution*". We can see that 68.2% of all measured data will be within the range of (average value $\pm\ \sigma$). The range: (average value $\pm\ 2\sigma$) encompasses 95.4 % of all the measured data and the range of (average value $\pm\ 3\sigma$ covers) 99.6% of all the measured data. Since the range (average value $\pm\ 6\sigma$) covers 99.9999998027 % of all the measured data, there is a process management/business management jargon called the *six-sigma management* that ensures that nearly 100% acceptable results (may be sales, may be error-free decisions …) are obtained using appropriate management practices.

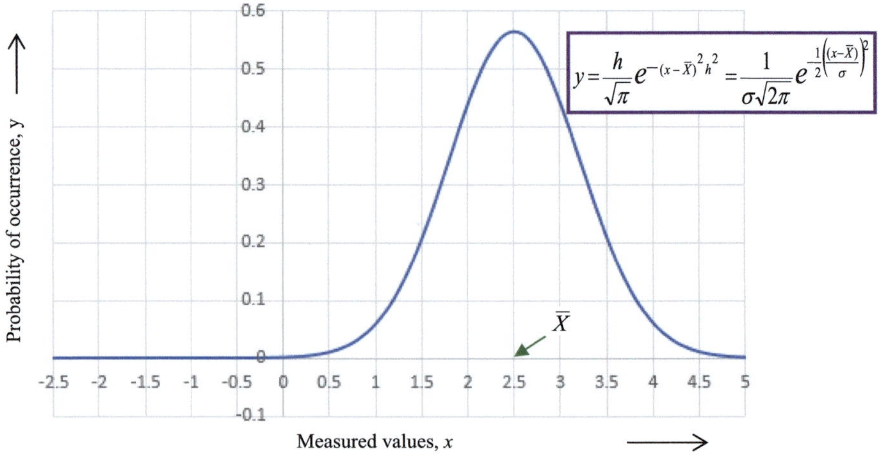

Fig. 3.16 Typical Gaussian function representation of measured data

3.3.7 *Propagation of Random Error*

In Sect. 3.1.3, we learnt how to estimate the limits of error in the value of a quantity calculated from two or more measured values that are afflicted with systematic errors… Similarly, we may use measured values possessing random errors and compute another quantity. The average value and standard deviation of the computed quantity will depend on the average values and standard deviations of the measured quantities, but their relative contributions to the final result will depend on the mathematical operations involved in the computation process.

3.3.7.1 Addition (or Subtraction) of Measured Quantities Possessing Random Errors

Let us assume that two voltages V_A and V_B are measured with random errors. Let V_A be measured N times as $V_{A1},\ V_{A2},\ V_{A3},\ \cdots\ V_{AN}$ with an average value of $\overline{V}_A$ and standard deviation σ_A. Similarly, let V_B be measured M times as $V_{B1},\ V_{B2},\ V_{B3},\ \cdots\ V_{BM}$ with an average value $\overline{V}_B$ and standard deviation σ_B. If we compute $V_C = V_A+V_B$, then the average value of V_C and its standard deviation can be obtained as given below. First, we can take V_{A1} and add to it $V_{B1},\ V_{B2},\ V_{B3},\ \cdots\ V_{BM}$ in turn, repeat the process with $V_{A2}, V_{A3,..}V_{AN}$ and thus obtain MN values of V_C. They can be represented as

$$V_C|_{i,j} = V_A|_i + V_C|_j \tag{3.55}$$

Then the average value $\overline{V}_C$ of V_C can be computed as

$$\bar{V}_C = \frac{1}{MN}\sum_{\substack{i=1.\\ j=1}}^{\substack{M\\ N}} V_{C_{i,j}} = \frac{1}{MN}\sum_{\substack{i=1.\\ j=1}}^{\substack{M\\ N}} \left(V_{A_i} + V_{B_j}\right) \Rightarrow \frac{M}{MN}\sum_{i=1.}^{N}\left(V_{A_i}\right) + \frac{N}{MN}\sum_{j=1}^{M}\left(V_{B_j}\right) = \bar{V}_A + \bar{V}_B \tag{3.56}$$

Equation (3.56) indicates that the "Average of a sum set is equal to the sum of the averages of the individual sets". Now standard deviation of V_C, say σ_C is

$$\sigma_c = \sqrt{\frac{1}{MN}\sum_{\substack{i=1.\\ j=1}}^{\substack{M\\ N}} \left(V - \overline{V}_C\right)^2} = \sqrt{\frac{1}{MN}\sum_{\substack{i=1.\\ j=1}}^{\substack{M\\ N}} \left(V_{A_i} + V_{B_j} - \overline{V}_A - \overline{V}_B\right)^2} = \sqrt{\frac{M}{MN}\sum_{i=1.}^{N}\left(V_{A_i} - \overline{V}_A\right)^2 + \frac{N}{MN}\sum_{j=1.}^{M}\left(V_{B_j} - \overline{V}_B\right)^2 + 2\frac{1}{MN}\sum_{i=1.}^{N}\left(V_{A_i} - \overline{V}_A\right)\left(V_{B_j} - \overline{V}_B\right)} \tag{3.57}$$

Since we have

$$\sum_{i=1}^{N}\left(V_{Ai} - \overline{V}_A\right)\sum_{j=1}^{N}\left(V_{Bj} - \overline{V}_B\right) = 0 \tag{3.58}$$

That is the sum of all deviations in each one of the terms in Eq. (3.58) is zero. Then

$$\sigma_C = \sqrt{\sigma_A^2 + \sigma_B^2}. \tag{3.59}$$

If we square Eq. (3.59), we get

$$\sigma_C^2 = \sigma_A^2 + \sigma_B^2 \tag{3.60}$$

The term σ_C^2 is called the variance of V_C, thus we have "The variance of a sum set is sum of the variances of the individual sets".

3.3.7.2 Formula for Propagation of Random Errors

Once again it would be worthwhile if we could derive a formula for obtaining the standard deviation or variance for a general case. To do that, let us take that we compute C as a function of measured quantities $A, B, X, Y \cdots$. Here, $A, B, X, Y \cdots$ have random errors with $\overline{A}, \overline{B}, \overline{X}, \overline{Y} \cdots$ as average values and $\sigma_A^2,\ \sigma_B^2,\ \sigma_X^2,\ \sigma_Y^2,\ \cdots$ as variances. Let C be

$$C = f(A,\ B,\ X,\ Y\ \cdots) \tag{3.61}$$

Then an incremental deviation ΔC in C can be expressed as

$$\Delta C = \left(\tfrac{\partial f}{\partial A}\Delta A\right) + \left(\tfrac{\partial f}{\partial B}\Delta B\right) + \left(\tfrac{\partial f}{\partial X}\Delta X\right) + \left(\tfrac{\partial f}{\partial Y}\Delta Y\right) + \cdots \tag{3.62}$$

where $\Delta A,\ \ \Delta B,\ \ \Delta X,\ \ \Delta Y \cdots$ are incremental deviations in $A, B, X, Y \cdots$ and $\frac{\partial f}{\partial A}, \frac{\partial f}{\partial B}, \frac{\partial f}{\partial X} \cdots$, etc., are the partial derivatives of the function f (with respect to the measured variable $A, B, X \cdots$, etc. The variance σ_C^2 in C can be derived as

$$\begin{aligned}\frac{1}{N_C}\sum_{N_C}\Delta C^2 &= \left(\left(\frac{\partial f}{\partial A}\right)^2\frac{1}{N_A}\sum_{N_A}\Delta A^2\right) + \left(\left(\frac{\partial f}{\partial B}\right)^2\frac{1}{N_B}\sum_{N_B}\Delta B^2\right)\\ &+\left(\left(\frac{\partial f}{\partial X}\right)^2\frac{1}{N_X}\sum_{N_X}\Delta X^2\right) + \cdots\end{aligned} \tag{3.63}$$

In the above expansion, we can omit the products of incremental variations of two unlike quantities (terms like $\Delta A_i\ \Delta B_j$) because the variations in one quantity are independent of the variations in the other (uncorrelated) and all such terms for a given pair of quantities add-up to zero in their summation. Thus, Eq. (3.63) reduces to

$$\sigma_C^2 = \left(\tfrac{\partial f}{\partial A}\right)^2\sigma_A^2 + \left(\tfrac{\partial f}{\partial B}\right)^2\sigma_B^2 + \left(\tfrac{\partial f}{\partial Y}\right)^2\sigma_Y^2 + \left(\tfrac{\partial f}{\partial X}\right)^2\sigma_X^2 \cdots \tag{3.64}$$

The variance of the calculated quantity is thus the weighted sum of the variances of the measured quantities, the weighting factor being the square of the partial derivative of the calculated quantity with respect to the related measured quantity.

3.4 Problems and Exercises

3.1 What reference conditions of temperature are to be adopted according to Indian Standard IS-1248 when no specific value is marked on an electrical indicating instrument?

3.2 A 300 V full-scale, class 1, PMMC voltmeter has a full-scale deflection of 100° and is graduated every 0.5°. Determine its resolution. If the meter reads 230 V, what is the worst-case error in its reading?

3.3 A 100 V full-scale, class 1 voltmeter having a sensitivity of 10 kΩ/V and a 200 V full-scale, class 0.5 voltmeter of sensitivity 15 kΩ/V are connected in series to measure 260 V. Determine the readings that will be obtained in each meter and the error in measuring 260 V. Determine the maximum voltage that can be measured using this series connected voltmeters and the error in that reading.

3.4 A multi range PMMC voltmeter has a sensitivity of 20 kΩ/V. The voltmeter is set at 10 V full scale to read the voltage across two points *a* and *b* in a circuit and a reading of 4.0 V is obtained. When the range is changed to 5 V full scale on the meter, a reading of 3.0 V is obtained. Determine the open circuit voltage across *a* and *b*.

3.5 A 6 A full-scale, class 1 ammeter having an internal resistance of 30 m Ω and a 5 A full-scale, class 2 ammeter having an internal resistance of 45 m Ω are paralleled to measure a current of 7.5 A. Determine the reading of each ammeter and the worst-case error in the reading of 7.5 A.

3.6 A current of 2 A DC is to be measured. Which one of the following PMMC ammeters you would choose and why?.

(a) 0–5 A, class 0.5
(b) 0–3 A, class 1.0

3.7 A 4 ½ digit DMM has accuracy specificities as DC Voltage: 0.2% FS + 2 counts, AC Voltage 0.5% FS + 10 counts and resistance 0.7% FS + 2 counts. Determine the worst-case error when the meter reads

(a) 12.345 V DC
(b) 500.0 mV DC
(c) 10.000 V AC
(d) 190.50 mV AC
(e) 1000.5 Ω
(f) 210.0 Ω

3.8 If, in the above problem, the percentage accuracies are specified as that of the reading, determine the errors.

3.9 In a Wheatstone bridge, the voltages across the two adjacent arms of the bridge are measured with a 5 V, class 0.5 m as 3.6 and 3.5 V and the difference voltage is then computed. Determine the worst-case error in computing the difference.

Determine the measurement error if a 120 mV FS, class 2.5 m is employed to measure the difference voltage and it reads 100 mV.

3.10 A student measures the height of his friend by asking him to stand against the doorway, which is known to be 200 cm high. He estimates that there is a 24 cm clearance over his friend's head and hence concludes the height of his friend to be 176 cm. If his estimation can be off by 2 cm (either direction), estimate the percentage error in the measurement.

3.11 The results of 14 separate measurements of the resistance of a resistor (in Ω) are:

2002.6	2002.8	2002.4	2001.6	2002.0	2002.2	2002.4
2002.6	2002.6	2002.4	2003.0	2002.2	2002.2	2002.0

(a) Find the best value of the resistor.
(b) Standard deviation of the measured values.
(c) Standard deviation of the mean (*This is not covered in the text*)

3.12 From p(x) in the equation given below, deduce the mean value and standard deviation of the distribution.

$$p(x)\frac{h}{\sqrt{\pi}}e^{-\left[h^2(x-\mu)^2\right]}, \text{ where } \mu \text{ is the mean of } x.$$

3.13 A resistor having a nominal value of 100 Ω is measured 100 times yielding the following data.

Resistance(Ω)	99.3	99.4	99.5	99.6	99.7	99.8	99.9	100	100.1	100.2	100.3	100.4
No. of times	0	2	3	7	12	17	22	17	12	5	3	0

a. Find the mean, median and mode (median and mode not covered in the text)
b. Compute the standard deviation σ
c. What percentage of the readings lie in the range ±σ and ±2σ from the mean.
d. How do the percentage in (c) compare with those expected for a normal distribution.

3.14 Legendre's principle of least square states that the most probable value of any observed quantity is such that the sum of the square of deviations of the observations from the value is the least. Using this principle, show that the most probable value of a measured quantity is the average of the observations.

3.15 A resistor is measured eleven times and the following readings in Ω are obtained. Find the best value of the resistor and its probable error. Assume that systematic error is absent, and that the distribution is normal.

(1) 200.2, (2) 220.0, (3) 200.0, (4) 199.4, (5) 200.0, (6) 200.2, (7) 200.2, (8) 200.6, (9) 199.6, (10) 199.8 and (11) 200.0

3.16 A student uses a digital counter to measure many times the frequency of the power supply. The mean of his readings is expressed as f = 49.6 Hz and standard deviation σ = 1 Hz. What fraction of the reading would you expect to lie within:

(a) 49.1Hz and 50.1Hz, (b) 48.6Hz and 50.6Hz (c) 47.6Hz and 51.6Hz
(d) 49.6Hz and 52.6Hz (e) 50.6Hz and 52.6Hz

Within what limits does 50% of the readings lie? Assume normal distribution.

3.17 You have on hand six resistors consisting of two 100 Ω resistors and four of 50 Ω resistors. The tolerance of 100 Ω and 50 Ω resistors are 1% and 2%, respectively. To build a resistor of 150 Ω with the least standard deviation which one of the connections would you use from the combinations (A), (B), (C), and (D) given below? What is the standard deviation of the connection of your choice?

3.18 It is proposed to build the circuits shown in Fig. P14a–c. All resistors used in the circuit have a nominal value of *R* Ω with a standard deviation σ. What is the effective value of resistance of each type of network and its standard deviation?

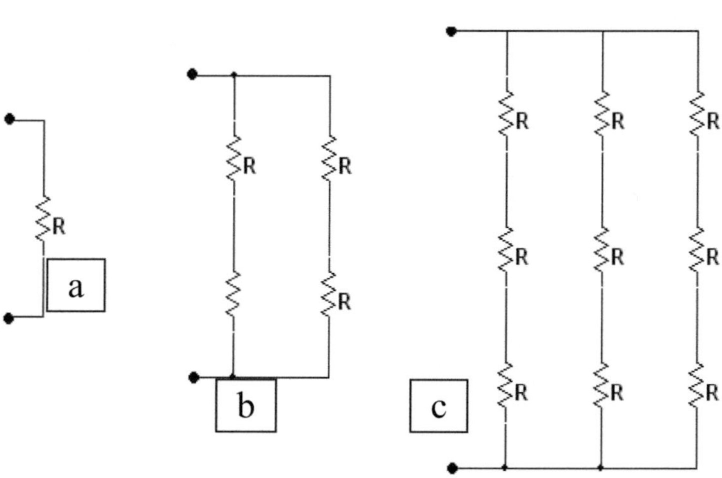

Fig. P14 Different ways of obtaining a resistance of value R

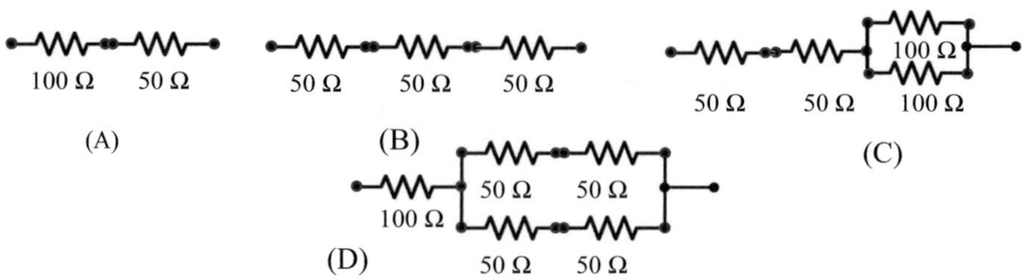

3.19 Using the formula "$y = 3x$", a quantity "y" is deduced from the value of x. Deduce the standard deviation of y, if σ_x is the standard deviation of x. However, a friend of yours puts $y = x + x + x$ and comes up with another answer for the standard deviation. What would be his answer and who is correct?

3.20 Using the formula $y = \sqrt{x}$, the quantity "y" is deduced from the value of x. Deduce the standard deviation of y, if σ_x is the standard deviation of x.

Chapter 4
The Potentiometers

The best and safest thing is to keep a perfect balance... If you can do that and achieve what you desire, then you really are a wise person.
—Euripides
The Greek Drama -406

4.1 Null Balance Methods of Measurement

The methods of measurement we learnt in Chap. 2 are called "*direct reading*" or "*direct indicating*" methods. The instruments described there measure a quantity by "reacting" to it and directly indicating its value on a scale. Hence, they are called "*direct reading instruments*". An alternative method available for determining the value of a measurand is the "*null-balance method*", wherein the measurement of an unknown quantity is accomplished by comparing it with a quantity whose value is known. We are all familiar with the 2-pan weighing balance of Fig. 4.1. In using this, we place the item we want to weigh on one of the pans and keep adding standard weights on the other till the beam supporting the two plates is horizontal with no

V. Jagadeesh Kumar, *Principles of Electrical Measurement and Measuring Instruments*,
https://doi.org/10.1007/978-981-96-5475-8_4

Fig. 4.1 Null balance weighing scale

tilt either way. Many times, we do the reverse as well. That is, we place a standard weight on one plate and keep adding more amounts of a quantity to the other till we get a "*balance*". In both cases, the weight of the quantity at balance will be equal to the sum of the standard weights placed on the other plate. Here the "balance" means that the two tilting forces in opposite directions exerted by the two weights on the supporting beam are equal, a state indicated by a central *balance indicator* becoming exactly vertical as seen in Fig. 4.1. Since we make a measurement by balancing two effects, the instrument itself is called "*the balance*".

Using the same balancing technique that we used for determining the weight of a substance, we can measure a DC voltage, say V_{ab} across terminals a and b in a DC circuit as indicated in Fig. 4.2. The symbol marked G in this figure is known as a *galvanometer*. The galvanometer, having a physical form shown in Fig. 4.3, is nothing but a sensitive center zero ammeter. The pointer of the galvanometer will be at the center of the scale (marked "0") when the current through G is zero. The pointer deflects to the right for positive values of current I_G and deflects towards the left if I_G is negative. In Fig. 4.2, V_s is a variable DC voltage source. To ascertain V_{ab}, we vary V_s till the current I_G is zero. When I_G is zero, we can deduce that $V_{ab} = V_s$. Since we know V_s, we also know V_{ab}. This type of measurement, wherein we compare a known quantity with an unknown quantity and vary the known quantity till the difference between the known and unknown values two is zero and obtain therefrom the value of the unknown quantity is called the *null-balance method of measurement*. There are two popular methods of null balance methods which have been developed. These are as follows:

1. The Potentiometric method of measurement and
2. The Bridge method of measurement.

The DC potentiometric method of measurement is applicable for DC voltage, current, and resistance. Similarly, the AC potentiometric method can be used to

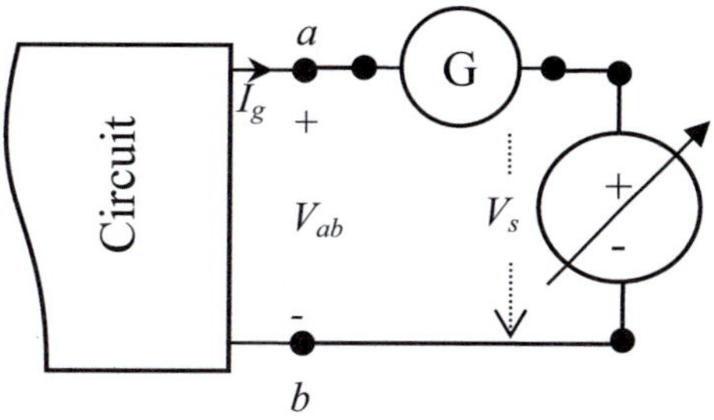

Fig. 4.2 Null balance method of measurement of a dc voltage

Fig. 4.3 The Galvanometer

measure AC voltage, current, impedance, and admittance. On the other hand, the bridge methods can be applied to measure resistance, inductance, and capacitance values and are not applicable for voltage and current measurements.

4.2 The DC Potentiometer

The DC potentiometer method of measurement shown in Fig. 4.4 was invented in 1841 by a German Physicist Johann Christian Poggendorff. Poggendorff, inferring correctly that a junction made of two conductors possessing differing work functions will generate an electromotive force (we call this as the thermo emf) proportional to the junction temperature, wanted to measure that thermo emf. We call such a sensor a *thermocouple.* Thermocouples are now widely used for the measurement of

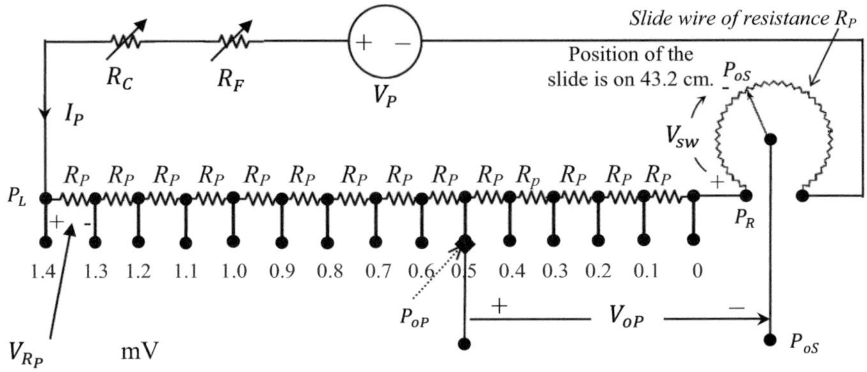

Fig. 4.4 The DC Potentiometer (Student Type: By Poggendorff)

temperature. At that point of time in 1841 only analog indicating instruments with a resolution of 1 in 100 or 1 in 200 of a volt were available and were incapable of measuring thermo emf. Hence, Poggendorff went ahead and invented the "*DC potentiometer*" and obtained a resolution of nearly 100 μV. However, the thermo emfs were two orders of magnitude lower than what he could measure and hence failed to demonstrate a *thermocouple*. That honor goes to Thomas Johann Seebeck who, also could not measure the thermo emf, but showed that thermo emf exists, through secondary effects (electromagnetic reaction).

4.2.1 The Student Potentiometer

Figure 4.4 gives the circuit diagram of a DC potentiometer invented by Johann Christian Poggendorff. Today, we call this device "*a student's DC potentiometer*" thus denying its originator his naming right. As can be seen in Fig. 4.4, the DC (student/Poggendorff) potentiometer consists of 14 resistances of equal value R_P (there can be more, say 20 resistors, as well), a wire with a slide on it (slide wire) also of value R_P and two variable resistances R_C and R_F in series with a DC source (battery) of value V_P. The values of R_C and R_F are adjusted such that a design value of current I_p flows in the circuit. The values of R_p and I_p are chosen to make the voltage drop V_{R_P} across each resistance of value R_P is

$$V_{R_P} = R_P I_P = 100.000\,\text{mV}.$$

For example, we can select $R_P = 10.000\ \Omega$ and $I_P = 10.000$ mA. With this condition, the voltage V_{oP} between terminal P_{oP} (assumed to be on the nth resistor from terminal with respect to terminal P_{oS} will be

Fig. 4.5 The 1 m slide wire fixed on a meter scale

$$V_{oP} = nV_{R_p} + V_{sw} = n(100.00\,\text{mV}) + V_{sw} = \left(n + \tfrac{xxx}{1000}\right)100\,\text{mV} \tag{4.1}$$

In Eq. (4.1), n is the number of resistors between points P_R and P_{oS}, and xxx is position of point on the slide wire of length 1 m, the voltage V_{sw} is the voltage on the slide wire between points P_R and P_{oS}. It is seen that pole P_{oP} is varied in steps of 100 mV. In the past, the slide wire was actually a 100 cm long resistive wire having a resistance R_P. The position of pole P_{oS} and hence the value of the voltage V_{sw} on the slide wire between points P_R and P_{oS} is read on a meter scale graduated in millimeters that is kept alongside the resistive wire. Figure 4.5 shows such an arrangement where the resistive wire is fixed on a meter scale and the slide is moved on the wire to obtain balance (zero reading on the galvanometer). Figure 4.5 shows part of the scale from about 40 cm mark to 60 cm mark. For example, if the slide (position of pole P_{oS}) is on the 512th mm mark ($xxx = 512$) on the meter scale, then the voltage $V_{sw} = \left(\frac{512}{1000}\right)100\,\text{mV} = 51.2$ mV. Presently, this type of DC potentiometer is used only for teaching the principles of operation of a DC potentiometer in an undergraduate laboratory class (hence the name *student potentiometer*). The slide wire is now realized using a ten-turn potentiometer shown in Fig. 4.6. We have seen that as we move the two poles P_{oP} and P_{oS}, the potentiometer output voltage V_{oP} will vary. For the positions of poles P_{oP} and P_{oS} shown in Fig. 4.4 the value of n is 5 and $xxx = 432$. Hence, we have

$$V_{oP} = \left(5 + \tfrac{432}{1000}\right)100 = 0.5432\,\text{V}. \tag{4.2}$$

To ensure the correctness of this reading we must make certain beforehand that $R_p I_p = 100.000$ mV. Thus, the resistances of value R_p must possess the exact designed value. In the example above $R_p = 10.000\ \Omega$. Then current I_P is set (adjusted) to be exactly 10.000 mA. The process of ensuring that $R_P I_P = 100.000$ mV by establishing the current in the potentiometer circuit to be exactly 10.000 mA is called: "*The Standardization of a potentiometer*".

4.2.1.1 Standardization of the Student Type DC Potentiometer

Before every measurement we make using a potentiometer, we must first standardize the potentiometer (That in Fig. 4.4, ensure that $R_p I_p = 100.0$ mV). For this, we need a standard voltage source, either the popularly called standard cell or any of the

(i) Digital readout (ii) Analog readout

Fig. 4.6 Ten-turn variable resistor

currently available electronic voltage reference standards. Let us assume that we have a standard cell of terminal voltage of 1.0186 V. We connect the standard cell as shown in Fig. 4.7. We then set the pole P_{oP} on the 1.0 mark (thus including ten resistors of R_P between points P_{oP} and P_R) and set the pole P_{oS} on the 18.6 mm mark on the slide wire (or 018.6 mark on the ten-turn variable resistor). This makes the potentiometer read the value of V_{oP} as 1.0186 V, made up of the voltage drops of 1.0000 V across 10 resistances of R_P, and 0.0186 V on the auxiliary (slide wire) resistor. Now we must make this reading a true representation by ensuring that $I_P =$ 10.00 mA. To this end, we first adjust the variable resistor R_C and then R_F so that the galvanometer G reads zero. Once the galvanometer reads zero, we then know that $V_{oP} = 1.0186$ V and hence $R_p I_p = 100.0$ mV. Now the potentiometer is ready for measurement. Here R_C is called the *coarse* variation resistor and R_F is called the *fine*-variation resistor. For the example we have chosen: ($R_p = 10.000\ \Omega$ and $I_p =$ 10.000 mA) The total resistance of the fourteen fixed resistances and the slide of value 10.00 Ω in series will be 150 Ω. If we have the value of V_P, say 3.0 V, then we can calculate the needed total resistance of the series circuit to make $I_p = 10.00$ mA as

$$\frac{3000\,\text{mV}}{10\,\text{mA}}\,\Omega = 300\,\Omega. \tag{4.3}$$

We already have 150 Ω of potentiometer resistors and thus, we need the value of $(R_C + R_F) = 150\ \Omega$. In this situation we choose a 150 Ω variable resistor as R_C and a 10 Ω variable resistor as R_F.

We assumed in the above discussion that we have a standard cell having a terminal voltage of 1.0186 V. This is the voltage of a standard cell known as the Weston Standard cell. The features of this cell and other voltage standards are presented next.

(i) ***The Standard Cell***.

We have seen that to standardize a DC potentiometer, we need a reference voltage source, whose terminal voltage is precisely known by calculation from fundamental

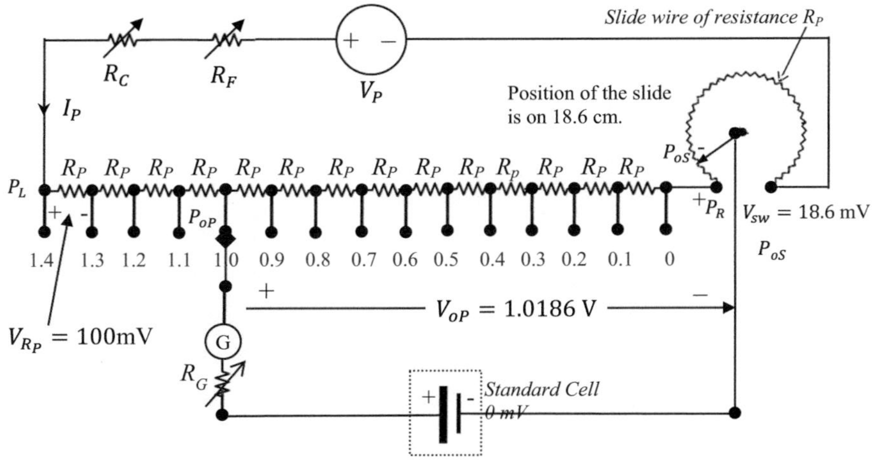

Fig. 4.7 Standardizing the DC Potentiometer

values. In the past, several types of standard cells were developed. However, after the invention of the "Weston Standard Cell" by Edward Weston in 1892, for nearly a century the Weston cell was employed for standardization. Figure 4.8 illustrates the schematic of a Weston standard cell, which is made of a H shaped glass structure. The positive electrode is made of mercury and the negative electrode is made of cadmium mercury amalgam. On the positive electrode, above the mercury pool, mercurous sulfate paste sandwiched between two porous spacers is placed. The electrolyte used is cadmium sulfate solution. Two possible forms of Weston standard cell can be prepared:

(a) **Saturated Cell:** Here the cadmium sulfate solution is saturated with excess cadmium sulfate crystals. Saturated cells provide long term stability of the terminal voltage but are susceptible for temperature variations.
(b) **Unsaturated Cell**: As the name indicates, this type of cell uses unsaturated cadmium sulfate solution. Its terminal voltage varies less with temperature in comparison with the saturated type. However, unsaturated cells have lower long term voltage stability.

In general, if the potentiometer is to be used in a controlled environment such as an air-conditioned laboratory with thermal regulation, saturated cells are preferred. On the other hand, unsaturated cells are preferred for field use, where temperature variations are expected. The typical terminal voltage of a Weston cell is 1.018632 V.

Today, we do not use the Weston Standard cell but use any one of three distinct types of voltage reference standards, namely, Josephson Reference, Bandgap reference and Zener reference. We do not call these standard cells but use the term "*voltage reference standards*". The term cell indicates that the emf is generated by converting chemical energy to electrical energy. On the other hand, the term reference standard

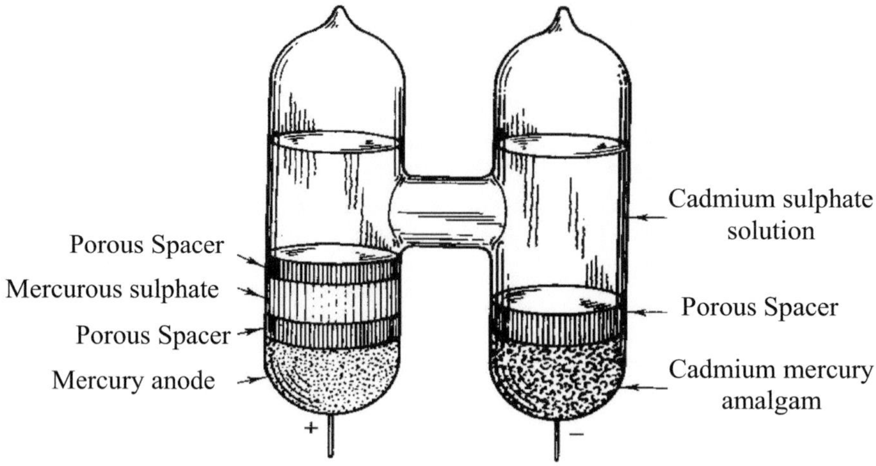

Fig. 4.8 The Weston standard cell

indicates that the required standard voltage is derived from an available reference voltage source.

(a) The Josephson Voltage Reference

In 1962, Brian Josephson of Cambridge University discovered that a junction created by inserting a thin insulator between two superconductors would exhibit discrete voltage levels across it, when irradiated with microwaves. If the microwave frequency is f, then one of the steps of constant voltage occurs at a value V_J given by

$$V_J = \frac{h}{2e} f. \quad (4.4)$$

With the symbols h and e representing Planck's constant and the elementary charge of an electron, equation (4.4) relates the junction voltage V_J to the value of a frequency f through fundamental constants. Furthermore, since frequencies can be precisely set and measured by reference to Cesium standards, the voltage of the Josephson standard can be determined with extraordinary accuracy and precision, with an uncertainty of about 0.01 ppm. The typical value of V_J at the frequencies employed being only in the order of a few micro volts, integrated circuit chips with series-connected arrays of thousands of Josephson junctions have been made to yield reference voltages in the order of a few volts. It is clear that Josephson junction-based voltage reference standards need special and expensive facilities for operation and maintenance. Hence, they are used only in National and International Standards Laboratories for high precision voltage measurements and calibration of other types of voltage references.

(b) The Bandgap Voltage Reference

The bandgap and hence the junction voltage of a *p-n* junction (under forward bias) has a negative temperature coefficient (under constant current). Hilbiber of Fairchild Semiconductor showed that two diode stacks biased at different current densities compensate the temperature effect and provide a temperature independent output voltage. Widlar also from Fairchild Semiconductor utilized this idea and developed the first bandgap voltage reference. Today several semiconductor manufacturers provide bandgap voltage reference integrated circuits (ICs).

Examples: (i) LT1009, LT1019, LT1389, and LT 1461 from Linear Technology; (ii) ADR380, ADR381, and AD680 from Analog Devices; (iii) REF3220 from Texas Instruments; and (iv) MAX8069 from Maxim Semiconductors.

(c) The "Buried Zener" Voltage Reference

A Zener diode, in reverse bias condition, provides near constant voltage drop across it. However, the Zener voltage changes when the exciting voltage or the operating temperature changes. Nevertheless, it is possible to design a voltage reference that uses a Zener and additional components so as to obtain a constant voltage with variations only at the "*parts per million*" level. Once again, such devices are available in IC form. Examples: AD 586, LT1021, LT1027, and LT1236. Figure 4.9 shows the circuit diagram of a simple Zener type voltage reference that uses an opamp.

(ii) *The Galvanometer*

For implementing a DC potentiometer, we need a DC galvanometer. The DC galvanometer is nothing but a center zero ammeter, as shown in Fig. 4.3. The overall accuracy and precision of a potentiometric measurement depend also on the quality of balance detection and hence on the resolution and sensitivity of the galvanometer. A typical low-cost galvanometer shown in Fig. 4.3 may have a sensitivity of 1 °/μA and a resolution of fraction of a microampere. On the other hand, a *light-spot* galvanometer

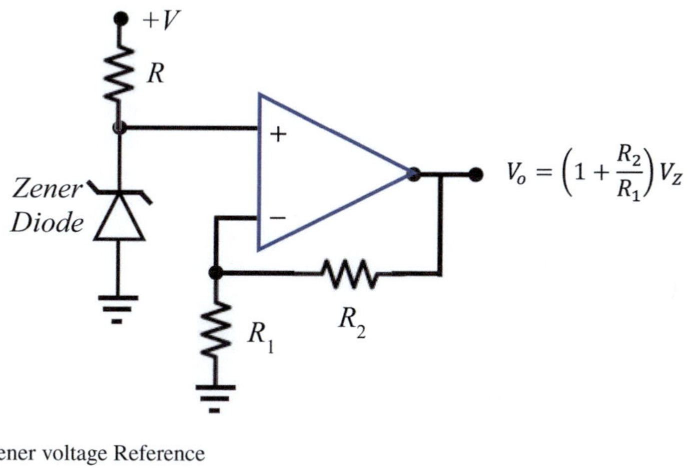

Fig. 4.9 Zener voltage Reference

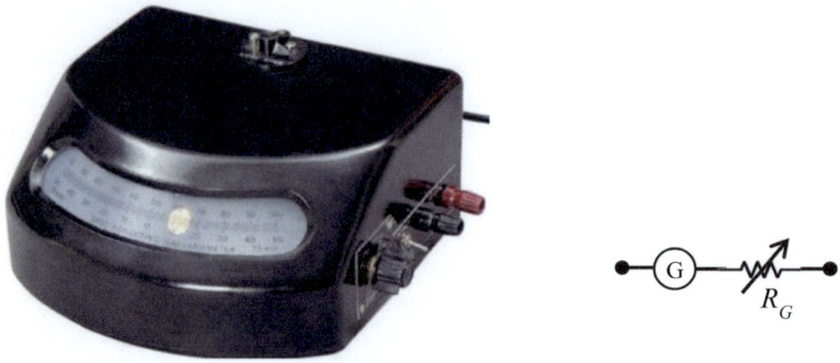

(a) Light Spot galvanometer (b) Circuit schematic of galvanometer

Fig. 4.10 The light-spot galvanometer

shown in Fig. 4.10a will have a resolution of 10 nA. At the start of balancing a potentiometer, the voltage across the galvanometer may be high and send a large current through the galvanometer, which may even damage the galvanometer. Hence all galvanometers have a variable series resistance R_G as shown in the schematic of Fig. 4.10b At the start we keep the voltage sensitivity of the galvanometer low with maximum possible series resistance R_G connected to the galvanometer and increase the voltage sensitivity by reducing the series resistance to zero as we approach the zero reading on the galvanometer.

4.2.1.2 Measurement of an Unknown DC Voltage Using the DC Potentiometer

(i) ***Measurement of unknown voltage that is less than the maximum voltage that can be set up on the potentiometer*** (for the student potentiometer shown in Fig. 4.11, the maximum voltage is 1.5 V) is straightforward. First, we standardize the potentiometer using a standard cell or a voltage reference standard as explained earlier (vide Fig. 4.7). After standardization of the DC potentiometer, we keep the variable resistances R_C and R_F undisturbed. We connect the unknown voltage and a galvanometer as shown in Fig. 4.11. We adjust poles P_{oP} and P_{oS} such that galvanometer G reads zero. Once G reads zero, we read off the value of the voltage from the positions of P_{oP} and P_{oS}. To illustrate, if P_{oP} is on the 0.9 V terminal and pole P_{oS} is on the 87.6 cm mark on the 1 m resistive wire of value R_p, the unknown voltage V_x is read as 0.9876 V.

(ii) ***Measurement of unknown voltage, whose value is more than the maximum known voltage*** that can be set up on the potentiometer is possible using a volt ratio box (VRB). In this method, we step-down the unknown voltage V_x by a known factor to a level compatible with the DC potentiometer. A volt ratio box, shown in Fig. 4.12, is nothing but a resistive divider using two resistances R_1

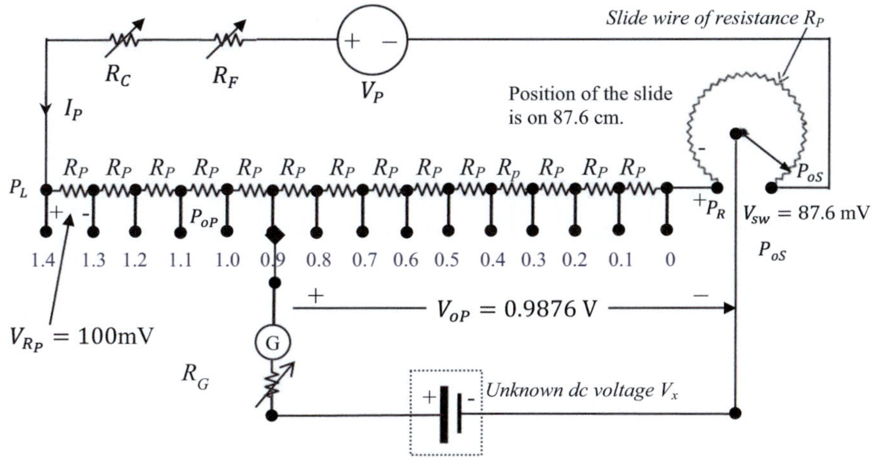

Fig. 4.11 Measurement of unknown dc voltage (Low value) with A DC Potentiometer

and R_2 as shown in Fig. 4.13. The values of R_1 and R_2 are so chosen that $\left(\frac{R_1}{R_1+R_2}\right)$ equals the selected ratio on the dial, such as 1/10 or 1/20 or 1/100 or 1/200 or 1/500. The unknown voltage is connected to the input of the VRB and the output of the VRB is connected to the DC potentiometer as shown in Fig. 4.13. Once again standardization of the DC potentiometer is first done using the resistances R_C and R_F and a standard cell as shown in Fig. 4.7. Thereafter R_C and R_F are undisturbed. With the connections as in Fig. 4.13, poles P_{oP} and P_{oS} adjusted such that galvanometer G reads zero. Once G reads zero, we read off the value from the positions of poles P_{oP} and P_{oS} and the value of V_x is computed by multiplying the reading by the ratio set on the VRB. Assuming that the balance is obtained with P_{oP} on the 0.6 mark, the slide P_{oS} on the 54.3 cm mark of the 1 m scale and the ratio set on the VRB as 100, the value of V_x will be 100*0.6543 = 65.43 V.

Fig. 4.12 Volt Ratio Box

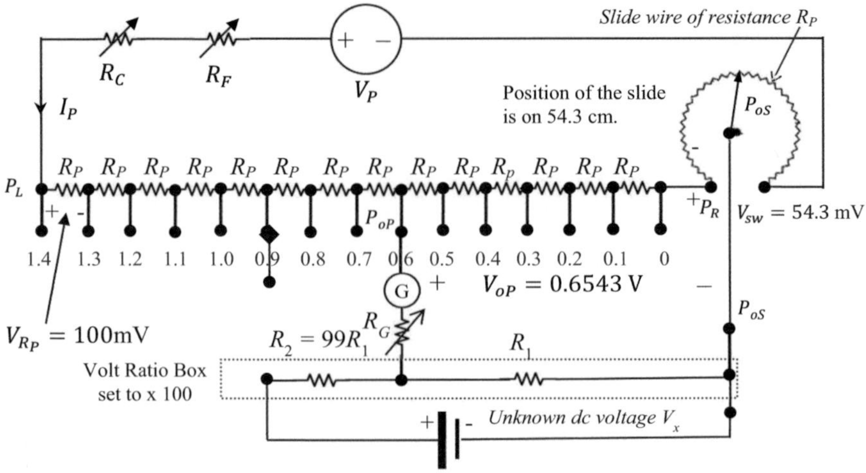

Fig. 4.13 Measurement of unknown dc voltage > 1.5 V using the Potentiometer

4.2.1.3 Measurement of an Unknown DC Current Using the DC Potentiometer

To measure a DC current using a DC potentiometer, the unknown current I_x is first converted into an equivalent DC voltage by passing it through a standard resistance (shown in Fig. 4.14) of known value R_S and measuring the voltage drop $R_S\ I_x$ across the resistance using the potentiometer, as shown in the Fig. 4.15. To ensure that contact resistances do not affect such a measurement, a four terminal standard resistance like the one illustrated in Fig. 4.14 is normally used. Current is led into and out of the four terminal standard resistance through the current terminals $\mathbf{C_1}$ and $\mathbf{C_2}$ and the voltage drop measured across the potential terminals $\mathbf{P_1}$ and $\mathbf{P_2}$. It is the resistance tapped between $\mathbf{P_1}$ and $\mathbf{P_2}$ that is adjusted to have exactly the specified value R_S of the standard resistance. Furthermore, no voltage drops occur across the contact resistances at these potential terminals as no current passes through them at the balance point. Thus, the voltage measured by the potentiometer at balance truly equals $R_S\ I_x$. If in the circuit of Fig. 4.15, balance is obtained with P_{oP} at 1.2 V and the pole P_{oS} on the 34.5 cm on the 1 m scale, the current I_x is computed as.

$$I_x = \left(\frac{1.2345\ \text{V}}{1\Omega}\right) = 1.2345\ \text{A}.$$

Fig. 4.14 Four-Terminal Standard Resistance

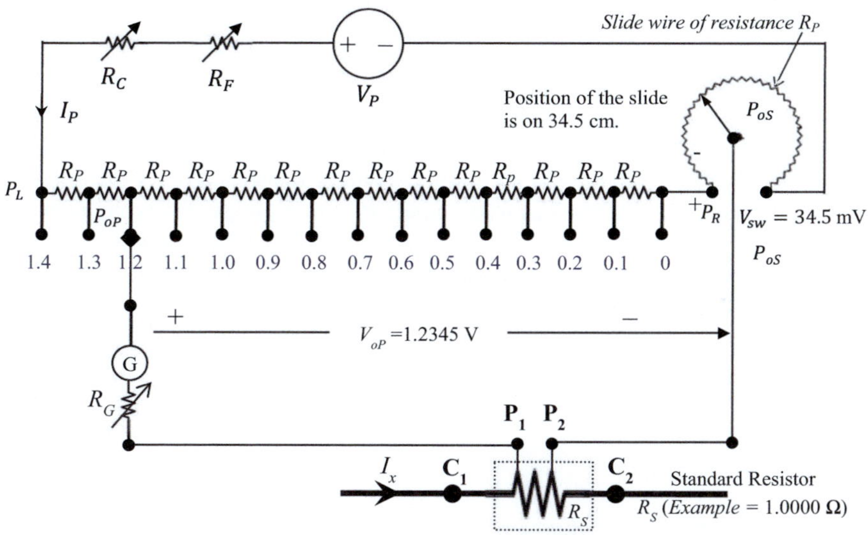

Fig. 4.15 Measurement of unknown dc current I_x using the Potentiometer

4.2.1.4 Measurement of an Unknown Resistance Using a DC Potentiometer

To measure the value of an unknown resistance R_x we make the circuit as shown in Fig. 4.16. Here, the unknown resistance R_x is connected in series with a standard resistance R_S and the series circuit is excited by a DC source so that a current I_x of

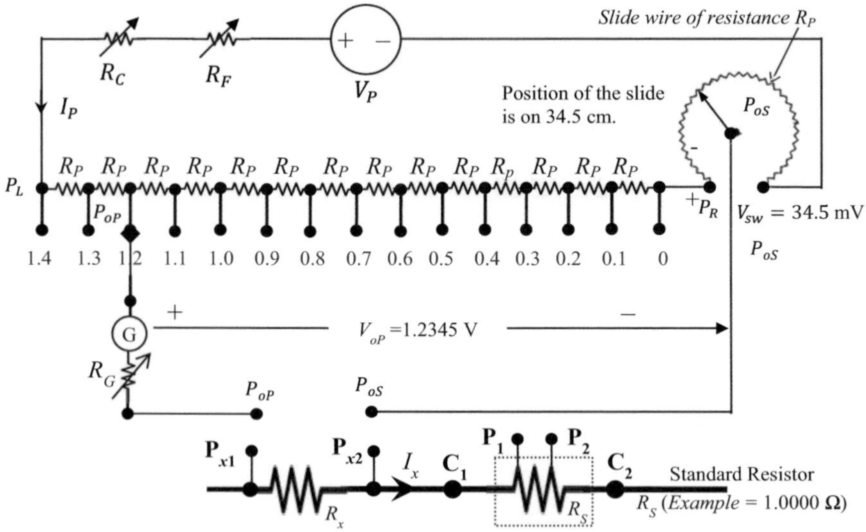

Fig. 4.16 Measurement of unknown Resistance R_r using the Potentiometer

suitable magnitude flows in it. Once again, we first standardize the potentiometer and then measure the voltage V_x across the unknown resistor R_x by connecting terminals P_{oP} *and* P_{oS} to terminals P_{x1} and $\boldsymbol{P_{x2}}$ respectively. Once the measurement of V_x is completed, then the voltage V_{Rs} across the standard resistance R_S is measured by connecting terminals P_{oP} and $\boldsymbol{P_{oS}}$ to terminals P_1 and $\boldsymbol{P_2}$ respectively. In each case described above, the potentiometer is first standardized and the voltages V_x and V_{Rs} measured, in turn, The current I_x through the unknown resistor is then computed as $I_x = \left(\frac{V_{Rs}}{R_S}\right)$. R_x is then calculated using V_x and I_x as

$$R_x = \left(\frac{V_x}{I_x}\right) = \left(\frac{V_x}{V_{Rs}}R_S\right). \tag{4.5}$$

4.2.1.5 Systemic Errors (Loading Effect) in a DC Potentiometer Type Measurement

In all the examples we have seen above, a measurement is made on the DC potentiometer after we get a zero reading on the galvanometer G. Thus, theoretically the terminals on which the voltage is measured are on open circuit. Thus, when we make a measurement using a potentiometer, systemic error (error due to loading) is zero. However, in practice what we read "zero" on the galvanometer, the current need not be zero. The zero indicated by a galvanometer really means that the current is smaller than the resolution of the galvanometer employed. Typically, the best possible resolution of a particularly good galvanometer is about 10 nA and thus measuring a voltage with a potentiometer is a sort of measuring with a voltmeter of internal resistance of 100 MΩ. With present day instrumentation, this can go to the level of 1 GΩ. Thus, one of the advantages of using a potentiometer is that the systemic (loading) errors can be made negligibly small. The foregoing advantage is available for the measurement of voltage, current and resistance directly with the DC potentiometer as shown in Figs. 4.11, 4.15, and 4.16, but is not available once we use a volt ratio box as given in 4.13.

The student potentiometer described above can only give a maximum of 4 decimal digits in any reading (5 significant digits). If we need a higher number of decimal/significant digits, we need to use different potentiometer configurations. Several different types were invented in the past, but, of these, only two have survived the test of time. They are the Wenner element and Kelvin-Varley type potentiometers. While both are current (that is being in use), the Kelvin-Varley is much more popular and is commercially available, even today.

4.2.2 The Wenner-Element Potentiometer

Figure 4.17 shows one decade of a Wenner element. Here, 20 resistances of value R_1 are connected on a 2-pole 10-way rotary switch. A resistor R_{11} is connected in between the ganged poles as seen in the Fig. 4.17. If we design such that $I_1 R_1 = 100$ mV, then the output voltage V_p will be $(n*100)$ mV where n is the number on which point h is placed (vide Fig. 4.17). The nice thing about Wenner element is that any number of elements can be cascaded to obtain the voltage in different elements variable in steps of, say, 1 V, 100 mV, 10 mV, 1 mV, 100 μV....

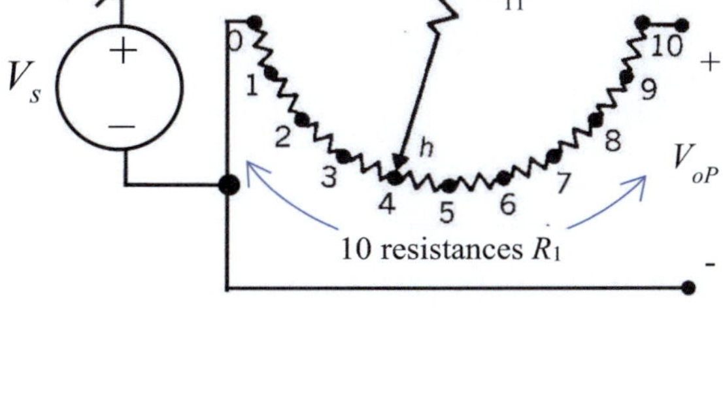

Fig. 4.17 Wenner element—One Decade

Figure 4.18 illustrates a four-decade Wenner DC potentiometer. Let the rotary switches of the four Wenner elements be set to N_1, N_2, N_3, and N_4. A careful analysis of the circuit shows that its input resistance as seen from the source side has the same value for all possible combinations of positions $\{N_1, N_2, N_3, N_4\}$ of the rotary switches. This means that once V_s, R_a, R_b, R_C, and R_F are set, I_1 does not change for any combination of the settings $\{N_1, N_2, N_3. N_4\}$. What is more, the values of I_2, I_3, and I_4 also do not change, as the switch positions are changed.

The design here is that

$$I_1R_1 = 1.0000\,\text{V}, I_2R_2 = 100.0\,\text{mV}, I_3R_3 = 10.0\,\text{mV}\,\textit{and}\,I_4R_4 = 1.0\,\text{mV}.$$

Accordingly, when the dials of the four elements are set as N_1, N_2, N_3, and N_4, the output voltage V_{oP} will be

$$\begin{aligned} V_{oP} =& N_1 + 0.1N_2 + 0.01N_3 + 0.001N_4 + 10R_1(I_2 + I_3 + I_4) \\ &+ 10R_2(I_3 + I_4) + 10R_3I_4 - V_{ab}\left(\frac{R_a}{R_a + R_b}\right) \end{aligned} \tag{4.6}$$

If we choose R_a and R_b such that

$$V_{ab}\left(\frac{R_a}{R_a+R_b}\right) = 10R_1(I_2 + I_3 + I_4) + 10R_2(I_3 + I_4) + 10R_3I_4 \tag{4.7}$$

we get

$$V_{oP} = (N_1 + 0.1N_2 + 0.01N_3 + 0.001N_4)V \tag{4.8}$$

It should be noted here that the value of V_{ab} is

$$V_{ab} = 10R_1(I_1 + I_2 + I_3 + I_4) + I_1R_{11}. \tag{4.9}$$

Thus, we can design, a priori, the values of R_a and R_b. If we make $I_1\ R_1 =$ 1.0000 V, $I_2R_2 =$ 100.000 mV, $I_3\ R_3 =$ 10.00 mV, and $I_4\ R_4 =$ 1.0 mV, then as

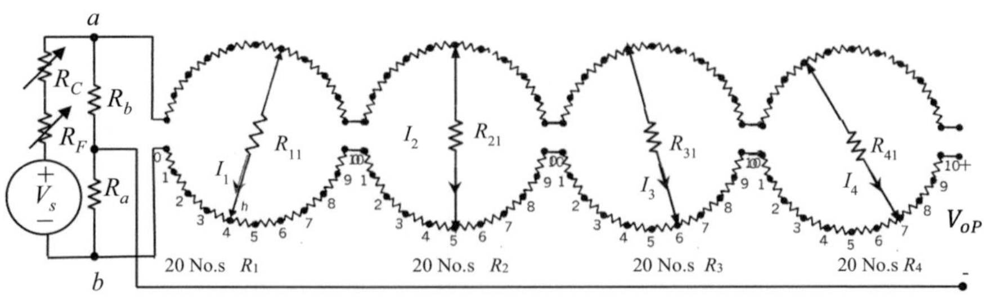

Fig. 4.18 Four-decade Wenner-element dc potentiometer

per Eq. (4.8), we see that the output voltage V_p will vary from 0.000 to 9.999 V as we vary (N_1 $N_2N_3N_4$) from 0000 to 9999. Of course, as is done in the student type potentiometer R_1, R_2 R_3, R_4, R_{11}, R_{21} R_{31}, R_{41} are designed and implemented using standard resistors. Here too, fulfillment of the values:

$$I_1R_1 = 1.00000\,\text{V}, I_2R_2 = 100.000\,\text{mV},\ I_3R_3 = 10.000\,\text{mV and}$$
$$I_4R_4 = 1.000\,\text{mV}$$

is obtained by standardization using a standard cell and a galvanometer as was done in the case of a student potentiometer. For the dial setting shown in Fig. 4.18, the voltage V_{oP} will be 4.567 V. Though the Wenner element-based potentiometer is quite elegant and flexible, the design becomes complex and requires almost double the number of precision resistors needed for the Kelvin-Varley type potentiometer. Hence, compared to the Wenner potentiometer, the Kelvin-Varley DC potentiometer is much more popular and sought after as it employs a simple design rule and requires slightly more than half the resistors of the Wenner potentiometer.

4.2.3 The Kelvin-Varley DC Potentiometer

Figure 4.19 illustrates the circuit schematic of a six-decade Kelvin-Varley type DC potentiometer. We see that each of the first five decades is made of eleven series-connected resistances of value R_1, R_2, R_3, R_4, and R_5 respectively. The sixth decade alone has ten resistances of R_6. Starting from the sixth decade, each decade is connected in parallel to a pair of adjacent resistances in the previous decade using a two-pole ganged switch as shown in Fig. 4.19. Now, the values of resistances R_1, R_2, R_3, R_4, and R_5 are chosen such that

$$10\,R_6 = 2\,R_5,\ 10\,R_5 = 2\,R_4,\ 10\,R_4 = 2\,R_3,\ 10\,R_3 = 2\,R_2\text{and}10R_2 = 2R_1.$$

With this design, decade 5 for example, has 11 resistances of value R_5, but a pair of these in series is shunted by an effective resistance of $2R_5$, arising from the 10 resistors of value R_6 in series. This makes the equivalent resistance of the combination to be R_5. Thus, even though there are 11 physical resistors of value R_5 the effective total resistance is $10\,R_5$ across terminals E and O_4. This is the same case with decades 1 to 4 as well. Each decade has 11 physical resistors, 2 of the resistors in series are paralleled with an equal-valued effective resistance, coming from the succeeding decade. Hence, each decade has, effectively, 10 resistors. Thus, if we assume that the voltage V_{AO} at node A with respect to node O is 10.00000 V, then, the voltage V_{BO1} will be 1.00000 V, the voltage V_{CO2} will be 0.10000 V, the voltage V_{DO3} will be 0.01000 V, the voltage V_{EO4} will be 0.00100 V. Thus, the value of voltage drop across each resistor of value R_1 is 1.0000 V. The exception is that 0.5 V each occurs across the two resistors of value R_1 that are spanned by the ganged switch. With

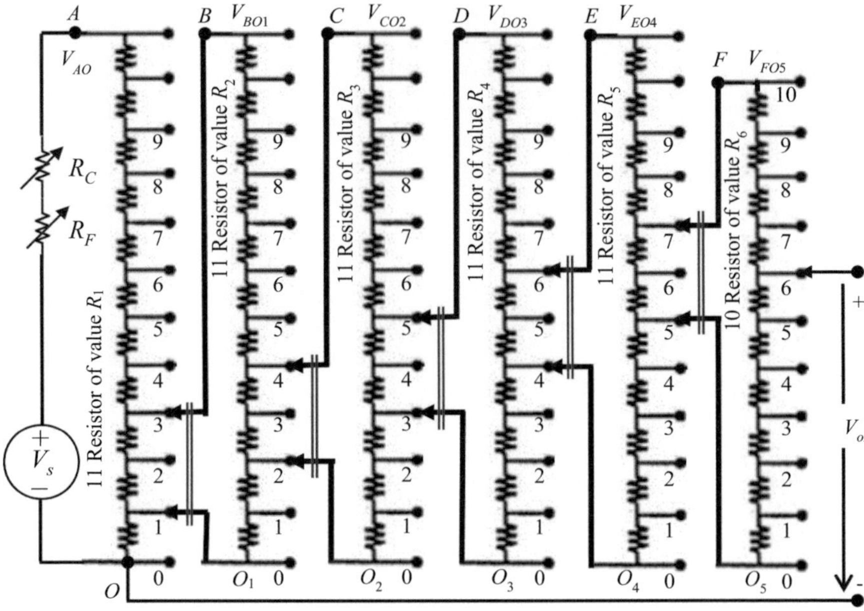

Fig. 4.19 Six-decade Kelvin-Varley dc potentiometer

similar exceptions in the succeeding decades, the voltages across each R_2, R_3, R_4 R_5, and R_6 are 0.10000 V, 0.0100 V, 1.00 mV, 0.10 mV, and 0.01 mV respectively. Thus, with the settings shown in Fig. 4.19, the output voltage V_{oP} is 1.23456 V. Once again, we first need to standardize the DC potentiometer to make $V_{AO} = 10.00000$ V. The scheme uses a standard Weston cell of 1.018638 V, and a galvanometer is shown in Fig. 4.20. Here, we set the dials of the Kelvin potentiometer at 101,864 as indicated in the figure and adjust R_C and R_F so as to get a zero reading on the galvanometer and thus standardize the potentiometer. Figure 4.21 shows a commercial seven-decade Kelvin-Varley unit from Fluke Corporation.

4.3 The AC Potentiometer

We learnt the technique of measurement of DC voltage, current, and resistance using a DC potentiometer. In a similar way, it is also possible to design and use the potentiometric method to measure an AC voltage, AC current, or AC impedance (or admittance). When we speak of the AC voltage $v(t)$ or AC current $i(t)$ we invariably take them to be sinusoidal functions of time, which have the general form:

$$v(t) = \sqrt{2}V\text{Cos}(\omega t + \varphi)V;\ and\ i\,(t) = \sqrt{2}I\text{Cos}(\omega t + \theta)A. \quad (4.10)$$

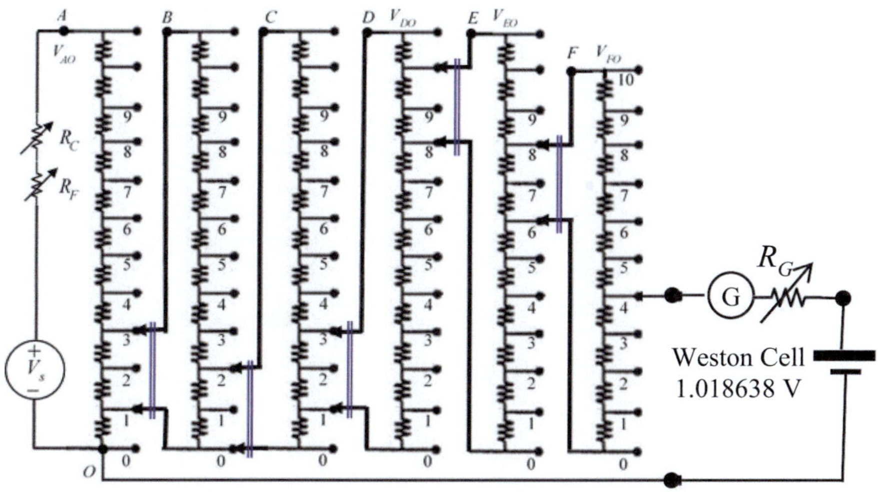

Fig. 4.20 Standardization of Kelvin-Varley dc potentiometer using Weston standard cell

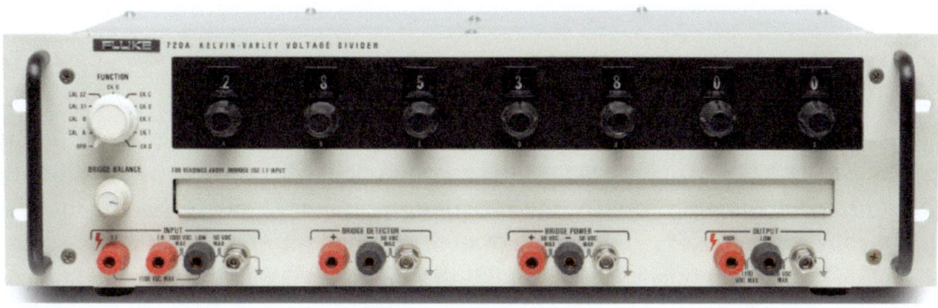

Fig. 4.21 Kelvin-Varley unit from Fluke Instruments

Here, V and I are the respective rms values and φ and θ are the phases of the voltage and current. In Eq. (4.10), $\omega = 2\pi ft$, with f being the frequency, and $f = \frac{1}{T}$ where T is the period of the sinusoidal signal. In order to process the AC voltage and current mathematically, it is convenient to use the phasor notation of the sinusoids instead of time functions. The process of converting sine functions to phasors and vice versa is depicted in the equations given below.

$$v(t) = \sqrt{2}V\,Cos(\omega t + \varphi)V \Leftrightarrow Re\left(\sqrt{2}Ve^{j\omega t}\right) \Leftrightarrow \underline{V} = V\angle\varphi$$

$$i(t) = \sqrt{2}I\,Cos(\omega t + \theta)A \Leftrightarrow Re\left(\sqrt{2}Ie^{j(\omega t+\theta)}\right) \Leftrightarrow \underline{I} = I\angle\theta \tag{4.11}$$

In Eq. (4.11), $\underline{V}$ and $\underline{I}$ are the respective voltage and current phasors, and being complex numbers, the voltage and current phasors are shown with underscores. Thus,

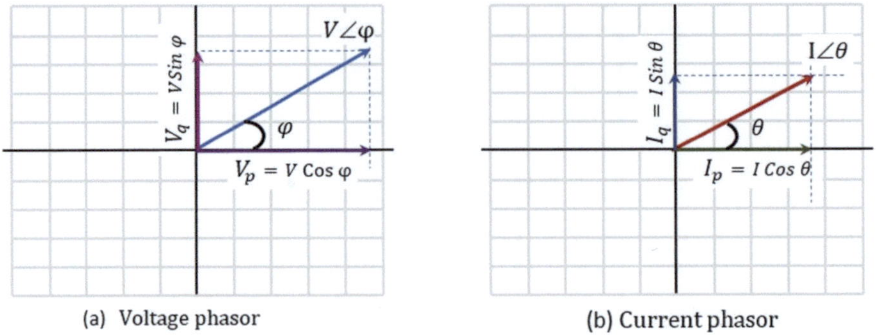

Fig. 4.22 Graphical form of voltage and current phasors

in Eq. (4.11) the phasors $\underline{V}$ and $\underline{I}$ are written in the "*polar form*" and is represented in graphic form in Fig. 4.22. From Fig. 4.22, we can see that we can also write each phasor as the sum of real and imaginary components:

$$\underline{V} = VCos(\varphi) + jVSin(\varphi) = V_p + jV_q \text{ and}$$

$$\underline{I} = I\,Cos(\theta) + I\,Sin(\theta) = I_p + jI_q \tag{4.12}$$

Here, $V_p = VCos(\varphi)$ is the *real* part of the phasor and is called the in-phase (zero phase) component of phasor $\underline{V}$. The component $V_q = VSin(\varphi)$ is called the *quadrature component*. Similarly, the phasor current $\underline{I} = I\theta$ possesses the in-phase component $I_p = I\,\mathrm{Cos}(\theta)$ and the quadrature component $I_q = I\,\mathrm{Sin}(\theta)$ as seen in Fig. 4.22. The convention used is to take the voltage $v(t)$ as the refence sinusoid $V0$ having zero phase (that is $\varphi = 0$). Thus, an AC (sinusoidal) current or voltage can be measured in the polar form (magnitude and phase angle) or in the coordinate form consisting of an in-phase component I_p (or V_p) and a quadrature component I_q (or V_q). Hence, we have two forms of AC potentiometers, namely, the "*polar type AC potentiometer*" that measures an AC voltage or current or impedance or admittance in the polar form and the "*coordinate type AC potentiometer*" that measures these quantities in the coordinate form. To clarify, we name these two methods in this manner because they directly measure the two components of the associated representation. We can always convert the result from a polar potentiometer to the coordinate form and vice versa. In any event, relative convenience and the accuracy of the final result need to be considered.

4.3.1 Polar Type AC Potentiometer

In the polar type of AC potentiometer, whose schematic is shown in Fig. 4.23, the magnitude of the voltage V_S and its phase θ_S are varied till the detector reads zero. Detector reading zero can be either the voltage across the detector or the current through the detector becoming zero. Once the detector reads zero, then:

$$v_{ab} = V_x \angle \theta_x \Leftrightarrow V_S \angle \theta_S. \tag{4.13}$$

With this $V_x = V_S$ and $\theta_x = \theta_S$. Here V_S can be varied using any type of DC potentiometer (student-Poggendorff or Wenner or Kalvin-Varley) and the phase angle varied using a phase shifter.

Originally, a wound rotor phase shifter was designed and used to vary θ. Later *R–C* phase shifters along with sinusoidal oscillators were used.

4.3.1.1 Standardization of AC Potentiometer

Figure 4.24 illustrates the implementation of the polar type of AC potentiometer with the base being a student-Poggendorff type of DC potentiometer. Since obtaining a standard AC source with the accuracy and precision of a DC standard cell or a DC voltage reference is not practical, all AC potentiometers are standardized using a DC standard cell and then the DC standardization achieved is transferred to the AC side. To accomplish this, in the early days, a typical AC potentiometer (polar type or coordinate type) had both DC and AC excitation sources as illustrated in Fig. 4.24, for the AC potentiometer. Whenever the DC excitation source is selected, the *balance* is detected using a galvanometer. Once the AC excitation is selected an AC *detector* is used to detect balance. To transfer the DC standardization to AC, an ammeter functioning as an AC-DC transfer standard TS is inserted in the current path as seen in Fig. 4.24. First, the potentiometer is set to DC excitation and standardized using a galvanometer. For this, switches S_1, S_2, and S_3 in Fig. 4.24 are set in position 1, thus selecting the DC source as excitation, the standard cell and the galvanometer inserted in the circuit. With the use of the standard cell and the galvanometer as shown in Fig. 4.24, the potentiometer is standardized, as described earlier for the DC potentiometer, by adjusting R_C and R_F till the galvanometer reads zero. Once

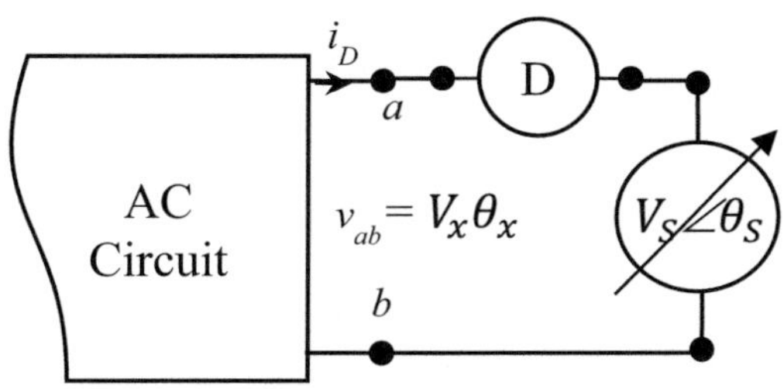

Fig. 4.23 Polar type AC Potentiometer

the galvanometer reads zero, thus achieving DC standardization, the reading on the AC-DC transfer standard TS is marked. Then S_1, S_2, and S_3 in Fig. 4.24 are set in position 2. Now the potentiometer is excited by the AC source $V\theta$. Once again R_C and R_F are varied till the indicator on the AC-DC transfer standard reads the same value that was marked earlier during dc standardization. By this process, the DC standardization using the DC standard cell is transferred to the AC side, provided:

(a) The AC-DC transfer indicates the same for DC and AC (rms) currents of identical values.
(b) The AC-DC transfer standard has good resolution near the designed standard current value of the potentiometer.

(i) ***The AC-DC Transfer Standard***

In the past, the AC-DC transfer was always implemented using a true rms ammeter, with a scale change to enlarge the portion around the design current of the AC potentiometer. To this end, the electrodynamic type of ammeter with a scale shaping option was employed as an AC-DC transfer standard for quite some time. Later, other methods (a prominent one is called the "*quarter-square*" method) were developed and used. Once the now popular thermal AC-DC transfer standard technique was developed, all other methods got eclipsed. Figure 4.25 illustrates the basic principle of a thermal AC-DC transfer standard. The two thermal transfer units A and B are identical and are made of

(1) A resistive heating element with resistance $R_{\theta A} = R_{\theta B} = R_{\theta}$ and
(2) A thermocouple (TC_A and TC_B respectively) made of dissimilar materials C and D.

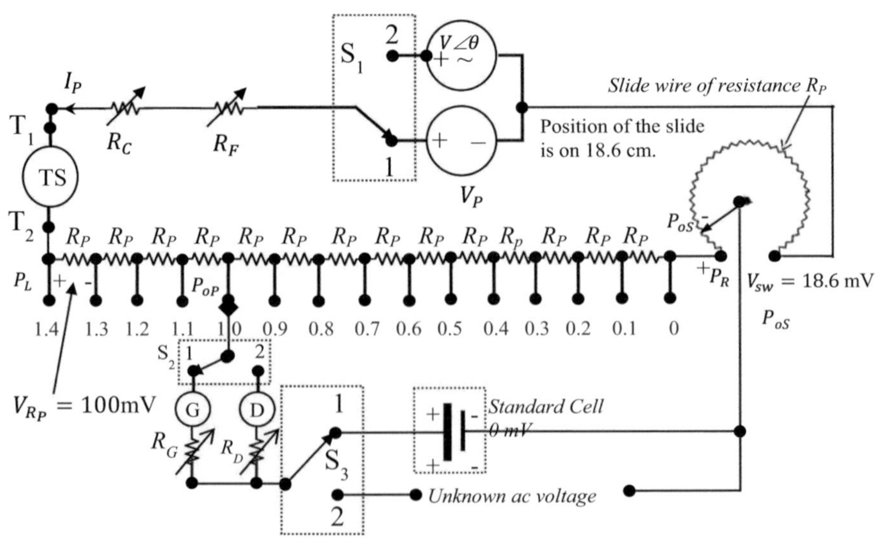

Fig. 4.24 Standardizing the AC Potentiometer with an AC-DC transfer standard TS

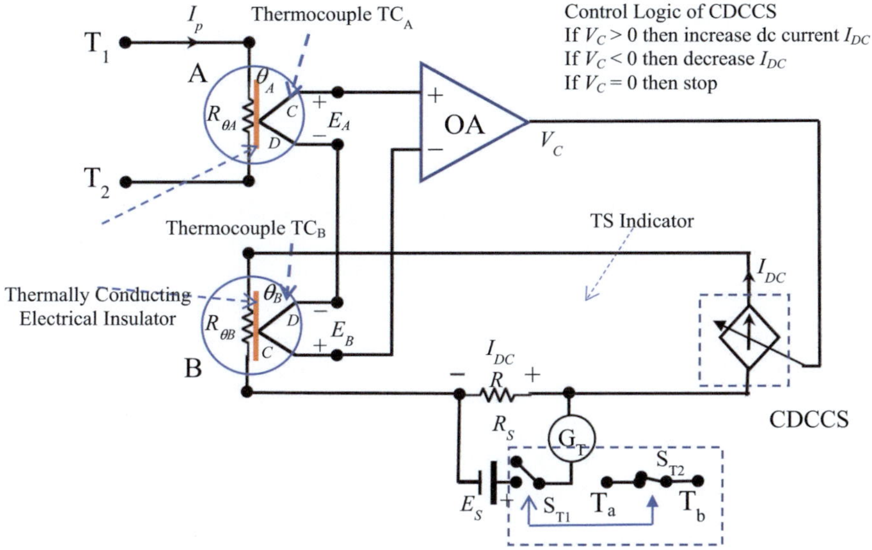

Fig. 4.25 Thermocouple based thermal type ac-dc transfer unit

The heating element and the thermocouple are encapsulated in individual *vacuum sealed* glass enclosure. The thermocouple senses the temperature θ (θ_A or θ_B as the case may be). In the modern version of AC potentiometer, using the thermal AC-DC transfer standard, the process of standardizing using DC excitation and transferring that to the AC is obtained in one step. For this, in the AC-DC transfer standard shown in Fig. 4.25, a standard resistance, a standard cell (or a voltage reference standard), ganged twin single pole single throw switch S_T and a DC galvanometer G_T are inserted on the DC side.

The Principle of AC-DC Transfer: It is seen from Fig. 4.25 that the AC current i_x with a rms value of I_x is sent into the resistor $R_{\theta A}$, dissipating a power $I^2{}_p\ R_{\theta A}$ thus elevating the temperature of $R_{\theta A}$ to θ_A, resulting in an increase in the thermo-emf E_A of TC_A. For this condition, the output V_c of the opamp OA will be positive and the DC current output of the controlled DC current source (CDCCS) will start increasing as per its control logic. Simultaneously θ_B will also increase due to the power dissipation in $R_{\theta B}$, resulting in the output E_B of TC_B to rise. Once $\theta_B = \theta_A$ then $E_B = E_A$, and V_c will be zero and the output of the CDCCS will stabilize with a current I_D, where $I^2{}_x\ R_{\theta A} = I^2{}_D\ R_{\theta B}$. With $R_{\theta A} = R_{\theta B}$ we get $I^2{}_p = I^2{}_{DC}$ and hence $I_{DC} = I_p$ Thus, the DC current output I_{DC} of the CDCCS will be equal to the rms value I_p of the AC current i_p of the AC potentiometer. Since we are generating a DC current I_{DC} that is equal to the rms value of an AC current using the thermal effects of these currents, and hence this method is called the "*Thermal AC-DC transfer*" standard.

With the modern AC-DC transfer standard, shown in Fig. 4.25, the DC standardization and the DC to AC transfer occurs in one step. Figure 4.26 shows a polar type of AC potentiometer redesigned using the modern AC DC transfer standard of Fig. 4.25 and only an AC source for excitation. Here for standardization, we close switch S_{T1} and automatically S_{T2} switch opens (as switches S_{T1} and S_{T2} are ganged switches) disconnecting the unknown voltage $V_x\theta_x$ from the potentiometer and connecting the galvanometer G_T inside the transfer standard. With switch S_{T1} in the closed position (and switch S_{T2} in open state), R_C first and R_F later varied till the galvanometer G_T in the AC-dc transfer standard reads zero. Once we have a zero reading on G_T, we get $I_{DC}\, R_S = E_S$ or $I_{DC} = \frac{E_S}{R_S} = I_p$, where I_p is the rms value of the potentiometer current Thus, choosing E_S and R_S appropriately we can ensure current I_p (which is the rms value of phasor current $\underline{I}_p$) to be exactly equal to the design value. For example, let the AC potentiometer of Fig. 4.26 be designed with $R_p = 10.000\ \Omega$ and current $I_p = 10.000$ mA. In the AC-DC transfer standard of Fig. 4.25 we must select E_S and R_S so that $I_{DC} = 10.000$ mA. If we select a Weston standard cell as E_S (E_S $= 1.018638$ V) then as R_S is selected as

$$R_S = \frac{1.018638}{10}\, k\Omega = 101.8638\ \Omega \tag{4.14}$$

Thus, as soon as the standardization in DC and transfer it to the AC is accomplished as well with switch S_{T1} closed and altering R_C first and R_F later to get zero reading on the galvanometer, switch S_{T1} is opened, and the AC potentiometer is returned in the measurement mode with switch S_{T2} closed and connecting $V_x\theta_x$ is connected to the AC potentiometer.

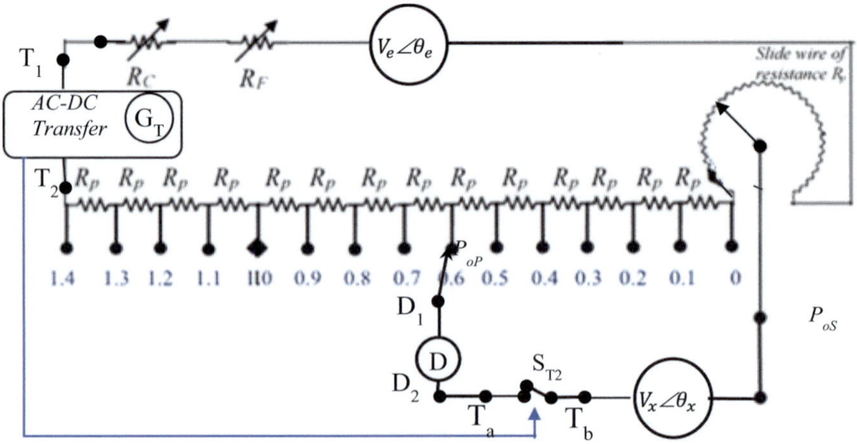

Fig. 4.26 Measuring an unknown ac voltage using a polar-type AC potentiometer

4.3.1.2 Measurement of an Unknown AC Voltage Using an AC Potentiometer

Once the standardization of the AC potentiometer is completed, it is ready for measurement of an unknown AC voltage as shown in Fig. 4.26. The unknown AC voltage is now connected to the output of the potentiometer using a series connected "*detector*". Just as a zero reading on the galvanometer ensures that the DC voltage output of the potentiometer equals the unknown DC voltage, a zero on the AC detector indicates that the AC voltage across the potentiometer exactly equals the unknown AC voltage, in magnitude and in phase. Thus, to measure the unknown AC voltage $V_x\theta_x$ the pointers P_{oP} and P_{oS} are varied as well as the phase θ_e of the excitation voltage V_e. Unlike balancing the DC potentiometer, balancing an AC potentiometer is little tedious as magnitude and phase needs to be adjusted, in repeated steps, to get a zero on the detector D. Once a zero is obtained in detector D, the magnitude V_x of the unknown voltage is read from the positions of P_{oP} and P_{oS} and the phase θ_x is taken as equal to θ_e at balance.

(i) ***The Detector***

The detector must indicate that there is zero current flowing through it (or zero voltage appearing across it). Initially a *vibrating galvanometer* was used as a detector. A vibrating galvanometer is a conventional indicating instrument (mostly PMMC type center zero ammeter) designed such that its natural frequency matches with the frequency of the excitation voltage of the AC potentiometer. Thus, the sensitivity of detecting zero AC current is made quite large. Later, an audio amplifier connected to a headphone served as the detector. After the invention of the oscilloscope, oscilloscopic detectors became the standard. An oscilloscopic detectors is nothing but an oscilloscope used in the *x*–*y* mode. The *x* input of the oscilloscope is connected to the AC excitation used in the potentiometer and the *y*-input is the voltage between terminals D_1 and D_2 of Fig. 4.26.

In the polar type of AC potentiometer, only the magnitude of the unknown voltage can be measured with 5 or 6 decades of significant digits. It is practically not possible to resolve the phase to better than 0.1°. Hence, today, the polar type of AC potentiometer is only of academic interest. On the other hand, the coordinate type potentiometer resolves both the in-phase and quadrature components of an unknown AC voltage, with six to nine significant digits (as per design).

4.3.2 Coordinate Type AC Potentiometer

We have already seen that an AC sinusoidal voltage, say, $\sqrt{2}V_x\text{Cos}(\omega t + \theta_x)$ can be resolved into $\sqrt{2}V_p\text{Cos}(\omega t)$–$\sqrt{2}V_q\text{Sin}(\omega t)$ or in phasor form $V_x\theta_x = V_{px} + jV_{qx}$. To determine the values of V_{px} and V_{qx}, we need two DC potentiometers converted to function as AC potentiometers as shown in Fig. 4.27. Once again for explaining the principle of operation, the basic student-type DC potentiometers are shown in Fig. 4.27. In practice, advanced Wenner or Kelvin-Varley type DC potentiometer configurations are employed. The excitation voltage of the in-phase component is taken as the reference voltage (one with zero phase angle). It is seen from Fig. 4.27 that while the in-phase component AC potentiometer is excited with the voltage $V_e0 = \sqrt{2}V_e\text{Cos}(\omega t)$ V, the quadrature component AC potentiometer is excited with $V_e90 = -\sqrt{2}V_e\text{Sin}(\omega t)$ V. These voltages can be obtained using.

(i) Mutual inductance (wound rotor, three-phase, phase shifter),
(ii) RC-Phase shifters, and
(iii) Quadrature oscillators.

Figure 4.28 illustrates the circuit diagram of a quadrature oscillator. With $v_1 = \sqrt{2}V_e\text{Sin}(\omega t)$, v_2 will be

$$v_2 = -\frac{\sqrt{2}V_e}{RC}\int Sin(\omega t)dt = \frac{\sqrt{2}V_e}{\omega RC}Cos(\omega t), \tag{4.15}$$

$$v_3 = -\frac{\sqrt{2}V_e}{\omega^2 R^2 C^2}\int Cos(\omega t)dt = -\frac{\sqrt{2}V_e}{\omega^2 R^2 C^2}Sin(\omega t) \text{ and} \tag{4.16}$$

$$v_1 = -\frac{R_1}{R_1}\left(-\frac{\sqrt{2}V_e}{\omega^2 R^2 C^2}Sin(\omega t)\right) = \frac{\sqrt{2}V_e}{\omega^2 R^2 C^2}Sin(\omega t). \tag{4.17}$$

Equation (4.17) indicates that what we assumed, namely, $v_1 = \sqrt{2}V_e\text{Sin}(\omega t)$ will be correct if and only if $\omega^2 R^2 C^2 = 1$ or $\omega = \frac{1}{RC}$. In other words, the circuit in the Fig. 4.28 will oscillate at the angular frequency $\omega = \frac{1}{RC}$ rad/s (frequency $f = \frac{1}{2\pi RC}$ Hz) providing the two outputs $\text{Sin}(\omega t)$ and $\text{Cos}(\omega t)$ required for a coordinate type AC potentiometer.

(i) ***Standardization of Coordinate Type AC Potentiometer***

As explained earlier, the standardization of an AC potentiometer (either polar or coordinate type) needs to be done in two steps. First the potentiometer is standardized

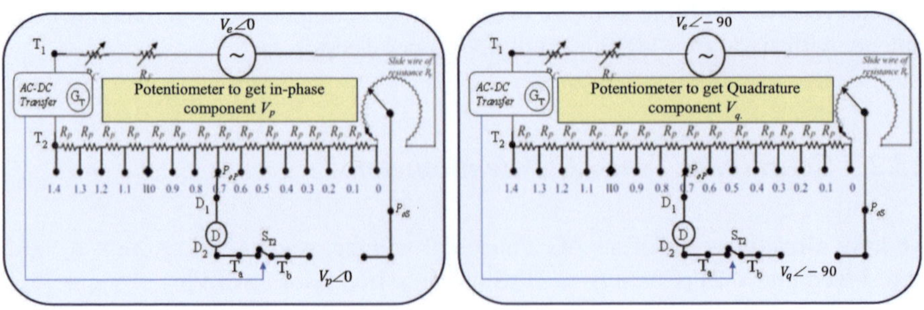

Fig. 4.27 Coordinate-type ac potentiometer

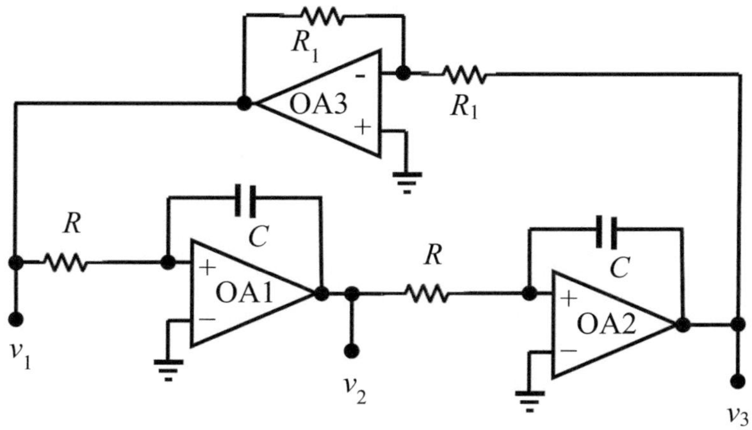

Fig. 4.28 Quadrature Oscillator to obtain sine and cosine voltages

using a DC standard cell (or using the single-step standardization of the present-day AC-DC transfer standard) and the standardization transferred using the AC-DC transfer unit. While the polar type of AC potentiometer has only one set to be standardized, for the coordinate type AC potentiometer the in-phase and quadrature potentiometers need to be individually standardized using a DC standard cell and the DC standardization transferred to AC using individual AC-DC transfer units (or using the single-step standardization of the present-day AC-DC transfer standard) as explained earlier.

(ii) ***Measurement of an Unknown AC Voltage Using a Coordinate Type AC Potentiometer.***

To measure an unknown AC voltage and the AC potentiometer is connected as shown in Fig. 4.29. The switching arrangement using two double pole double throw (DPDT) switches shown in Fig. 4.29 enables obtaining a voltage phasor $\underline{V}_P = \left(\pm V_p \pm jV_q\right)$. Thus phasor $\underline{V}_P$ can lie in the entire 360° range. When DPDT switch S1 is in position 1, the $+\,V_p$ option is selected. Similarly, when DPDT switch S2 is in position 1, the $+\,V_q$ option is selected. On the other hand, if S1 (S2) is in position 2 the voltage option $-V_p\,(-V_q)$ is selected. The two potentiometers (in-phase and quadrature) are adjusted to get zero reading on the detector. Once the detector reads zero, we have.

$$V_x\angle\theta_x = \pm Vp \pm jVq = V_{xp} + jV_{xq} \tag{4.18}$$

Equation (4.17) indicates that we measure the in-phase and quadrature components V_{xp} and V_{xq} of the unknown AC sinusoidal voltage $V_x\theta_x$ with the appropriate signs. Hence

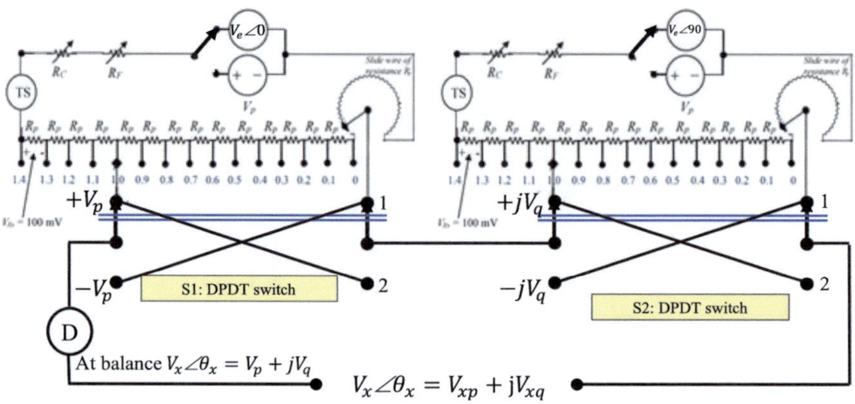

Fig. 4.29 Measurement of unknown ac voltage using the coordinate type of ac potentiometer

$$V_x = \sqrt{\left(V_{xp}^2 + V_{xq}^2\right)} \; and \; \theta_x = Tan^{-1}\left(\frac{V_{xq}}{V_{xp}}\right). \tag{4.19}$$

(iii) ***Measurement of AC Current Using a Coordinate Type AC Potentiometer.***

We have learnt the method of measurement of an AC voltage using either the polar or coordinate type AC potentiometer. Following the method used with a DC potentiometer, we can measure an AC current by sending the unknown current through a standard AC resistance, and measure the voltage drop across it, using either the polar or coordinate type AC potentiometer. Figure 4.30 shows such an arrangement for the measurement of AC current using the coordinate type AC potentiometer.

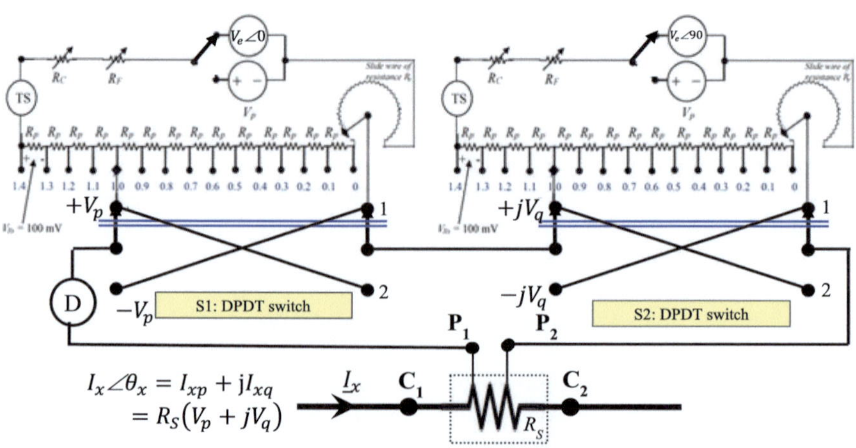

Fig. 4.30 Measurement of unknown ac current using the coordinate type of ac potentiometer

(iv) ***Measurement of Impedance or Admittance Using an AC Potentiometer***

Similarly, the value of an unknown resistance R_x under AC excitation can also be measured, by measuring the current I_x through and voltage V_x across the unknown resistance and calculating R_x as $= V_x/I_x$, assuming R_x to be purely resistive. An added advantage of an AC potentiometer is that we can measure the real part (resistance R_x) and imaginary part (reactance X_x) of an impedance $\underline{Z_x}$ by measuring the phasors of voltage across and current through it. If we use the polar type of AC potentiometer and measure the voltage (phasor) across $\underline{V_x}$ as: $\underline{V_x} = V_x\theta_v$ and $\underline{I_x} = I_x\theta_i$ then we can calculate $\underline{Z_x}$ as

$$Z_x = R_x + jX_x = \frac{V_x}{I_x}Cos(\theta_v - \theta_i) + j\frac{V_x}{I_x}Sin(\theta_v - \theta_i). \tag{4.20}$$

If required, we can also calculate from the measured values of $\underline{V_x}$ and $\underline{I_x}$, the conductance G_x and susceptance B_x of an admittance Y_x as

$$Y_x = G_x + jB_x = \frac{I_x}{V_x}Cos(\theta_i - \theta_v) + j\frac{I_x}{V_x}Sin(\theta_i - \theta_v). \tag{4.21}$$

On the other hand, if we use a coordinate type AC potentiometer and measure V_x as $V_R + jV_X$ and I_x as $I_R + jI_X$, then we calculate Z_x as

$$Z_x = \sqrt{\frac{V_R^2+V_X^2}{I_R^2+I_X^2}}Cos\emptyset + j\sqrt{\frac{V_R^2+V_X^2}{I_R^2+I_X^2}}Sin\emptyset, \tag{4.22}$$

where $\emptyset = \left[\mathrm{Tan}^{-1}\left(\frac{V_X}{V_R}\right) - \mathrm{Tan}^{-1}\left(\frac{I_X}{I_R}\right)\right]$. Similarly, we can calculate admittance $\underline{Y_x}$ as

$$Y_x = \sqrt{\frac{I_R^2+I_X^2}{V_R^2+V_X^2}}Cos\varphi + j\sqrt{\frac{I_R^2+I_X^2}{V_R^2+V_X^2}}Sin\varphi, \tag{4.23}$$

where

$$\varphi = \left[\mathrm{Tan}^{-1}\left(\frac{I_X}{I_R}\right) - \mathrm{Tan}^{-1}\left(\frac{V_X}{V_R}\right)\right]. \tag{4.24}$$

Another application of an AC potentiometer is in the measurement of the errors (ratio error and phase error) of an instrument transformer (current transformer and potential/voltage transformer). This application will be described in detail in the sequel (under the topic on voltage and current scaling).

4.4 Problems Exercises

E4.1. A potentiometer of the student type is designed to have a dial of 15 steps, 0.1 V and 10 Ω each, and a 10 Ω slide wire. Because of a mistake, the resistance of the step between the zero and the 0.1 V dial contact is 10.1 Ω. This potentiometer is standardized against a standard cell of 1.0190 V and then is used to measure an unknown voltage that gives balance for a reading of 0.2150 V on the dials. What is the true value of the unknown? Express the error in the observed reading as a percentage.

E4.2. A potentiometer of the student type has 15 steps of 0.1 V (R = 10 Ω) each and a slide wire of 11 Ω. The working battery has a voltage of 3.0 V.

(a) With the dials set at 1.2460 V, what is the internal (Thevenin) resistance of the potentiometer circuit as seen from the galvanometer position?
(b) If the potentiometer is standardized against a standard cell marked 1.01920 but the standard cell voltage has dropped 1.01884 V, what dial reading be obtained when balancing a source of 0.6362 V?
(c) The potentiometer is as described above, except that the 0.1 to 0.2 V step is 10.2 Ω, and the 0.9 to 1.0 V step is 9.6 Ω. Standardization is carried out in the regular manner against a correct 1.01900 V standard cell. What reading will be obtained when balancing a 0.2466 V source?

E4.3. The student type potentiometer has a main dial with 15 steps, each 0.1 V and 10 Ω, and a slide wire of 20 Ω, shunted to give it a range of 0.1 V. The potentiometer is used with a 3 V battery and a galvanometer of 30 Ω resistances and a sensitivity of 0.25 μA/mm.

(a) Find the resistance of the slide wire shunt.
(b) The working current is standardized correctly, and the dials are set at 0.45 V when measuring an emf of 0.499 V with negligible internal resistance. Find the galvanometer deflection.

E4.4. A potentiometer has 15 steps, each of 0.1 V and 5 Ω, and a slide wire of 5.5 Ω, in series with the working battery (emf = 2.20 V, r = 0.1 Ω) and a rheostat. The galvanometer has a sensitivity of 0.03 μA/mm and a resistance of 50 Ω.

(a) Find the resistance in the rheostat for standardization.
(b) If the slide has 11 turns, 200 divisions per turn, and the reading can be estimated to one fifth of a division, to what fraction of a volt can the reading be taken?
(c) The potentiometer is being used to measure an emf of 1.20 V with negligible internal resistance. What error of setting (in volts) from the true balance is necessary to deflect the galvanometer spot one millimeter?

E4.5. A potentiometer of the student type has a main dial with 15 steps, each 50 Ω, and a slide wire with a resistance of 75 Ω. The working current is adjusted so that each main-dial step represents 0.1 V. The working battery has an emf of 3.0 V. The galvanometer has a resistance of 50 Ω and a protective series resistance of 10 kΩ. The standard cell has an emf of 1.0186 V and an internal resistance of 250 Ω.

(a) Find the working current and the voltage range covered by the slide wire
(b) With the working current properly adjusted, the standard cell is connected to the circuit, but with the dials set by accident at 1.1186 V instead of 1.0186 V. Find the galvanometer current with protective resistance included?
(c) Same, with protective resistance

E4.6. A student type potentiometer has a main dial with 15 steps, each 10 Ω and 0.1 V and a slide covering 0.11 V. The battery has an emf of 3.0 V and a resistance of 0.1 Ω. The standard cell has an emf of 1.0186 V and a resistance of 250 Ω. The galvanometer has a resistance of 50 Ω and a sensitivity of 0.05 μA/mm. With the working current adjusted correctly and the dials set for 0.600 V, the switch is thrown to standardization.

(a) Find the galvanometer current, with no protective resistance
(b) Find the protective series resistance to limit the deflection to 50 mm

Chapter 5
Bridge Methods

Balance is the perfect state of still water. Let that be our model.
—Confucius

5.1 DC and AC Bridges

When two circuit branches (with each branch made of two circuit elements in series) in parallel are "bridged" by a third branch (that is by joining the mid points of the circuit branches as shown in Fig. 5.1) we get a "Bridge Circuit". The bridge circuit precedes the PMMC meter which is the very first analog indicating instrument. Thus, the oldest method of measurement is the "bridge method". Quite a few bridge circuits are still used for teaching as well as for the practical measurement of passive components such as resistance, inductance, and capacitance. Bridge methods of measurement can be broadly classified into

1. DC bridges and
2. AC bridges.

V. Jagadeesh Kumar, *Principles of Electrical Measurement and Measuring Instruments*,
https://doi.org/10.1007/978-981-96-5475-8_5

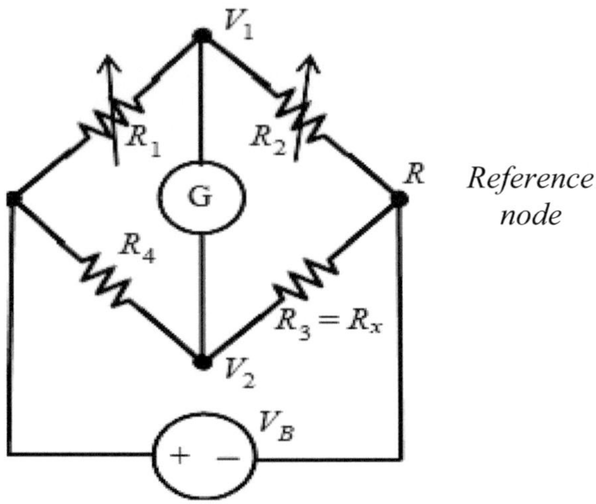

Fig. 5.1 The Wheatstone bridge

When a bridge circuit is excited with a DC source, we term that bridge as a DC bridge. On similar lines, if a bridge is exited with an AC source (normally sinusoidal AC source), then we term that bridge as an AC bridge. Obviously, DC bridges can be used to measure only resistances. AC bridges can be used to measure all three components, namely, resistance, inductance, and capacitance.

5.2 DC Bridges

Electrical engineering started with electrical power sources and devices providing unidirectional voltage and current. Today we call this "*direct current*" (DC) system. Thus, the null balance method, called the Wheatstone bridge, is the very first null balance technique developed to measure the value of an unknown DC resistance R_x. *Fun fact*: The Wheatstone bridge was originally invented by Samuel Hunter Christy in 1833 and improved and popularized by Sir Charles Wheatstone in 1843. The Wheatstone bridge was invented, not for the measurement of resistors but for soil analysis (compressibility measurement) through the measurement of soil resistance!

5.2.1 The Wheatstone Bridge

The circuit schematic of the Wheatstone bridge is shown in Fig. 5.1. Let R be the reference node, i.e., one having zero voltage. Here if the galvanometer reads zero then we have $V_1 = V_2$. Hence, we have

$$V_1 = V_B\left(\frac{R_2}{R_1+R_2}\right) = V_2 = V_B\left(\frac{R_3}{R_3+R_4}\right). \tag{5.1}$$

Rearranging Eq. (5.1) we get

$$R_2(R_3 + R_4) = R_3(R_1 + R_2). \tag{5.2}$$

Evaluating Eq. (5.2), we get

$$R_2R_4 = R_1R_3R_3 = R_x = \left(\frac{R_2R_4}{R_1}\right) \tag{5.3}$$

Equation (5.3) indicates that if we balance the Wheatstone bridge (balancing a Wheatstone bridge is to obtain zero on the galvanometer) then one of the resistors (we have chosen the resistor R_3) is expressed in terms of the other three resistors (R_1, $R_{,2}$ and R_4). The resistors R_2 and R_4 together are called the *product arm* as they appear as a product (multiplication) in Eq. (5.3). Equation (5.3) can also be written as

$$R_3 = R_x = R_4\left(\frac{R_2}{R_1}\right) = R_2\left(\frac{R_4}{R_1}\right), \tag{5.4}$$

and hence the resistances R_1 and R_2 (or resistors R_1 and R_4) constitute the *ratio arm.*

The Wheatstone bridge is good to measure resistors in the range of a few tens of ohms to few hundreds of kilo ohms. If the resistance to be measured (R_3) is very small or very large, then the errors in measuring them using the Wheatstone bridge becomes high. For example, Fig. 5.2 shows the wiring of a Wheatstone bridge. The wiring introduces additional resistances as seen in the expanded Wheatstone bridge circuit configuration shown in Fig. 5.3. We see that the connecting wires introduce the wiring resistances R_{w1} to R_{w4}. In general, the connecting wires we use in the laboratory (we call them patch chords) will have about 60 mΩ/m resistivity. With this connection when we calculate R_3 in terms of the other three resistors with galvanometer connected between terminals *b* and *f*, as

$$R_3 + R_{w3} = (R_4)\left(\frac{R_2+R_{w2}}{R_1+R_{w1}+R_{w4}}\right). \tag{5.5}$$

Let $R_{w3} = 100$ mΩ for purposes of illustration. Hence if we try to measure, say a resistance of value 1 Ω, we will end up with an error of 100 mΩ in 1 Ω and that is an error +10% error. We may say 'Ha! we can connect the bottom terminal of galvanometer G in Fig. 5.3 to point *e* as shown by the dashed line in Fig. 5.3 instead of *f*. Then we will have

$$R_3 = R_x = (R_4 + R_{w3})\left(\frac{R_2+R_{w2}}{R_1+R_{w1}+R_{w4}}\right). \tag{5.6}$$

Again, Eq. (5.6) shows that we still have errors in the determination of R_3. When we determine the values of the bridge resistances, it is possible to design R_1 and R_2

Fig. 5.2 Wiring a Wheatstone bridge

Fig. 5.3 Wiring schematic of Wheatstone bridge

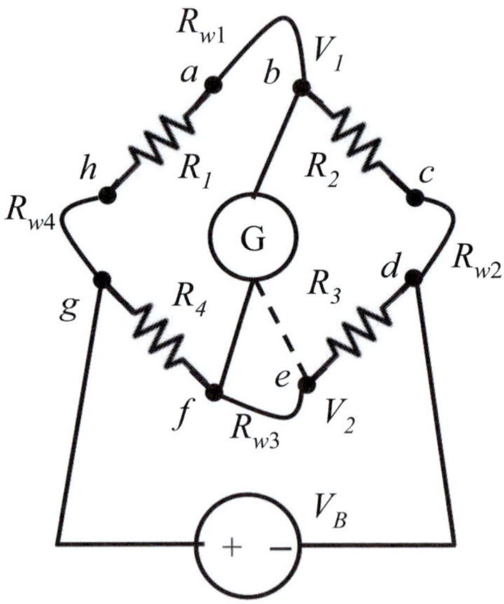

to be large-valued so that R_{w1}, R_{w2}, and R_{w4} are negligibly small and thus do not affect the measurement of small-valued resistance (R_3). However, the effect of R_{w3} will still exist.

Similarly, when we need to measure large resistance, say R_3 having the value of a few megaohms, Wheatstone bridge measurements will incur large error. Thus, we need special bridge forms to measure:

(i) Low valued resistors and
(ii) High value resistors.

To measure low value resistors Kelvin invented a modified Wheatstone bridge, which today we call the *Kelvin Bridge*. Similarly, we use the Megohm bridge (once again a modified Wheatstone bridge) for the measurement of resistances of extremely high value (Insulation) resistances.

5.2.2 The Kelvin Bridge

Kelvin bridge was invented by William Thomson (later he became Baron Kelvin) specifically for the measurement of low valued resistances. The circuit schematic of the Kelvin bridge is shown in Fig. 5.4. Here R_3 is the low-valued resistance to be measured, R_4 a fixed low-valued standard resistance and R_L is the resistance of their interconnecting link, which is made as low as convenient. R_1, R_2, R_A, and R_B are variable standard resistances with values in the kilo ohm range. The bridge is balanced (galvanometer current is made zero) by varying the ratio R_1/R_2, simultaneously making R_A/R_B also to have the same ratio.

To analyze the bridge, let us convert the delta connected resistances R_A, R_B, and R_L with nodes *a*, *b*, and *c* into a star connected set of resistances R_x, R_y, and R_z as in Fig. 5.5. Here:

$$R_x = \left(\frac{R_A R_B}{R_A + R_B + R_L}\right) \tag{5.7}$$

$$R_y = \left(\frac{R_A R_L}{R_A + R_B + R_L}\right) \tag{5.8}$$

$$R_z = \left(\frac{R_B R_L}{R_A + R_B + R_L}\right) \tag{5.9}$$

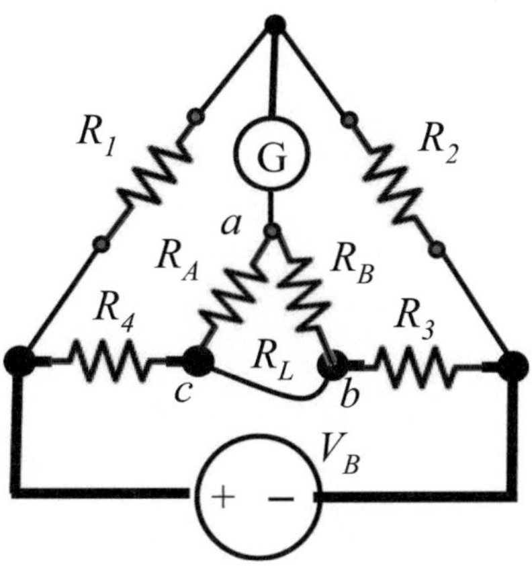

Fig. 5.4 Kelvin bridge

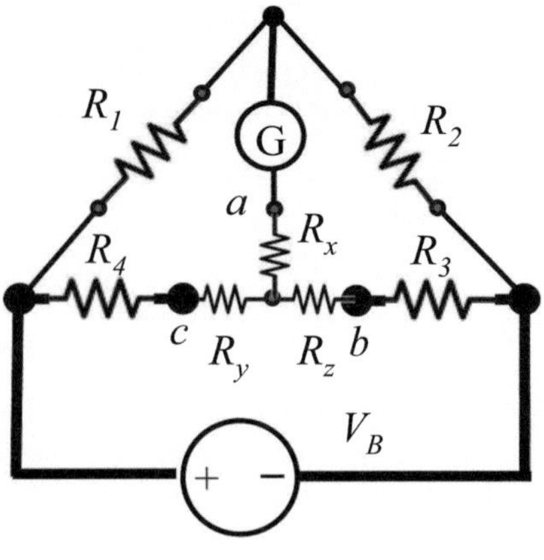

Fig. 5.5 Kelvin bridge—redraw

Once the galvanometer reads zero there will be zero current in the resistor R_x and hence has no effect, once the bridge is balanced. Hence, we have

$$R_3 + R_z = (R_4 + R_y)\left(\frac{R_2}{R_1}\right) \tag{5.10}$$

Substituting R_y and R_z from Eqs. (5.7) and (5.8) in Eq. (5.5), we get

$$R_3 = \left(R_4 + \frac{R_A R_L}{R_A+R_B+R_L}\right)\left(\frac{R_2}{R_1}\right) - \left(\frac{R_B R_L}{R_A+R_B+R_L}\right). \tag{5.11}$$

Rearranging Eq. (5.11) we get

$$R_3 = R_4\left(\frac{R_2}{R_1}\right) + \frac{R_2}{R_1}\left(\frac{R_A R_L}{R_A+R_B+R_L}\right) - \frac{R_B R_L}{R_A+R_B+R_L}. \tag{5.12}$$

Equation (5.12) is simplified as

$$\mathrm{R_3} = \mathrm{R_4}\left(\frac{\mathrm{R_2}}{\mathrm{R_1}}\right) + \left(\frac{\mathrm{R_A\ R_L}}{\mathrm{R_A+\ R_B+\ R_L}}\right)\left[\left(\frac{\mathrm{R_2}}{\mathrm{R_1}}\right) - \left(\frac{\mathrm{R_B}}{\mathrm{R_A}}\right)\right]. \tag{5.13}$$

If we ensure

$$\left(\frac{R_2}{R_1}\right) = \left(\frac{R_B}{R_A}\right)$$

then Eq. (5.13) becomes

$$\mathrm{R_3} = \mathrm{R_4}\left(\frac{\mathrm{R_2}}{\mathrm{R_1}}\right). \tag{5.14}$$

We note that

$$\left(\frac{\mathrm{R_A R_L}}{\mathrm{R_A + R_B + R_L}}\right) < \mathrm{R_L},$$

which has a small value. Hence, even if there is a small error in making the ratios,

$$\left(\frac{R_2}{R_1}\right) = \left(\frac{R_B}{R_A}\right),$$

exactly equal, the error made in evaluating R_3 through the formula in Eq. (5.14) is the product of two small quantities and hence becomes negligibly small. The above analysis indicates that if we ensure that by design $\left(\frac{R_2}{R_1}\right) = \left(\frac{R_B}{R_A}\right)$ and the values of R_1, R_2, R_A, and R_B are in the range of few kilo ohms then however small the value of R_3, its value can be measured with very low errors. When the value of R_3 is small (in milli ohms) then we need to send large enough current so that the voltage drops across R_3 and R_4 are sufficiently large to overcome effects of noise and provide adequate emf to cause sufficient deflection in the galvanometer for an unbalance. Hence in Figs. 5.4 and 5.5 the wiring from the excitation to the resistors R_3 and R_4 are shown by thick lines. Conventionally, a few amperes to a few hundred amperes may be required. A look at the Kelvin bridge of Fig. 5.4 suggests that there is a "*bridge inside another bridge*" and hence the Kelvin bridge is sometimes called as "*Kelvin double bridge*". Since there is only "one bridge" in effect, it is better to avoid using the term "*Kelvin double bridge*" and use the correct term Kelvin Bridge. Similar to the Kelvin bridge which is meant for the measurement of low valued resistors, the megohm bridge is employed for measuring resistances of values higher than 10^6 ohms.

5.2.3 *The Megohm Bridge*

When the unknown resistance has a large value, two problems arise: (i) the current through the unknown resistance will become small and thus balancing with a galvanometer (a current operated device) will become problematic, if not impractical and (ii) the current through the unknown resistor will become close to leakage currents in external parasitic elements.

In the megohm bridge shown in Fig. 5.6 the first issue is solved by (i) increasing the excitation voltage and employing a voltage detector rather than a galvanometer. In the past Vacuum Tube Voltmeter (VTVM) was employed as the null detector. Presently, chopper stabilized DC amplifier and sensitive voltmeter are employed.

(ii) The issue of leakage current is solved using:

   (a) a guard electrode and
   (b) a shield as indicated in Fig. 5.6.

The shield is kept at potential V_G. and the guard ring is tied to potential V_1. By doing this it is seen from Fig. 5.7, the two ends of the parasitic leakage resistance

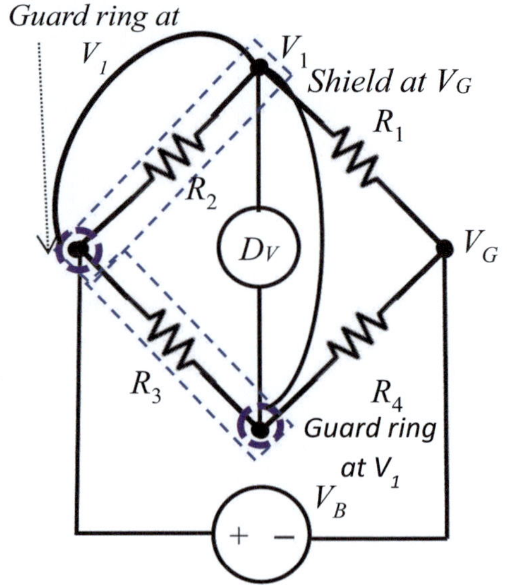

Fig. 5.6 The Megohm bridge

R_{3b} will be at the same potential at balance and thus will not affect the measurement. Since by design, the values of R_1, R_2, and R_4 are kept small compared to the leakage resistance R_{3a}, its effect on the measurement of R_3 will also be small. Another difference between a Wheatstone bridge and the megohm bridge is that the bridge excitation will be 500 V or 1000 V.

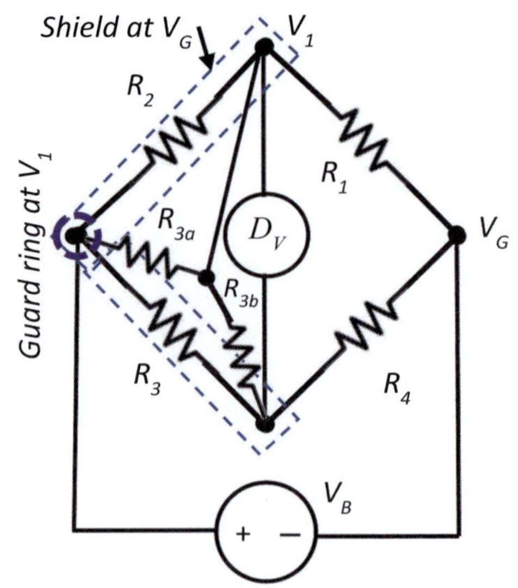

Fig. 5.7 The Megohm bridge with leakage resistances

5.3 The AC Bridge Method of Measurement

AC bridge methods are developed mainly to measure the values of inductors (coils) and capacitors. When inductors and capacitors are excited with sinusoidal sources (voltage or current sources), then the ratio of the voltage across and current through the element (inductor or capacitor) will be a complex number, called an impedance $\mathbf{Z}_x$, expressed as

$$\mathbf{Z_x} = \mathrm{Z_x}\theta_\mathrm{x} = \frac{\underline{\mathrm{V}}_\mathrm{x}}{\underline{\mathrm{I}}_\mathrm{x}} = \frac{\mathrm{V_x}\theta_\mathrm{v}}{\mathrm{I_x}\theta_\mathrm{i}} = \frac{\mathrm{V_x}}{\mathrm{I_x}}(\theta_\mathrm{v} - \theta_\mathrm{i}) = \mathrm{R_x} \pm j\mathrm{X_x} \tag{5.15}$$

In Eq. (5.15), the impedance is indicated as $\mathbf{Z_x}$ (with a bold face font) to show that the impedance is a complex number. Here, $\underline{V}_x$ is the voltage phasor that appears across $\mathbf{Z}_x$ and $\underline{I}_x$ is the phasor current through $\mathbf{Z}_x$. R_x is the resistive part of the impedance $\mathbf{Z}_x$ and X_x is the reactance part ($= \omega L$ for inductance and $= -1/\omega C$ for the capacitance). Similarly, we have learnt that the reciprocal of the impedance $\mathbf{Z}_x$ gives us the admittance $\mathbf{Y}_x$, once again a complex number.

$$\mathbf{Y_x} = \frac{\underline{\mathrm{I}}_\mathrm{x}}{\underline{\mathrm{V}}_\mathrm{x}} = \frac{1}{\mathbf{Z_x}} = \mathrm{G_x} \pm j\mathrm{B_x}. \tag{5.16}$$

In Eq. (5.16), G_x is the conductance and B_x is the susceptance of admittance $\boldsymbol{Y}_x$. For practical capacitors and inductors, the resistance part of the impedance and the conductance part of the admittance will always be a positive number. The reactance will be positive for inductors and negative for capacitors. To measure the impedance, we once again construct an AC bridge as shown in Fig. 5.8. We measure the unknown impedance $\mathbf{Z}_x = \mathbf{Z}_3$ with the known values of $\mathbf{Z}_1$, $\mathbf{Z}_2$, and $\mathbf{Z}_4$. In Fig. 5.8, the bridge excitation $\underline{V_B} = \sqrt{2}V_B\,\mathrm{Cos}(\omega t)$ V. Once the values of $\mathbf{Z}_1$, $\mathbf{Z}_2$, and $\mathbf{Z}_4$ are adjusted such that the detector D reads zero, then we have

$$\underline{\mathrm{V}}_1 = \underline{\mathrm{V}}_\mathrm{B}\left(\frac{\mathbf{Z}_2}{\mathbf{Z}_1+\mathbf{Z}_2}\right) = \underline{\mathrm{V}}_2 = \underline{\mathrm{V}}_\mathrm{B}\left(\frac{\mathbf{Z}_3}{\mathbf{Z}_3+\mathbf{Z}_4}\right). \tag{5.17}$$

Rearranging Eq. (5.17), we get

$$\mathbf{Z}_3(\mathbf{Z}_1 + \mathbf{Z}_2) = \mathbf{Z}_2(\mathbf{Z}_3 + \mathbf{Z}_4). \tag{5.18}$$

From Eq. (5.18), we get

$$\mathbf{Z}_2\mathbf{Z}_4 = \mathbf{Z}_1\mathbf{Z}_3\mathbf{Z}_3 = \left(\frac{\mathbf{Z}_2\mathbf{Z}_4}{\mathbf{Z}_1}\right). \tag{5.19}$$

Making precision variable resistors is easier than manufacturing precision variable capacitors or inductors. Hence, the optimal design of the bridge shown in Fig. 5.8 requires to be accomplished with as many resistive elements as possible. For example, if

$\mathbf{Z}_3 = R_3 \pm \mathrm{j}X_3$ then we can make $\mathbf{Z}_1$ and $\mathbf{Z}_2$ as resistors of R_1 and R_2, resulting in

Fig. 5.8 General AC bridge

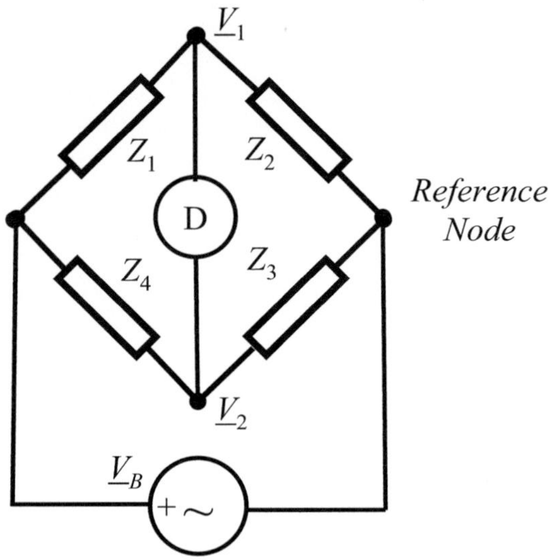

$$\mathbf{Z}_3 = \mathrm{R}_3 \pm j\mathrm{X}_3 = \left(\frac{\mathrm{R}_2}{\mathrm{R}_1}\right)\mathbf{Z}_4 = \frac{\mathrm{R}_2}{\mathrm{R}_1}\left(\mathrm{R}_4 \pm \mathrm{j}\mathrm{X}_4\right). \tag{5.20}$$

From Eq. (5.20), we deduce that the unknown impedance is measured as series combination of a resistance R_x and reactance X_x as indicated in Fig. 5.9.

$$\mathrm{R}_\mathrm{x} = \mathrm{R}_3 = \left(\frac{\mathrm{R}_2}{\mathrm{R}_1}\right)\mathrm{R}_4 \; \textit{and} \; \mathrm{X}_\mathrm{x} = \mathrm{X}_3 = \left(\frac{\mathrm{R}_2}{\mathrm{R}_1}\right)\mathrm{X}_4. \tag{5.21}$$

Thus, the AC bridge designed with $\mathbf{Z}_1$ and $\mathbf{Z}_2$ as resistors of R_1 and R_2 can measure an inductor with another known inductor or an unknown capacitor with a known capacitor. Since Z_1 and $\mathbf{Z}_2$ appear as a ratio of two resistors in computing the value of the unknown impedance, the bridge is classified as "*Ratio-Arm Real bridge*".

We can also design a bridge with $\mathbf{Z}_2$ and $\mathbf{Z}_4$ as resistors R_2 and R_4 and then we will get

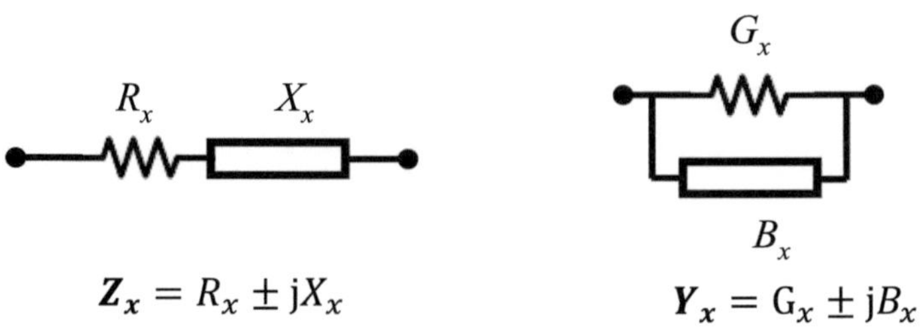

Fig. 5.9 The structure of impedance Z_x

$$\mathbf{Z}_3 = \mathrm{R}_3 \pm \mathrm{jX}_3 = (\mathrm{R}_2\mathrm{R}_4)\mathrm{Y}_1 = (\mathrm{R}_2\mathrm{R}_4)\left(\mathrm{G}_1 \pm \mathrm{jB}_1\right). \tag{5.22}$$

From Eq. (5.22), we see that the unknown impedance $\mathbf{Z}_3$ is measured using the known admittance $\boldsymbol{Y}_1 = G_1 \pm jB_1$, which means that we measure inductance using capacitor (The converse scheme is never attempted as creating variable standard inductance is not practical). We can also rewrite Eq. (5.22) as

$$\mathbf{Y}_3 = \mathrm{G}_3 \pm \mathrm{jB}_3 = \left(\frac{\mathbf{Z}_1}{\mathbf{Z}_2}\right)\mathbf{Y}_4. \tag{5.23}$$

In this case, we measure the unknown element as an admittance (conductance G_x in parallel with susceptance B_x as shown in Fig. 5.9). We can design that $\mathbf{Z}_1$ and $\mathbf{Z}_2$ as resistors of R_1 and R_2 then:

$$\mathbf{Y}_3 = \mathrm{G}_3 \pm \mathrm{jB}_3 = \left(\frac{\mathrm{R}_1}{\mathrm{R}_2}\right)\left[\mathrm{G}_4 \pm \mathrm{jB}_4\right]. \tag{5.24}$$

This bridge will come under the category "*Ratio Arm Real*" bridge. We can choose $\mathbf{Z}_1$ to be a capacitor and $\mathbf{Z}_2$ as resistance R_2 and thus Eq. (5.19) becomes

$$\mathbf{Z}_3 = \mathrm{j}\omega\mathrm{C}_1\ \mathrm{R}_2\mathbf{Z}_4\mathrm{R}_3 = \omega\mathrm{C}_1\mathrm{R}_2\mathrm{X}_4\ \textit{and}\ \mathrm{X}_3 = \omega\mathrm{C}_1\mathrm{R}_2\mathrm{R}_4. \tag{5.25}$$

It should be noted that this type of bridge will come under the "*Ratio Arm Imaginary*". Similarly, we can design a bridge to be "*product arm imaginary*". The number and types of AC bridges developed is quite large. However, only a few are historically or academically relevant today and a sub-set of these are currently in use as listed in Table 5.1. All AC bridges measure either the impedance $R_x \pm jX_x$ or the admittance $G_x \pm jB_x$ as indicated in Fig. 5.9.

Table 5.1 Typical AC bridges and their use

Sl. no	AC bridge name	For the measurement of
1	Maxwell-Wien Bridge	Low Q Inductor
2	Owens Bridge	Low Q Inductor
3	Anderson Bridge	High Q Inductor
4	Hays Bridge	Inductors
5	Schering Bridge	Capacitors
6	Wien Bridge	Standard Capacitor, Frequency
7	Heaviside Campbell Bridge	Mutual Inductance
8	Transformer Ratio Bridge	Capacitor and Inductor
9	Active (source) AC bridge	Capacitor and Inductor

5.3.1 Measurement of Inductance

Unlike practical resistors, which can be built to be purely resistive to a close degree, practical inductors do not have just the property of inductance. Depending on the frequency of excitation, a practical inductor will possess additional elements as illustrated in Fig. 5.10. Figure 5.10a shows an ideal inductor and the corresponding voltage and current phasors. When we wind a coil (without an iron core) the AC resistance of the wire with which the coil is wound will come into the equivalent circuit as shown in Fig. 5.10b. It should be noted here that the circuit shown in Fig. 5.10b for an air-cored coil is applicable only for low and medium frequencies. At high frequency excitation, the inter-turn and parasitic capacitances of the wound coil become appreciable, and we need to use the equivalent circuit of Fig. 5.10c. Irrespective of the equivalent circuit, we see from the phasor diagram that the current phasor $\underline{I}_{La}$ lags the voltage phasor $V_L 0$ by an angle θ. Hence the impedance $\mathbf{Z}_L$ of the coil at that frequency will be

$$\mathbf{Z}_L = \frac{V_L}{I_{La}}\left[\mathrm{Cos}(\theta) + j\,\mathrm{Sin}(\theta)\right], \tag{5.26}$$

resulting in the impedance to be

$$\mathbf{Z}_L = (R_s + j\omega L_s),$$

where R_s is the equivalent series resistance and L_s is the equivalent series inductance at that frequency. Most inductance measuring AC bridges measure this R_s and L_s. Once we use a soft ferromagnetic core, apart from the losses in the winding resistance, there will be losses in the core due to hysteresis and eddy current phenomena. That

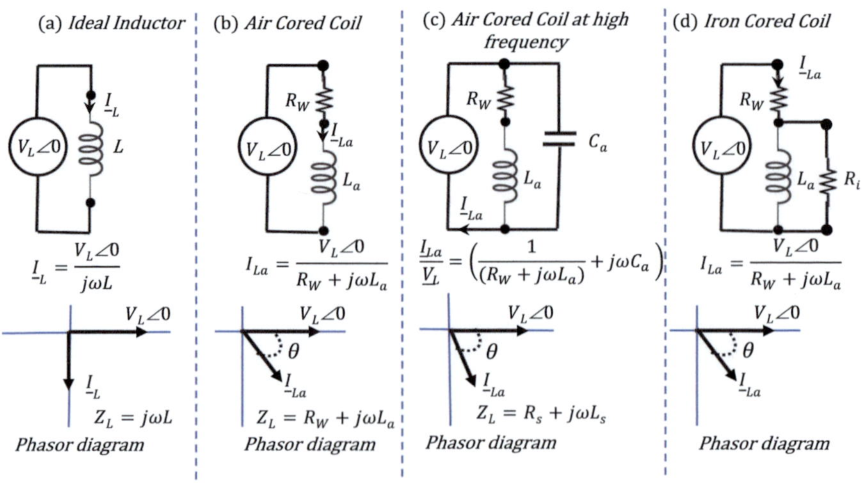

Fig. 5.10 Equivalent circuit of a practical air-cored inductor

would introduce a parallel resistor R_i across the inductor as shown in Fig. 5.10d. In practice, this circuit has a restricted validity since resistor R_i that represents the iron losses in the core is nonlinear and its value changes with the excitation. It is seen from the phasor diagrams of different equivalent circuits, that irrespective of the equivalent circuit we may have to choose for an inductor (depending on whether air-cored or iron cored or on the frequency of excitation), the current drawn by the inductor lags the voltage applied by an angle θ and hence we can write:

$$\boldsymbol{Z}_L = \frac{\underline{V}_L}{\underline{I}_L} = R_s + j\omega L_s, \tag{5.27}$$

Where in R_s and L_s are also functions of ω in general. Most inductance measurement methods measure the values of R_s and L_s at a particular desired frequency of excitation. It is possible to represent a practical inductor as a combination of an inductance L_p in parallel with a resistance R_p, thus giving:

$$\boldsymbol{Y}_L = \frac{\underline{I}_L}{\underline{V}_L} = \frac{1}{R_p} + \frac{1}{j\omega L_p}. \tag{5.28}$$

However, we rarely use the parallel combination of resistance and inductance to represent a practical inductor. Another important factor associated with a practical inductor (coil) is the quality factor Q, given by

$$Q = \frac{\omega L_s}{R_s}. \tag{5.29}$$

Some AC bridges measure the quality factor Q directly.

5.3.1.1 Maxwell Bridge

Figure 5.11 gives the circuit schematic of an AC bridge for the measurement of inductors possessing low quality factor. Popularly known as Invented by James Clark Maxwell in 1873, it is popularly known as Maxwell Bridge. This bridge is also called Maxwell Inductance-capacitance bridge or Maxwell-Wien bridge. The bridge elements must satisfy the following relations as per Eq. (5.30) for its balance condition:

$$\mathrm{R}_3 + j\omega \mathrm{L}_3 = \frac{\mathrm{R}_2\,\mathrm{R}_4}{\boldsymbol{Z}_1} = \mathrm{R}_2\mathrm{R}_4\left(\frac{1}{\mathrm{R}_1} + \mathrm{j}\omega \mathrm{C}_1\right). \tag{5.30}$$

This leads to

$$\mathrm{R}_\mathrm{X} = \mathrm{R}_3 = \frac{\mathrm{R}_2\,\mathrm{R}_4}{\mathrm{R}_1}\ \textit{and}\ \mathrm{L}_\mathrm{X} = \mathrm{L}_3 = \mathrm{R}_2\mathrm{R}_4\mathrm{C}_1 \tag{5.31}$$

Furthermore, the Q factor of the coil $Q_x = Q_3$ can be calculated as

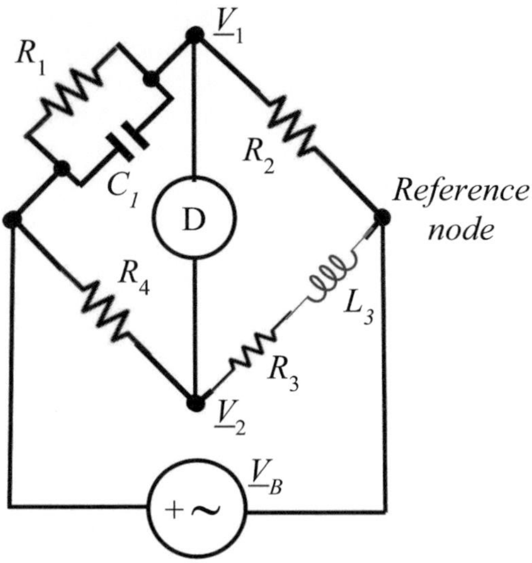

Fig. 5.11 The Maxwell bridge

$$Q_x = Q_3 = \frac{\omega L_3}{R_3} = \frac{\omega R_1 R_2\ R_4 C_1}{R_2\ R_4} = \omega R_1 C_1. \tag{5.32}$$

The bridge is well suited for low Q (<50) coils. When Q is large, R_3 becomes small compared to ωL_3 and hence the value of R_1 required for balancing the bridge becomes impractical. For large Q inductors, we can use the Anderson bridge shown in Fig. 5.12.

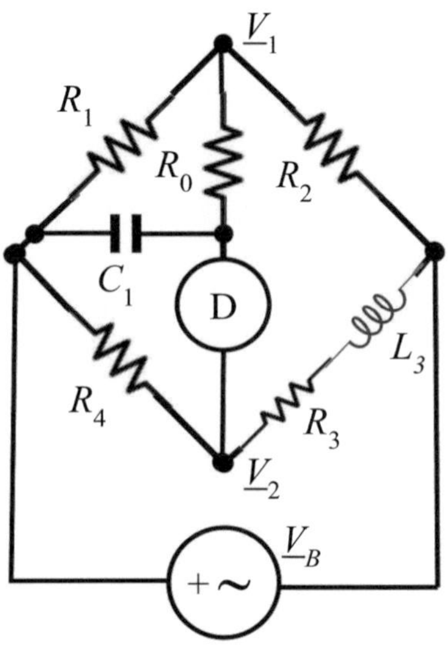

Fig. 5.12 The Anderson bridge

5.3.1.2 The Anderson Bridge

Comparing the Maxwell bridge of Fig. 5.11 and the Anderson bridge of Fig. 5.12, we see that a resistor R_0 is added in series with capacitor C_1 and that the detector is now connected at the junction of R_0 and C_1. We convert the star circuit made of R_0, R_1, and R_2 in Fig. 5.12 to delta as shown in Fig. 5.13. Then:

$$R_A = \frac{R_0R_1 + R_1R_2 + R_2R_0}{R_2}. \tag{5.33}$$

$$R_B = \frac{R_0R_1 + R_1R_2 + R_2R_0}{R_1}. \tag{5.34}$$

Since R_C comes in parallel to the source, it does not enter the balance requirements and hence we need not write its value. Thus, at balance, we have

$$R_3 + j\omega L_3 = R_BR_4\left(\frac{1}{R_A} + j\omega C_1\right). \tag{5.35}$$

Substituting Eqs. (5.33) and (5.34) in Eq. (5.35), we get

$$R_3 = \frac{R_2R_4}{R_1} \textit{ and } L_3 = R_BR_4C_1. \tag{5.36}$$

In general, the Anderson bridge is best suited for inductors of Q value > 20.

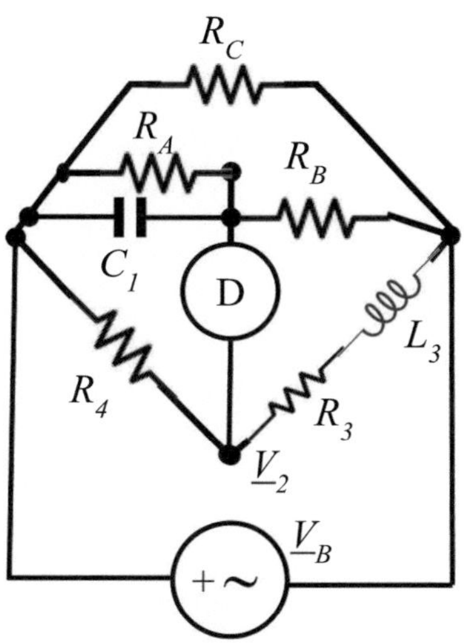

Fig. 5.13 Equivalent of Fig. 5.12

5.3.2 Measurement of Capacitance

Unlike an inductor, a practical capacitor can be manufactured to have only the property of capacitance at normal operating voltages and frequencies. Especially standard capacitors can be built with air or nearly ideal insulators (such as mica) and can be made to function as "pure" capacitors as shown in Fig. 5.14a. Here, the capacitor's current phasor $\underline{I}_c$ leads the capacitor voltage phasor $\underline{V}_c$ by 90°. However, when either the capacitor needs to have large values, such as hundreds or thousands of microfarads or must withstand high excitation voltage (HV capacitors) or to have small size, then the dielectric employed cannot be optimized for ideal (perfect insulator) characteristics. Such capacitors will possess leakage current I_{Lx} and hence the equivalent circuit of such capacitors will have additional resistance R_L added in parallel to C_p to represent the internal leakage current I_L as shown in Fig. 5.14b. When large-valued capacitors are made, the resistance of the plates and the connecting wire becomes appreciable, resulting in the series resistance R_p added to the capacitor C_{px} and parallel resistor R_{Lx} as shown in Fig. 5.14c.

Once the capacitor C_x is excited with a voltage $\underline{V}_x$, the current $\underline{I}_x$ flows through the capacitor resulting in the phasors as shown in Fig. 5.14b, c. Thus, at that frequency, the parallel capacitance C_{px} and parallel resistor R_{px} equivalent circuit of Fig. 5.14d is adequate to represent the phasors $\underline{V}_x$ and $\underline{I}_x$. Though it is rare, the series equivalent circuit of Fig. 4.14e is also used to represent a practical capacitor. It is seen both in Fig. 5.14b, c that the current phasor deviates by an angle δ from the expected 90° of an ideal capacitor of Fig. 5.14a. This angle δ is called the *loss angle* of the capacitor and the value *tan* δ (= equivalent series resistance/equivalent series reactance =

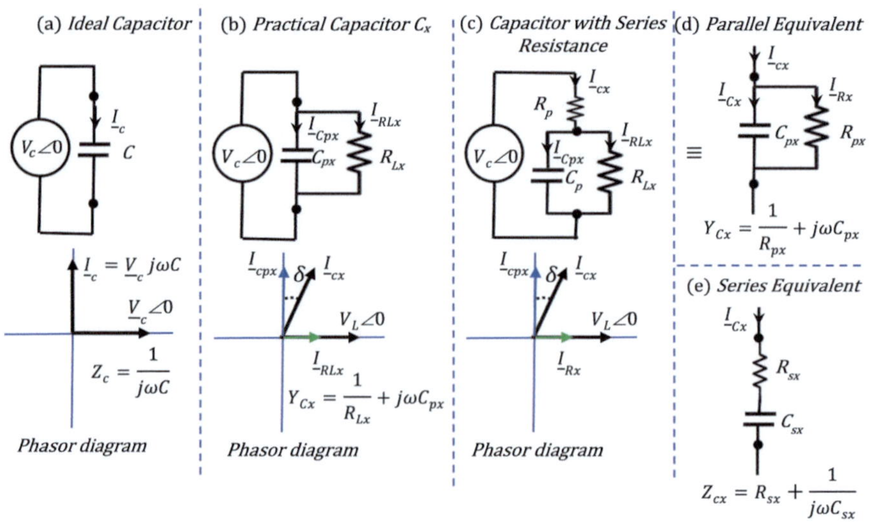

Fig. 5.14 Equivalent circuit of a practical capacitor

$j\omega C_{sx}R_{sx}$) is called the loss tangent and also as the dissipation factor D of the capacitor. A popular AC bridge that was developed long ago for the measurement of the value of the capacitance C_{sx} using the series equivalent circuit of Fig. 5.14 and *tan* δ of the capacitor and is still in use is the Schering bridge.

5.3.2.1 The Schering Bridge

The circuit schematic of the Schering bridge is shown in Fig. 5.15. The unknown capacitance represented by C_x in series with R_x is compared with a standard capacitance C_s. Once the bridge is balanced (detector D reading zero), we get

$$Z_2 = \left(\frac{Z_1 Z_3}{Z_4}\right) \rightarrow R_x + \frac{1}{j\omega C_x} = \left(\frac{R_3}{j\omega C_s Z_4}\right) = \frac{R_3}{j\omega C_s}\left(\frac{1}{R_4} + j\omega C_4\right) \tag{5.37}$$

Resolving Eq. (5.37) we get

$$R_x = R_3\left(\frac{C_4}{C_s}\right) \textit{ and } C_x = C_s\left(\frac{R_4}{R_3}\right) \tag{5.38}$$

From Eq. (5.38), we can get

$$\tan\delta = j\omega C_x R_x = \omega C_4 R_4. \tag{5.39}$$

Thus, using the Schering bridge, we can get the series equivalent as well as the capacitance and loss tangent (*tan* δ) of an unknown capacitor. An added advantage of a Schering bridge is that we can, by design, keep the voltage V_B to be large and at the same time V_1 and V_2 to be small so that the test capacitor can be excited using a high voltage and the operating parts (Normally we make R_3, R_4, and C_4 as variable elements) are at a safe, low voltage for an operator to handle without safety issues. Such an arrangement is called the high-voltage Schering Bridge. It is also possible

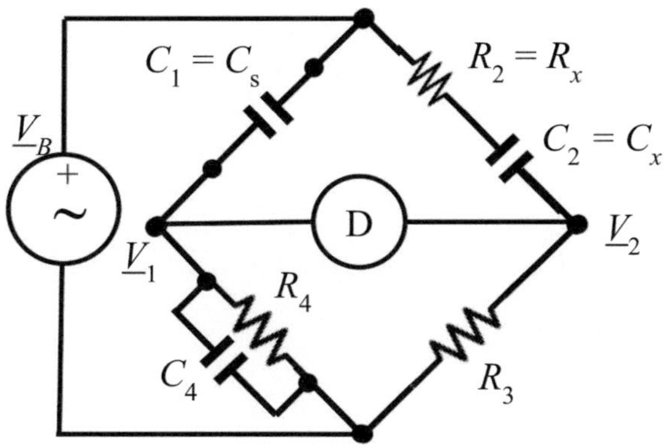

Fig. 5.15 The Schering bridge

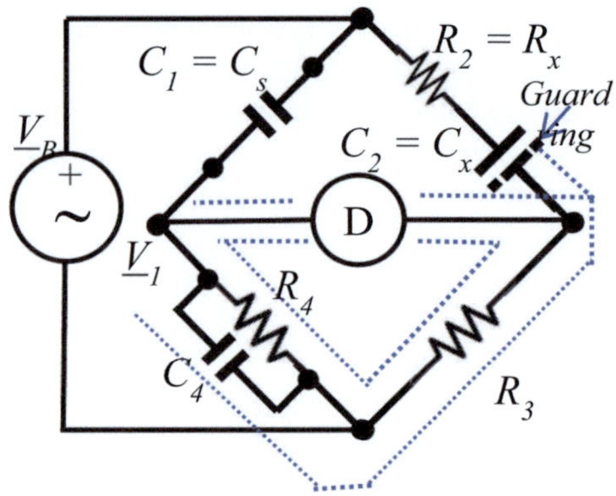

Fig. 5.16 Shielded Schering bridge

to shield the Schering bridge as well as measure capacitors that possess a guard ring as shown in Fig. 5.16.

5.3.2.2 The Transformer Ratio Bridge

The transformer ratio bridge, as the name suggests, uses a transformer with two variable-turn secondary sections, whose voltage ratio is adjustable, and which constitute the ratio arms of an AC bridge. The two secondary sections of the transformer in Fig. 5.17 produce output voltages with phasors $\underline{V}_{AO}$ and $\underline{V}_{OB}$ as

$$\underline{\mathrm{V}}_{\mathrm{AO}} = \mathrm{N_x}\underline{\mathrm{V}}_{\mathrm{P}},\ and\ \underline{\mathrm{V}}_{\mathrm{OB}} = \mathrm{N_s}\underline{\mathrm{V}}_{\mathrm{P}}, \tag{5.40}$$

where N_x and N_s are the respective turns ratios with respect to the primary). Here, N_x and N_s can be individually varied from 0.000000 to 1.00000. The six (or even up to 9) significant digits variation is obtained using multiple cores and windings along with electronics as illustrated in the sequel. Figure 5.17 indicates how a transformer ratio bridge can be used for the measurement of an unknown capacitor C_x by comparing C_x with a standard variable capacitor C_s in parallel to a standard variable resistor R_s. Once the detector D reads zero, we get the currents: $\underline{I}_{cs} + \underline{I}_{rs} = \underline{I}_{cx}$. Hence, we have

$$\mathrm{N_x}\underline{\mathrm{V}}_{\mathrm{p}}(\mathrm{G_x} + \mathrm{jB_x}) = \mathrm{N_s}\underline{\mathrm{V}}_{\mathrm{p}}(\mathrm{G_s} + \mathrm{jB_s}). \tag{5.41}$$

In Eq. (5.41) $G_s = \frac{1}{R_s}, B_s = \omega C_s, G_x = \frac{1}{R_{xp}}$ and $B_x = \omega C_{xp}$. Simplifying Eq. (5.41) and equating the respective real and imaginary parts on the two sides of the equation, to, we get

$$\mathrm{R_{xp}} = \frac{\mathrm{N_x}}{\mathrm{N_s}}\mathrm{R_s}\ and\ \mathrm{C_{xp}} = \frac{\mathrm{N_s}}{\mathrm{N_x}}\mathrm{C_s}. \tag{5.42}$$

Here we adopted the parallel equivalent circuit of Fig. 5.14d for the unknown capacitor and assumed the standard capacitor to be ideal (without power loss). Transformer

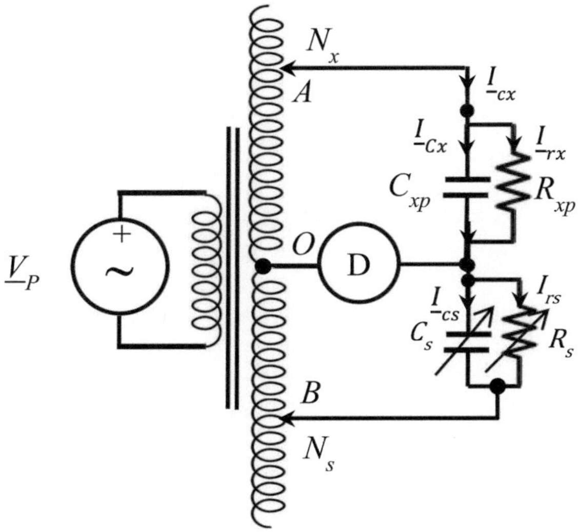

Fig. 5.17 The transformer ratio bridge: capacitor—parallel equivalent

ratio bridge can also be used to measure the series equivalent circuit of an unknown capacitor as shown in Fig. 5.18 once the detector D reads zero, we get the currents: $\underline{I}_{cs} = \underline{I}_{cx}$. Hence, we have

$$\frac{N_x \underline{V}_p}{(R_{xs} - jX_x)} = \frac{N_s \underline{V}_p}{(R_s - jX_s)}. \tag{5.43}$$

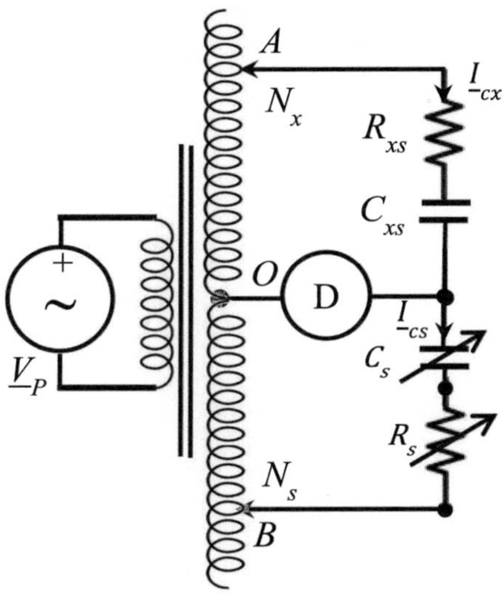

Fig. 5.18 The transformer ratio bridge—capacitor: series equivalent

In Eq. (5.43), $X_s = \frac{1}{\omega C_s}$ and $X_x = \frac{1}{\omega C_{xs}}$. Inverting Eq. (5.43) and equating the respective real and imaginary parts on the LHS to the respective real and imaginary parts on the RHS, we get

$$\mathrm{R_{xs}} = \frac{\mathrm{N_x}}{\mathrm{N_s}}\mathrm{R_s}\ and\ \mathrm{C_{xs}} = \frac{\mathrm{N_s}}{\mathrm{N_x}}\mathrm{C_s} \tag{5.44}$$

It is seen that using a TRB, we can directly measure the parameters of either the series equivalent or the shunt equivalent of an unknown capacitor. In a TRB we have four variables namely, N_x, N_s, C_s, and R_s to balance the bridge. Thus, it is possible to optimize the variable elements. In such a case either the standard capacitor or standard resistor alone can be a variable element. Normally obtaining a variable standard resistor is more convenient, in a TRB, standard capacitor is varied in larger steps (in the 1, 10, 100 … sequence) and the standard resistance is varied in very fine steps.

5.3.2.3 The Current Comparator for the Measurement of Capacitance

It is also possible to use a current comparator scheme to measure capacitors and obtain a parallel or series equivalent model of the unknown capacitor. A current comparator, shown in Fig. 5.19, looks like the transformer ratio bridge but is quite different in design and construction. Here, the transformer is no longer acting as the ratio arm of a bridge but functions as a comparator of currents, in this case comparing I_{cx} the current of the unknown capacitor C_x with the sum $\underline{I}_{cs}$ of currents through the standard capacitor C_s and the standard resistor R_s in parallel. While in the transformer ratio bridge, the secondaries of the transformer act as arms of an AC bridge, here both the windings act as the primaries carrying the currents of the unknown and standard capacitors. Here when the detector reads zero, it indicates that the net mmf needed to set up the flux in the core is zero, i.e., $N_x\underline{I}_x = N_s\underline{I}_s$. We then have

$$\mathrm{N_x}\underline{\mathrm{V}}_\mathrm{p}(\mathrm{G_x} + \mathrm{jB_x}) = \mathrm{N_s}\left(\underline{\mathrm{V}}_\mathrm{p}\mathrm{G_s} + \mathrm{jB_s}\right), \tag{5.45}$$

resulting in

$$\mathrm{R_{xp}} = \frac{\mathrm{N_x}}{\mathrm{N_s}}\mathrm{R_s}\ and\ \mathrm{C_{xp}} = \frac{\mathrm{N_s}}{\mathrm{N_x}}\mathrm{C_s}. \tag{5.46}$$

We can also use the current comparator to obtain the series equivalent of a capacitor using the scheme shown in Fig. 5.20. Once again, at balance, $N_x\underline{I}_x = N_s\underline{I}_s$, resulting in

$$\frac{\mathrm{N_x}\underline{\mathrm{V}}_\mathrm{p}}{(\mathrm{R_{xs}} - \mathrm{jX_x})} = \frac{\mathrm{N_s}\underline{\mathrm{V}}_\mathrm{p}}{(\mathrm{R_s} - \mathrm{jX_s})}. \tag{5.47}$$

With $X_s = \frac{1}{\omega C_s}$ and $X_x = \frac{1}{\omega C_{xs}}$, we get

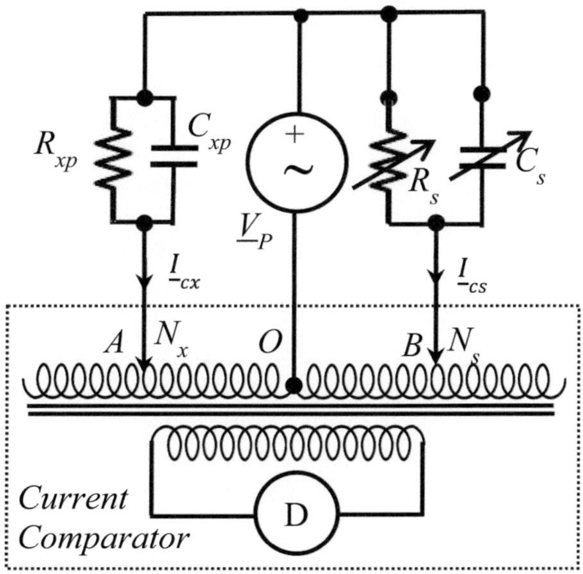

Fig. 5.19 Capacitance measurement using current comparator—parallel equivalent

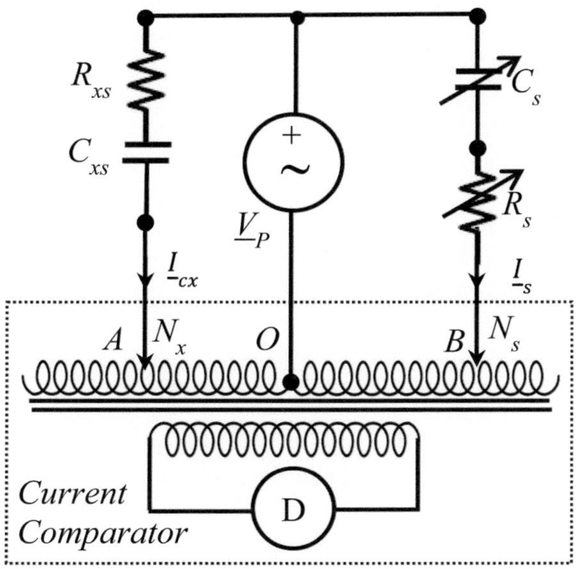

Fig. 5.20 Capacitance measurement using current comparator—series equivalent

$$R_{XS} = \frac{N_x}{N_s} R_S \textit{ and } C_{XS} = \frac{N_s}{N_x} C_S. \tag{5.48}$$

5.3.2.4 The Wien Bridge

Another bridge that can be used for the measurement of capacitance is the Wien Bridge shown in Fig. 5.21, where Arm 3 is the shunt equivalent of the capacitor under test. Balancing a Wien bridge is different from the other AC bridges. Here, one of the variables to be adjusted is the excitation frequency. Once the bridge is balanced, we have

$$\omega^2 = \frac{1}{R_2 R_3 C_2 C_3}. \tag{5.49}$$

$$\frac{C_3}{C_2} = \frac{R_1}{R_4} - \frac{R_2}{R_3}. \tag{5.50}$$

Rearranging Eq. (5.50), we get

$$C_3 = C_x = \frac{1}{\omega}\sqrt{\frac{R_4}{(R_1 R_3 - R_4 R_2) R_2}}. \tag{5.51}$$

From Eq. (5.51) we see that in order to measure a high-quality capacitor, we do not need another accurately known standard capacitor. The right-hand side of Eq. (5.51) contains only values of frequency and resistances. Both these variables can be determined with several significant digits and excellent accuracy. Thus, Wien bridge is useful in determining the value of standard capacitances to a high degree of precision and accuracy. Another use of the Wien bridge is to measure the frequency. If we ensure $C_2 = C_3 = C$ and $R_2 = R_3 = R$, then at balance we get

$$\omega = \frac{1}{RC}. \tag{5.52}$$

Apart from measurements on standard capacitance and frequency, the Wien bridge topology is also used to generate a sine wave using positive feedback. A typical

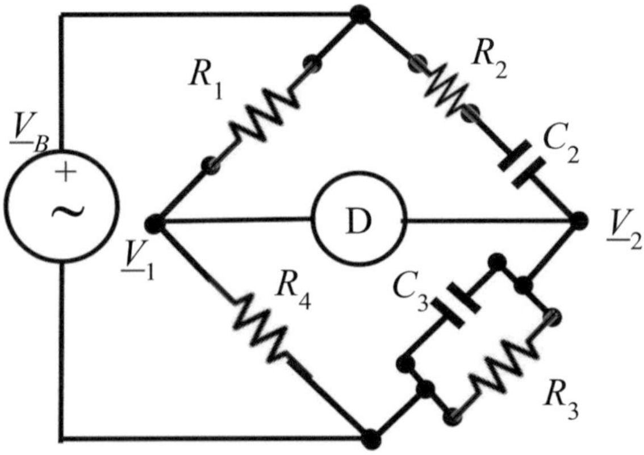

Fig. 5.21 The Wien bridge

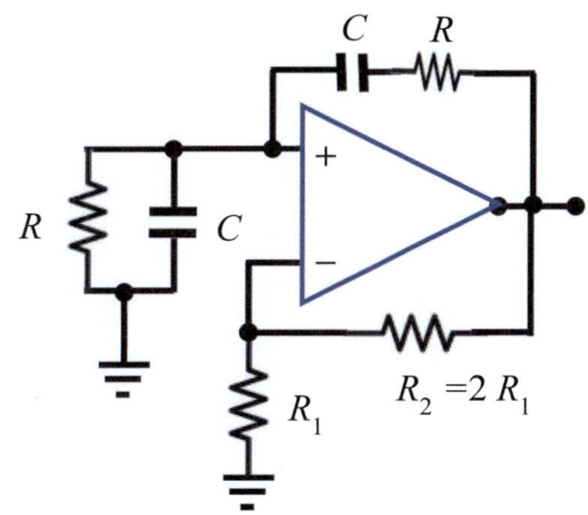

Fig. 5.22 The Wien bridge oscillator

Wien bridge oscillator using an opamp is shown in Fig. 5.22. The topology shown in Fig. 5.22 is amenable to building an amplitude stabilized sinewave oscillator.

5.3.3 *Universal AC Bridge: Active Source Type*

Universal AC bridges are those that can be used to measure a capacitor or an inductor. Of the universal bridges, the active (source type) AC bridge is the popular one. Unlike the bridge methods we have seen so far wherein only passive elements (R and C) were varied, in the modern AC bridges variable AC voltage sources are employed and are aptly called *Active AC bridges*. These bridges can measure the impedance of the unknown element as ($R_{xs} \pm \mathrm{j}X_s$) in its general form or its admittance as ($G_{xp} \pm \mathrm{j}B_p$). Figure 5.23 shows the schematic of a typical active AC bridge. It is seen in Fig. 5.23 that apart from the unknown element Z_X, the bridge is made of a standard resistance R_S and two AC sources $\underline{V}_{s1}$ and $\underline{V}_{s2}$. The magnitude and phase of the AC voltages $\underline{V}_{s1}$ and $\underline{V}_{s2}$ are adjusted till the detector D reads zero. Once the detector D reads zero (that is the active AC bridge is balanced), we have

$$\mathbf{Z_x} = \mathrm{R_x} \pm j\mathrm{X_x} = Re\left(\frac{\underline{\mathrm{V}}_{s2}}{\underline{\mathrm{V}}_{s1}}\mathrm{R_S}\right) + Im\left(\frac{\underline{\mathrm{V}}_{s2}}{\underline{\mathrm{V}}_{s1}}\mathrm{R_S}\right) \tag{5.53}$$

or

$$\mathbf{Y_x} = \mathrm{G_x} + j\mathrm{B_x} = Re\left(\frac{\underline{\mathrm{V}}_{s1}}{\underline{\mathrm{V}}_{s2}\mathrm{R_S}}\right) + Im\left(\frac{\underline{\mathrm{V}}_{s1}}{\underline{\mathrm{V}}_{s2}\mathrm{R_S}}\right). \tag{5.54}$$

In Eqs. (5.53) and (5.54), the term *Re* indicates the real part and *Im* indicates the imaginary part of the associated complex number. It is easily seen that at balance the current phasor:

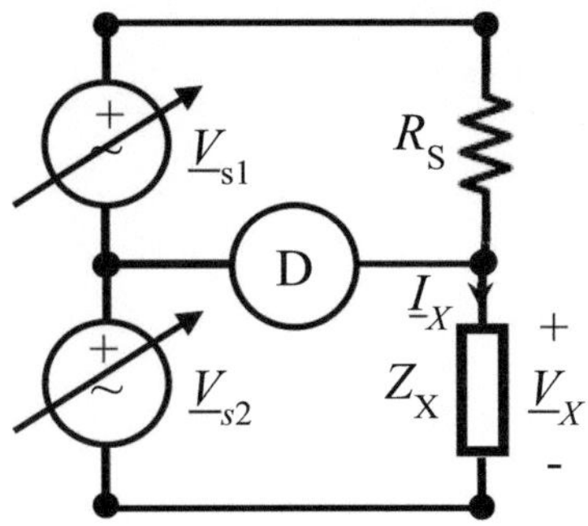

Fig. 5.23 Active AC bridge

$$\underline{I}_x = \left(\frac{\underline{V}_{s1}}{R_S}\right).$$

It is possible to fix $\underline{V}_{s1} = V_1 0$ and thus fix the current through the unknown impedance Z_x (OR admittance Y_x) as a constant and measure the impedance at that constant current. With this arrangement, only source $\underline{V}_{s2}$ needs to be made as a variable source. On the other hand, it is possible to fix $\underline{V}_{s2} = V_2 0$ and vary $\underline{V}_{s1}$ to obtain balance. In such a situation, the measurement of the unknown impedance $\boldsymbol{Z_x}$ (OR admittance $\boldsymbol{Y_x}$) is obtained with constant voltage excitation. Thus, the unknown element can be measured either

(i) at a specified current or
(ii) under a specified voltage excitation.

In the constant current excitation, we vary only $\underline{V}_{s2}$ and if we express $\underline{V}_{s2} = V_{p2} + \mathrm{j}V_{q2}$ at balance, then Eq. (5.53) can be written as

$$\mathbf{Z_x} = \mathrm{R_x} \pm j\mathrm{X_x} = \frac{\mathrm{V_{p2}}}{\mathrm{V_1}}\mathrm{R_S} + j\frac{\mathrm{V_{q2}}}{\mathrm{V_1}}\mathrm{R_S}. \tag{5.55}$$

Hence $R_x = \frac{V_{p2}}{V_1}R_S$ and $X_x = \frac{V_{q2}}{V_1}R_S$. Similarly, in the constant voltage excitation, we vary only $\underline{V}_{s1}$ and if at balance, $\underline{V}_{s1} = V_{p1} + \mathrm{j}V_{q1}$, then Eq. (5.54) can be written as

$$\mathbf{Y_x} = \mathrm{G_x} \pm j\mathrm{B_x} = \frac{\mathrm{V_{p1}}}{\mathrm{V_2R_S}} + j\frac{\mathrm{V_{q1}}}{\mathrm{V_2R_S}}. \tag{5.56}$$

The variable voltages can easily be generated using a coordinate AC potentiometer setup described earlier and thus the value of an impedance or admittance can be determined with several significant digits. Another universal bridge that can measure both capacitance and inductance is called the "Quasi-balance bridge".

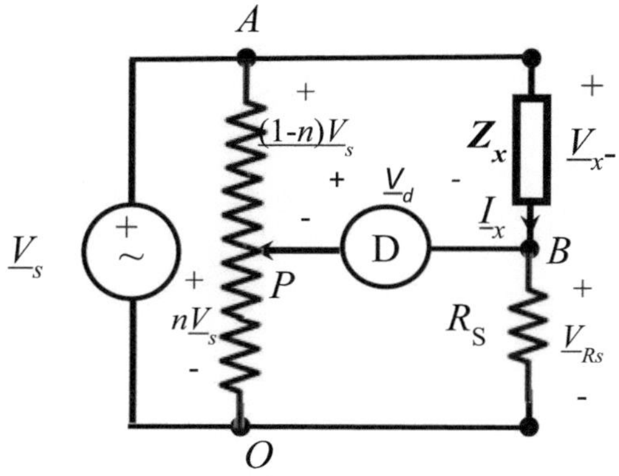

Fig. 5.24 Quasi-balance active AC bridge

5.3.4 Universal AC Bridge: Quasi-Balance Type

We have already remarked that manufacturing precision resistors (fixed or variable) is much easier and cheaper than manufacturing fixed or variable precision capacitors or inductors. Thus, we can say the best form of an AC bridge is the one which has the three arms (other than the element whose value is being measured) to be resistors as shown in Fig. 5.24. However, the problem now is that it is not possible to get a null balance. That is the voltage $\underline{V}_d$ sensed by detector D in Fig. 5.24 cannot be made zero as we can only vary the magnitudes V_{PO} and V_{AP} of the respective voltages and not their phases However, it is possible to use some other criterion and deduce the unknown impedance $\mathbf{Z}_x$ therefrom. A bridge that operates on such a criterion (not exactly *balance*) is called the "*Quasi-Balance*". Since a quasi-balance is obtained with a "nonzero" option, it becomes necessary to obtain a couple of quasi-balances to measure inductors and capacitors as explained next.

5.3.4.1 Quasi-Balance Bridge for the Measurement of Inductors

The Quasi-balance bridge for the measurement of an inductor (coil) is shown in Fig. 5.25a. Here we have used the series equivalent circuit of a practical inductor (coil) and represented its impedance $\mathbf{Z}_{Ls}$ as coil resistance R_{Ls} in series with the reactance ωL_{xs} of the coil. Detector D is a high impedance voltage indicator and is assumed not to draw any current from the circuit, thereby permitting us to treat the test inductor and the standard resistor R_s to be virtually in series. The current phasor $\underline{I}_{xL}$ lags the voltage phasor $\underline{V}_s$ by an angle θ, where

$$\left[\theta = Tan^{-1}\left(\frac{\omega L_{xs}}{R_{Ls} + R_S}\right)\right].$$

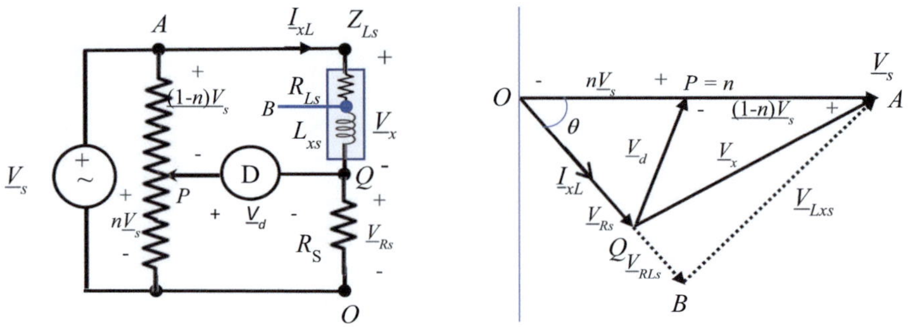

(a) Circuit diagram (b) Phasors of the QBB in (a)

Fig. 5.25 Quasi-balance AC bridge for measuring an inductor

as shown in Fig. 5.25b. The phasors of the applied voltage $\underline{V}_s$, the voltage drop $\underline{V}_{Rs}$ across R_S, and the voltage $\underline{V}_x$ across $\mathbf{Z}_{Ls}$ $(=R_{Ls}+j\omega L_{xs})$ are also shown in Fig. 5.25b. As we vary the potentiometer variable point P, the voltages V_{PO} and V_{AP} will be nV_s and $(1–n)\ V_s$, where n is a variable fraction. If we use the Poggendorff or Wenner or Kelvin-Varley type of potentiometer, then n can have several significant digits. We can also use an inductive voltage divider (IVD) instead of a resistive potentiometer to realize voltages V_{PO} and V_{AP}. As we move the point P on the potentiometer, the difference voltage $\underline{V}_{PQ}=\underline{V}_d$ changes. Now we bring in the two quasi-balance conditions necessary to measure R_{Ls} and L_s of the coil under test as

(i) The first quasi quasi-balance is obtained by moving the point P to a position marked as P_1 (value of $n=n_1$) such that the phasors $\underline{V}_d\big|_{n_1}$ and $\underline{V}_s$ are at right angle as shown in Fig. 5.26a. Here we have $\underline{V}_{P_1O}=n_1\underline{V}_s$ with n_1 being a fraction.

(ii) After we have completed the first quasi-balance, we obtain the second quasi-balance, once again by moving point P to point P_2 (value of $n=n_2$) such that the phasors $\underline{V}_d\big|_{n_2}$ and $\underline{V}_{Rs}$ the voltage across R_s are at right angle as shown in Fig. 5.26b. In this case, $\underline{V}_{P_2O}=n_2\underline{V}_s$, with n_2 being a fraction.

In Fig. 5.26b, we have $\angle QOP_1=\angle QOP_2=\theta$ and $\angle OP_2Q=\angle OAB=(90–\theta)$ in the right-angled triangles OQP_2 and OBA. Hence triangles OQP_2 and OBA are similar triangles. With this, we have

$$\frac{\mathrm{BA}}{\mathrm{QP_2}}=\frac{\mathrm{OA}}{\mathrm{OP_2}}\rightarrow\frac{\omega L_s\,\mathrm{I_{xL}}}{\mathrm{V_d}|_{\mathrm{n_2}}}=\frac{\mathrm{V_s}}{\mathrm{n_2\,V_s}}\rightarrow\omega L_{\mathrm{S}}=\frac{\mathrm{V_d}|_{\mathrm{n_2}}}{\mathrm{n_2 I_{xL}}}. \tag{5.57}$$

We also have we have $\angle QOP_1=\angle QOP_{2=}\ \theta$ and $\angle OP_2Q=\angle OQP_1=(90–\theta)$ in the right-angled triangles QP_1O and QP_1P_2. Hence triangles QP_1O and QP_1P_2 are similar triangles. With this we have we have

$$\frac{\mathrm{QP_2}}{\mathrm{OQ}}=\frac{\mathrm{P_1P_2}}{\mathrm{QP_1}}=\rightarrow\mathrm{V_d}|_{\mathrm{n_2}}=\frac{\mathrm{R_S I_L(n_2-n_1)V_s}}{\mathrm{V_d}|_{\mathrm{n_1}}}. \tag{5.58}$$

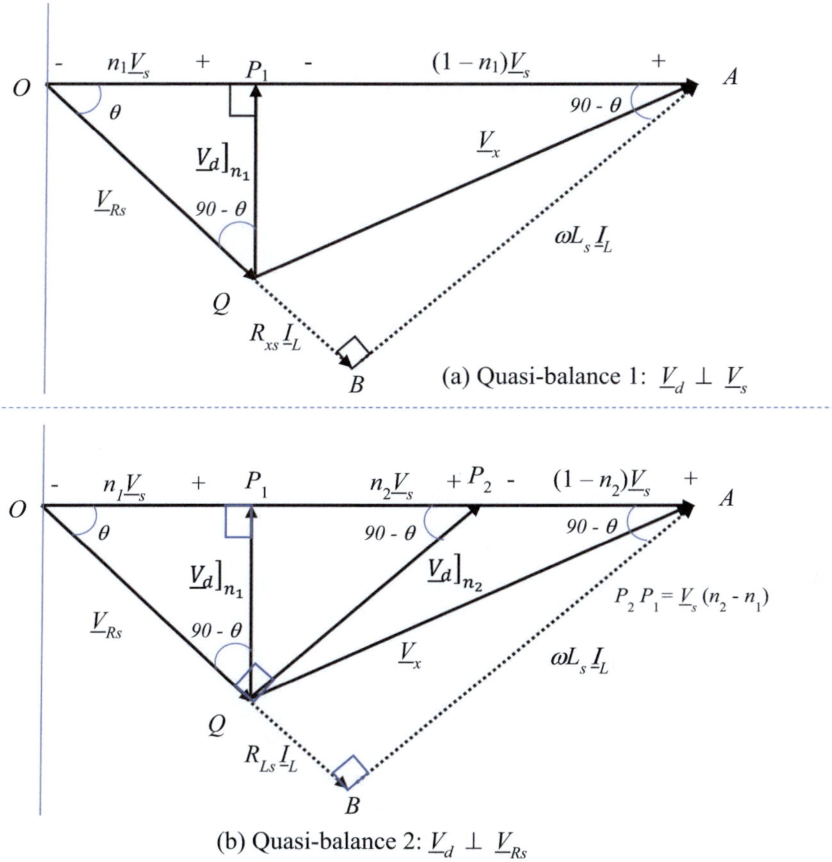

Fig. 5.26 Phasor diagram of Quasi-balances for inductor measurement

From similar triangles QP_1O and QP_1P_2, we can write

$$\frac{\mathrm{QP_1}}{\mathrm{P_1P_2}} = \frac{\mathrm{OP_1}}{\mathrm{QP_1}} \rightarrow (\mathrm{QP_1})^2 = \left(\underline{\mathrm{V}}_{\mathrm{d}}\Big|_{\mathrm{n_1}}\right)^2 = \mathrm{n_1(n_2 - n_1)V_s^2} \tag{5.59}$$

Simplifying Eq. (5.59), we get

$$\underline{\mathrm{V}}_{\mathrm{d}}\Big|_{\mathrm{n_1}} = \mathrm{V_s}\sqrt{\mathrm{n_1(n_2 - n_1)}}. \tag{5.60}$$

Substituting Eqs. (5.60) into (5.58), we get

$$\mathrm{V_d}\Big|_{\mathrm{n_2}} = \frac{\mathrm{R_S I_{xL}(n_2 - n_1)V_s}}{\mathrm{V_s}\sqrt{\mathrm{n_1(n_2 - n_1)}}} = \frac{\mathrm{R_S I_{xL}}}{\mathrm{V_s}}\sqrt{\frac{\mathrm{(n_2 - n_1)}}{\mathrm{n_1}}}\mathrm{V_s} = \mathrm{R_S I_{xL}}\sqrt{\frac{\mathrm{(n_2 - n_1)}}{\mathrm{n_1}}} \tag{5.61}$$

Substituting Eqs. (5.61) into (5.57), we get

$$L_S = \frac{R_S}{\omega n_2}\sqrt{\frac{(n_2-n_1)}{n_1}}. \tag{5.62}$$

From similar triangles OQP_2 and OBA we get

$$\frac{OQ}{OB} = \frac{OP_2}{OA} \rightarrow \frac{R_s\, I_L}{(R_s + R_{Ls})I_L} = \frac{n_2 V_s}{V_s} \rightarrow R_{Ls} = \frac{(1-n_2)}{n_2} R_S. \tag{5.63}$$

Equation (5.63) indicates that using two quasi-balances, we can determine the series equivalent inductance of a coil (inductor). Equation (5.63) indicates that with the second quasi-balance we can obtain the series equivalent resistance of a coil (inductor). If we need the value of the Q factor Q_L of the coil (inductor), the same can be obtained as

$$Q_L = \frac{\omega L_s}{R_{Ls}} = \frac{\frac{\omega R_S}{\omega n_2}\sqrt{\frac{(n_2-n_1)}{n_1}}}{R_S\frac{(1-n_2)}{n_2}} = \frac{1}{(1-n_2)}\sqrt{\frac{(n_2-n_1)}{n_1}}. \tag{5.64}$$

5.3.4.2 Quasi-Balance Bridge for the Measurement of Capacitors—Series Equivalent Circuit

The Quasi-balance bridge for the measurement of a capacitor, shown in Fig. 5.27a, is like the one used for inductor in Fig. 5.25a if we use the series equivalent circuit representation of the test capacitor C_x as a resistance R_{cs} in series with the reactance $(1/\omega C_s)$ of the series equivalent capacitor C_s. The phasors of the applied voltage $\underline{V}_s$, the voltage drop $\underline{V}_{Rs}$ across R_S, and the voltage $\underline{V}_x$ across Z_{cs} $(= R_{cs} + (1/j\omega C_{xs})$ will be as shown in Fig. 5.14b. Here, the capacitor current phasor $\underline{I}_{xc}$ leads the voltage phasor $\underline{V}_s$ by an angle φ, where

$$\left[\varphi = Tan^{-1}\left(\frac{1}{\omega C_{xs}(R_{cs}+R_S)}\right)\right] \tag{5.65}$$

And the capacitor voltage $\underline{V}_{xc} = \underline{V}_{Rcs} + \underline{V}_{cs}$, as shown in Fig. 5.27b. As we vary the potentiometer (variable point P is moved up and down), the voltage $\underline{V}_{PO} = m\underline{V}_s$ and the voltage V_{AP} will be (1–m) $\underline{V}_s$, where m is a fraction. Once again, if we use the Poggendorff or Wenner or Kelvin-Varley type of potentiometer, then m can have several significant digits. Here too we can use an inductive voltage divider (IVD) instead of a resistive potentiometer to realize voltages $\underline{V}_{PO}$ and $\underline{V}_{AP}$. As we move the point P on the potentiometer, the difference voltage $\underline{V}_{PQ} = \underline{V}_d$ changes. Now we bring in the two quasi-balance conditions necessary to measure R_{cs} and C_{xs} of the capacitor under test as

(i) The first quasi quasi-balance is obtained by moving the point P to a position marked as P_1 such that the phasors $\underline{V}_d\big|_{m_1}$ and $\underline{V}_s$ are at right angle as shown in Fig. 5.28. Here we have $\underline{V}_{P_1O} = m_1\underline{V}_s$, with m_1 being a fraction. (Underline V in $\underline{V}_{P1O}$).

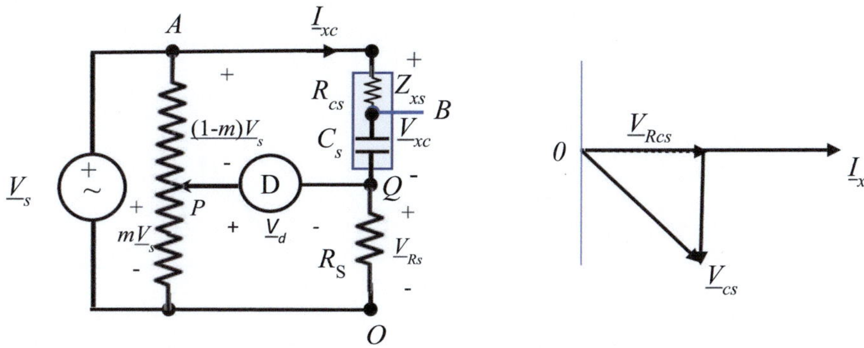

Fig. 5.27 Quasi-balance AC bridge for measuring capacitor (series equivalent circuit)

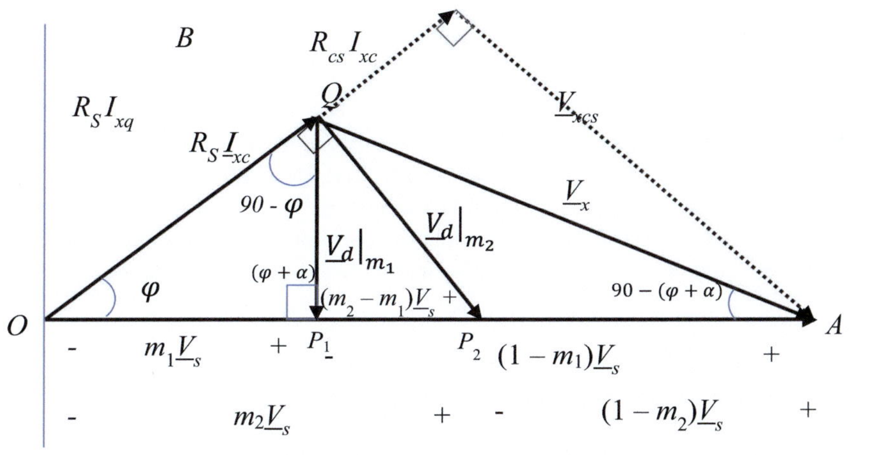

Fig. 5.28 Phasor diagram of quasi-balances for Capacitor (Series equivalent circuit)

(ii) After we have completed the first quasi-balance, we obtain the second quasi-balance, once again by moving point P to point P_2, such that the phasors $\underline{V}_d\big|_{m_2}$ and $\underline{V}_{Rs}$ the voltage across R_s are at right angle as shown in Fig. 5.28. In this case, $\underline{V}_{P2O} = m_2\underline{V}_s$, with m_2 being a fraction.

The values m_1 and m_2 so obtained and the value of the standard resistance R_s enable us to determine C_s and R_{cs}, using a similar analysis as in Sect. 5.2.4.1 for the case of an inductor. The relevant formulas to be used are

$$\mathrm{C_{xs}} = \frac{\mathrm{m_2}}{\omega R_s}\sqrt{\frac{\mathrm{m_1}}{\mathrm{m_2}-\mathrm{m_1}}}\,and \tag{5.66}$$

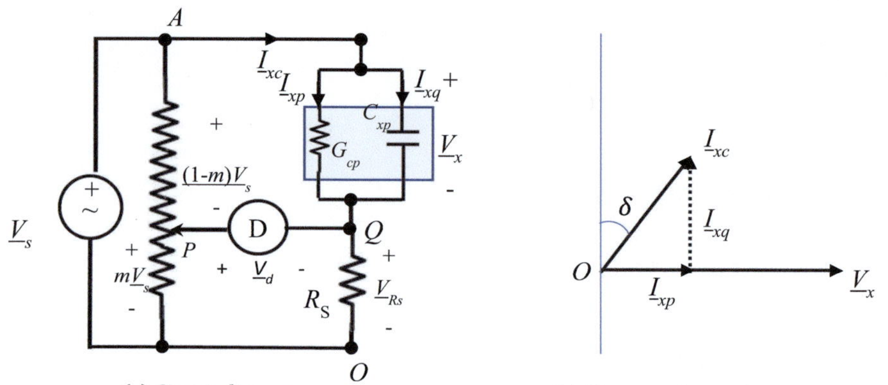

(a) Circuit diagram (b) Phasors of capacitor current

Fig. 5.29 Quasi-balance AC bridge for measuring capacitor (Parallel Equivalent Circuit)

$$R_{cs} = \frac{(1-m_2)}{m_2} R_S. \tag{5.67}$$

The two equations above can be used to determine the series equivalent resistor R_{cs} and series equivalent value C_{xs} of a capacitor. However, when we deal with capacitors, the parallel equivalent circuit as shown in Fig. 5.29a is preferred than the series equivalent circuit. We can determine the parallel equivalent conductance G_{cp} and parallel equivalent capacitor C_{xp} also using quasi-balance as detailed next.

5.3.4.3 Quasi-Balance Bridge for Measurement on Capacitors—Parallel Equivalent Circuit

The quasi-balance bridge for determining the parallel equivalent conductance G_{cp} and capacitance C_{xp} is shown in Fig. 5.29a. The phasors of the currents $\underline{I}_{xp}$ and I_{xq} through the equivalent parallel conductance G_{cp} and capacitance C_{xp} respectively are shown in Fig. 5.29b. Here,

$$\mathrm{G_{cp}} = \frac{\mathrm{I_{xp}}}{\mathrm{V_x}} and \tag{5.68}$$

$$\omega \mathrm{C_{xp}} = \frac{\mathrm{I_{xq}}}{\mathrm{V_x}}. \tag{5.69}$$

From Fig. 5.30 that illustrates the phasors of the QBB we see that

$$\mathrm{G_{cp}} = \frac{\mathrm{I_{xp}}}{\mathrm{V_x}} = \frac{\mathrm{QB}}{\mathrm{R_S QA}} and \tag{5.70}$$

$$\omega \mathrm{C_{xp}} = \frac{\mathrm{I_{xq}}}{\mathrm{V_x}} = \frac{\mathrm{OB}}{\mathrm{R_S QA}}. \tag{5.71}$$

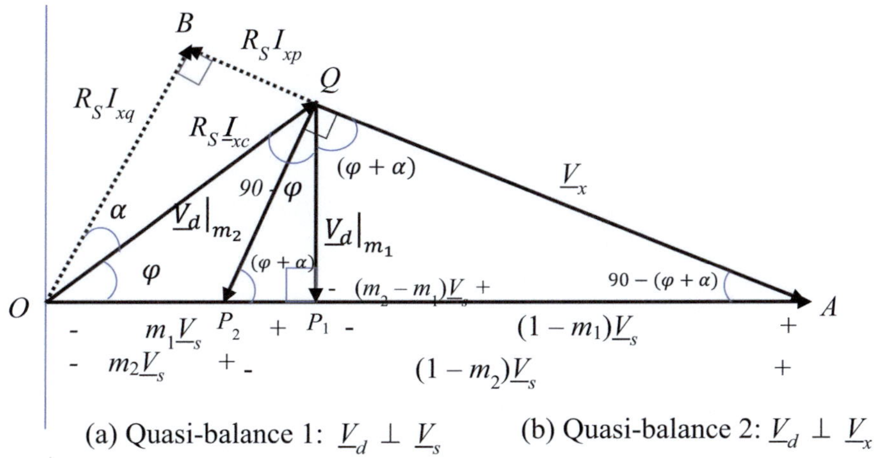

Fig. 5.30 Phasor diagram of quasi-balances for Capacitor (parallel equivalent circuit)

Here too we perform two quasi-balances. First, we obtain the quasi-balance such that $\underline{V}_d \perp \underline{V}_s$ with potentiometer variable point P at P_1 (with $OP_1 = m_1\underline{V}_s$) and the second quasi-balance is obtained with $\underline{V}_d \perp \underline{V}_x$ with potentiometer variable point P at P_2 (with $OP_2 = m_2\underline{V}_s$). We note that the right-angled triangles OBA and P_2QA in Fig. 5.30 are similar and hence:

$$\frac{AB}{AQ} = \frac{AO}{AP_2} \rightarrow \frac{V_x + R_S I_{xp}}{V_x} = \frac{V_s}{(1-m_2)V_s} \rightarrow \frac{R_S I_{xp}}{V_x} = \left(\frac{1}{(1-m_2)} - 1\right). \tag{5.72}$$

Simplifying Eq. (5.72), we get

$$G_{xp} = \frac{I_{xp}}{V_x} = \frac{m_2}{R_S(1-m_2)}. \tag{5.73}$$

From similar triangles OBA and P_2QA we also get

$$\frac{OB}{P_2Q} = \frac{AO}{AP_2} \rightarrow \frac{R_S I_q}{\underline{V}_d\big|_{m_2}} = \frac{V_s}{(1-m_2)V_s} \rightarrow I_q = \frac{\underline{V}_d\big|_{m_2}}{R_S(1-m_2)}. \tag{5.74}$$

In the right-angled triangles P_2P_1Q and AP_1Q we have $\angle Q\, P_2P_1 = \angle P_2QA = (\varphi + \alpha)$

$$\frac{QP_2}{QP_1} = \frac{QA}{AP_1} \rightarrow \frac{\underline{V}_d\big|_{m_2}}{\underline{V}_d\big|_{m_1}} = \frac{V_x}{(1-m_1)V_s} \rightarrow \underline{V}_d\big|_{m_2} = \frac{V_x\, \underline{V}_d\big|_{m_1}}{(1-m_1)V_s}. \tag{5.75}$$

We also have

$$\frac{QP_1}{P_2P_1} = \frac{AP_1}{QP_1} \rightarrow (QP_1)^2 = \left(\underline{V}_d\big|_{m_1}\right)^2 = (m_1 - m_2)(1 - m_1)V_s^2, \tag{5.76}$$

resulting in

$$\underline{V}_d\Big|_{m_1} = V_s\sqrt{(m_1 - m_2)(1 - m_1)}. \tag{5.77}$$

Substituting Eqs. (5.77) in (5.75), we get

$$\underline{V}_d\Big|_{m_2} = \frac{V_x\, V_s\sqrt{(m_1-m_2)(1-m_1)}}{(1-m_1)V_s}. \tag{5.78}$$

Substituting Eqs. (5.78) in (5.74), we get

$$I_q = \frac{V_x\sqrt{(m_2-m_1)(1-m_1)}}{R_S(1-m_2)(1-m_1)} \rightarrow \frac{I_q}{V_x} = \omega C_{xp} = \frac{1}{R_S(1-m_2)}\sqrt{\left(\frac{m_1-m_2}{1-m_1}\right)}. \tag{5.79}$$

From the above equation we see that with the standard resistance R_s and the two quasi-balance potentiometer values m_1 and m_2, we can determine the susceptance ωC_{xp}. Once again, the number of significant digits and the accuracy depends on the significant digits and accuracy in R_s, m_1 and m_2. Like the Q factor for inductor, the $Tan\ \delta$ of a capacitor is also an important characteristic to be determined. We can obtain this value using the following formula.

$$Tan\delta = \frac{1}{\omega C_{xp}\, R_{xp}} = m_2\sqrt{\left(\frac{1-m_1}{m_1-m_2}\right)}. \tag{5.80}$$

5.3.4.4 Implementation of a Quasi-Balance Bridge

Though quasi-balance bridge theory was developed in 1950, the practical implementation was a challenge as the quasi-balance requires the fine detection of two phasors being orthogonal to each other. The invention of the ubiquitous phase sensitive detector in 1970 resulted in the practical implementation of the quasi-balance bridge. A phase sensitive detector resolves a phasor $\underline{V}_i = (V_i\theta)$ either to the in-phase component $V_{ip}0°$ or quadrature component $V_{iq}90°$ with respect a reference, say $V_R0°$, as shown in Fig. 5.31. A phase sensitive detector that provides V_{ip} as the output with $V_i\theta$ and V_R0 as inputs is called the in-phase phase sensitive detector (IPSD) and a phase sensitive detector that provides the quadrature component V_{iq} with $V_i\theta$ and V_R0 as inputs is called the quadrature phase sensitive detector (QPSD). A phase sensitive detector (also popularly known as *synchronous detector* and *lock-in-amplifier*) can be implemented in two distinct forms, namely,

(a) The multiplier type PSD and
(b) Synchronous (switching) type PSD.

(a) **The Multiplier Type Phase Sensitive Detector**

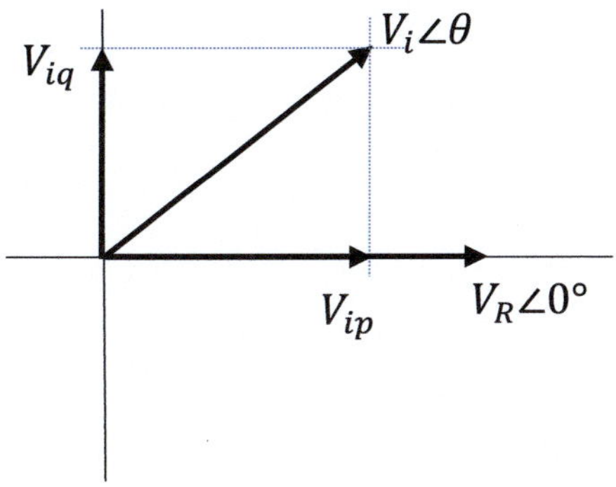

Fig. 5.31 Components of a phasor

Figure 5.32 shows the block schematic of a multiplier type phase sensitive detector. As the name indicates the method consists of a two-input analog multiplier that multiplies signals v_i (popularly known as input) and v_R (called the reference signal). The output v_m of the multiplier is

$$\mathrm{v_m} = \mathrm{v_i v_R}. \tag{5.81}$$

The output of the multiplier is fed as the input to an integrator, whose averaged output (the DC component of the output) is

$$\mathrm{v_o} = Avg(\mathrm{v_i v_R}) = \int \mathrm{v_i v_R} dt. \tag{5.82}$$

By choosing the appropriate reference phasor, we can implement the IPSD or QPSD as illustrated in Fig. 5.32a, b respectively.

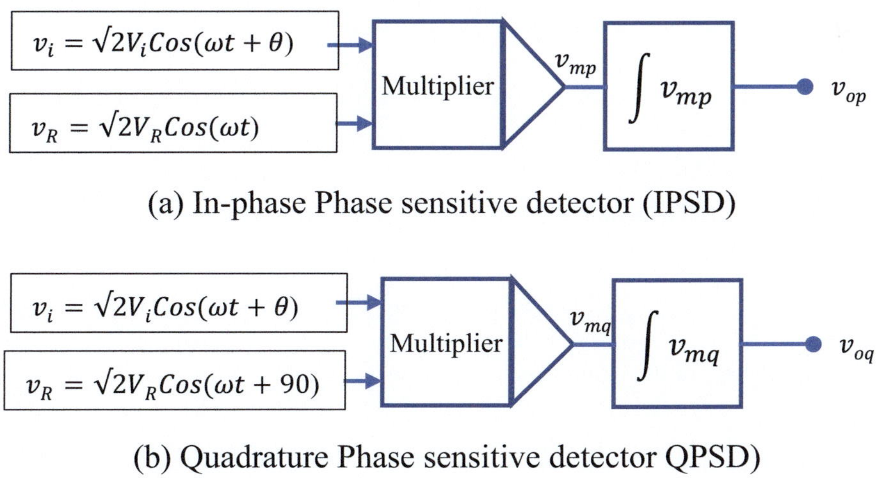

Fig. 5.32 Multiplier type Phase sensitive detector

(a) **Multiplier Type IPSD**

The multiplier type IPSD configuration is shown in Fig. 5.32a. Here:

$$\mathrm{v_i} = \sqrt{2}\mathrm{V_i}Cos(\omega t + \theta \; and \; \mathrm{v_R} = \sqrt{2}\mathrm{V_R}Cos(\omega t) \tag{5.83}$$

The output of the multiplier will then be

$$\mathrm{v_{mp}} = \mathrm{v_i v_R} = \sqrt{2}\mathrm{V_i}Cos(\omega t + \theta)\sqrt{2}\mathrm{V_R}Cos(\omega t). \tag{5.84}$$

The, the output v_{op} of the integrator is

$$\mathrm{v_{op}} = \frac{1}{\mathrm{T}}\int_0^{\mathrm{T}} \mathrm{v_i v_R dt} = \frac{1}{\mathrm{T}}\int_0^{\mathrm{T}} \sqrt{2}\mathrm{V_i}\mathrm{Cos}(\omega t + \theta)\sqrt{2}\mathrm{V_R}\mathrm{Cos}(\omega t)\mathrm{dt}. \tag{5.85}$$

Here,

$$T = \frac{1}{f} = \frac{2\pi}{\omega}.$$

Equation (5.85) can be simplified using the identity:

$$Cos(\mathrm{A})Cos(\mathrm{B}) = \frac{\mathrm{Cos(A-B)}}{2} + \frac{\mathrm{Cos(A+B)}}{2}. \tag{5.86}$$

Resulting in the output of the integrator as

$$\mathrm{v_{op}} = \frac{2\mathrm{V_i V_R}}{2\mathrm{T}}\int_0^{\mathrm{T}}[\mathrm{Cos}(\theta) + \mathrm{Cos}(2\omega t + 90^\circ)]\mathrm{dt}. \tag{5.87}$$

Evaluating the integral in Eq. (5.87) results in

$$\mathrm{v_{op}} = \mathrm{V_i V_R}Cos(\theta). \tag{5.88}$$

Multiplier Type QPSD

If we use $v_R = \sqrt{2}V_R\mathrm{Cos}(\omega t + 90^\circ)$ as indicated in Fig. 5.32b, the output v_{mq} of the multiplier is

$$\mathrm{v_{mq}} = \mathrm{v_i v_R} = \sqrt{2}\mathrm{V_i}Cos(\omega t + \theta)\sqrt{2}\mathrm{V_R}Cos(\omega t + 90^\circ). \tag{5.89}$$

The output v_{oq} of the integrator is

$$\mathrm{v_{oq}} = \frac{1}{\mathrm{T}}\int_0^{\mathrm{T}} \mathrm{v_i v_R dt} = \frac{1}{\mathrm{T}}\int_0^{\mathrm{T}} \sqrt{2}\mathrm{V_i}\mathrm{Cos}(\omega t + \theta)\sqrt{2}\mathrm{V_R}\mathrm{Cos}(\omega t + 90^\circ)\mathrm{dt}. \tag{5.90}$$

Equation (5.90) can be simplified as

$$v_{oq} = V_i V_R Cos(\theta - 90°) = V_i V_R Sin(\theta). \quad (5.91)$$

(b) **The Synchronous (Switching) Type Phase Sensitive Detector**

Like the multiplier type phase sensitive detector, the synchronous (switching) type sensitive detector also extracts the in-phase component v_{op} and the quadrature component v_{oq} of a phasor $V_i\theta$. Figure 5.33 shows the block schematic of a synchronous (switching) type phase sensitive detector. As the name indicates the synchronous (switching) type phase sensitive detector consists of a single pole double-throw (SPDT) analog switch, a zero-crossing detector, an inverting unity gain amplifier and an integrator. Here too by choosing the refence phasor as V_R0 we obtain an IPSD and with V_R90 we realize a QPSD.

Synchronous (Switching) type IPSD

The schematic of a synchronous (switching) type IPSD configuration is shown in Fig. 5.33a. Here depending on the polarity of v_R the SPDT selects v_i or $-v_i$.

If $v_R \geq 0$ then v_i is selected otherwise $-v_i$ is selected. Thus, the output of the switch v_{sp} can be mathematically represented as $v_{sp} = Sgn[v_R]$, with the $Sgn[v_R]$ function having a value +1 if $v_R \geq 0$ and −1 if $v_R < 0$. With $v_i = \sqrt{2}V_i\text{Cos}(\omega t + \theta)$ and $v_R = \sqrt{2}V_R\text{Cos}(\omega t)$, the output of the integrator is

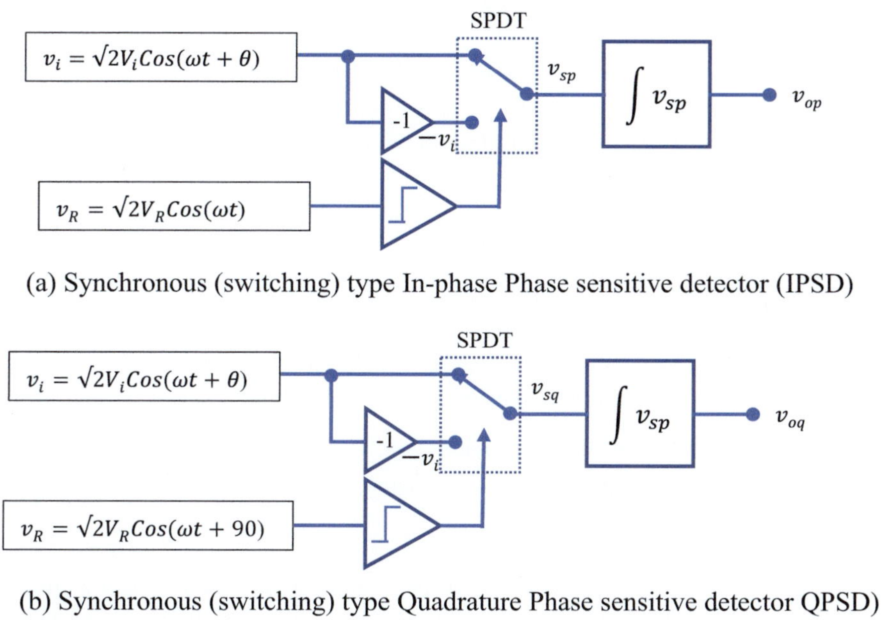

Fig. 5.33 Synchronous (switching) type Phase sensitive detector

$$\mathrm{v_{op}} = \frac{1}{\mathrm{T}}\int_{0}^{\mathrm{T}} \sqrt{2}\mathrm{V_i}\mathrm{Cos}(\omega t + \theta)\ \mathrm{Sgn}\left[\sqrt{2}\mathrm{V_R}\mathrm{Cos}(\omega t)\right] \mathrm{dt}. \tag{5.92}$$

Evaluating the integral in Eq. (5.92), we get

$$\mathrm{v_{op}} = \frac{\sqrt{2}\mathrm{V_i}}{\omega T}\left[\mathrm{Sin}(\omega t + \theta)]_{0}^{\frac{T}{4}} - \mathrm{Sin}(\omega t + \theta)]_{\frac{T}{4}}^{\frac{3T}{4}} + \mathrm{Sin}\,(\omega t + \theta)]_{\frac{3T}{4}}^{T}\right]. \tag{5.93}$$

Evaluating Eq. (5.93) and substituting $\omega T = 2\pi$, we get

$$\mathrm{v_{op}} = \frac{\sqrt{2}\mathrm{V_i}}{2\pi}[-\mathrm{Sin}(\theta) + \mathrm{Cos}(\theta) + \mathrm{Cos}(\theta) + \mathrm{Cos}(\theta) + \mathrm{Cos}(\theta) + \mathrm{Sin}(\theta)]. \tag{5.94}$$

$$\mathrm{v_{op}} = \frac{2\sqrt{2}\mathrm{V_i}}{\pi} Cos(\theta). \tag{5.95}$$

Synchronous (Switching) Type QPSD

The schematic of a synchronous (switching) type IPSD configuration is shown in Fig. 5.33b. With $v_i = \sqrt{2}V_i\mathrm{Cos}(\omega t + \theta)$ and $v_R = \sqrt{2}V_R\mathrm{Cos}(\omega t + 90)$, the output of the integrator is

$$\mathrm{v_{oq}} = \frac{1}{\mathrm{T}}\int_{0}^{\mathrm{T}} \sqrt{2}\mathrm{V_i}\mathrm{Cos}(\omega t + \theta)\ \mathrm{Sgn}\left[\sqrt{2}\mathrm{V_R}\mathrm{Cos}(\omega t + 90^\circ)\right] \mathrm{dt}. \tag{5.96}$$

Evaluating the integral in Eq. (5.96), we get

$$\mathrm{v_{oq}} = \frac{\sqrt{2}\mathrm{V_i}}{\omega T}\left[-\mathrm{Sin}(\omega t + \theta)]_{0}^{\frac{T}{2}} + \mathrm{Sin}(\omega t + \theta)]_{\frac{T}{2}}^{T}\right]. \tag{5.97}$$

Evaluating Eq. (5.97) and substituting $\omega T = 2\pi$, we get

$$\mathrm{v_{oq}} = \frac{\sqrt{2}\mathrm{V_i}}{2\pi}[\mathrm{Sin}(\theta) + \mathrm{Sin}(\theta) + \mathrm{Sin}(\theta) + \mathrm{Sin}(\theta)] = \frac{2\sqrt{2}\mathrm{V_i}}{\pi} Sin(\theta). \tag{5.98}$$

Of the two (multiplier/synchronous) phase sensitive detectors, the synchronous (switching) type phase sensitive detectors are more popular than the multiplier type. Semiconductor integrated circuits that implement the circuit schematic of a PSD shown in Fig. 5.33a are available. An example is the AD630 from the Analog Devices. For "synchronous detection and synchronous demodulation", "*Lock-in amplifier*" employs both IPSD and QPSD and provides both v_{op} and v_{oq} as well as $\sqrt{\left(v_{op}^2 + v_{oq}^2\right)}$ are designed and are now commercially available.

PSD for quasi-balance

For implementing the quasi-balance to measure inductance all we need is a simple IPSD. To get the first quasi-balance, we feed the phasor $\underline{V}_d$ as input and phasor $\underline{V}_s$ as reference. We adjust the point *P* till we get a zero at the output of the IPSD. Since we know that neither $\underline{V}_d$ or $\underline{V}_s$ is zero, then the output v_{op} of IPSD can only be zero when $Cos(\theta) = 0°$ and hence $\theta = 90°$.

Thus, for the Measurement on Inductors: To obtain the series equivalent resistance R_L and inductance L_L of an inductor, we use the quasi-balance bridge of Fig. 5.25 and connect the phasor $\underline{V}_d$ reference to an IPSD of Fig. 5.33a as shown in Fig. 5.34. First the phasor $\underline{V}_s$ is selected as input by setting the single pole double throw (SPDT) switch in position "0", adjust P to get zero (on the galvanometer) on the output v_{op} of the IPSD, get m_1. Once the first quasi-balance is obtained, SPDT is set in position "1", thus select $\underline{V}_{Rs}$ as the input to the IPSD. Once again, we obtain the second quasi-balance by adjusting point P till the output of IPSD to be zero and get m_2. Once m_1 and m_2 are obtained the values of R_L, L_L, and Q_L are computed using Eqs. (5.62), (5.63), and (5.64) respectively.

For the Measurement on Capacitors

To obtain the parallel equivalent resistance conductance G_{cp} and Capacitance inductance C_{cp} of capacitor, we use the quasi-balance bridge of Fig. 5.29 and connect the phasor $\underline{V}_d$ reference to an IPSD of Fig. 5.33a as shown in Fig. 5.35. First the phasor $\underline{V}_s$ is selected as input by setting the single pole double throw (SPDT) switch in position "0", adjust P to get zero (on the galvanometer) on the output V_{op} of the IPSD, get m_1. Once the first quasi-balance is obtained, SPDT is set in position "1", thus select $\underline{V}_x$ as the input to the IPSD. Once again, we obtain the second quasi-balance by adjusting point P till the output of IPSD to be zero and get m_2. Once m_1 and m_2 are obtained the values of G_{cp}, C_{cp}, and $Tan\ \delta$ are computed using Eqs. (5.73), (5.80), and (5.81) respectively.

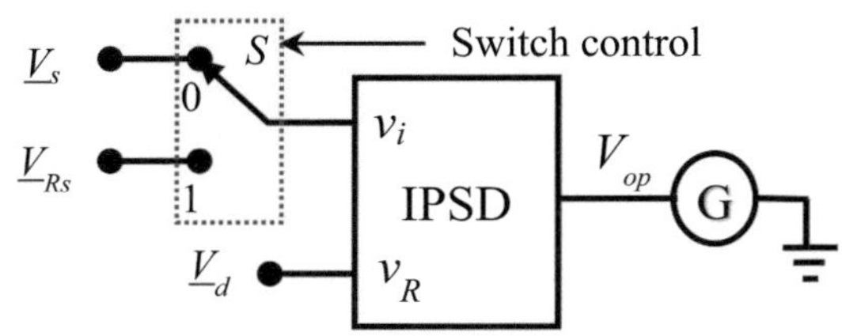

Fig. 5.34 Use of IPSD for quasi-balance detection for inductance

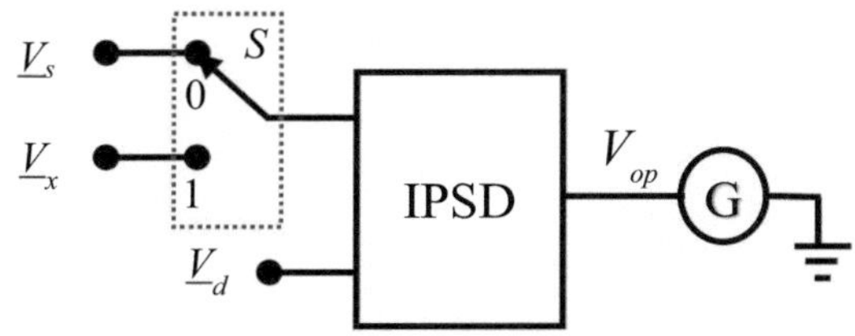

Fig. 5.35 Use of IPSD for quasi-balance detection for capacitor

5.4 The Multi-decade Transformer Ratio Device

In measurements using the transformer ratio bridge as well as for the quasi-balance bridge, we need the ratio of the output to input to be varied with several significant digits. We can obtain the variation using a potentiometer. It is also possible to obtain such a variation using a transformer. Such an arrangement is popularly called the inductive voltage divider (IVD). A simplified schematic of a five-decade ratio-adjustable transformer (also called the Inductive voltage divider or simply IVD) is shown in Fig. 5.36. Here each stage is identical to every other stage. The secondary taps are at steps of (1/10)th of the primary. Hence if the first stage is excited with a 10 V (rms) source then the first stage outputs will vary in steps of one volt, the second stage taps will be 0.1 V per step, the third stage 0.01 V and so on. However, the winding resistance and most importantly the excitation current drawn by the primary winding in each stage will introduce errors in the division (will not be exactly equal to (1/10)th of the applied voltage in each sub section of the secondary). To reduce this error:

(i) The windings are made on toroidal core of extremely high permeability μ_r thus reducing the magnetizing current and
(ii) A passive feedback scheme as indicated in Fig. 5.37 is used. This shifts the magnetizing current away from the primary winding P_{11}–P_{12} to an auxiliary winding P_{21}–P_{22}. As seen in Fig. 5.37, windings (P_{11}–P_{12}) and (P_{21}–P_{22}) are identical ($10N_1$ turns) and are paralleled and excited. Now winding (P_{11}–P_{12}) and the secondary (with taps at every N_1 turns) encompass both core 1 and core 2, while winding (P_{21}–P_{22}) encompasses core 2 alone. If we assume that the flux in core 1 is φ_1 and that of core 2 is φ_2 then the induced emf's e_{p1} across P_{11} and P_{12} and e_{p2} across P_{21} and P_{22} will be

$$e_{p1} = 10N_1 \frac{d(\varphi_1+\varphi_2)}{dt} \tag{5.99}$$

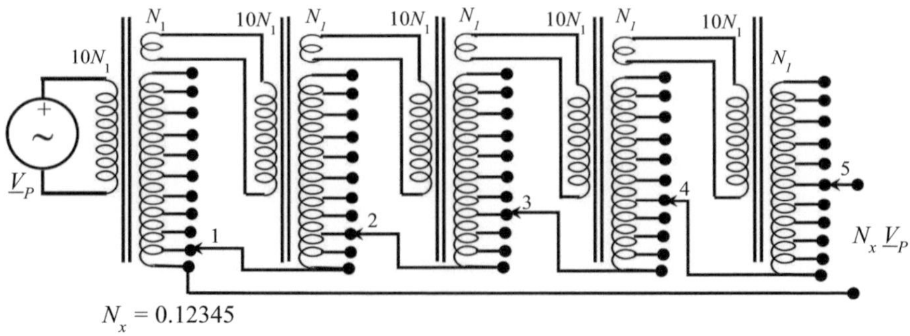

Fig. 5.36 Five decade transformer ratio device

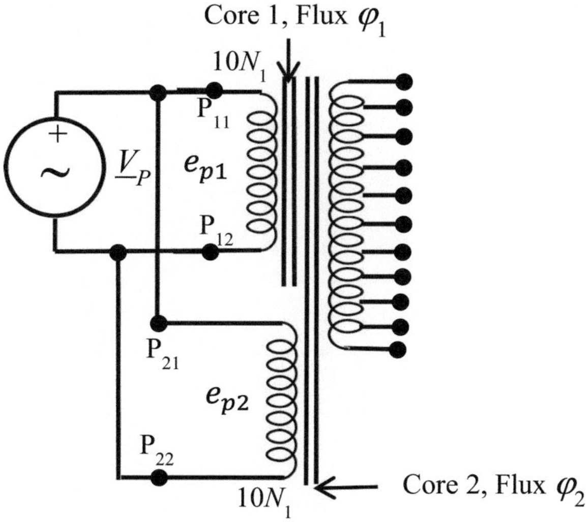

Fig. 5.37 Two-core, two-primary inductive voltage divider

$$e_{p2} = 10N_1 \frac{d\varphi_2}{dt}. \quad (5.100)$$

Since the windings (P_{11}–P_{12}) and (P_{21}–P_{22}) are paralleled, we have

$$e_{p1} = 10\,N_1 \frac{d(\varphi_1+\varphi_2)}{dt} = e_{p2} = 10\,N_1 \frac{d\varphi_2}{dt}. \quad (5.101)$$

Equation (5.101) indicates that

$$\frac{d(\varphi_1 + \varphi_2)}{dt} = \frac{d\varphi_2}{dt}.$$

This is possible if and only if,

$$\frac{d\varphi_1}{dt} = 0.$$

Hence core 1 will have zero AC flux and winding (P_{11}–P_{12}) will have zero excitation current. Thus, the secondary winding sub sections will have on open circuit, 1/10th of the primary applied voltage. However, in practice, the windings (P_{11}–P_{12}) and (P_{21}–P_{22}) may not be identical and that will introduce a small error.

(iii) It is also possible to use additional active feedback compensation as shown in Fig. 5.38. In Fig. 5.38, the induced emf e_d in the detector winding of N_d turns is:

$$e_d = N_d \frac{d\varphi_1}{dt}. \quad (5.102)$$

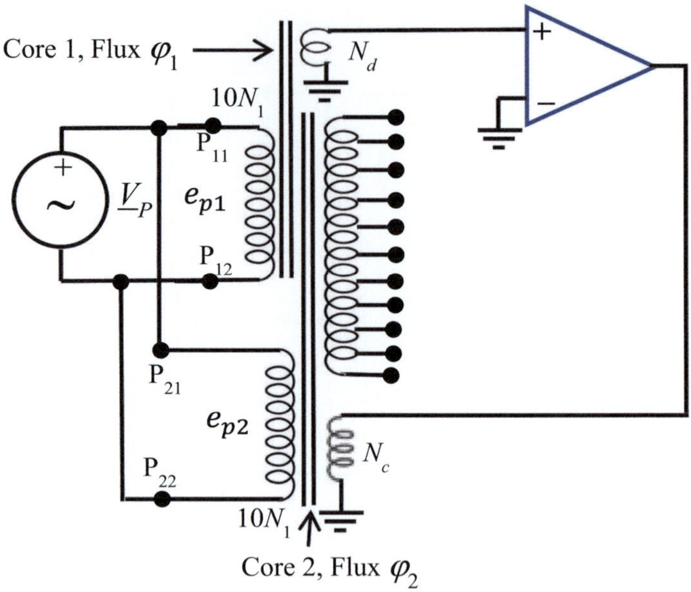

Fig. 5.38 Two-primary inductive voltage divider with active electronic error compensation

If we assume that the opamp in Fig. 5.38 is ideal, then the voltage e_d at its input will be zero. Or in other words:

$$e_d = N_{\mathrm{d}} \frac{d\varphi_1}{dt} = 0.$$

Since N_d is nonzero, $\frac{d\varphi_1}{dt} = 0$. The opamp OA will drive a current through the compensating winding of N_c turns to ensure that the AC flux φ_2 in core 2 is increased to such a level that flux φ_1 in core 1 is nearly zero. Figure 5.39 shows a commercial IVD. It should be noted here that the schemes of Figs. 5.37 and 5.38 depict only one stage of Fig. 5.36. With the electronic compensation scheme of Fig. 5.38, it is possible to design and fabricate up to a 9-decade variable ratio transformer (also known as Inductive Voltage Divider, IVD) of extremely low errors. An IVD is distinctly different from a current comparator. While the windings in an IVD sustain a voltage, the windings in a current comparator carry a current. More importantly, under balance the flux in the core of a current comparator will be zero.

5.5 Problems and Exercises

E5.1. A resistance of approximately 3000 Ω is needed to balance a Wheatstone bridge. It is obtained using a five dialed box (steps of 1000 Ω, 100 Ω, 10 Ω, 1 Ω, and 0.1 Ω) guaranteed to 0.1%. For this accuracy, how many of these dials would it be worth adjusting?

E5.2. A coil of copper wire is measured on a Wheatstone bridge and specified by the operator to have a resistance of 255.2 Ω. The measurement was made at a temperature of about 20 °C. To what degree of exactness must the temperature be known if all figures in the resistance value are to be regarded as significant? ($\alpha = 0.00393$/°C for copper at 20 °C).

E5.3. The ratio arms of a Kelvin double bridge are 1000 Ω each. The standard resistance is set at 0.1000 Ω; the unknown resistance is 0.1002 Ω. The galvanometer has a resistance of 500 Ω, and a sensitivity of 0.005 μA/mm. A current of 10 A is passed through S and X from a 2.2 V battery in series with a rheostat. The resistance of the yoke may be considered negligible.

(a) What galvanometer deflection is produced for this condition?
(b) From the results of (a), what is the resistance unbalance to give a galvanometer deflection of 1 mm?
(c) The ratio arms of a Kelvin double bridge are 1000 Ω each. The galvanometer has a resistance of 100 Ω, and a sensitivity of 0.003 μA/mm. A current of 10 A is passed through bridge from a 2.2 V battery in series with a rheostat. The resistance of the yoke is negligible. The standard resistance is set at 0.1000 Ω, The galvanometer gives a deflection of 25 mm. By how much does the unknown differ in resistance from the standard?

E5.4. A Kelvin double bridge has 2 ratio arms available of each of following values, 100 Ω, 500 Ω, 1000 Ω, 10,000 Ω, guaranteed to ± 0.05%. The adjustable standard has 9 steps of 0.001 Ω each (±0.05%), and a calibrated bar of 0.0011 Ω with a 110 division scale (any setting guaranteed to ± 2 μΩ). A DC ammeter shunt, rated 50 mV, 200 A full scale is to be measured on this bridge. (a) Select the ratio arms you would use (b) Make a complete circuit diagram, showing all equipment needed for the test. Assume that a 2.2 V storage battery is available.

E5.5. For the settings you have selected, find the possible uncertainty in X caused by the uncertainty in the standard, and expressed a percentage of X. What additional error may be caused by errors in the ratio arms? (Consider that the 2 arms of each value are matched to a negligible difference).

E5.6. A bridge has the following constants: arm AB, R = 800 Ω in parallel with C = 0.159 μF; BC, R = 1000 Ω; CD, R = 500 Ω; DA, C = 0.636 μF in series with an unknown resistance. Find the frequency for which the bridge is in balance and the value of R in DA to produce balance.

E5.7. A bridge has in arm AB, R = 800 Ω in parallel with C = 0.5 μF; BC, R = 400 Ω in series with C = 1 μF; CD, R = 1000 Ω; DA, Pure R of unknown value.

(a) Find the frequency for which this bridge is in balance. (b) Find the resistance required in DA to produce this balance.

Chapter 6
Voltage and Current Scaling

"All I have done is to try experiments in as vast a scale as I could".
—Mahatma Gandhi

6.1 Instrument Transformers

To meet today's electricity demands, the present-day power systems must handle hundreds of, sometimes even thousands of megawatts. To cater to these levels of power, the system voltage and current levels must also increase. Typically, voltage levels as high as 400 kV and currents of several hundred to several thousand amperes are needed. This in turn necessitates suitable devices to measure high voltages and large currents. To ensure safety of the personnel and practicality of measuring instruments, the high voltages and high currents must be scaled down (stepped down) using *scaling devices*. The popular scaling devices are the *Current Transformer* (acronym CT) for scaling the current and the *Voltage Transformer* also called *Potential Transformer*, (acronym VT or PT) for scaling the voltage. Combined, the current and voltage transformers are called "*Instrument Transformers*" and are indispensable in

V. Jagadeesh Kumar, *Principles of Electrical Measurement and Measuring Instruments*,
https://doi.org/10.1007/978-981-96-5475-8_6

today's power system infrastructure. Additionally, a phasor measurement unit (PMU) has also become an essential instrument for the measurement of relevant parameters on an electrical power system.

6.2 The Current Transformer

We learnt that an ammeter's range can be extended (increased) by simply adding a shunt resistance R_{sh} across the ammeter. For example, let us say that we have a 1 mA full-scale DC ammeter having an internal resistance of 30 Ω. Then we can extend the range (full-scale) to 1 A DC by adding a shunt resistance as shown in Fig. 6.1a. The value of the shunt must be such that when one ampere flows into the parallel circuit of the shunt and the ammeter, one milliampere flows through the meter and the rest 999 mA flows through the shunt resistor. Hence the value of the shunt resistance

$$R_{sh} = \frac{30\ \text{mV}}{999\ \text{mA}} = 30.03\ \text{m}\Omega. \tag{6.1}$$

The power dissipated in the shunt resistor is 30 mW. Now if we try and extend the range of this 1 mA full scale meter to 300 A full-scale as indicated in Fig. 6.1b, then the value of the shunt resistor is

$$R_{sh} = \frac{30\ \text{mV}}{299.999\ \text{A}} = 0.1\ \text{m}\Omega. \tag{6.2}$$

The power rating of the required 1 mΩ shunt to extend the current measurement range of the 1 mA full-scale ammeter to 300 A full-scale (is $299.999^2 * 0.0001$) $= 9$ W. Such a shunt resistor will be bulky and needs to be placed outside of the ammeter as shown in Fig. 6.2. Thus, extending the range of an ammeter using shunt resistance has limitations. Extending the range of ammeters for AC currents will be more complex as the inductance of the ammeter also needs to be considered. Since power systems operate with sinusoidal voltages and currents at the frequency of 50 Hz or 60 Hz, we can use a transformer to step down the 300 A current to 1 A and measure the 1 A current using a simple 1 A full-scale ammeter as shown in Fig. 6.3.

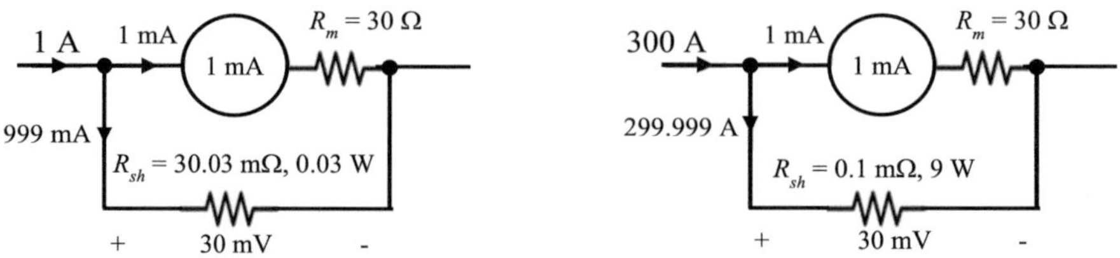

(a) Shunt for 1 A range extension

(b) Shunt for 300 A range extension

Fig. 6.1 Extending the full-scale of an ammeter

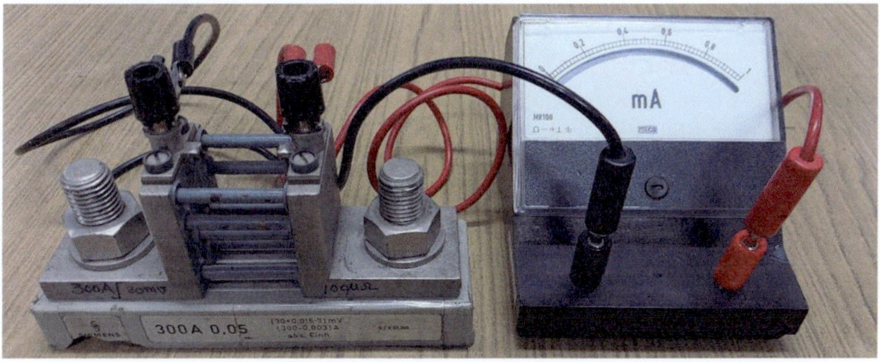

Fig. 6.2 Shunt (0.1 mΩ) for 300 A full-scale ammeter

Fig. 6.3 Ideal current transformer

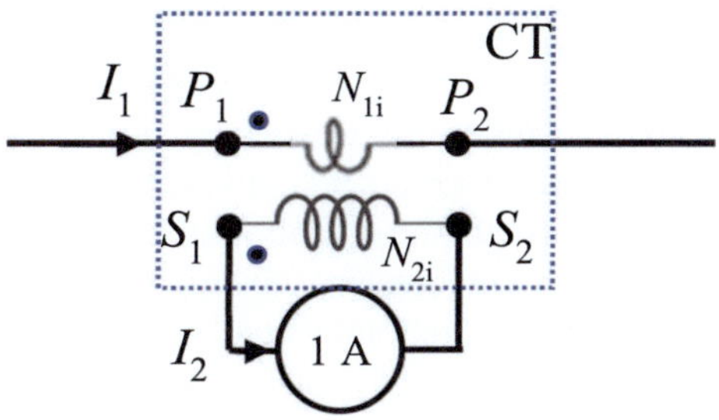

Let the current transformer's primary winding P_1:P_2 be wound with N_{1i} turns and the secondary winding S_1:S_2 with N_{2i} turns. If the current transformer (CT) of Fig. 6.3 is assumed to be ideal, then we ha

$$\underline{I}_1 N_{1i} = \underline{I}_2 N_{2i}. \tag{6.3}$$

Equation (6.3) is popularly known as "*ampere turn balance*", that is in an ideal current transformer, the secondary ampere turns are balanced by equal primary ampere turns. Rearranging Eq. (6.3) we get

$$\underline{I}_1 = \left(\frac{N_{2i}}{N_{1i}}\right)\underline{I}_2. \tag{6.4}$$

In Eq. (6.4), $\underline{I}_1$ is the primary current phasor, $\underline{I}_2$ is the phasor of the secondary current. Normally we design a current transformer by setting the maximum value I_{1FS} of $\underline{I}_1$ and the corresponding maximum value of $\underline{I}_2$ is I_{2FS}. Then the "*nominal ratio*" K_N of the current transformer is defined as

$$K_N = \left[\frac{I_{1FS}}{I_{2FS}}\right]_{\text{By design}}. \tag{6.5}$$

For example, we may design a current transformer to step down the primary current in the range of 0 A to 100 A to the secondary current in the range of 0 A to 1 A. Then the nominal ratio of the current transformer is (100/1) = 100. It may be noted that in practice K_N is the constant we use to multiply the current I_2 read on the ammeter to get the value of the primary current I_1. In short, we call this CT as a 100/1 A (to be read as hundred bar one ampere) current transformer, with a nominal ratio K_N of 100.

6.2.1 The Practical Current Transformer

When we manufacture a practical current transformer, the primary and secondary windings are wound with a wire (normally insulated copper wire) that possesses a nonzero resistivity (resistivity ρ of copper is 168 μΩ m). Hence, the primary and secondary windings (coils) will possess *winding resistances* R_1 and R_2, respectively. To obtain the condition that all the flux generated by the primary winding is coupled to the secondary winding (so that coupling coefficient $\approx$1), the primary and secondary windings are wound on a soft ferromagnetic core possessing high permeability. To establish a flux Ø in the soft ferromagnetic core, the primary winding will draw a magnetizing current $\underline{I}_m$. Once we have the flux Ø in the core, hysteresis and eddy current power losses will occur in the soft ferromagnetic core. The combined hysteresis and eddy current losses are called the *iron losses* of the core. To cater to the iron losses, the primary winding will also draw a current $\underline{I}_w$ to supply the power required to sustain the losses in the core.

Figure 6.4 shows the equivalent circuit of a practical current transformer that considers all the non-idealities described above. In Fig. 6.4, R_1 is the resistance of the primary winding, R_2 is the resistance of the secondary winding, R_i is the equivalent resistance that represents the iron losses in the soft ferromagnetic core, L_m is the equivalent inductance that represents the flux generated in the core, L_1 is the equivalent inductance that represents the leakage flux on the primary side (flux generated by $\underline{I}_1$ that is not coupled to the secondary), and L_2 represents the leakage flux of the secondary winding (that is flux generated by $\underline{I}_2$ that is not coupled to the primary). With these non-idealities, the use of a current transformer to scale (mostly step down) a current will, in practice, introduce errors in the scaling factor and hence in the measurement of the primary current. The impedance $\boldsymbol{Z}_B$ in Fig. 6.4 is the internal impedance of the ammeter A_m and is called the burden connected to the current transformer. Normally, we term the impedance connected on the secondary side of a transformer as "*load on the transformer*". However, use of the same terminology, viz. *load*, for the impedance connected on the secondary side of a current transformer (or a voltage transformer) will create confusion as there is an actual load in the main circuit carrying the primary current $\underline{I}_1$. . Therefore, the impedance connected on the

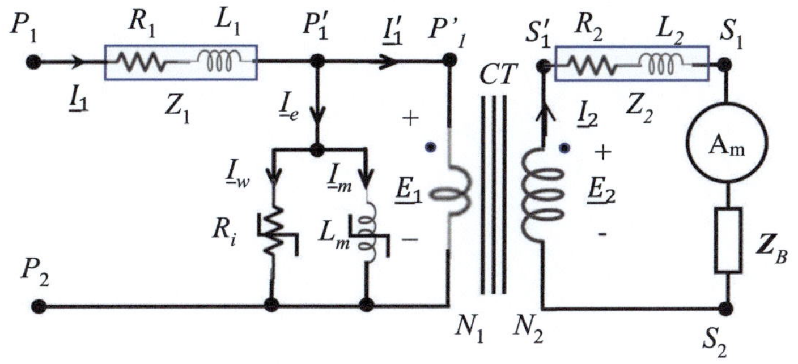

Fig. 6.4 Equivalent circuit of a practical current transformer

secondary side of a current transformer (or the voltage transformer) is called by a different name, *burden* and the subscript "*B*" is used. It should be noted that L_m and R_i vary nonlinearly with the flux Ø in the core. Hence, $\underline{I}_m$ and $\underline{I}_w$ vary nonlinearly with respect to the primary current $\underline{I}_1$. Moreover, $\underline{I}_m$ and $\underline{I}_w$ will also be non-sinusoidal (will contain harmonics), mainly due to the nonlinear *B*-*H* characteristics of the core. In Fig. 6.4, the quantity $\underline{I}_1'$ is the reflected secondary current $\underline{I}_2$ on the primary side. Once we have identified and segregated the non-idealities, then the transformer CT (Primary terminals P_1' : P_2 and secondary terminals S_1' : S_2) in Fig. 6.4 is an ideal one. Hence, we have

$$\underline{I}_1' = \frac{N_2}{N_1}\underline{I}_2 = K_t\underline{I}_2. \tag{6.6}$$

where

$$K_t = \tfrac{N_2}{N_1}. \tag{6.7}$$

Here, K_t is called the turns ratio. In most designs $K_t = K_N$ and we will assume this to be the case for now. Normally, for drawing the phasor diagram of a transformer's currents and voltages, the flux $Ø_c$ in the core is chosen as the reference phasor $\underline{Ø}_c$. The secondary induced emf $\underline{E}_2$ will then lead the flux by 90° and the secondary current $\underline{I}_2$ will lag the secondary induced emf (for inductive burden) by an angle δ. Similarly, on the primary side, the primary induced emf leads the core flux by 90° and the primary current will lag the primary induced emf for an inductive burden by an angle α. **Thus, all the phasors of the primary and secondary currents and voltages will occupy one quadrant, namely, the first quadrant**. Such a phasor diagram will be cluttered and will confuse the reader as it will then be very difficult to visually segregate the primary and secondary quantities. ***In order to enhance the visualization,*** **in the phasor diagram of a current transformer shown in** Fig. 6.5, ***the secondary quantities are shown in quadrant 3*** **(*that is: the secondary voltage and current phasors are rotated by*** **180°)**. Here, the burden is assumed to be inductive.

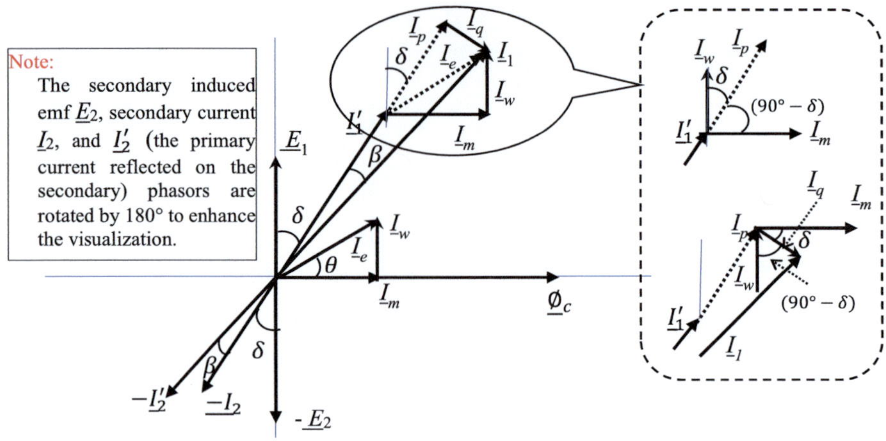

Fig. 6.5 Current transformer: phasors (lagging power factor burden)

In most practical cases this is true, as practical ammeters will possess resistance and inductance. On the primary side, the excitation current $\underline{I}_e$ will be

$$\underline{I}_e = \left(\underline{I}_w + \underline{I}_m\right). \tag{6.8}$$

From Fig. 6.4, we see that

$$\underline{I}'_1 = \underline{I}_1 - \underline{I}. \tag{6.9}$$

Thus, the excitation current $\underline{I}_e$ is taken off the primary current $\underline{I}_1$ and thsse balance current $\underline{I}'_1$ is transformed to the secondary as $\underline{I}_2$ as seen in the phasor diagram shown in Fig. 6.5. If the current transformer is ideal, then we would have obtained $\underline{I}'_2$, which is the reflected primary current $\underline{I}_1$ on the secondary side as

$$\underline{I}'_2 = \frac{\underline{I}_1}{K_N}. \tag{6.10}$$

It must be noted that the phasor diagram of Fig. 6.5 **is not drawn to scale**. Compared to $\underline{I}_1$ and $\underline{I}'_1$, the current phasors $\underline{I}_p$, $\underline{I}_q$, $\underline{I}_e$, $\underline{I}_m$, and $\underline{I}_w$ will be very small (a few percentage points of $\underline{I}_1$ or $\underline{I}'_1$) and thus β will also be very small. Hence, the phasor diagram in Fig. 6.5 is for illustration only and does not represent the magnitudes or phase in appropriate proportion.

6.2.2 *Errors of a Current Transformer*

In current-transformer usage, we take the primary current to be K_N times the measured value of the secondary current I_2. Here, I_2 is the rms value (magnitude) of the phasor current $\underline{I}_2$. We see from Fig. 6.5 that this result viz. $I_1' = K_N I_2$ differs from the actual value I_1 of the primary current. Again, I_1 and I_1' are the magnitudes of phasors $\underline{I}_1$ and $\underline{I}_1'$. Therefore, we have an error in the measurement, equal to the measured value I_1' *minus* the actual value I_1. This error is called the "*ratio error*". The reason that this error is named "*the ratio error*" is that the actual ratio K_A of the two currents,

$$K_A = \frac{I_1}{I_2}. \tag{6.11}$$

differs from the nominal ratio K_N used in the measurement process. We also see that the phase of the secondary current phasor $\underline{I}_2$ deviates by a small angle β from that of the actual primary current $\underline{I}_1$ reflected on the secondary, namely $\underline{I}_2'$ as shown in Fig. 6.5. Here, β is called *the phase error* (some national standards call this β as *phase displacement*), normally expressed in terms of minutes (1/60th of a degree) or in centiradians. Thus, a practical current transformer possesses a ratio error and phase error. While ratio error affects the measurement of the current, the phase error will affect the measurement of power and energy in a power system when power and energy are measured using current and voltage transformers in conjunction with a wattmeter and energy meter respectively.

6.2.2.1 Ratio Error of a Current Transformer

As already mentioned, we determine the primary current I_1 by multiplying the measured secondary current I_2 by the nominal ratio K_N of the CT. For example, if we use a 100/1 A CT and measure the secondary current as 0.56 A, then we take the primary current to be (100*0.56) = 56 A. Thus, the measured value of the primary current $I_{1m} = I_1' = K_N I_2$ and the actual (true) value is $I_1 = K_A I_2$, where

$$K_N = \frac{I_{1m}}{I_2}. \tag{6.12}$$

is the actual ratio of the primary current to the secondary current. Hence, the error in the ratio of a practical current transformer can be expressed as

$$\text{Error in the ratio} = \frac{K_N - K_A}{K_A} 100\%. \tag{6.13}$$

The term that is most popular is "*Ratio Error*" and not the "*Error in the Ratio*". Substituting the values of K_A and K_N from Eqs. (6.11) and (6.12), we get the ratio error in terms of I_1 and I_2 as

$$\text{Ratio error } = \frac{I_{1m} - I_1}{I_1} 100\%. \quad (6.14)$$

It should be noted that in the present discussion we consider only the rms values of the currents, i.e., the magnitudes of the respective phasors and not the phasors themselves. We now drop a perpendicular from the phasor $\underline{I}_1$ to $\underline{I}_1'$ as shown in Fig. 6.5 resulting in the phasors $\underline{I}_p$ and $\underline{I}_q$. From the phasor diagram of Fig. 6.5 and noting that I_q is very small compared with $(I_1' + I_p)$, we see that

$$I_1 \approx (I_1' + I_p)$$

Furthermore, we know from the property of an ideal transformer that $I_1' = K_t I_2$. Substituting these values in Eq. (6.14) we get

$$\text{Ratio error } = \frac{K_N I_2 - (K_t I_2 + I_p)}{(K_t I_2 + I_p)} 100\%. \quad (6.15)$$

Rearranging Eq. (6.15) we get

$$\text{Ratio error } = \left(\frac{(K_N - K_t) I_2}{(K_t I_2 + I_p)} - \frac{I_p}{(K_t I_2 + I_p)} \right) 100\%. \quad (6.16)$$

In a practical design, we must make the ratio error as small as possible and hence $I_p \ll K_t I_2$. With this assumption, we can simplify Eq. (6.16) as

$$\text{Ratio error } \approx \left(\left(\tfrac{K_N - K_t}{K_t} \right) - \tfrac{I_p}{I_1} \right) 100\%. \quad (6.17)$$

From the phasor diagram shown in Fig. 6.5, (see the insets), we can deduce

$$I_p = I_w Cos\delta + I_m Sin\delta, \quad (6.18)$$

and

$$I_q = I_m Cos\delta - I_w Sin\delta. \quad (6.19)$$

In Eqs. (6.18) and (6.19), δ is the secondary power factor angle. Substituting Eq. (6.18) in Eq. (6.17), we get

$$\text{Ratio error } = \left(\left(\frac{K_N - K_t}{K_t} \right) - \frac{I_w Cos\delta + I_m Sin\delta}{I_1} \right) 100\%. \quad (6.20)$$

From Eq. (6.20), we see that if we make $K_t < K_N$, then the first term becomes positive, and thus it is possible in design to obtain

$$\left(\frac{K_N - K_t}{K_t}\right) - \frac{I_w Cos\delta + I_m Sin\delta}{I_1} = 0,$$

resulting in zero ratio error in the current transformer. Such a current transformer with K_t intentionally made to be different from (less than) K_N is called the "*Turns ratio compensated current transformer*". Alternatively, the term *turns compensated current transformer* is also used. However, such a design to be implemented practically is fraught with problems as I_w and I_m do not vary in proportion to I_1. Hence "turns compensation" is generally used to "trim the ratio error" to a very small extent and that too at a particular value of the primary current.

6.2.2.2 Ratio Correction Factor

Some national standards (especially ANSI, the United States of America's national standard) specify what is known as "*Ratio Correction Factor*" (RCF) instead of percentage ratio error. The RCF is defined as "*the factor by which the nominal ratio* K_N (as marked on the nameplate) *of a current transformer must be multiplied to obtain the actual ratio* K_A''.

$$\text{Ratio Correction Factor}(RCF) = \left(\frac{K_A}{K_N}\right). \tag{6.21}$$

In Eq. (6.21), we already know that K_N is the nominal ratio of a current transformer and K_A is the actual ratio of the current transformer. From the definition of the percentage ratio error as in Eq. (6.13) we get

$$\text{Ratio error} = \left(\frac{K_N - K_A}{K_A}\right)100 = \left(\frac{K_N}{K_A} - 1\right)100 = \left(\frac{1}{RCF} - 1\right)100\% \tag{6.22}$$

We can rearrange Eq. (6.22) and get

$$RCF = \left(\frac{100}{100 + \%RE}\right). \tag{6.23}$$

Note: The RCF should not be confused with the *reading correction factor* and *Transformer correction factor* (TCF). These are not part of any standard but are used in practice. The reading correction factor is the factor that when multiplied with the reading obtained on current transformer ($=K_N I_2$) will give the actual primary current (that is without any error). Similarly, the transformer correction factor is that factor that needs to be multiplied with the wattmeter reading obtained using current and potential transformers to get the correct system power. Whether it is RCF or reading/transformer correction factor, the factor is not a constant but varies with the operating current/voltage/power. In general, ratio error and phase error (or RCF or

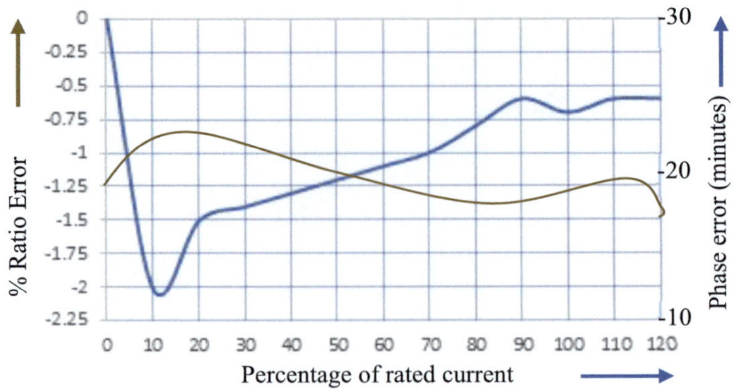

Fig. 6.6 Ratio error and phase errors of a current transformer

reading/transformer correction factor) are normally shown by curves as shown in Fig. 6.6.

6.2.2.3 Phase Error of a Current Transformer

In Fig. 6.5, we see that the reflected secondary current $\underline{I}_1'$ is not only not equal in magnitude to the primary current, but also has a phase shift of value β with I_1. Hence a current transformer also introduces a phase error. From Fig. 6.5, we can deduce that

$$Tan\beta = \frac{I_q}{I_1' + I_p} = \frac{I_q}{I_1}. \tag{6.24}$$

and we have

$$I_q = (I_m Cos\delta - I_w Sin\delta). \tag{6.25}$$

From Eqs. (6.24) and (6.25), we get

$$\text{Phase error}\beta = Tan^{-1}\left(\frac{I_q}{I_1}\right) = Tan^{-1}\left(\frac{I_m Cos\delta - I_w Sin\delta}{I_1}\right). \tag{6.26}$$

As per the international system of units and standards (SI), the phase error is expressed using the unit centiradians (crad). However, traditionally the phase error was expressed using the unit *minutes*. Today, we can use either centiradians or minutes. Thus, phase error β can be obtained as

$$\text{Phase error}\beta = Tan^{-1}\left(\frac{I_m Cos\delta - I_w Sin\delta}{I_1}\right) 100\,\text{crad}. \tag{6.27}$$

or

$$\text{Phase error } \beta = Tan^{-1}\left(\frac{I_m Cos\delta - I_w Sin\delta}{I_1}\right)3438\text{ min.} \tag{6.28}$$

Since in most cases, the value of β will be small, we can approximate $Tan\beta \approx \beta$ and hence Eqs. (6.27) and (6.28) can be modified as

$$\text{Phase error } \beta \approx \left(\frac{I_m Cos\delta - I_w Sin\delta}{I_1}\right)100\text{ crad.} \tag{6.29}$$

or

$$\text{Phase error}\beta \approx \left(\frac{I_m Cos\delta - I_w Sin\delta}{I_1}\right)3438\text{ min.} \tag{6.30}$$

The sign of the phase error β is taken as positive if the reflected secondary current on the primary side $\underline{I}_1'$ leads the primary current $\underline{I}_1$. Hence, the phase error will be negative if the secondary current reflected on the primary side $\underline{I}_1'$ lags the primary current $\underline{I}_1$. Figure 6.7 shows the phasor diagram of a current transformer with the secondary power factor being unity (that is $\delta = 0$). In this figure too, the phasors of the secondary side of the current transformer are shown in the third quadrant of the graph intentionally to improve the viewing (This is achieved by intentionally adding 180° to each phasor). From Fig. 6.7, we can deduce that the ratio error depends only on the watt-loss component I_w of the excitation current, and the phase error is dictated only by the magnetizing component I_m of the excitation current when $\delta = 0$. Equations (6.31) and (6.32) illustrate the ratio and phase errors when the secondary power factor is unity (UPF) as $Cos\delta = 1$ and $Sin\delta = 0$.

$$\text{Ratio error} = \left(\left(\tfrac{K_N - K_t}{K_t}\right) - \tfrac{I_w}{I_1}\right)100\% \tag{6.31}$$

$$\text{Phase error } \beta \approx \left(\frac{I_m}{I_1}\right)3438\text{ min.} \tag{6.32}$$

Figure 6.8 shows the phasors of a current transformer with the overall secondary operating under leading power factor. Such a situation is rare in practice but can happen with modern electronic instruments connected on the secondary side.

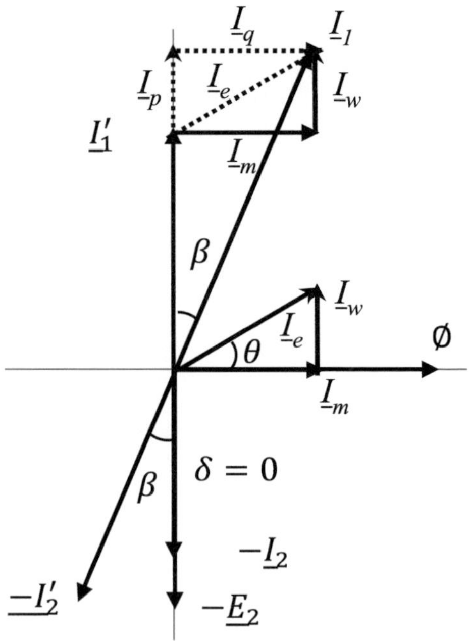

Fig. 6.7 Current transformer: Phasors for unity power factor burden

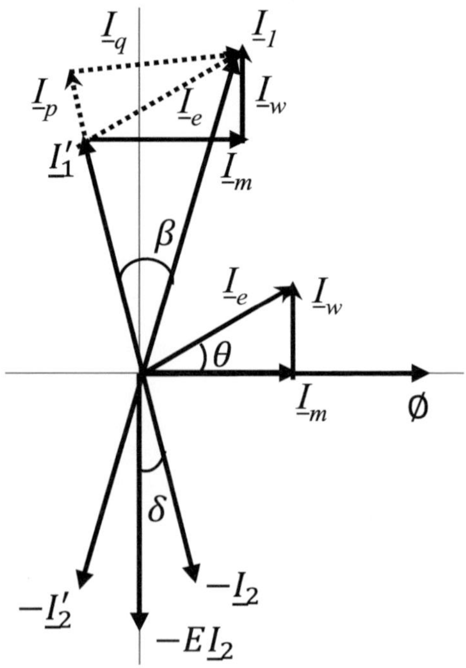

Fig. 6.8 Current transformer: Phasors (leading power factor burden)

6.2.3 *Types of Current Transformers*

In a power system, current transformers are used to fulfill the following needs:

(i) To measure or meter currents.
(ii) To monitor a current continuously and if it exceeds a certain safe value, trip (switch off) the power system to protect it from irreparable damage. This aspect is called *overload protection* or *over current protection* or *short circuit protection.*
(iii) To be used as a standard for testing other current transformers.
(iv) To measure non-sinusoidal currents with or without DC, especially when power electronic compensators such as STATCOM (static var compensator) or UPFC (unified power factor controller) are employed.

To meet the above requirements, five different types of current transformers are used in a power system, namely,

(i) The *measuring* (or *metering*) *current transformers*,
(ii) The *protection current transformers,*
(iii) The *Standard Current transformer*,
(iv) The *hall-effect type current transformer, and*
(v) The *Fiber-optic current transformer.*

6.2.3.1 Metering (Measuring) Current Transformers

As the name indicates, the metering current transformers are used to measure the current and thus must have as small a ratio error and phase error as required. Metering current transformers are classified in terms of "*accuracy class*" as per measurement standards such as

(i) The international standard IEC 61,869-2:2012 (Old IEC 60,044-1),
(ii) The British Standard BS3938 and
(iii) The Indian standard IS2705. Table 6.1 lists the classification of metering current transformer as per the Indian Standard IS2705. Here, the error limit specifies the maximum permissible value of ratio error and phase error at that primary current level. For example, the ratio error limit for a class 1 current transformer at 5% of the rated primary current is ±3.0%. Table 6.1 also indicates that the metering current transformers are expected to perform normally (within the set ratio and phase error limits) up to 120% of the full-scale current.

Table 6.1 Accuracy class of current transformer (CT) as per IS2705

Sl. no	Accuracy class	% Ratio error limit				Phase error limit in min			
		Percentage of full-scale current				Percentage of full-scale current			
		5	20	100	120	5	20	50	100
1	0.1	0.4	0.2	0.1	0.1	15	8	5	5
2	0.2	0.75	0.35	0.2	0.2	30	15	10	10
3	0.5	1.5	0.75	0.5	0.5	90	45	30	30
4	1.0	3.0	1.5	1.0	1.0	180	90	60	60

Apart from the four accuracy classes shown in Table 6.1, we also have class 3 and class 5 current transformers that are not popular and are recommended for deployment in situations that need casual monitoring/metering of current. These (class 3 and class 5) current transformer specifications set no limit for the phase error but specify just the ratio error limit applicable from 50 to 120% of the rated current as

(a) Class 3 current transformer: The percentage ratio error limit is ±3% from 50 to 120% of full-scale current.
(b) Similarly, for the Class 5 current transformer: The percentage ratio error limit is ±5% from 50 to 120% of full-scale current.

Table 6.2 illustrates the accuracy class and the corresponding ratio and phase error specifications of measuring (metering) current transformers as per IEC 61,869-2 specifications. IEC is the International Electrotechnical Commission that sets standards for instrumentation and is recognized by 172 countries. We see that in the IEC specification, the phase error is also given in centiradians. Apart from the ratio and phase errors, metering current transformer specifications must include the following parameters.

I. Rated burden,
II. Insulation level, and
III. Instrument safety (or security) factor (ISF).

These are explained next.

(I) Rated Burden.

In the equivalent circuit of Fig. 6.4, the induced emf $\underline{E}_2 = (Z_2 + Z_B)\underline{I}_2$ and hence $\underline{E}_2$ is dependent on the value of the burden. Higher the burden, larger will be the induced emf resulting in a proportional increase in the core flux as

$$e_2 = N_2 \frac{d\emptyset}{dt}. \tag{6.33}$$

Table 6.2 Accuracy class of current transformers as per IEC 61,869-2

Accuracy class	Ratio error limit (%)				Phase error							
					±min				±Centiradians			
	Percentage of rated current				Percentage of rated current				Percentage of rated current			
	5	20	100	120	5	20	100	120	5	20	100	120
0.1	0.40	0.20	0.10	0.10	15	8	5	5	0.45	0.24	0.15	0.15
0.2	0.75	0.35	0.20	0.20	30	15	10	10	0.90	0.45	0.30	0.30
0.5	1.50	0.75	0.50	0.50	90	45	30	30	2.70	1.35	0.90	0.90
1.0	3.00	1.50	1.00	1.00	180	90	60	60	5.40	2.70	1.80	1.80

In Eq. (6.33), e_2 is the time domain representation of the phasor $\underline{E}_2$. An increase in e_2 will increase the excitation current $\underline{I}_e\left[= \left(\underline{I}_m + \underline{I}_w\right)\right]$ and hence the ratio and phase errors will increase. Thus, the errors of a current transformer depend on the value of the burden connected to its secondary terminals. Hence, a manufacturer ascertains the maximum value of the burden that can be connected to the current transformer and still obtain the ratio and phase errors within the set limits. That level of burden, which if crossed, results in errors exceeding the desired limits, is called the "*Rated Burden*". Burden is normally indicated in terms of apparent power using the unit "*voltampere* (VA)" and power factor. Examples of rated burden:

(a) 5 VA, UPF (to be read as: Five voltamperes, at unity power factor).
(b) 10 VA, 0.8 lag.

(To be read as: Ten voltamperes at an inductive power factor of 0.8)

(II) Insulation Level.

As is seen in Fig. 6.9, the primary of the current transformer is inserted in the power line and hence the primary winding (Terminal P_1 and P_2) will be at the system line voltage level. The secondary is near a person or persons reading the secondary current on an ammeter. Hence, the insulation between the primary and the secondary must withstand the system voltage level. Thus, one of the design criteria of a current transformer is the insulation to be provided between the primary and secondary windings. A current transformer designed for, say, 400 V system cannot and should not be used on an 11 kV line. A current transformer designed for an 11 kV system can be used on a 400 V system. However, a current transformer designed and fabricated to withstand a 11 kV insulation level will be more expensive than a current transformer designed and fabricated for a 400 V system. Hence, for proper use of a current transformer, the insulation level is (must be) specified.

(III) Instrument Safety/Security Factor (ISF).

When a fault occurs on the power line, depending on the design criteria and the nature of the fault, the fault current can reach very high values. The secondary current, if not limited, will also rise to that level. For example, if we are using a 100/1 A

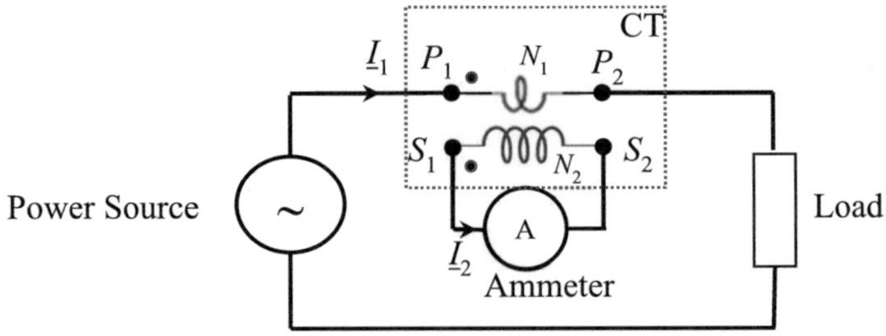

Fig. 6.9 Connection-diagram for a current transformer

current transformer then the ammeter in Fig. 6.9 will be a 1 A full-scale ammeter. During a fault, say, the primary fault current is 1000 A. Then the secondary current will rise to 10 A. Sending a 10 A current through a 1 A full-scale ammeter will irreversibly damage the ammeter and may cause injuries to persons nearby. Hence, for measuring or metering current transformers the maximum value the secondary current can reach must be limited by design. The ratio of the maximum limit permitted on the secondary current to the nominal full-scale value of the secondary current is called the "*Instrument Safety Factor*" or ISF. The nomenclature of *Instrument Security Factor* (ISF) instead of the *instrument safety factor* is also employed but means the same. For example, in a 100/1 A current transformer with an ISF of 2.5, whatever be the primary (fault) current, the secondary current will not exceed 2.5 A. Examples of typical specifications for a measuring or metering current transformer are given below.

(a) 100 /1 A, 5VA, UPF, LT CT with ISF3.0
(b) 1000/5 A, 10 VA, 0.8 lag, 11 kV, bar primary CT with ISF 3.5.

In example (a), we have a 100 bar 1 A CT with insulation level of 400 V (LT means low tension and in India the low-tension voltage is 400 V) which is expected to be connected to a nominal burden of 5 VA rating at unity power factor (that is burden is purely resistive of value 5 Ω). We see that the example (b) has a new term "*bar primary*". Stepping down 1000 A to 5 A requires the turns ratio:

$$\frac{N_2}{N_1} = 200.$$

This can be obtained by using myriad pairs of values for N_1 and N_2. For example, we can keep N_1 as 2 turns and N_2 as 400 turns or keep N_1 as 1 turn and N_2 as 200 turns. Whenever we make N_1 as 1 turn, it simply means that the primary is just a bar of conductor that encompasses (goes through) the core. Normally, the primary that carries hundreds of amperes is manufactured as a bar having rectangular cross section (to handle *skin effect*) as seen in Fig. 6.10b and hence the nomenclature "*Bar Primary*". Thus, *a bar primary current transformer* has $N_1 = 1$. Figure 6.10 illustrates both wound primary and bar primary current transformers.

(a) Multi range wound primary CT

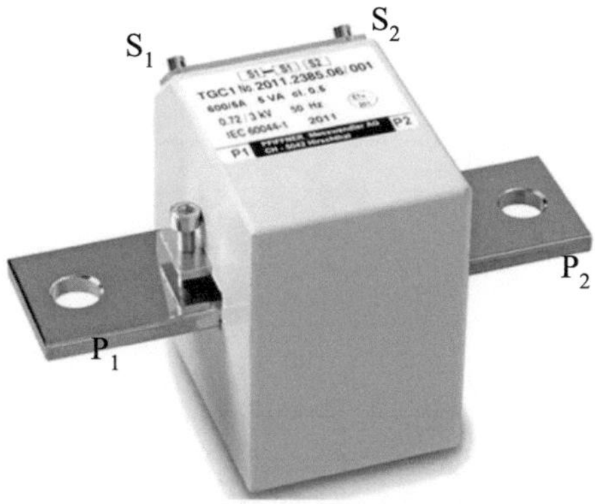

(b) Bar Primary Current Transformer

Fig. 6.10 Practical measuring current transformers

6.2.3.2 Protection Current Transformers

As the name indicates these current transformers are used to monitor the current, switch off the power when the system is overloaded, or large currents flow due to a fault (short circuit). Hence, protection current transformers are expected to work even when the primary current exceeds several times (say, y times) the rated full-scale current. Because we are interested only in the magnitude of the primary current, that too during fault conditions, only the composite error (say as x %) is specified for the protection current transformer at the expected maximum current (y times the normal full-scale operating current of the system) and phase error at this current is not mentioned as it is not required. However, some national standards require the protection type current transformers satisfy phase error limit at the normal maximum operating current of the power system. From the foregone discussion, we understand that the working range of a protection current transformer is specified by a factor y derived as

$$y = \frac{\text{Maximum Current the CT is expected to handle}}{\text{Full} - \text{scale current of the power system}}.$$

For example, if we have 200/1 A protection current transformer and is expected to work till a maximum current of 1000 A, then:

$$y = \frac{1000}{200} = 5.$$

Since the protection CT must handle very large currents (even if for a limited time), in general, protection CTs will be bulkier than the measuring (metering) CTs. Thus, for the protection type current transformer, as per IS/IEC specifications, the limit of error is given as xPy. The letter P indicates that the CT is a protection type

Table 6.3 Protection type CT—limits of error

Sl. no	Class of protection	Error at rated current I_N		Composite error at current limit (y I_N) (%)
		Ratio error (%)	Phase error	
1	5Py	±1	±60′	±5
2	10Py	±3	Not specified	±10
3	15Py	±5	Not specified	±15

current transformer. Here x is the accuracy class (limit of the composite error of the protection current transformer). As per IS2705-Part 3 standard there are only three accuracy classes for the protection type current transformers, namely 5Py, 10Py, and 15Py. Table 6.3 provides the accuracy limits of the three classes of protective current transformers.

The 5Py class must have the ratio error ≤±1% and phase error ≤±60' at the nominal full-scale value of current and the composite error cannot be more than ±5% at the limiting current. For example, the composite error of the 5P10 class must be ≤±5% at the maximum (10 times the nominal full-scale) operating current the CT is designed for.

The 10Py class must have the composite error ≤±3% at the rated full-scale current (*Note*: phase error is not specified, can be any value) and the composite error ≤±10% at the maximum (y times the nominal full-scale) current.

The15Py class of protection CTs must have the composite error ≤±5% (*Note*: phase error is not specified, can be any value) at the rated full-scale current and the composite error ≤±15% at the maximum (y times the nominal full-scale) current.

Typical specification examples of few protective current transformers are given below.

(a) 500/5 A, 15 VA, 0.8 lag, 5P10, LT CT.

Expansion: Nominal (operating) primary current is 500 A full-scale with corresponding full-scale secondary current as 5 A. Rated burden is 15 VA at 0.8 lag power factor. Limit of ratio error ≤±1% and phase error ≤±60' at 500 A. The CT will work till the primary current is 3 times the nominal value (500*15= 5000 A) and must have a composite error ≤±5% up to the current is 5000 A. Insulation level is 400 V.

(b) 100/5 A, 50 VA, 0.8 lag, 10P15, 11 kV CT.

Expansion: Nominal (operating) primary current is 100 A full-scale with corresponding full-scale secondary current as 5 A. Rated burden is 50 VA, 0.8 lag power factor. Limit of accuracy at 100 A: Composite error ≤±3%. The CT will work till the primary current is 5 times the nominal value (100*15 = 1500 A) and must have the composite error to be ≤±10% up to the current is 1500 A.

(c) 500/5 A, 50 VA, 0.8 lag, 15P5, 110 kV.

Expansion: Nominal (operating) primary current is 500 A full-scale with corresponding full-scale secondary current as 5 A. Rated burden is 50 VA, 0.8 lag power factor. Limit of accuracy at 500 A : Composite error $\leq \pm 5\%$. The CT will work till the primary current is 10 times the nominal value ($500*5 = 2500$ A) and must have the composite error to be $\leq \pm 15\%$ in the range < 500 A to 2500 A>. The CT is rated for a 110 kV system.

6.2.3.3 Standard Current Transformers

The standard current transformer, as the name indicates, is used as a "*standard*" with which a current transformer under test is compared and the ratio and phase errors of the test current transformer is determined. Such testing is called "*comparison method of testing*". Testing of current transformers using the *comparison method* is dealt with in the sequel. A standard current transformer is designed and built to have very small ratio and phase errors. To obtain very small ratio and phase errors, we need to

(a) Reduce the excitation current or
(b) Compensate for the excitation current.

(a) **Standard CT with Very Low Excitation Current**

This type of standard current transformers that are designed to have very small excitation current by design are called *passive standard current transformers.* From the equivalent circuit of a current transformer shown in Fig. 6.4, we can deduce that the excitation current $\underline{I}_e$ is

$$\underline{I}_e = \frac{\underline{E}_1}{R_i} - j\frac{\underline{E}_1}{\omega L_m}. \tag{6.34}$$

We also have

$$\underline{E}_1 = \frac{N_1}{N_2}\underline{I}_2((R_2 + R_B) + j\omega(L_2 + L_B))\text{and}L_m = \frac{\mu_o\mu_r A_c N_1^2}{l_m}. \tag{6.35}$$

In Eq. (6.35), μ_o is the permeability of free space, μ_r is the relative permeability of the core, A_c is the cross-sectional area of the core, and l_m is the mean magnetic path length of the path the flux goes through. From Eqs. (6.34) and (6.35), we realize that to reduce the excitation current we need to

I. **Reduce the Induced EMF $\underline{E}_1$:** Reducing $\underline{E}_1$ can be achieved by reducing: (a) the secondary leakage impedance ($R_2 + j\omega L_2$) and (b) the burden Z_{BS} on the standard current transformer.
II. **Increase the Value L_m:** Increasing L_m can be achieved by increasing the number of turns (of course if we increase primary turns N_1, we need to increase the secondary turns N_2 as well). L_m can also be increased by using a soft ferromagnetic core of very high permeability. Thus, standard current transformers are designed to operate at high ampere turns and the core is made of thin insulated strips of very high permeability material such as permalloy (80% nickel, 14%

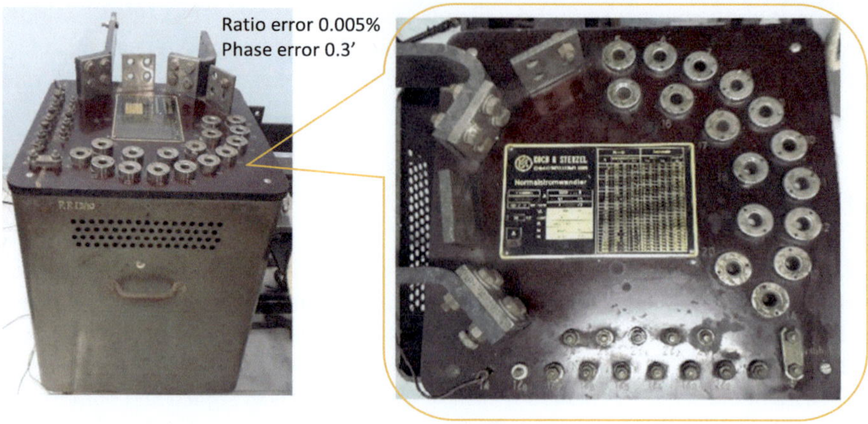

(a) Multi range wound primary standard CT (b) Primary and secondary connectors

Fig. 6.11 Multi-ratio standard current transformer (Passive)

iron, 4% molybdenum and small amounts of silicon, copper, and carbon). The permalloy 80 has a relative permeability of 300,000 at moderate flux densities. Such standard current transformers will obviously be bulky and expensive.

III. **Increase the Value of R_i:** To reduce excitation current $\underline{I}_e$, the watt loss component $\underline{I}_w$ of the excitation current also needs to be reduced. Thus, apart from increasing L_m, we also need to increase the value of R_i. Increasing R_i can be achieved with the use of soft ferromagnetic cores that possess low hysteresis loss. Of course, to reduce eddy current loss (i) we need to use a core with high resistivity and (ii) assemble the core using insulated thin strips of the core. Figure 6.11 shows such a passive, multi ratio current transformer with ratio error <0.005% and phase error <0.3′.

(b) Compensate for the Excitation Current in Current Transformers

Standard current transformers that use the technique of compensation of excitation current to reduce the ratio and phase errors to insignificant levels are called "***Electronically compensated reference current transformers***" due to the reason that the excitation current is compensated using electronics. They are also called active standard current transformers or electronic standard current transformers. All such techniques of electronic compensation of the excitation current operate on the simple principle that if the core flux is reduced to zero or nearly zero, then primary current need not supply the excitation current, thus reducing the errors. These schemes fall under two categories, namely,

(A) Single-core scheme and
(B) Two-core scheme.

These schemes are illustrated next.

(A) *Electronically Error Compensated Current Transformer—Single Core Scheme.*

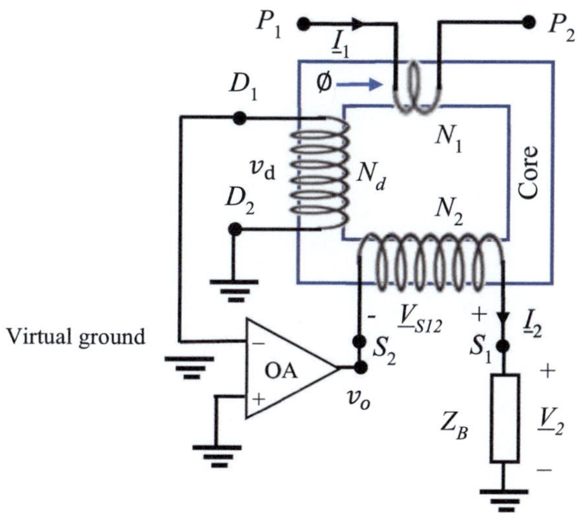

Fig. 6.12 Single core three-winding electronically compensated standard current transformer

Figure 6.12 illustrates the technique wherein apart from the primary winding of N_1 turns and secondary winding of N_2 turns, a third detector winding of N_d turns is wound on the core. The output $v_{\mathrm{d}}\ \left(= N_{\mathrm{d}}\frac{d\emptyset}{dt}\right)$ of the detector winding is connected to the inverting input of the opamp OA as shown in Fig. 6.12. The non-inverting input of OA is grounded. The opamp's output v_{o} is connected to terminal S_2 of the secondary. The burden is connected between terminal S_1 of the secondary and ground. With the connections as shown, the inverting terminal will be at virtual ground and hence $v_{\mathrm{d}} = 0$, resulting in

$$N_{\mathrm{d}}\frac{d\emptyset}{dt} = 0. \tag{6.36}$$

Since N_{d} is not zero, we will have $\frac{d\emptyset}{dt} = 0$. That is, we have *zeroACacflux*. Zero AC flux results in zero excitation current $\underline{I}_e$. Hence $N_1\underline{I}_1 = N_2\underline{I}_2$ giving rise to zero ratio and phase errors.

However, the scheme in Fig. 6.12 is a negative feedback system, with the feedback achieved through the core flux. In practice, due to stability issues associated with the negative feedback systems, we will have a gain margin beyond which the feedback system will be unstable. Hence, we will not be able to use an opamp with very large gain (10^6 or higher) but employ an amplifier with finite gain A_{OA} (the gain A_{OA} must be less than the gain margin) to ensure stable operation. If A_{OA} is the maximum gain possible to ensure stability, then we have

$$v_o = A_{OA}N_d\frac{d\emptyset}{dt} = -N_2\frac{d\emptyset}{dt} + v_2. \tag{6.37}$$

Rearranging equation (6.37), we get

$$\frac{d\emptyset}{dt}=\frac{v_2}{(A_{OA}N_d+N_2)}. \tag{6.38}$$

Assuming the burden to be inductive ($\boldsymbol{Z_B}=(R_B+j\omega L_B)$, we have

$$v_2=i_2(R_2+R_{\text{B}})+\tfrac{di_2}{dt}(L_2+L_{\text{B}}). \tag{6.39}$$

If we did not have the compensation scheme, then the core flux would have been $\emptyset'$ and we have the secondary voltage v_2' without the compensation and with the resulting flux $\emptyset'$ as

$$v_2'=N_2\tfrac{d\emptyset'}{dt}. \tag{6.40}$$

With $(A_{OA}N_d+N_2)\gg 1$, then we have $v_2\approx N_2\frac{d\emptyset'}{dt}$ and Eq. (6.38) reduces to

$$\tfrac{d\emptyset}{dt}=\tfrac{d\emptyset'}{dt}\left(\tfrac{N_2}{(A_{OA}N_d+N_2)}\right)=\tfrac{d\emptyset'}{dt}\left(\tfrac{1}{\left(A_{OA}\frac{N_d}{N_2}+1\right)}\right). \tag{6.41}$$

Equation (6.41) indicates that operating the compensation scheme of Fig. 6.12 with a finite gain A_{OA}, we will reduce the AC flux in the core by a factor:

$$\left(1+\frac{A_{OA}N_{\text{d}}}{N_2}\right)$$

and thus reduce the errors of the current transformer. In the absence of the amplifier, terminal S_2 will be grounded and the induced emf E_2' due to $\emptyset'$ will supply the burden of $E_2'I_2$ VA.

Once we connect the amplifier, then its output v_{o} must be $\approx E_2'$ and supply the current I_2. Thus, the scheme illustrated in Fig. 6.12 has one major drawback, namely, the opamp must supply the entire voltampere requirement $[I_2(\mathbf{Z}_{\text{B}}+\mathbf{Z}_2)]$ of the secondary side. Hence, the scheme in Fig. 6.12 is not popular for building standard current transformers. However, this scheme is adopted as current to voltage converter for "solid-state" energy meters, wherein the secondary voltampere demand is small and in the range of milli voltampere.

(b) The compensation schemes that employ two cores instead of one, shown in Figs. 6.13 and 6.14, reduce the voltampere demand on the opamp OA by nearly two orders of magnitude and thus overcomes this lacuna. Hence most commercial electronically compensated standard current transformers use one of the schemes shown in Figs. 6.13 and 6.14 or a similar scheme with small variation.

Two-Core Electronically Error Compensated CT—Scheme 1

Figure 6.13 shows an electronically compensated standard current transformer that uses two cores, namely, Core A and Core B. As seen in Fig. 6.13, the primary winding of N_1 turns and the secondary winding of N_2 turns are wound such that they enclose

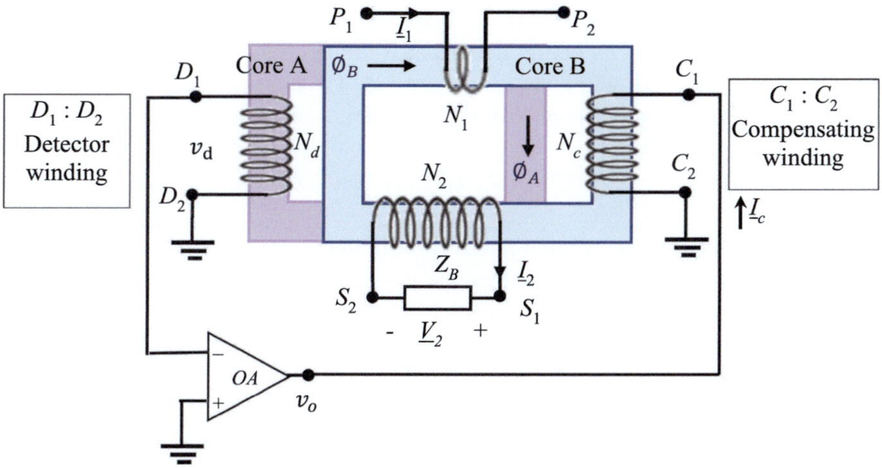

Fig. 6.13 Two core electronically compensated standard current transformer—Scheme 1

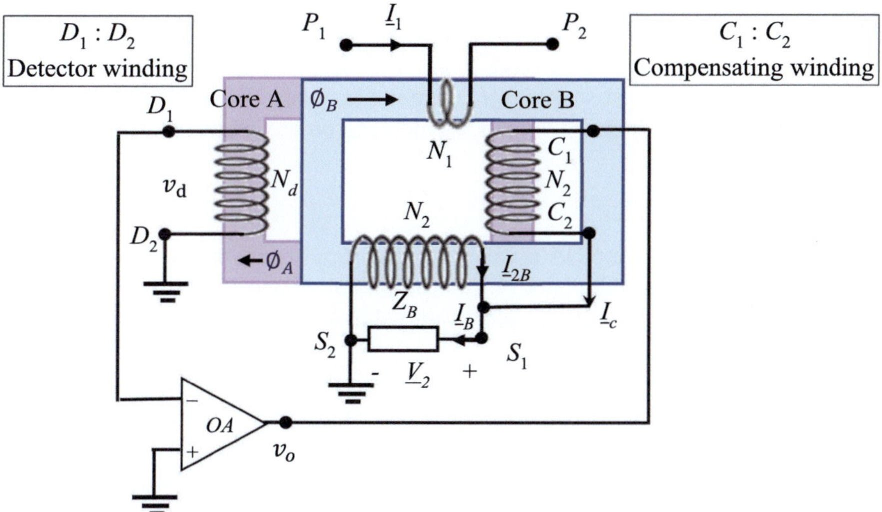

Fig. 6.14 Two Core electronically compensated standard current transformer—Scheme 2

both cores A and B. The detector winding of N_d turns is wound on the Core A and the compensating winding of N_c turns is wound on Core B. The output $v_d \left(= N_d \frac{d\emptyset_A}{dt}\right)$ of the detector winding is connected to the inverting input of the opamp OA as shown in Fig. 6.13. The non-inverting input of OA is grounded. The opamp's output v_o is connected to terminal C_1 of the compensating winding and terminal C_2 of the compensating winding is grounded. The burden Z_B is connected between terminals S_1 and S_2 of the secondary of the current transformer.

Let the flux in Core A be Ø_A and that in Core B is Ø_B. With the connections as shown and with opamp OA being ideal (that is the open loop gain $A_{OA} = \infty$), the inverting terminal of OA will be at virtual ground. Thus, terminal D_1 of the detector winding will be at virtual ground while its terminal D_2 is grounded. Hence, we have

$$v_\text{d} = N_\text{d}\frac{d\text{Ø}_A}{dt} = 0. \tag{6.42}$$

Since N_d is not zero, we should have $\frac{d\text{Ø}_A}{dt} = 0$. Once again, the AC flux in core A is zero resulting in ampere turn balance for Core A. Hence, we get

$$N_1\underline{I}_1 = N_2\underline{I}_2. \tag{6.43}$$

Applying ampere turn balance to core B, we get

$$N_1\underline{I}_1 - N_1\underline{I}_e + N_c\underline{I}_c = N_2\underline{I}_2\,. \tag{6.44}$$

Equations (6.43) and (6.44) together yield

$$N_c\underline{I}_c = N_1\underline{I}_e\underline{I}_c = \frac{N_1\underline{I}_e}{N_c}. \tag{6.45}$$

Here, the opamp OA must supply $V_\text{o}I_\text{c}$ voltampere, where V_o is the rms voltage of the opamp output v_o. Thus, the voltampere demand on the opamp is

$$V_\text{o}I_\text{c} = \frac{N_1}{N_2}V_2I_\text{e}. \tag{6.46}$$

Here, we see that the opamp OA is called upon to supply only the excitation voltampere requirement. Thus, the major portion of the voltampere demand of the burden comes from the primary, only the excitation voltampere demand of the core is met by the amplifier OA.

Here too, the scheme in Fig. 6.13 is a negative feedback system, with the feedback achieved through the flux in the cores A and B. In practice, due to stability issues associated with the negative feedback systems, we will have a gain margin beyond which the feedback system will be unstable. Hence OA cannot be a high gain opamp but an amplifier of gain A_{OA} to ensure stable operation of the feedback scheme. With the finite gain A_{OA}, the output v_o of the opamp OA in Fig. 6.13 will be

$$v_o = A_{OA}N_d\frac{d\text{Ø}_A}{dt} = N_c\frac{d\text{Ø}_B}{dt}. \tag{6.47}$$

Rearranging equation (6.44), we get

$$\frac{d\text{Ø}_A}{dt} = \frac{N_c}{A_{OA}N_d}\frac{d\text{Ø}_B}{dt}. \tag{6.48}$$

Equation (6.48) shows that the flux in core A will be a fraction of the required flux for the transformer and the reduction factor is

$$\frac{A_{OA} N_{\mathrm{d}}}{N_{\mathrm{c}}}. \tag{6.49}$$

Hence, core A will have reduced flux and thus have reduced excitation current drawn from the primary current. Hence, the errors of the current transformer will be reduced by the loop gain of the feedback compensation scheme of Fig. 6.13.

Two-Core Electronically Error Compensated CT—scheme 2

The scheme in Fig. 6.13 is a voltage-controlled voltage source (VCVS) type of negative feedback scheme. It is possible to design a voltage controlled current source (VCCS) feedback scheme as shown in Fig. 6.14 for obtaining a standard current transformer with very small errors. Here, we see that the compensating winding is put on core A itself and has N_2 turns. The burden current $\underline{I}_{\mathrm{B}}$ is obtained by adding the compensating winding current $\underline{I}_{\mathrm{c}}$ to the secondary current $\underline{I}_{2\mathrm{B}}$. Hence, we get:

$$\underline{I}_B = \underline{I}_{2\mathrm{B}} + \underline{I}_c. \tag{6.50}$$

The ampere turns balance for core A (with $v_{\mathrm{d}} = 0$) is

$$N_1 \underline{I}_1 = N_2 \left(\underline{I}_{2B} + \underline{I}_c\right) = N_2 \underline{I}_B \Rightarrow \underline{I}_1 = \left(\frac{N_2}{N_1}\right) \underline{I}_B. \tag{6.51}$$

Equation (6.51) indicates that the burden current multiplied by the ratio of the current transformer correctly indicates the primary current without (ratio or phase) errors.

Here too, Eq. (6.51) derived with opamp OA in Fig. 6.14 is ideal and thus we realized an ideal current transformer with zero ratio and phase errors. However, this scheme is also a negative feedback scheme and hence will possess a gain margin. That is gain higher that the gain margin A_{OA} for OA will make the scheme unstable. With finite gain A_{OA} of OA, the output v_{o} of the amplifier OA is (OA is no longer an opamp):

$$v_o = A_{OA} N_d \frac{d\text{Ø}_A}{dt}. \tag{6.52}$$

As with the schemes of Figs. 6.12 and 6.13, the scheme in Fig. 6.14 also is a negative feedback scheme with a loop gain:

$$\left(1 + \frac{A_{OA} N_{\mathrm{d}}}{N_2}\right).$$

Hence the errors of the CT will reduce by that much. The difference between the schemes of Figs. 6.13 and 6.14 is that in the scheme 1, the core flux Ø_B is altered such that core flux Ø_A is reduced (to zero theoretically) and thus a magneto motive force of intensity of nearly $N_1 \underline{I}_e$ is added to core B magnetic circuit. Thus, the flux in the core B is increased to a level to ensure $N_1 \underline{I}_1 = N_2 \underline{I}_2$. With this, the secondary current $\underline{I}_2$ is directly altered such that $N_1 \underline{I}_1 = N_2 \underline{I}_2$ (theoretically, but practically

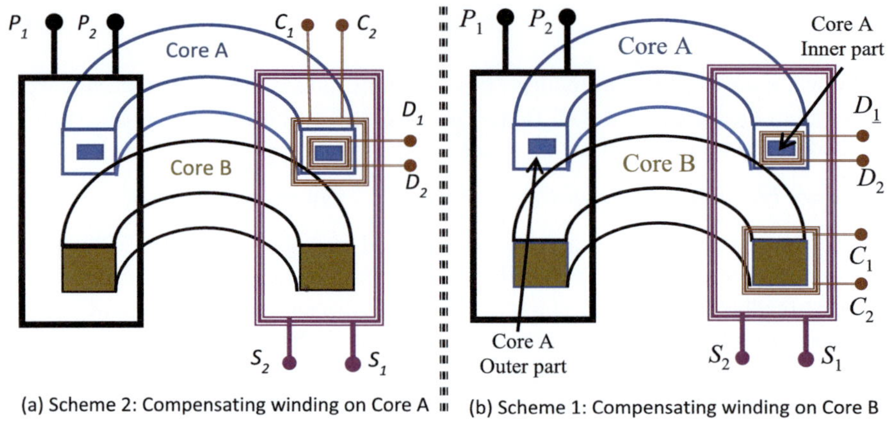

Fig. 6.15 The winding arrangement of electronically compensated standard current transformer: Schemes 1and 2

$N_1\underline{I}_1 \approx N_2\underline{I}_2$). In the second scheme shown in Fig. 6.14, the core flux $\emptyset_A$ is reduced (to zero theoretically) by introducing additional compensating ampere turns N_2I_c to act on core A itself. In that process a current of value:

$$\frac{N_1}{N_2}\underline{I}_e$$

is added directly to the burden current $\underline{I}_B$ and hence $\underline{I}_B$ is altered to get reduced errors. As a negative feedback system, the scheme of Fig. 6.14 has better stability characteristics as the output is "short-circuit-stable" and the feedback deals with only one core flux, namely, the flux in core A.

Practical Implementation

In a practical implementation, the core A is split into two parts as shown in Fig. 6.15, namely the inner part and the outer part. The inner part of core A is wound with the detector winding. Then the outer part is assembled. Only for scheme 2, the compensating winding is then wound on the outer part of core A as shown in Fig. 6.15a. For scheme 1, the compensating winding is wound on core B as seen in Fig. 6.15b. Once the detector and compensating windings are wound, the two cores are assembled (stacked one over the other) and then the secondary is wound on the combined core, again as shown in Fig. 6.15. Finally, the primary is wound, once again encompassing both the cores.

This way of splitting the core A into two parts and embedding the detector winding improves stability as close coupling is ensured with the core. Moreover, the detector winding is provided with shielding that is necessary to avoid external interference affecting the detector, as the detector winding will be sensing nearly zero (very low) flux. It must be noted that core A must have large relative permeability near zero flux condition for the two-core schemes to operate as explained here.

6.2.3.4 Hall-Effect Based AC/DC Current Transformers

As the name indicates, the current transformer uses the "hall Effect". Dr. Edwin Hall in 1879, a doctoral candidate at Johns Hopkins University, in Baltimore was experimenting the theory of electron flow proposed by Kelvin in 1849. During his experiments, Hall found that when a magnet was placed with its field perpendicular to one face of a thin rectangle of gold through which a current is flowing, a difference in potential appeared at the opposite edges. He found that this voltage is proportional to the current flowing through the conductor, and the flux density or magnetic induction perpendicular to the conductor. This effect is now known as Hall Effect.

The Hall-Effect Magnetic Flux Sensor

For several decades Hall effect had no practical application and was found useful only in the realm of theoretical physics. The reason is quite simple, if the material is a conductor such as metal, then cross currents will result due to the Hall voltage and will immediately reduce the voltage developed, more importantly the cross currents will heat up the sensor. Since we cannot send a current through an insulator, Hall Effect cannot be utilized there too. The Hall voltage will be sustained if and only if the conductivity of the material is optimal, as in a semiconductor. Hence it is not a surprise that all Hall effect sensors are invariably made of semiconductor material. Let us consider a block of material of rectangular cross section with length "*l*", width "*w*", and thickness "*t*" through which a current I_R is flowing as shown in Fig. 6.16. The material is subjected to a magnetic field (flux density B) in a direction perpendicular to the flow of current. If we assume that the current is caused by the flow of electrons (the material is made of either a metal or N-type semiconductor) then the electrons will interact with the magnetic field. Due to the Lorentz force, the electrons will be pushed to one side (upwards for the directions of current and flux as indicated in Fig. 6.16). Hence if we look at the charge density in the "*w*" direction, the charge (electrons) density will be more on the upper side compared to the lower side. Hence there will be a potential difference between terminals 1 and 2 as shown in Fig. 6.16. This voltage is the Hall voltage V_H, and can be derived as

$$V_H = R_H \frac{B_z.I_x}{t}, \tag{6.53}$$

where R_H is the Hall coefficient of the material and can be easily derived as

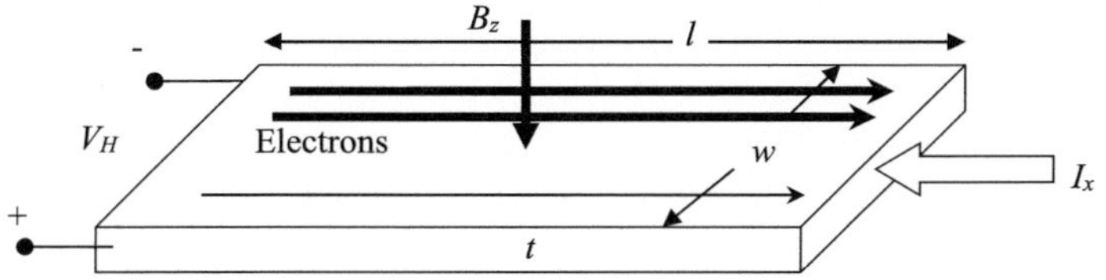

Fig. 6.16 The hall effect

$$R_H = \frac{1}{nq} \tag{6.54}$$

It is easily seen from Eq. (6.54) that the Hall voltage is inversely proportional to the charge carrier density n and the thickness of the sensor. Hall sensors can be used to measure current I_x if we make B_z constant and B_z if we make I_x constant and n if we make $B_z.I_x$ a constant. Since Hall sensors are always made very thin to increase the Hall voltage it is seldom used for the direct measurement of currents. On the other hand, these types of sensors are quite popular for the measurement of flux density B. By measuring the flux density B produced by a current flowing in a circuit, it is also possible to measure large currents I. With the development of power electronic circuits, the need for the measurement of currents (and voltages) that have both DC and AC parts, that too the AC part containing harmonics has arisen. Conventional current transformers work on the principle of magnetic induction and hence will not function at all for DC currents and will function poorly for currents possessing either low frequency (<25 Hz) or high frequency (> 200 Hz) components. For measuring DC currents, it is possible to design a DC current transformer comprising a common primary winding and a pair of windings connected in series opposition and powered by an AC source. Such current transformers, though they enable DC current measurement, will not be useful if the current to be measured possesses both DC and AC, as normally encountered in power electronic circuitry. The Hall effect sensor based current transformer, popularly known as "Hall effect CT" is best suited for such an application. Thus, the hall-effect current transformer (some time called the Hall effect current transducer) was developed, specifically for measuring currents in power electronic circuits. There are two types of Hall effect type current sensors, namely, the open loop type and the feedback (closed loop) type.

(a) The Open Loop Type Hall Effect Current Sensors:

Figure 6.17 illustrates the open loop type Hall effect type current sensor. These types of sensors are not popular as the output is dependent on the operating temperature and on the current I_H.

(b) The Feedback Type Hall Effect Type Current Transformer:

The feedback or closed loop type Hall effect type current transformer as the name indicates employs the feedback principle. Figure 6.18 provides the circuit schematic of a typical Hall-effect current transformer (transducer). Like a conventional current transformer, here too we have the primary winding of N_1 turns and the secondary winding of N_2 turns wound on a soft ferromagnetic core as shown in Fig. 6.18. A flux sensor (in this case a Hall-effect sensor) is inserted into the core assembly to sense the flux Ø in the core. The output of the flux sensor is given as the input to the opamp OA. The opamp OA drives the secondary winding. With the connections as given in Fig. 6.18, the output v_s of the flux sensor will be zero. The output v_s of the flux sensor will be zero if and only if flux $Ø = 0$ and that is possible if and only if $N_1 i_1 = N_2 i_2$. Hence we have

$$i_1 = \left(\frac{N_2}{N_1}\right) i_2 = K_N i_2 \,. \tag{6.55}$$

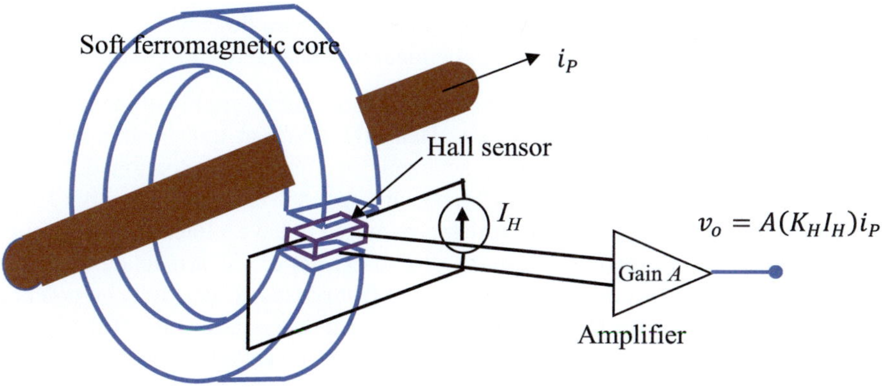

Fig. 6.17 Open loop type Hall effect current Sensor

Equation (6.55) indicates that current i_2 will be a faithful replica of the current i_1 with the scaling factor K_N. This type of Hall Effect current transformer has several advantages as listed below.

(i) The bandwidth of the current transformer is dictated only by the bandwidth of the opamp.
(ii) The Hall sensor current and the Hall coefficient of the Hall sensor do not directly affect the performance of the current transformer.

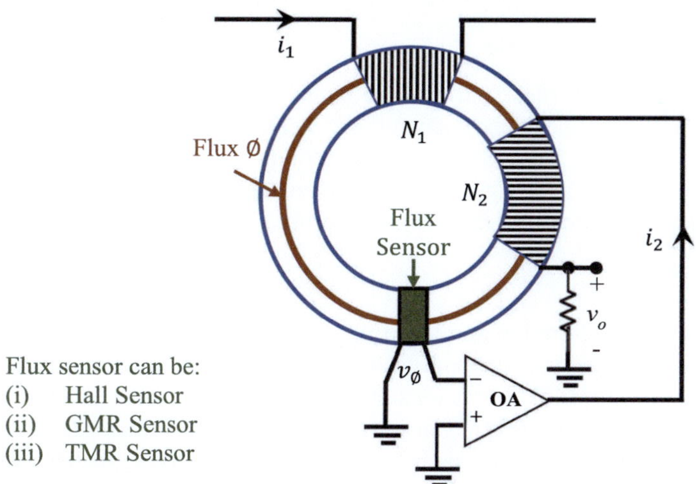

Fig. 6.18 Hall-effect current transformer

(iii) The Hall sensor must have good sensitivity near zero flux level.
(iv) The core needs to have only large permeability near zero flux condition.

Hence commercial Hall effect type current sensors are invariably of the feedback compensated type. In the feedback type Hall effect current transformer, the Hall sensor must have very good sensitivity near zero flux, as already mentioned. Hence, the Hall sensor used in Hall-effect CT is specially tuned to have high sensitivity near zero unlike the regular Hall sensor, which is expected to have a linear characteristic. Today, we can use Tunneling magneto resistive (TMR) or Giant magneto resistive (GMR) flux sensor instead of Hall effect sensor that possess a sensitivity an order of magnitude higher than the Hall effect sensor. But still these transformers (that is AC/DC current transducers using TMR or GMR sensors) are called "Hall effect current transformer".

6.2.3.5 Fiber-Optic Current Transformers

The fiber optic current transformer (popularly called the fiber optic current transducer) senses the current using the Faraday Magento optic effect. Though the Faraday Magento optic effect is known for more than a century, its use for current measurements, especially in high voltage environments, become feasible after the advent of single mode polarization preserving optical fibers. Due to their high cost, the fiber optic current transformer becomes economically viable only for high voltage systems. Today we have commercial fiber optic current transformers available for high voltage power systems (DC or AC) up to 800 kV.

(a) **Light and Its Properties**

Light has several properties, starting from duality. Certain aspects of its behavior such as travel in a straight line can be understood by treating light as being made of particles called "*photons*". On the other hand, certain other aspects like interference can be explained only by treating light as an electromagnetic wave. As an electromagnetic traveling wave, light has the following properties:

(i) Frequency,
(ii) Phase,
(iii) Amplitude (referred to as intensity),
(iv) Speed or velocity (is a constant in a homogeneous medium. In vacuum, the speed of light is 300 Mm/s (exactly 299,792,458 m/s). On the other hand, the velocity drops to 200 Mm/s inside a glass of refractive index 1.5),
(v) Wavelength (depends on the frequency of light and the refractive index of the medium light is traveling).

Light can be classified as "*coherent light*" and "*non coherent light*". In coherent light all the electromagnetic waves have the same frequency and phase. Light can also be classified as

(i) Randomly Polarized,
(ii) Linearly polarized,
(iii) Circularly polarized, or
(iv) Elliptically polarized.

Figure 6.19 shows a single electromagnetic wave. Here the y-axis is the electric field, and the z-axis is the magnetic field. The x-axis can be either time or distance. If the x-axis is marked in time, then one cycle of the wave is the period T of the electromagnetic wave and hence the frequency f of the wave is $f = \frac{1}{T}$. If the x-axis is marked in distance, then one cycle gives one wavelength $= \frac{v}{f}$, where v is the velocity of propagation. If all the electromagnetic waves in the light have the electric field, say in the y-axis, then we call the light linearly polarized light (polarized in the vertical direction). It is also possible that the electric fields of individual waves in the light is oriented at an angle θ, then we call the light as polarized light at an angle θ. If we combine two polarized lights, one with 0° polarization and the other at 90° (say one is vertically polarized and the other horizontally polarized) of equal magnitude (intensity) then we get a circularly polarized light. On the other hand, if we combine two polarized lights at 90° with differing magnitudes or having equal magnitudes but a difference of polarization angles other than 90° or integral multiples of 90°, we get an elliptically polarized light. If the polarizations of individual waves in the beam of light are not correlated, then we get an unpolarized light or randomly polarized light. Though polarization is not dependent on the frequency, in practice polarized light can be created using only coherent light. In a coherent light all electromagnetic waves have the same frequency (single color) and phase. The polarized light that we use most today is generated using light amplification by stimulated emission of radiation (LASER) technique. It may be noted that on a day-to-day basis, the light we see are all noncoherent, randomly polarized waves.

(b) **Faraday Magento Optic Effect**

The polarization of a linearly polarized light under the influence of a magnetic field undergoes rotation. This effect was first studied by Michael Faraday and hence is

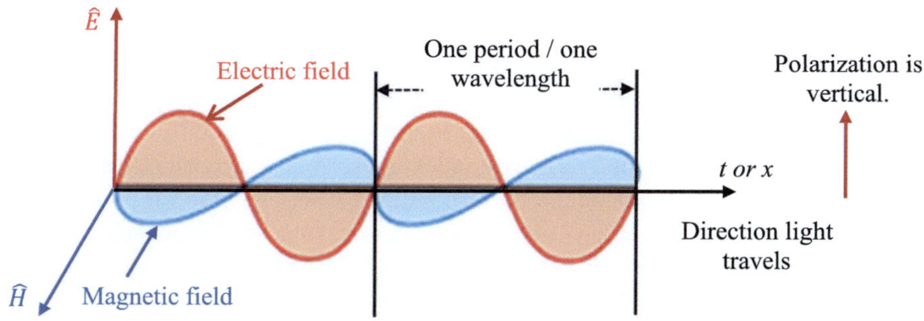

Fig. 6.19 Electromagnetic wave

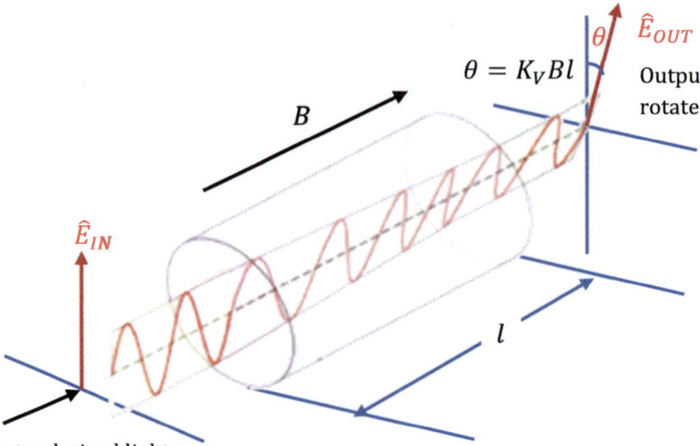

Fig. 6.20 Faraday magneto optic effect

called the Faraday Magneto optic effect. Figure 6.20 illustrates this phenomenon. The angle of rotation of polarization is given by

$$\theta = K_V \int_0^l B dl = K_V B l. \tag{6.56}$$

$\theta = K_V B l$ in equation is valid if the light travel and the flux direction are collinear and B is uniform along the length of travel as shown in Fig. 6.20. In Eq. (6.56), l is the length of interaction of the magnetic flux with the light and K_V is a constant, called the Verdet constant, which depends on the material in which the light is traveling. The use of Faraday magneto optic effect for the measurement of current became popular after the advent of the polarization preserving optical fiber (also called single mode fiber).

(c) **Optical Fibers**

An optical fiber as the name indicates is a cylindrical wire (few micrometers to few hundred micrometers cross section) made of dielectric materials such as glass (more specifically silicon glass) or transparent plastic. An optical fiber consists of an inner cylinder called the core with a refractive index n_{cr} surrounded by an outer cylinder (called the cladding) possessing a refractive index n_{cl}. As shown in Fig. 6.21. If we have $n_{cr} > n_{cl}$ then the core acts like an optical waveguide and transmits light (of course certain wavelengths only) without refraction. An Optical fiber that has a core-cladding such that at the interface of the core-cladding, the refractive index jumps from n_{cr} to n_{cl}, is called "*step-index*" optical fiber. It is also possible to fabricate and use optical fibers where the change of refractive index from n_{cr} to n_{cl} occurs gradually from core to cladding, resulting in "*graded index*" optical fibers. Optical

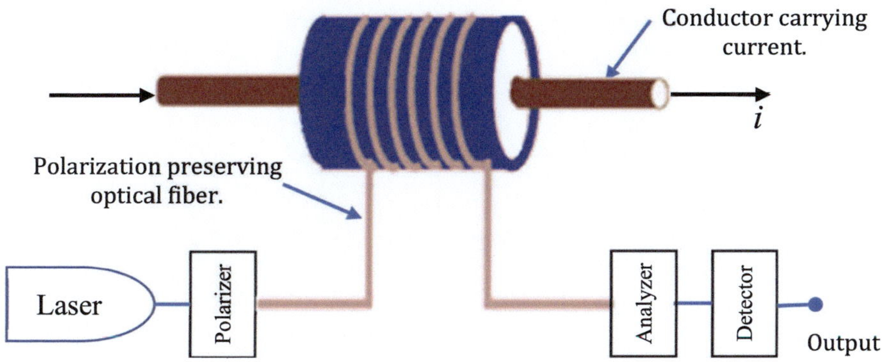

Fig. 6.21 Fiber optic current measurement

fibers that sustain only the fundamental mode of transmission of light is called the "*single mode*" fibers and those that sustain more than one mode of propagation are called the "*multi-mode*" fibers. Today's communication systems predominantly use optical fibers instead of coaxial cables.

(d) Fiber Optic Current Transformers

Here, a polarization preserving optical fiber is wound on to the current carrying conductor as shown in Fig. 6.21. A polarized light is launched into the fiber using a laser (conventional or laser diode) and a polarizer. The output of the fiber is sent through an analyzer. Analyzer is nothing but another polarizer, whose polarization axis adjusted such that its output is zero when current i is zero. The output light from the analyzer is converted into a voltage using a photodiode detector. The scheme shown in Fig. 6.21 is called the fixed optical fiber current measurement system, as once the fiber is wound on the current carrying conductor, the entire system gets finalized. It is also possible to keep one end of the optical fiber to be free as seen in Fig. 6.22. Here, in the free-standing optical fiber current sensor, after the polarizer, a beam-splitter is introduced, and the output of the beam-splitter (direct output) is fed as input to one end of the optical fiber. The other end of the optical fiber is fitted with a mirror. The other output of the beam splitter (90° beam) is analyzed using an analyzer and the output of the analyzer detected using a photodiode to get an output proportional to current i. Unlike the fixed optical fiber current measurement system, the fiber in the free-standing optical fiber current sensor can be wound and unwound on the conductor carrying the current to be measured. Moreover, with proper adjustments, the output from the free-standing optical fiber current sensor can be twice that of the fixed optical fiber current measurement system.

The optical fiber current sensor has the advantages of

(i) Can sense sinusoidal, non-sinusoidal currents with or without DC,
(ii) Unlike conventional current transformers that require one current transformer to measure and another for protection purposes, a single optical fiber current sensor can satisfy the measurement and protection needs,

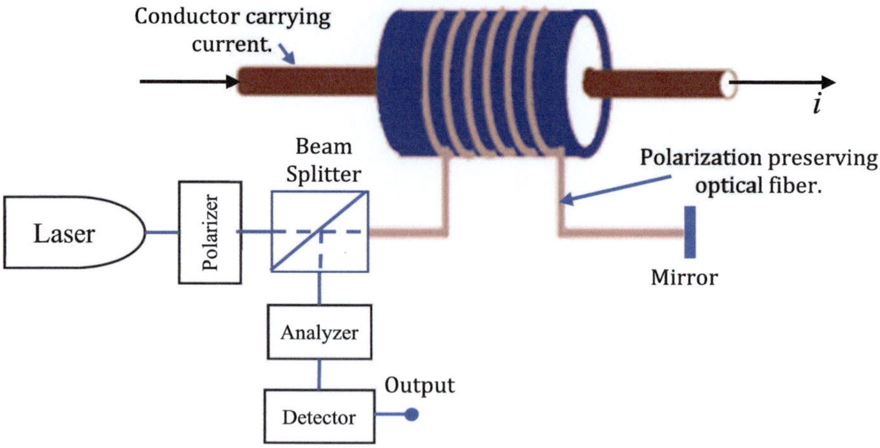

Fig. 6.22 Free-standing fiber optic current measurement

(iii) Easily interfaceable to present day digital instruments, and
(iv) Most importantly use of optical fiber reduces the insulation cost. Thus, optical fiber CTs are best suited for measurement of current in high voltage systems.

However, the cost of the system is very high and hence the use of optical fiber current sensor either for measurement or for protection purposes is very limited.

6.2.4 Design and Construction of Current Transformers

We already know that the ratio and phase errors of a current transformer increase with an increase in the secondary induced emf $\underline{E}_2 = (\mathbf{Z}_2 + \mathbf{Z}_B)\underline{I}_2$. While $\mathbf{Z}_B$ is determined by the user, the value of $\mathbf{Z}_2$ is determined by the design. Hence to lower the ratio and phase errors, $\mathbf{Z}_2$ must be kept as small as possible. Since $\mathbf{Z}_2 = R_2 + j\omega L_2$, both the winding resistance and secondary leakage reactance must be optimized. L_2 can be kept to a minimum if we employ a toroidal core shown in Fig. 6.23. Due to this reason nearly 100% of the current transformers in use today employ a toroidal core. Once we use a toroidal core, a single conductor inserted into the toroid becomes a "turn" ($N_1 = 1$), thus permitting forming of a "bar primary" current transformer. We also have

$$e_2 = N_2 \frac{d\emptyset}{dt} \emptyset = \frac{1}{N_2} \int e_2. \tag{6.57}$$

Hence flux Ø can be reduced by increasing N_2. Of course, increasing secondary turns N_2 increases the length of the wire used for winding the secondary and thus

Fig. 6.23 Toroidal core with winding

increases R_2, resulting in increased errors. Thus, these parameters must be optimized at the design stage itself. Even when the primary current is sinusoidal, the secondary current becomes non-sinusoidal due to the nonlinear B-H relationship. Figure 6.24a illustrates the typical B-H loops of a soft ferromagnetic core at three different maximum flux densities B_1, < B_2 < B_3. Here B_3 is assumed to be greater than the saturation flux level of the core.

For the design of a current transformer, the B-H loop is not useful. For that we use the dynamic B-H curve shown in Fig. 6.24b for a cold rolled grain-oriented (CRGO) silicon steel, hot-rolled silicon steel and iron. Figure 6.24b is obtained by joining all the tips of the individual B-H loops shown in Fig. 6.24a. Apart from the above constraint the ISF of metering current transformers necessitates that the operating flux at the limit of ISF be kept approximately at the knee point of the dynamic B-H curve of the soft ferromagnetic core. On the other hand, for a protection current transformer, the knee point of the dynamic B-H curve must be far below the flux level

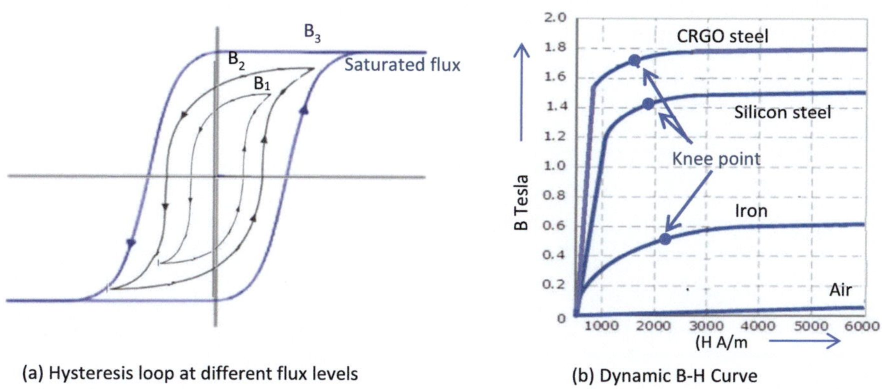

Fig. 6.24 B-H characteristics of soft ferromagnetic core

when the primary current is at the limit of specification (5 times or 10 times). The knee point is defined as the point on the dynamic B-H curve where a 10% increase in B will result in a 50% increase in H. From the forgone discussions, we see that the design of a current transformer is a complex task and many times a designer may have to do several iterations to arrive at an optimal design. Another inference we can arrive at is that a given current transformer will exhibit least errors if the value of $\mathbf{Z}_\mathrm{B}$ is zero (short circuited secondary current transformer).

6.2.5 *Current Transformer on Open Circuit*

When the secondary side of a current transformer gets open circuited when the primary current is nonzero, the secondary will develop a high voltage across the secondary terminals and such a situation is not safe. The equivalent circuit of a current transformer with secondary on open circuit is shown in Fig. 6.25a. This situation can be thought of as though the current transformer is excited using an AC current source as shown in Fig. 6.25b, since $\underline{I}_2$ and $\underline{I}'_2$ are zero. Hence the primary current $\underline{I}_1$ becomes the excitation current $\underline{I}_e$ and that saturates the core even if the primary current is a small fraction of the full-scale (nominal) current. The core getting saturated (that is core flux is no longer changing) results in the $\frac{d\emptyset}{dt}$ near zero crossing rising to a large value. Hence the secondary induced emf will be as shown in Fig. 6.26, with the peak voltage even reaching several kilovolts. Such a high value of voltage can cause

(a) flashover and
(b) dielectric breakdown.

Both conditions will cause irreversible damage not only to the current transformer but also to interconnected and nearby instruments as well. Hence, under no circumstances can the secondary of a current transformer be on open circuit. It is a good engineering practice that the secondary is short circuited even when a current transformer is in storage.

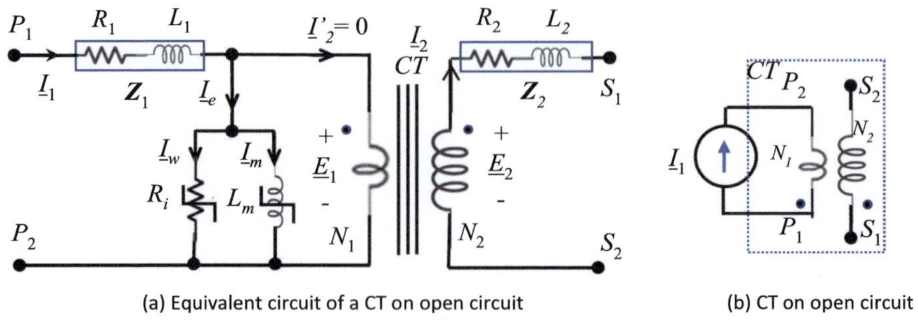

(a) Equivalent circuit of a CT on open circuit (b) CT on open circuit

Fig. 6.25 Equivalent circuit of a current Transformer with secondary on open circuit

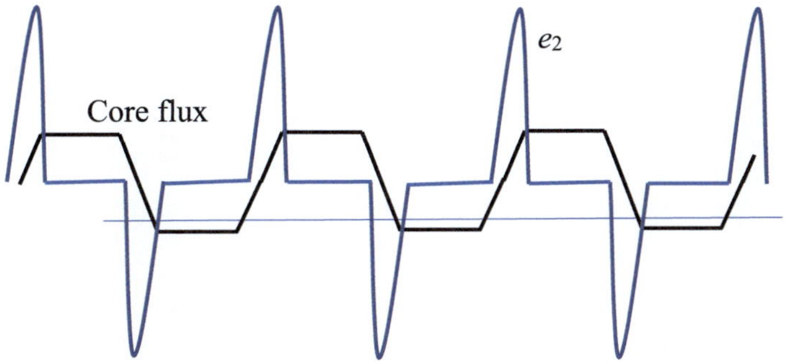

Fig. 6.26 Core flux and secondary voltage of a current Transformer with secondary on open circuit

6.2.6 *High Voltage Current Transformer*

A current transformer to be used on a high voltage power line must have the required heavy insulation between the primary and secondary terminals. In general, a current transformer to be used on a high voltage (HVCT) will be placed on top of an insulation stack as shown in Fig. 6.27a and the secondary terminals are brought down. In few instances (especially for protection type of current transformer, where the weight of the current transformer is quite large), the current transformer is mounted at the bottom of the insulation stack and the primary line is brought down from the transmission lines and taken back up to the tower as illustrated in Fig. 6.27b. Rarely, a two-stage current transformer design is also used for realizing a high voltage CT (A primary CT cascaded to a secondary CT). As in any high voltage transformer, the insulation can be dry type or oil (wet) type.

6.2.7 *Measurement of Errors of Current Transformers*

In general, the errors of any measuring apparatus can be determined using:

(a) The Absolute method or
(b) The Comparison method.

However, the absolute method of testing a current tarnsformer (or voltage transformer) is very tedious and is fraught with practical problems. Hence, for determining the ratio and phase errors of a current transformer (or voltage transformer), we normally employ the comparison method. The schematic for the comparison method of determining the ratio and phase errors of a current transformer is shown in Fig. 6.28. Here, the errors of the test transformer X is determined by comparing it with a standard current transformer S that has the same nominal ratio but nearly zero ratio and phase errors. The primary windings of current transformer under test X

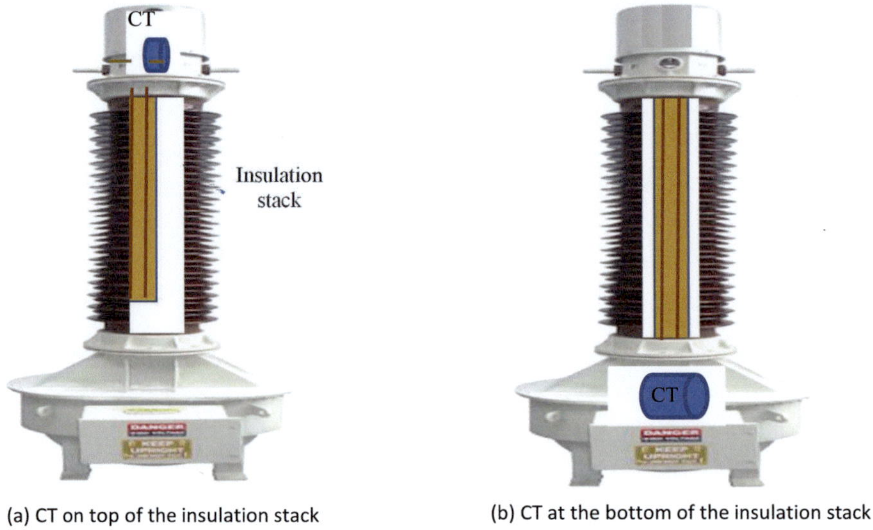

Fig. 6.27 HV current transformers

and the standard current transformer S are connected in series. The secondary windings of X and S are also connected in series as shown in Fig. 6.28. Here $\boldsymbol{Z}_{BS}$ is the rated burden of the standard current transformer and $\boldsymbol{Z}_{BX}$ is the rated burden or test burden of the transformer whose ratio and phase errors are to be determined. With the connections as given in Fig. 6.28, the difference current $\underline{I}_d$ obtained by subtracting $\underline{I}_{2S}$ from $\underline{I}_{2X}$ as: $\underline{I}_d = \left(\underline{I}_{2X} - \underline{I}_{2S}\right)$ and made to flow through the impedance $\boldsymbol{Z}_d$. The phasors of the currents, $\underline{I}_{2S}$, $\underline{I}_{2x}$ and $\underline{I}_d$ in the circuit of Fig. 6.28 are as shown in Fig. 6.29. From the fact that β is usually very small, the errors of the test C.T. are easily shown to be given by the following *approximate* expressions.

Fig. 6.28 Comparison method of testing a current Transformer

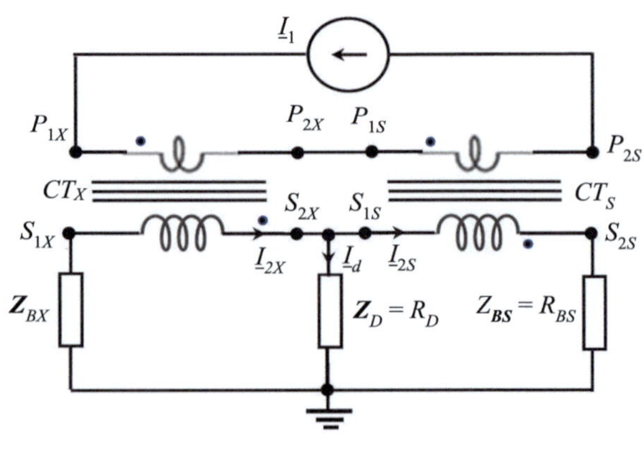

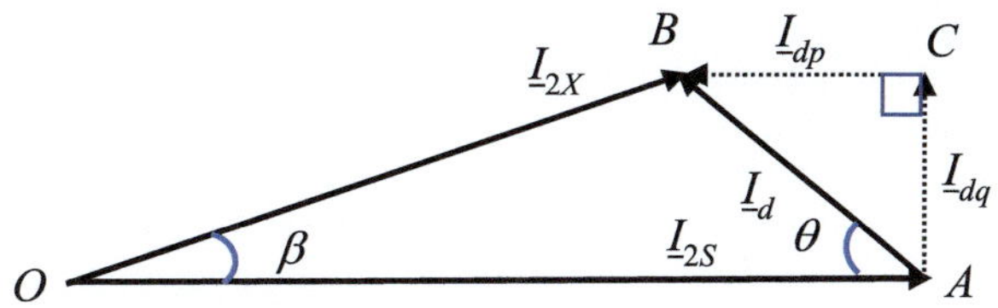

Fig. 6.29 Comparison method of testing a current Transformer-Phasors (Not to scale)

$$\text{Ratio error} \approx -\left(\frac{I_d Cos\theta}{I_{2S}}\right)100 = -\left(\frac{I_{dp}}{I_{2S}}\right)100\,\%. \tag{6.58}$$

and

$$\text{Phase error } \beta \approx \left(\frac{I_d Sin\theta}{I_{2S}}\right)3438 = \left(\frac{I_{dq}}{I_{2S}}\right)3438 \text{ min.} \tag{6.59}$$

or

$$\text{Phase error } \beta \approx \left(\frac{I_d Sin\theta}{I_{2S}}\right)100 = \left(\frac{I_{dq}}{I_{2S}}\right)100 \text{ centiradians.} \tag{6.60}$$

Measurement of the ratio and phase errors of a test CT by implementing Eqs. (6.58) and (6.59) or (6.60) was accomplished with an AC bridge.

6.2.7.1 Bridge Method of Measurement of the Errors of a Current Transformer

Of the several bridge forms that were developed for the measurement of the errors of a current transformer, the "Hohle" bridge, the "Pitch-Elliot" bridge and the Arnold bridge were used to build instrument transformer testing/calibration units and sold commercially. Figure 6.30 shows the schematic of an Arnold bridge for testing current transformers. In the Arnold bridge, the currents $\alpha \underline{I}_{2S}$ (as a voltage drop across a variable resistance and $\gamma \underline{I}_{2S}\angle 90°$ (Using the principle that the induced emf in a mutual inductance is shifted in phase by 90°from the inducing current) are generated and balanced against $\underline{I}_d$ so that $\underline{I}_d = \alpha\left(\underline{I}_{2S} + \gamma \underline{I}_{2S}\angle 90°\right)$. Once α and γ are adjusted to get a balance (that is detector D reads zero) then we have $I_{dp} = \alpha I_{2S}$ and $I_{dq} = \gamma I_{2S}$. Hence, we can compute the ratio and phase errors of a current transformer as

$$\text{Ratio error } = -\left(\frac{I_{dp}}{I_{2S}}\right)100 = -\alpha 100\,\%. \tag{6.61}$$

$$\text{Phase error } \beta = \left(\frac{I_{dq}}{I_{2S}}\right)3438 = \gamma\ 3438 \text{ min.} \tag{6.62}$$

Or

$$\text{Phase error } \beta = \left(\frac{I_{dq}}{I_{2S}}\right)100 = \gamma\,100 \text{ centiradians.} \tag{6.63}$$

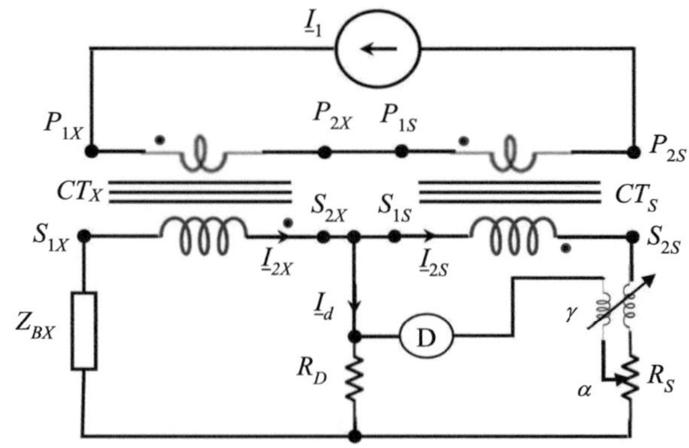

Fig. 6.30 Arnold's bridge for testing a current Transformer

Figure 6.30 is a very simplistic version of the Arnolds bridge drawn only to illustrate the principle of measurement. The actual practical Arnolds bridge is more complex than the one shown in Fig. 6.30. In a typical Arnold bridge the design of the bridge itself is done such that the dial that varies α itself is marked in terms of percentage ratio error and the dial that varies γ is marked in terms of phase error (minutes or centiradians as the case may be). However, the bridge methods of measurements are no longer in use. The methods that are current use the direct measurement of ratio and phase errors using the ubiquitous phase sensitive detector we learnt in an earlier chapter.

6.2.7.2 Phase Sensitive Detector Based Measurement of the Errors of a Current Transformer

Figure 6.31 shows the block diagram of the phase sensitive detector (PSD) method of measurement of the errors of a current transformer. The currents $\underline{I}_{2S}$ and $\underline{I}_{d}$ are converted into equivalent voltages $\underline{E}_{2S}$ and $\underline{E}_{d}$ using standard resistances R_{BS} and R_{DS} respectively as shown in Fig. 6.31. Three phase sensitive detectors, one quadrature PSD (QPSD) and two in-phase PSDs (IPSD 1 and IPSD 2) are employed. The reference input V_r of all three PSDs is $\underline{E}_{2S}$. The input of the IPSD 1 is set as $A_1\underline{E}_{d}$ using an amplifier of gain A_1 as shown in Fig. 6.31. The input V_i of the QPSD is set as $A_2\underline{E}_{d}$ using an amplifier of gain A_2 and the reference input of the IPSD 2 is set as $\underline{E}_{2S}$. With these connections, the outputs V_{op}, V_{oq}, and V_{os} of IPSD 1, QPSD, and IPSD 2 are

$$V_{op} = \tfrac{2\sqrt{2}}{\pi} A_1 R_{DS} I_d Cos\theta. \tag{6.64}$$

$$V_{oq} = \tfrac{2\sqrt{2}}{\pi} A_2 R_{DS} I_d Sin\theta. \tag{6.65}$$

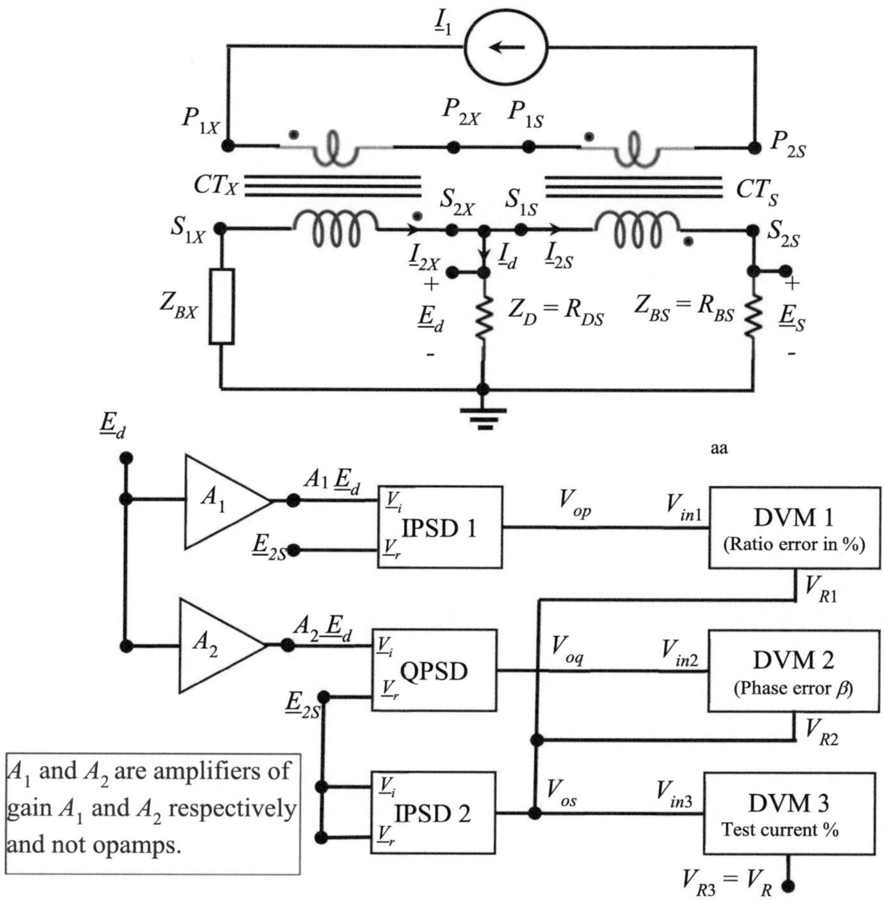

Fig. 6.31 Testing a current transformer using PSD

$$V_{os} = \frac{2\sqrt{2}}{\pi} R_{BS} I_{2S}. \tag{6.66}$$

The outputs V_{op}, V_{oq}, and V_{os} of the three PSDs are tied to the inputs V_{in1}, V_{in2}, and V_{in3} respectively of three individual digital voltmeters DVM 1, DVM 2, and DVM 3 as seen in Fig. 6.31. While the reference inputs V_{R1} and V_{R2} of DVM 1 and DVM 2 are tied to the output V_{os} of IPSD 2 as shown in Fig. 6.31, the reference input V_{R3} of DVM 3 is set as V_R. If we use 4½ digit DVMs, the DVMs will read:

$$DVM\,\text{Reading} = 10000\frac{V_{in}}{V_{ref}},$$

where V_{in} is the input to the DVM and V_{ref} is the reference input of that DVM. Thus, the readings of the three DVMs will be

$$\text{Reading}_{DVM1} = 10000\frac{A_1 R_{DS}}{R_{BS}}\left(\frac{I_d Cos\theta}{I_{2S}}\right). \tag{6.67}$$

$$\text{Reading}_{DVM2} = 10000\frac{A_2 R_{DS}}{R_{BS}}\left(\frac{I_d Sin\theta}{I_{2S}}\right). \tag{6.68}$$

$$\text{Reading}_{DVM3} = 10000\left(\frac{\frac{2\sqrt{2}}{\pi}I_{2S}}{V_R}\right). \tag{6.69}$$

By design if we make

$$\left[10000\left(\frac{A_1 R_{\text{DS}}}{R_{\text{BS}}}\right)\right] \text{to be 1.0000 when} \left(\frac{I_d Cos\theta}{I_{2S}}\right) = 0.01,$$

then DVM 1 will read the ratio error of the test CT with a full scale of 2.000%. Similarly, by design if we make

$$\left[10000\left(\frac{A_{21} R_{\text{DS}}}{R_{\text{BS}}}\right)\right] = 3438 \text{ when} \left(\frac{I_d Sin\theta}{I_{2S}}\right) = 1$$

then DVM 2 will read phase error in minutes. On the other hand, if we make

$$\left[10000\left(\frac{A_{21} R_{\text{DS}}}{R_{\text{BS}}}\right)\right] = 10000 \text{ when} \left(\frac{I_d Sin\theta}{I_{2S}}\right) = 1,$$

then DVM 2 will read the phase error in centiradian, with a full-scale of 200 centiradians. If we make

$$\left(\frac{2\sqrt{2}R_{BS}}{V_R \pi}[I_{2S}]_{FS}\right) = 100.00,$$

then DVM 3 will read the percentage of the testing current. Here, $[I_{2S}]_{FS}$ is the full-scale secondary current (nominally either 5 A or 1 A, but can be any standard value)

6.2.7.3 Digital Method of Measurement of the Errors of a Current Transformer

In the digital method of testing current (and voltage) transformers too, phase sensitive detection is employed, but with digital samples. The voltages $\underline{E}_d$ and $\underline{E}_s$ are converted to digital samples using high resolution (24 bit) simultaneous sampling (time synchronized sampling) analog to digital converters and fed to a digital processing unit. If we take N uniform samples per cycle of voltages $\underline{E}_d$ and $\underline{E}_s$ in Fig. 6.31 as E_{dj} and E_{sj} then we can compute the ratio and phase errors as

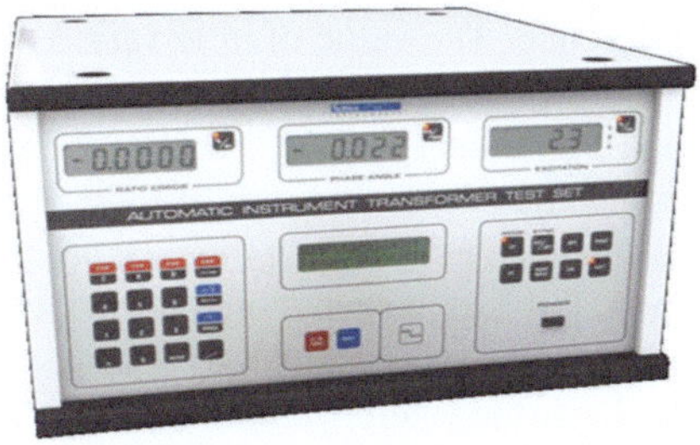

Haefely Tettex CT test set

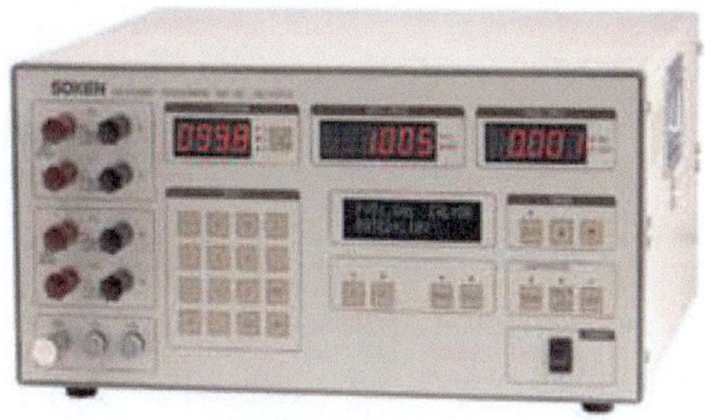

Soken CT test set

Fig. 6.32 Digital current transformer test sets

$$\text{Ratio error} = \frac{\sum_1^N (E_{dj}E_{Sj})}{\sum_1^N (E_{Sj})^2}\frac{R_{DS}}{R_{BS}}100\%. \tag{6.70}$$

$$\text{Phase error}\beta = \frac{\sum_1^N \left(E_{dj}E_{S\left(j+\frac{N}{4}\right)}\right)}{\sum_1^N (E_{Sj})^2}\frac{R_{BS}}{R_{DS}}3438\,\text{min}. \tag{6.71}$$

or

$$\text{Phase error}\beta = \frac{\sum_1^N \left(E_{dj}E_{S\left(j+\frac{N}{4}\right)}\right)}{\sum_1^N (E_{Sj})^2}\frac{R_{BS}}{R_{DS}}100\,\text{centiradians}. \tag{6.72}$$

Figure 6.32 shows a couple of instrument transformer testing units, one manufactured by Haefely-Tettex and the other by Soken.

6.2.7.4 Quasi-Balance Method of Measurement of the Errors of a Current Transformer

We can also use the quasi-balance principle to measure the errors of a current transformer. Figure 6.33 illustrates the principle of operation of the quasi-balance technique for the measurement of the errors of a current transformer. As can be seen in Fig. 6.33, here too, the test current transformer's errors are determined by comparing it with a standard current transformer. As shown in Fig. 6.33a, the difference current $\underline{I}_d$ is converted into a voltage $\underline{V}_d = R_{DS}\underline{I}_d$, represented by the phasor $\underline{V}_{DC}$ in Fig. 6.33b by the current to voltage converter (*CVC*). The secondary current $\underline{I}_{2S}$ of the standard current transformer is converted into a voltage using the standard resistive potentiometer R_{BS} as shown in Fig. 6.33a. Line *OC* in the phasor diagram of Fig. 6.33b represents the phasor $R_{BS}\underline{I}_{2S}$. A fraction $-mR_{BS}\underline{I}_{2S}$ (where $m < 1$) of this voltage is subtracted from the voltage $\underline{V}_{CV}$ to obtain the phasor $\underline{V}_a$. As we vary *m*, the voltage $\underline{V}_a$ changes. Like in the impedance measurement, here too, with the use of the IPSD shown in Fig. 6.33a, two quasi-balances are obtained. It is seen that the input to the IPSD in Fig. 6.33b is the phasor $\underline{V}_a$. The first quasi-balance is obtained with phasor $R_{BS}\underline{I}_{2S}$ selected as the reference. The value of m is varied till we get a zero on the output V_P of the IPSD.

If we employ a synchronous switching type PSD, then $V_P = KV_a Cos(\beta)$. Since V_a is not zero, the output V_P of the IPSD will become zero if and only if $Cos(\beta)$ is zero. $Cos(\beta)$ will go to zero if and only if the phase angle β between the phasors $\underline{V}_a$ and $R_{BS}\underline{I}_{2S}$ is 90°, as indicated by point E in Fig. 5.25b. Let the value of m for the first quasi-balance be denoted as m_1. The second quasi-balance is obtained by setting the switch SW in position 2 and thus selecting $R_{DS}\underline{I}_d$ as the reference and adjusting m till we get a zero at the output V_P of the IPSD. Once V_P is zero we get the condition that $\underline{V}_a R_{DS}\underline{I}_d$ marked by point F. Let the value of m for the first quasi-balance be denoted as m_2. From Fig. 6.33b we get

$$R_{DS}I_d Cos\theta = EC = m_1 R_{BS} I_{2s}. \tag{6.73}$$

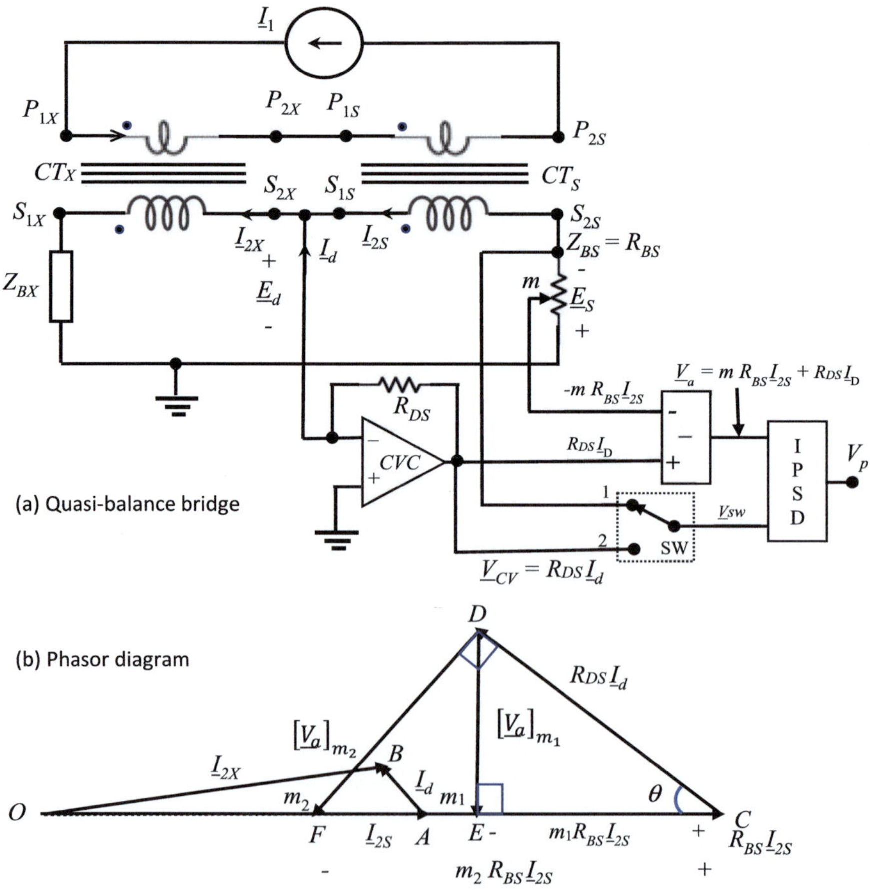

Fig. 6.33 Testing a current transformer using Quasi-Balance

Rearranging Eq. (6.73) we get

$$\%\text{Ratio error} = \frac{I_d Cos\theta}{I_{2S}} * 100 = \frac{m_1 R_{BS}}{R_{DS}} * 100. \tag{6.74}$$

In Fig. 6.30b, we have right-angled triangles DEC and DEF are similar. Therefore:

$$\frac{DE}{FE} = \frac{EC}{DE} DE^2 = FE * EC. \tag{6.75}$$

Substituting $DE = R_{DS} I_d Sin\theta$, $EC = m_1 R_{BS} I_{2s}$ and $FE = (m_2 - m_1) R_{BS} I_{2s}$ in Eq. (6.75) we get

$$(R_{DS} I_d Sin\theta)^2 = ((m_2 - m_1) R_{BS} I_{2s})(m_1 R_{BS} I_{2s}). \tag{6.76}$$

From Eq. (6.76), we get

$$\begin{aligned} \text{Phase error } \beta &= Tan^{-1}\left(\frac{I_d Sin\theta}{I_{2S}}\right) \\ &= Tan^{-1}\left(\sqrt{m_1(m_2 - m_1)}\frac{R_{BS}}{R_{DS}}\right) * 3438 \text{ min}. \end{aligned} \tag{6.77}$$

In Eq. (6.77), if we multiply by 100 instead of 3438 we will get the phase error in centiradians.

6.3 The Potential (Voltage) Transformer

In a high voltage power system, mainly for safety and secondly to provide galvanic isolation, measurement of the system voltage is carried out using a voltage transformer VT (also called the potential transformer PT). Like the current transformer here too we have a primary of N_1 turns and a secondary of N_2 turns as indicated in Fig. 6.34a. If the voltage transformer is ideal, then we have

$$\frac{V_1}{N_1} = \frac{V_2}{N_2}. \tag{6.78}$$

Equation (6.66) is known as volts per turn balance. However, a practical voltage transformer once again will possess nonidealities as shown in Fig. 6.34b. Once again here, R_1 is the resistance of the primary winding, R_2 is the resistance of the secondary winding, R_i is the equivalent resistance that represents the iron losses in the soft ferromagnetic core, L_m is the equivalent inductance that represents the flux generated in the core, L_1 is the equivalent inductance that represents the leakage flux on the primary side, and L_2 represents the leakage flux of the secondary winding. With these nonidealities present in a practical voltage transformer, the secondary voltage will not be as given in Eq. (6.78). The impedance Z_B in Fig. 6.34 is the internal impedance of the voltmeter V_m. Once again, we term the impedance connected on the secondary side of a voltage transformer as *burden* to avoid confusion with the

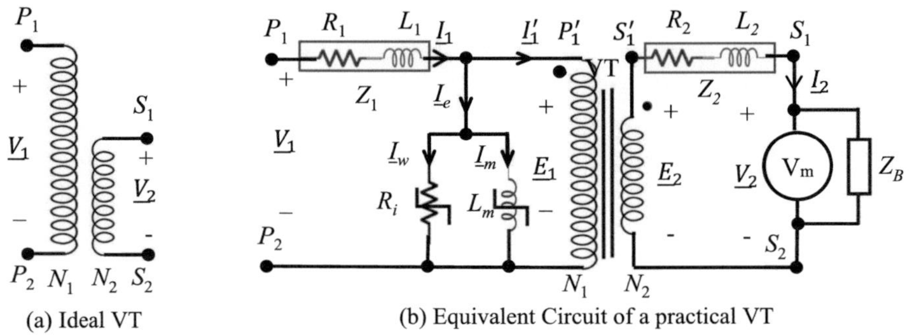

Fig. 6.34 Voltage transformer

"*load*" on the power system. Here too L_m and R_i vary nonlinearly with the flux Ø in the core and hence with respect to the primary voltage $\underline{V}_1$. Moreover $\underline{I}_m$ and $\underline{I}_w$ will also be non-sinusoidal (will contain harmonics). In Fig. 6.34, the quantity $\underline{I}'_1$ is the reflected secondary current $\underline{I}_2$ on the primary side and quantity $\underline{I}'_2$ is the reflected primary current $\underline{I}_1$ on the secondary side. Once we have identified and segregated the nonidealities, then the transformer VT (Primary terminals $P'_1 : P_2$ and secondary terminals $S'_1 : S_2$) in Fig. 6.34 is an ideal one. Figure 6.35 shows the phasor diagram of a VT whose equivalent circuit is as shown in Fig. 6.34b. Unlike the current transformer, here the excitation current and the burden (secondary) current along with the winding resistances and reactances cause voltage drops and thus contribute to the fact that the secondary voltage deviates from the expected value of $\underline{V}_2 = \frac{N_2}{N_1}\underline{V}_1$ and thus introduce errors in the measurement. Figure 6.36 illustrates the phasors of all the quantities referred to the primary side. From Fig. 6.36 we get

$$R_p = R_1 + R'_2 \text{and} X_p = X_1 + X'_2. \tag{6.79}$$

In equation R'_2 and X'_2 are the secondary resistance R_2 and reactance X_2 reflected on the primary side, respectively.

$$V_p = V_1 Cos\beta = K_t V_2 + I'_1 \left(R_p Cos\delta + X_p Sin\delta\right) + I_e (R_1 Cos\gamma + X_1 Sin\gamma). \tag{6.80}$$

and

$$V_q = V_1 Sin\beta = I'_1 \left(R_p Sin\delta - X_p Cos\delta\right) + I_e (R_1 Sin\gamma - X_1 Cos\gamma). \tag{6.81}$$

Since β is normally small, we can take Cos$\beta \approx 1$ and $\gamma \approx \theta$. Hence,

$$I_e R_1 Cos\gamma = I_e R_1 Cos\theta = I_w R_1 and I_e X_1 Sin\gamma = I_e X_1 Sin\theta = I_m X_1. \tag{6.82}$$

With this, we have from Eq. (6.80):

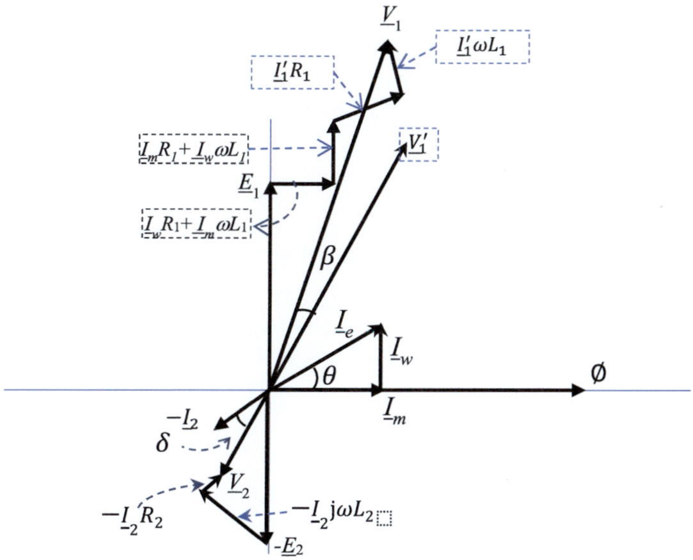

Fig. 6.35 Voltage transformer primary and secondary phasors (Not to scale)

$$V_2' = \left(\frac{N_1}{N_2}\right) V_2 = K_t V_2$$

$$R_2' = \left(\frac{N_1}{N_2}\right)^2 R_2 = K_t^2 R_2$$

$$X_2' = \left(\frac{N_1}{N_2}\right)^2 \omega L_2 = K_t^2 X_2$$

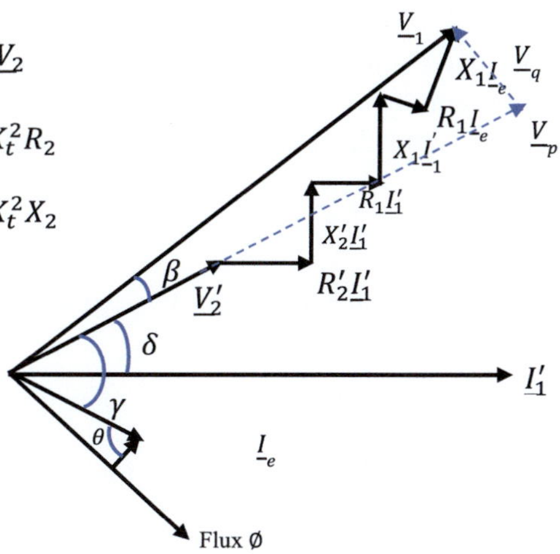

Fig. 6.36 Voltage transformer primary and secondary phasors

$$\frac{V_1}{V_2} = K_v = K_t + \frac{I_1'\left(R_p Cos\delta + X_p Sin\delta\right) + (I_w R_1 + I_m X_1)}{V_2}. \tag{6.83}$$

In Eq. (6.83), K_v is the ratio actual of the VT. With K_N being the nominal ratio (design value) of the VT, the percentage ratio error is

$$\text{Ratio error} = \left(\frac{K_N - K_v}{K_v}\right)100 \approx \left(\frac{K_N - K_v}{K_t}\right)100\%. \tag{6.84}$$

$$\text{Ratio error} = \left[\left(\frac{K_N - K_t}{K_t}\right) - \left(\frac{I_1'(R_p Cos\delta + X_p Sin\delta) + (I_w R_1 + I_m X_1)}{V_1}\right)\right] 100\%. \tag{6.85}$$

If we make $K_N = K_t$, then Eq. (6.85) reduces to

$$\text{Ratio error} = -\left(\frac{I_1'(R_p Cos\delta + X_p Sin\delta) + (I_w R_1 + I_m X_1)}{V_1}\right) 100\%. \tag{6.86}$$

If we have the secondary on open circuit, then I_1' is zero and hence ratio error of an open circuit VT will be

$$\text{Ratio error} = -\left(\frac{I_w R_1 + I_m X_1}{V_1}\right) 100\%. \tag{6.87}$$

From the phasor diagram, we get

$$\beta = Tan^{-1}\left(\frac{V_q}{V_1}\right) \text{radians}. \tag{6.88}$$

From Eqs. (6.70) and (6.71), we get the phase error β as

$$\beta = Tan^{-1}\left(\frac{I_1'(R_p Sin\delta - X_p Cos\delta) + (I_m R_1 - I_w X_1)}{V_1}\right) radians. \tag{6.89}$$

Once again if we multiply Eq. (6.89) with 3148, we get the phase error β in minutes and multiplying Eq. (6.89) with 100 gives us phase error β in milliradians. Since β is normally small, we can take $Tan\beta \approx \beta$ and hence we have

$$\beta \approx \left(\frac{I_1'(R_p Sin\delta - X_p Cos\delta) + (I_m R_1 - I_w X_1)}{V_1}\right) radians. \tag{6.90}$$

The phase error β of an open circuit VT will be

$$\beta \approx \left(\frac{(I_m R_1 - I_w X_1)}{V_1}\right) radians. \tag{6.91}$$

Instead of ratio error, American standard uses a slightly different notation, namely ratio correction factor (RCF) also called the voltage ratio factor (VRF) to represent the error of a VT, where

$$VRF = RCF = \left(\frac{K_v}{K_N}\right). \tag{6.92}$$

From Eqs. (6.84) and (6.92) we get

$$\%\text{Ratio error} = 100\left(\frac{1}{RCF} - 1\right). \tag{6.93}$$

In normal usage, the burden on a VT will typically be of unity power factor and hence the ratio and phase errors of a VT with unity power factor burden are

$$\text{Ratio error} = \left[\left(\frac{K_N - K_t}{K_t}\right) - \left(\frac{I_1'(R_p Cos\delta) + (I_w R_1 + I_m X_1)}{V_1}\right)\right]100\%. \tag{6.94}$$

$$\text{Phase error } \beta = Tan^{-1}\left(\frac{(I_m R_1 - I_w X_1) - I_1' X_p Cos\delta}{V_1}\right)\text{radians.} \tag{6.95}$$

6.3.1 Three-Phase Voltage Transformer

Depending on the three-phase system, different configurations of three-phase VTs are employed.

(I) Three-Phase Four-Wire System

Here three-single phase VTs are used with the primaries of the three VTs connected in star and the secondaries are also connected in star as shown in Fig. 6.37. It is also possible to connect the primaries in delta and secondaries in star resulting in a factor $\frac{1}{\sqrt{3}}$ in the transformation ratio. However, both primaries and secondaries in delta-delta connection are seldom used. A practical three-phase VT is shown in Fig. 6.37b. Instead of using three individual VTs, in practice, we use a single core with three limbs to wind the primary and secondaries of the three-phase transformer, with one limb per phase, as indicated in Fig. 6.37c.

(II) Three-Phase Three-Wire System

It is also possible to use only two single phase VTs and obtain a three-phase VT as shown in Fig. 6.38. This type of connection is called the open V connection and s not preferred.

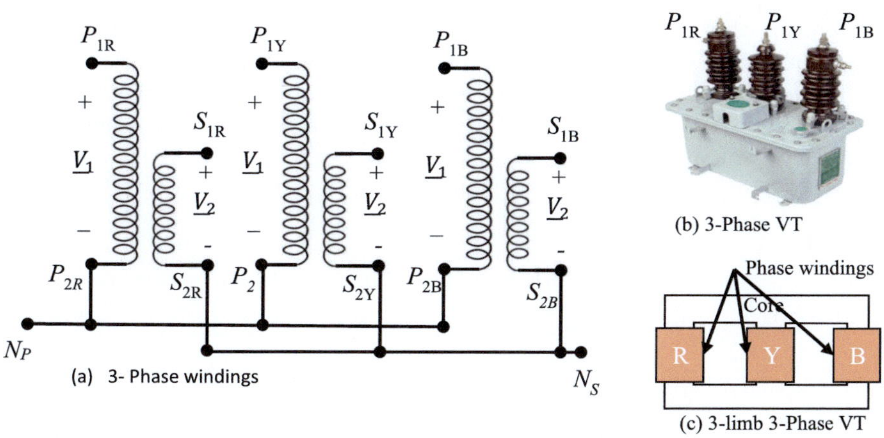

Fig. 6.37 Three-phase four wire voltage transformer

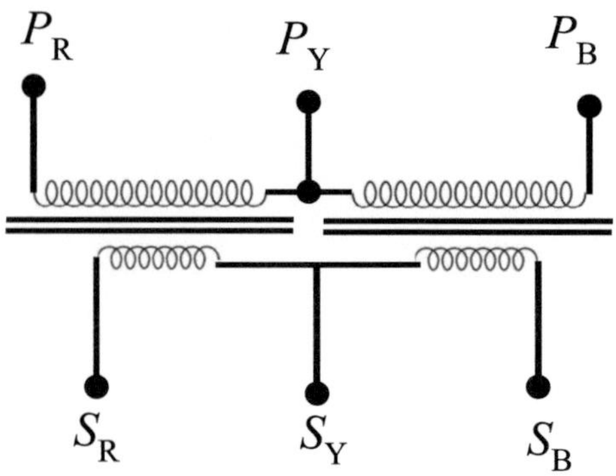

Fig. 6.38 V-connection for 3-phase VT

Unlike current transformers, voltage transformers are classified into two categories only, namely, Measuring (metering) VTs and Standard VTs.

6.3.2 *The Measuring (Metering) VT*

A voltage transformer intended to be used for measurement is again classified based on the accuracy that is ratio and phase errors. Table 6.4. indicates different accuracy classes of voltage transformers as per IS3156. Measuring / metering voltage transformers are expected to work within the accuracy limits up to 120% of the nominal value.

Typical specification of a VT is as given below.

(i) 11 kV / 110 V, 10 VA, UPF, Class 1, single phase, oil cooled.
(ii) $\frac{110}{\sqrt{3}}$ kV / $\frac{110}{\sqrt{3}}$ V, 15 VA, UPF, Class 0.5, single phase, Dry type.

From the above, it is seen that there are two types of insulation, namely, oil insulation and the dry (normally epoxy) type insulation. Apart from the above, the specification may also include "indoor type" and "outdoor" type. While the indoor type insulation VTs must be housed in a protected environment, the outdoor type units can be kept in the open (yard).

Table 6.4 Accuracy class of VT (IS3156)

Sl. no	Accuracy class	% Ratio Error limit	Phase error limit in min
1	0.1	±0.1	±5
2	0.2	±0.2	±10
3	0.5	±0.5	±20
4	1.0	±1.0	±40
5	3.0	±3.0	Not applicable

6.3.3 The Capacitive Voltage Transformer

When the system voltage level is a few hundred kilovolts, the cost of insulation of the transformer becomes substantially high. To offset this high cost, a capacitive voltage transformer is employed. The circuit schematic of a typical CVT is given in Fig. 6.39. Here, C_1 is rated for the system voltage level and C_2 is rated for a fraction of the system voltage level. The open circuit voltage e_o at point O will be

$$e_o = e_{1A}|_{withP_1:P_2 \text{ on open circuit}} = \frac{C_1}{(C_1 + C_2)} v_1. \tag{6.96}$$

The transformer T_r is normally an isolation (1:1) transformer but can also be a (step-down by a factor α) transformer, where α is normally not large. Even with the values of C_1 and C_2 properly chosen Eq. (6.84) will be valid under loaded condition if and only if the capacitive divider has zero output impedance. Thus, when we connect the capacitive divider with a step-down transformer and load the unit at its secondary terminals S_1 and S_2 ratio and phase errors will be introduced. To reduce the ratio and phase errors an additional compensating inductor L_c is introduced. Figure 6.40a shows the CVT with the transformer primary represented by primary leakage inductance L_1 and magnetizing inductance L_m (neglecting primary winding resistance and iron loss). Now we look at the Thevenin equivalent of the circuit of Fig. 6.40a to the left of points a and b given in Fig. 6.40b. Here it should be noted that the Thevenin voltage v_{th} across a-b is as given in Eq. (6.84). Now if we make the series L-C circuit made of the parallel combination of C_1 and C_2 in series with inductances L_c and L_1, resonate at the power frequency by adjusting L_c, then the Thevenin equivalent circuit of Fig. 6.40b reduces to the one shown in Fig. 6.40c. Thus, once the value of L_c is adjusted to obtain resonance, the primary leakage inductance of the transformer is compensated. The errors will now be dependent only on the (i) primary and secondary winding resistances and (ii) the secondary

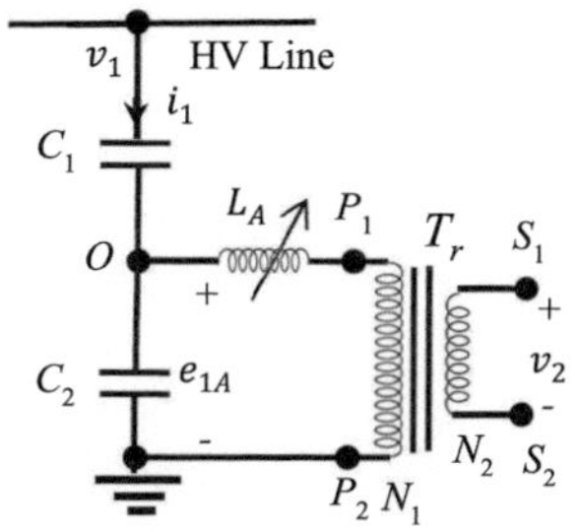

(a) Capacitive voltage transformer (CVT)

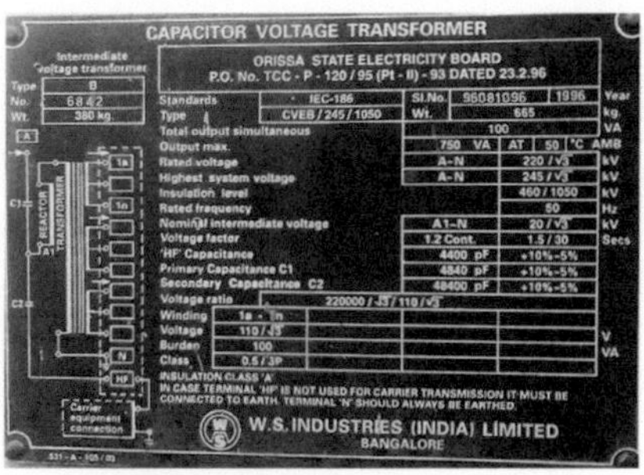

(b) Name plate of a CVT

Fig. 6.39 Capacitive voltage transformer

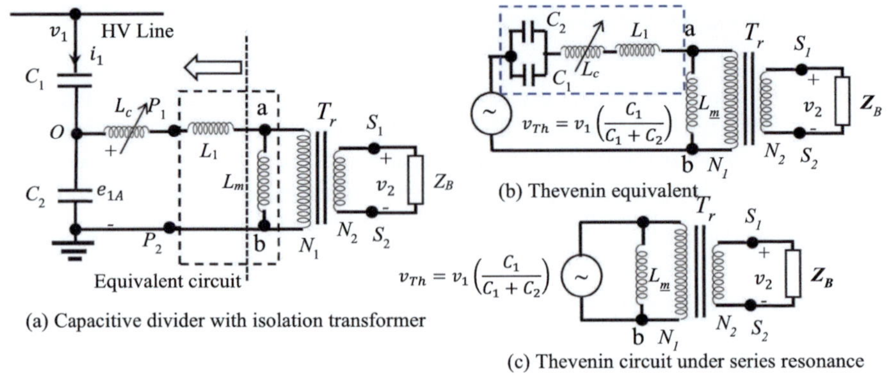

Fig. 6.40 Equivalent circuit of a CVT

Fig. 6.41 Capacitive voltage transformer

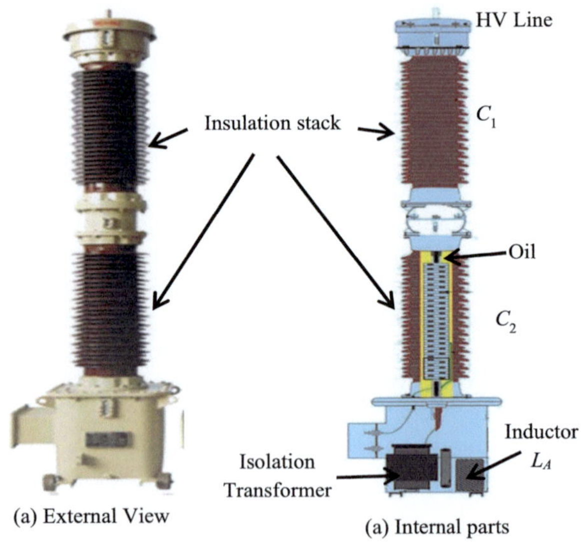

leakage reactance. Thus, the secondary voltage v_2 will have reduced ratio and phase errors. Figure 6.41 portrays a commercial CVT.

6.3.4 *The Standard VT*

As in the case of current transformers, the ratio and phase errors of a voltage transformer are determined using the comparison method. That is, the ratio and phase errors of a voltage transformer under test are obtained by comparing it with a standard voltage transformer of the same nominal ratios. The standard voltage transformer is expected to possess a very low ratio and phase errors. Here again, we have

(i) Passive standard voltage transformers that are designed such that they possess very low ratio and phase errors. Figure 6.42 portrays such a standard VT that has a ratio error of $\pm 0.01\%$ and a phase error of $\pm 1'$. It is also possible to use a 2-stage configuration shown in Fig. 6.43 with three windings to obtain reduced errors. Here the primary windings (P_{1A}–P_{2A}) and (P_{1B}–P_{2A}) are identical with N_1 turns each. While primary winding (P_{1A}–P_{2A}) encloses both cores A and B, the primary windings (P_{1B}–P_{2B}) encompasses only core B. If we assume ideal conditions, then we must have the induced emf e_{1A} in the winding (P_{1A}–P_{2A}) to be equal to the induced emf e_{1B} in the winding (P_{1B}–P_{2B}), resulting in

$$e_{1A} = N_1 \frac{d(\emptyset_A + \emptyset_B)}{dt} = e_{1B} = N_1 \frac{d\emptyset_B}{dt}. \tag{6.97}$$

Equation (6.85) indicates that $\frac{\mathrm{d}\emptyset_A}{\mathrm{dt}} = 0$. Thus winding ($P_{1A}$–$P_{2A}$) is relieved of the need to carry exciting current and the voltage drop it entails. However, it is difficult to make the two windings to be identical, hence in practice the flux reduction in core A will not be very substantial.

(ii) Once again it is possible to use electronics to obtain voltage transformer having very small ratio and phase errors. Of the several schemes available, the one shown in Fig. 6.44 is optimal as this scheme provides isolation for the electronic circuit and places the minimum voltampere demand on the amplifier OA. Here, core A is wound with the detector winding of N_D turns and core B is wound with the compensating winding of N_C turns.

(a) Name plate of standard VT

(b) The passive standard VT

Fig. 6.42 Passive standard VT

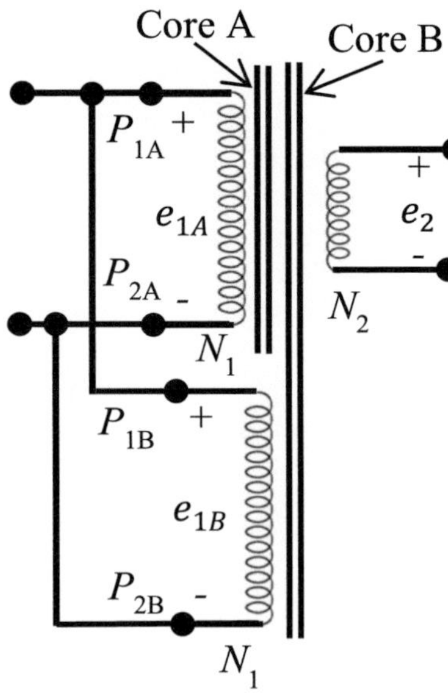

Fig. 6.43 Two core standard VT

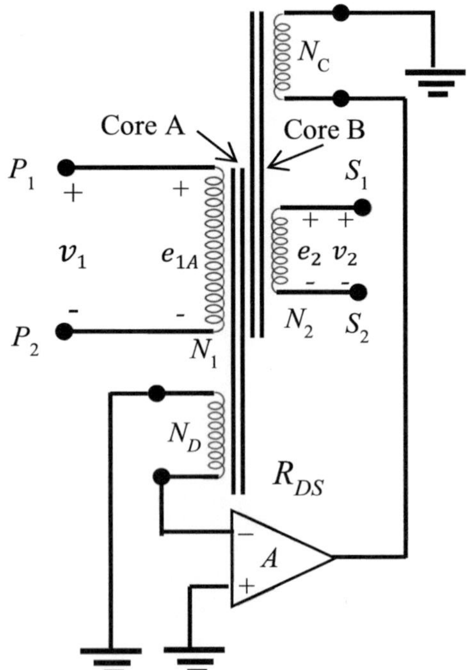

Fig. 6.44 Two core electronic standard VT

The primary of N_1 turns and the secondary of N_2 turns are wound encompassing both the cores. The output of the detector winding is given as input to the amplifier A and the output of the amplifier drives the compensating winding. If we assume the amplifier to be an ideal opamp then we have

$$v_D = N_D \frac{d\varnothing_A}{dt} = 0. \tag{6.98}$$

Hence, we will have

$$\frac{v_1}{N_1} = \frac{v_2}{N_2}. \tag{6.99}$$

Once again, the scheme in Fig. 6.44 is negative feedback and hence will be stable if and only if the gain of the amplifier is lower than the gain margin. Thus, the scheme will help in reducing the errors of the voltage transformer without the application of the scheme by a factor of the closed loop gain of the feedback scheme.

6.3.5 Testing of a Voltage Transformer

As in the case of current transformers the testing of voltage transformers is also accomplished using the comparison method of testing. The ratio and phase errors of a VT are determined by comparing it with a standard VT of similar ratio as shown in Fig. 6.45a. With this connection, we get the phasors of the secondary voltage $\underline{V}_{2X}$ of VT under test, secondary voltage $\underline{V}_{2S}$ of standard VT and the difference voltage represented $\underline{V}_d = (\underline{V}_{2X} - \underline{V}_{2S})$. as in Fig. 6.45b. With this, we can easily derive that

$$\text{Ratio error} = \left(\frac{V_{dp}}{V_{2S}}\right)100\%. \tag{6.100}$$

$$\text{Phase error}\beta \approx \left(\frac{V_{dq}}{V_{2S}}\right)3438\ \text{min}. \tag{6.101}$$

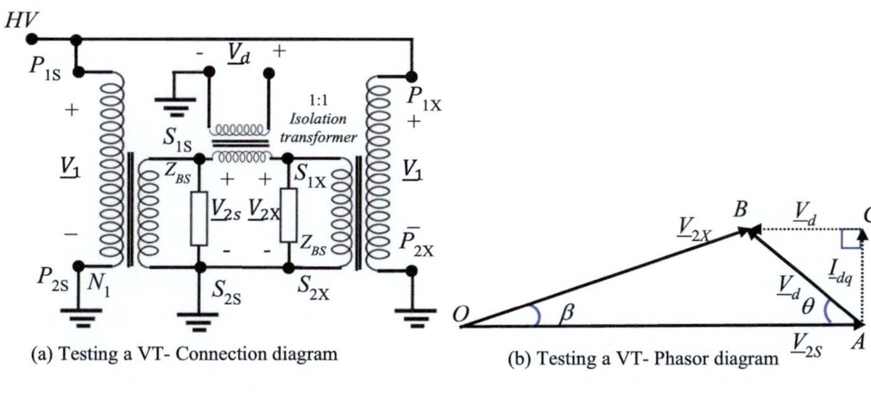

Fig. 6.45 Testing of voltage transformer

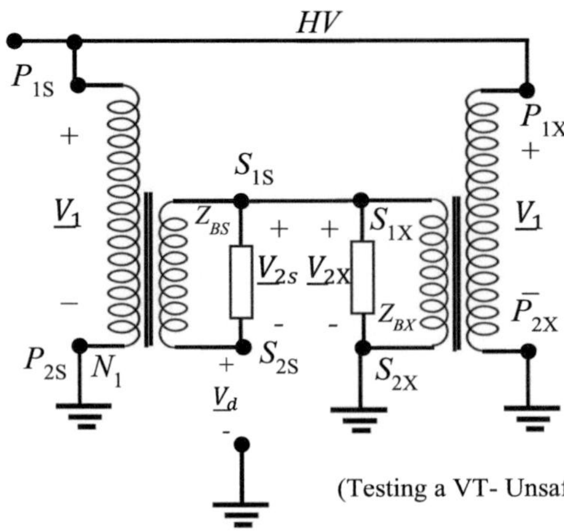

Fig. 6.46 Testing of voltage transformer without isolation transformer

In Eqs. (6.88) and (6.89), we have V_{dp} is the in-phase component of the difference voltage $\underline{V}_d$ with respect to $\underline{V}_{2S}$ and V_{dq} is the quadrature component of $\underline{V}_d$ with respect to $\underline{V}_{2S}$. Once again, instead of multiplying $\left(\frac{V_{dq}}{V_{2S}}\right)$ by 3438, we multiply by 100, we get the phase error in centiradians. We see that the equations are very similar to the ones we obtained for a current transformer and hence using the same method we learnt for testing a current transformer, we can determine the errors of a voltage transformer. Hence for testing a VT, we can use

(i) The bridge method,
(ii) The Quasi-balance bridge method,
(iii) The PSD based method, and
(iv) Inductive voltage divider-based method.

We see in Fig. 6.45 that the difference voltage $\underline{V}_d$ is not referenced to ground and hence we need to use a 1:1 isolation transformer to get the voltage referenced to ground. It maybe we can get the same difference voltage referenced to ground without a need for an isolation transformer by not connecting the secondary terminal S_{2s} not connected to ground as shown in Fig. 6.46. But in high voltage connections, we never keep a terminal floating due to safety concerns. Hence, the connection as shown in Fig. 6.46 is never used as this way of connection is not safe.

6.4 Problems and Exercise

E6.1 A 1000/1 A, 5 VA UPF current transformer (CT) shows an excitation current of (8 + j6) A at rated primary current (Reference phasor is the flux and the primary current leads the flux by 45°). Determine the secondary current phasor and hence determine the ratio and phase errors (Do not use formulae).

E6.2 A 100/1 A, 5 VA, UPF CT with primary turns $N1 = 10$ exhibits ratio and phase errors of -0.5 % and -30 min respectively at half the rated current at 50 Hz. Determine the error of the CT at full rated current and 60 Hz (Assume linear B-H relationship in this region).

E6.3 A 100/1 A, 5 VA, UPF, bar primary CT draws an exciting current of 2.4545° A, when the primary current is $100\angle 0°$ A. Determine the ratio and phase error, as well as the iron losses of the core (Neglect secondary winding resistance and leakage reactance).

E6.4 A 500/5 A, 5 VA, UPF, Class 1 CT has 99 secondary turns and zero ratio error. The phase error is + 30 min. If the burden is changed to 0.8 PF, determine the ratio and phase error of the CT. Neglect the secondary winding resistance.

E6.5 A 100/1 A, 5 VA, UPF, Class 1 CT has 496 secondary turns and zero ratio error and a phase error of +30 min. Determine the number of primary turns. Determine the excitation current drawn by the CT. Also determine the ratio error of the CT if the burden is changed to 2.5 VA UPF (Neglect the secondary winding resistance and assume linear B-H relationship).

Chapter 7
Graphical Methods of Measurement

"Use a picture. It's worth a thousand words."
—Arthur Brisbane

7.1 Need for Graphical Methods

While we generally agree that a picture is worth more than a thousand words, at a first glance the importance of "*visual*" inputs for measurements and instrumentation is not apparent. While graphical methods are not warranted for measurements, recording the measurements for post-analysis or for safe keeping requires graphical methods of measurements, especially in the past when digital storage was not invented or was very expensive. Hence, the first graphical instruments developed were for continually recording the measurements. These instruments are popularly called *recording meters*. Figure 7.1 shows a typical recording multimeter. This type of recording meter was available for voltage, current, power, temperature, humidity, and wind velocity. In such a typical meter, a pen is attached to the end of the pointer, and the paper on which the pen writes is made to move with respect to time, thus creating on the paper a continuous record in the form of a graph of the concerned parameter as the

V. Jagadeesh Kumar, *Principles of Electrical Measurement and Measuring Instruments*,
https://doi.org/10.1007/978-981-96-5475-8_7

y-amplitude with respect to time represented by the x-axis. Hence these instruments were called the y–t recorders. In these recorders, the input (called the y-input) is recorded on a paper as an x–y graph, with the x-axis being time (seconds, minutes, hours, or days). Following this, the x–y recorder shown in Fig. 7.2 was developed. In this recorder, both x- and y-inputs can be any two physical quantities (of course converted into an equal voltage) enabling us to obtain their relative variations as an x vs y record. After the advent of AC power generation, transmission, and distribution, instruments to display the variations in voltage and or current as a function of time became the need of the hour, The first such instrument that was developed was the "Oscillograph" that used a moving light spot that recorded the variation on a photographic plate. The next development was what is known today as the Cathode Ray Oscilloscope or simply CRO. CROs are no longer manufactured but they metamorphosized into today's digital storage oscilloscope (DSO). Though the DSO is distinctly different from a CRO, most of the measurement features of the CRO are extended in a DSO. Thus, from a user point of view, operating a CRO and a DSO are very similar. The heart of a CRO is the "*cathode ray tube*" popularly called the CRT. CRT was the display, not only in CRO but in several other instruments that needed a display, such as "Radar", Patient monitors (to display the Electrocardiogram (ECG), temperature and blood pressure of a patient), Television, and computer terminal, to name a few. However, today the CRT is no longer used for any of these applications and is replaced with a "*Liquid Crystal Display*" (LCD). Since understanding the working principle of a CRT leads to a better and easier understating of an oscilloscope (CRO or DSO), we will discuss the CRT-based oscilloscope here and the digital storage (sampling) oscilloscope in the sequel.

7.2 The Cathode Ray Tube

The name cathode ray oscilloscope or CRO came about as this instrument used a cathode ray tube (popularly called the CRT) to display the input voltage connected to the instrument on a screen. The cathode ray tube or CRT was invented to build an instrument to visualize time-varying parameters but became very popular as an entertainment device as it became the heart of yesteryear's televisions. Figure 7.3 shows two versions of CRTs used in an oscilloscope. As seen in Fig. 7.3, the CRT is made of a glass enclosure with a long cylindrical tube that expands to house a screen at the front. Initially, the screen was circular in structure as shown in Fig. 7.3a but was later changed to a rectangular form as shown in Fig. 7.3b. Inside the glass tube, the screen is coated with a phosphorescent material that emits light if an electron beam meets it. The internal structure of a CRT is illustrated in Fig. 7.4. Before looking at the details of the internal parts, let us first learn the working principle of the CRT. As seen in Fig. 7.4, an electron gun kept at the cylindrical end (farthest point from the screen) generates free electrons and these electrons are then focused by the focusing electrodes to form an *electron beam*. When the electron beam hits the screen, the energy possessed by the electrons is transferred to the phosphorescent material, and

Fig. 7.1 Recording meter

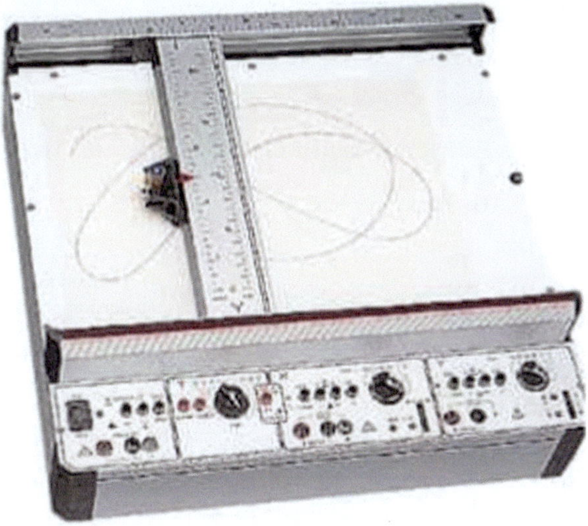

Fig. 7.2 *X–Y* recorder

a spot of light normally in the visible green or bluish-green wavelength is generated. While this was maintained for CRO, for the television the color of the spot can be red, blue, green, or a combination of these three primary colors thus giving rise to a dot of any chosen color. The entire glass enclosure is evacuated and vacuum sealed. A vacuum is required for two reasons.

(i) The electrons should not collide with stray atoms/molecules and should be free to travel from the cathode to the anode without any hindrance before reaching the screen.
(ii) For obtaining a high bandwidth of operation of the CRT, the anode (screen) voltage must be a high voltage (>100 kV) and the vacuum provides high dielectric breakdown and low leakage characteristics needed for such a high voltage level.

The main components that are housed inside the CRT as seen in Fig. 7.4 are:

I. **The electron generator**: To generate electrons, a tungsten filament is heated. This (the heated tungsten filament) also serves as the cathode. If the cathode

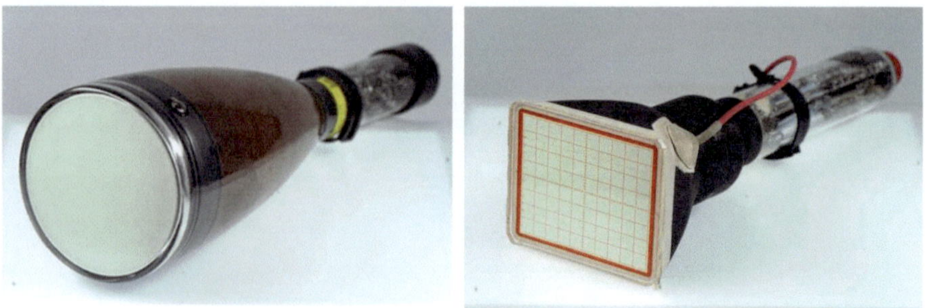

(a) Early Cathode Ray Tube (b) The later version CRT

Fig. 7.3 The cathode ray tube

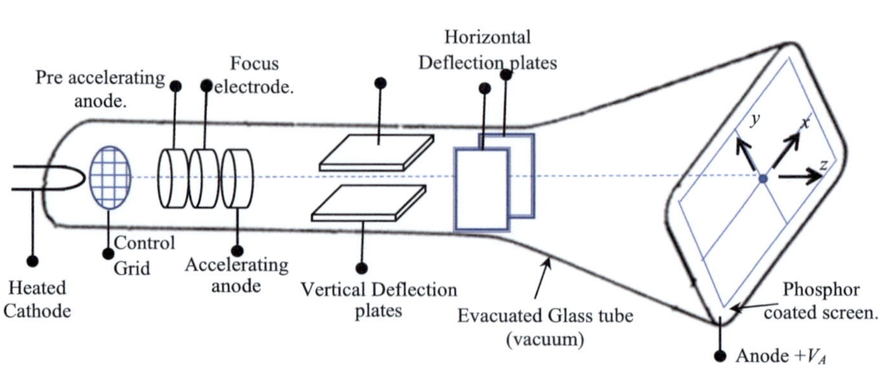

Fig. 7.4 The cathode ray tube internal electrodes

terminal is in galvanic contact with the heating filament, then we call the arrangement as: *directly heated cathode.* If the cathode is galvanically insulated from the tungsten filament, we get an *indirectly heated cathode* or *isolated cathode.*

II. **The anode (screen)**: The front face of the glass tube that serves as the screen to view the waveform is coated with a phosphorescent material (on the inner side of the glass) and is excited with a high voltage of positive polarity (with reference to the cathode) and thus serves as the anode. The field gradient between the anode and the cathode propels the electrons generated by the cathode toward the anode. The design is such that the electrons get accelerated to attain a steady-state velocity v_z before they reach the x and y-deflection plates kept inside the CRT. The phosphorescent material luminesces (emits light at a wavelength) when an electron meets it. Initially, most of the phosphorescent materials that were developed emitted light in the green color region. Later multicolor phosphorescent materials were developed and used, especially in yesteryear's color television.

III. **The control grid**: At a short distance from the cathode, a metallic grid is positioned and is termed the "*control grid*". A negative voltage applied to this electrode controls the number of electrons that pass through the control grid and reach the anode (screen). When the number of electrons hitting the screen increases, then more light will be generated by the screen. Thus, by controlling the magnitude of the negative voltage applied to this electrode, the intensity of the light spot on the screen is controlled. If we keep increasing the grid voltage in the negative direction, at a particular level the grid will reflect all the electrons back to the cathode and thus will extinguish the beam and hence the light spot on the screen. This voltage level that stops all the electrons is called the "*Blanking voltage*", this being the minimum magnitude of voltage that blanks the screen.

IV. **The focusing electrodes**: Three metallic hollow cylinders kept inline after the control grid in the central axis of the cylindrical part of the CRT form the focusing set of electrodes. Figure 7.5 shows a cutaway view of the focusing electrode set. The cylinder that comes next to the control grid is called the "*pre-accelerating*" anode and is applied a potential $+V_{PA}$. The middle cylinder is called the focusing anode and is applied with a voltage $+V_{FA}$. The cylinder is next to the focus anode on the screen side is called the accelerating anode and is applied with a voltage $+V_{AA}$. The voltages $+V_{PA}$ and $+V_{AA}$ are set and $+V_{FA}$ is made variable. As shown in Fig. 7.5 the focus electrode voltage is intentionally made to be less than that of the pre-accelerating and accelerating anodes. This establishes the electric field lines as shown in Fig. 7.5. Electrons that travel at the center along the axis of the cylindrical focusing set of anodes proceed in the same direction, un-hindered. An electron that enters the electric field at an angle gets deflected so that the electron crosses the center line at a distance. By adjusting the voltage difference applied between the focusing anode and the accelerating anodes, the electric field strength is varied and the point at which all the electrons cross the center line can be made to fall right on the screen.

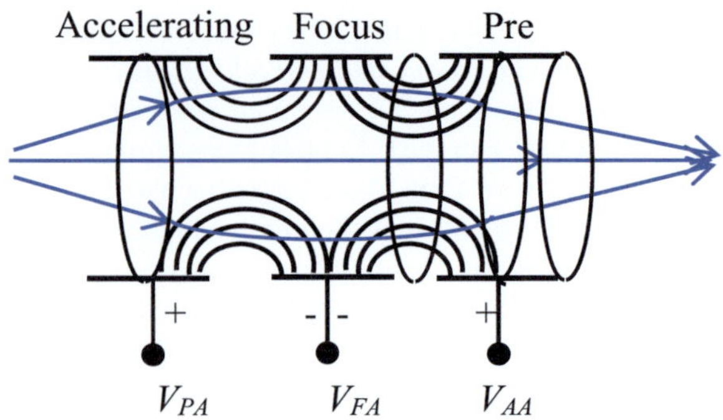

Fig. 7.5 Electrostatic focusing with three electrodes

Hence whenever we adjust the focus knob on a CRO, we change the voltage applied to the middle electrode of the three focusing electrodes set.

V. **The deflection plates**: There are two pairs of parallel plates that serve to deflect the electrons generated by the electron gun. One pair is kept vertically. Ironically, these vertically kept plates serve as "*horizontal deflection electrodes*"—vide the definition for the *x*, *y*, and *z* directions as indicated in Fig. 7.4 and the other pair of plates that are kept horizontally serve as *vertical deflection electrodes.*

We have seen that electrons are generated by the directly or indirectly heated cathode, controlled by the control grid, accelerated by the anode, and focused by the focusing electrodes to hit the screen and display a dot as shown in Fig. 7.4. The stream of electrons that pass through is called an "*electron beam*". By exciting the *x*- and *y*-deflection plates with voltages of appropriate magnitude and polarity, we can deflect (move) this beam right/left and up/down and hence move the dot or light spot displayed on the screen to left/right and up/down.

7.2.1 The y-Deflection

Figure 7.6 shows the schematic (cutaway section) of the *y*-deflection plates (the *y*-deflection plates are kept horizontally inside the tube) applied with a voltage of magnitude V_{Dy} and with the polarity as indicated. With the application of the voltage V_{Dy} there will be a vertical force on the electrons traveling with a velocity of v_z in the *z*-direction. With the polarity as shown in Fig. 7.6, the electrons will experience an upward thrust and move up as indicated in Fig. 7.6. Once the electrons leave the deflection plate at the point marked as *S*, the electrons will travel thereafter in a straight line and hit the screen and display a dot (light spot) at the point marked *R* as shown in Fig. 7.6. If we have $V_{Dy} = 0$, then the electron beam would have traveled straight and displayed a light spot at point *Q*. As illustrated in Fig. 7.6, the length of the deflection plate is l_y, the distance between the plates is D_y, and the distance between the midpoint *O* of the deflection plate and the screen is L_y. With $+V_A$ as the anode voltage (at the screen), the kinetic energy of the electron must be equal to

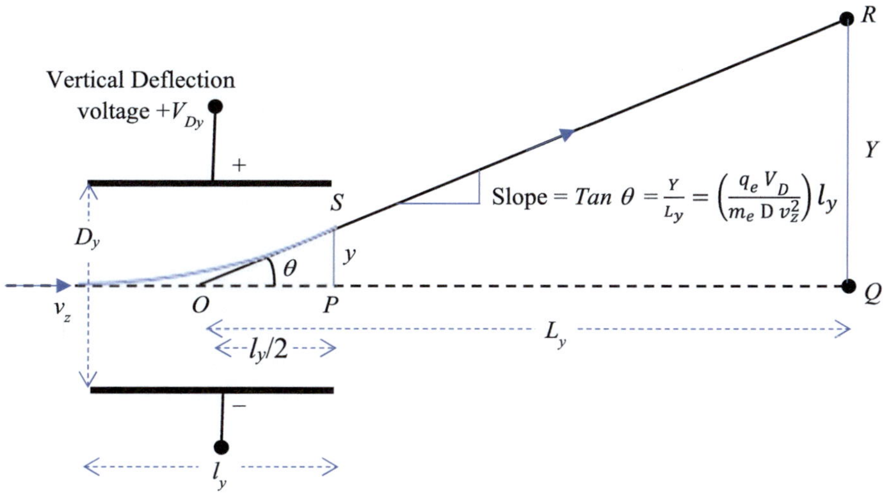

Fig. 7.6 Electrostatic deflection in the cathode ray tube

the potential energy applied and hence, we have

$$\tfrac{1}{2} m_e v_z^2 = q_e V_A \rightarrow v_z^2 = \tfrac{2q_e}{m_e} V_A \tag{7.1}$$

In Eq. (7.1), m_e and q_e are the mass and charge, respectively, of an electron and v_z is the velocity of the electron in the z-direction as indicated in Fig. 7.6. The electrons have zero velocity in the y-direction at the entry point of the y-deflection plates and as they move in the z-direction, the velocity of the electron in the upward (y) direction v_y will increase and can be derived as

$$v_y = a_y t. \tag{7.2}$$

In Eq. (7.2), a_y is the acceleration of the electron in the y-direction. The electrostatic field applied to the electron when it is under the y-deflection electrodes is $E_y = \frac{V_{Dy}}{\mathrm{D}_y}$ V/m and hence the electrons when they are in this field experience a force. Since force F_y on the electron is the field E_y times the charge q_e of the electron.

$$F_y = \tfrac{q_y V_{Dy}}{D_y}. \tag{7.3}$$

Sine force = mass times the acceleration, we get

$$F_y = m_e a_y. \tag{7.4}$$

From Eqs. (7.3) and (7.4) we obtain

$$a_y = \frac{q_e V_{Dy}}{m_e D_y}. \tag{7.5}$$

For the motion of the electron in the y-direction, we have $y = Y_o + v_{yo}t + \frac{1}{2}a_y t^2$. Since the initial displacement Y_o and initial velocity v_{yo} in the y-direction are zero we have

$$y = \frac{1}{2}a_y t^2 = \frac{q_e V_{Dy}}{2m_e D_y} t^2. \tag{7.6}$$

For the motion of the electron in the z-direction, we have $z = Z_o + v_z t + \frac{1}{2}a_z t^2$. Since the initial displacement $Z_o = 0$ and if we assume the electron had obtained the final velocity v_z by the time it reaches the deflection plates (normally this is ensured by designing a long initial cylindrical section as shown in Figs. 7.3 and 7.4) then $a_z = 0$. Thus, we have $z = v_z t$ and hence:

$$t = \frac{z}{v_z}. \tag{7.7}$$

Substituting this value of t in Eq. (7.7), in Eq. (7.6) we get

$$y = \frac{q_e V_{Dy}}{2m_e D_y} t^2 = \left(\frac{q_e V_{Dy}}{2m_e D_y v_z^2}\right) z^2. \tag{7.8}$$

Thus, the deflection in the y-direction within the length of the deflection plates will be a parabola as shown in Fig. 7.6. Now the slope of the path taken by the electron in the y-direction will be

$$\text{Slope} = \frac{dy}{dz} = \left(\frac{q_e V_{Dy}}{m_e D_y v_z^2}\right) z. \tag{7.9}$$

The slope of the electron path when $z = l_y$ is

$$\left.\frac{dy}{dz}\right|_{z=l_y} = \left(\frac{q_e V_{Dy}}{m_e D_y v_z^2}\right) l_y = \text{Tan}\,\theta. \tag{7.10}$$

Since

$$\text{Tan}\,\theta = \frac{Y}{L_Y},$$

we have

$$Y = \left(\frac{q_e V_{Dy}}{m_e D_y v_z^2}\right) l_y L_Y. \tag{7.11}$$

From Fig. 7.5 we see that $OP = \left(\frac{y}{\text{Tan}\,\theta}\right)$ resulting in:

$$OP = \frac{\left(\frac{q_e V_{Dy}}{2m_e D_y v_z^2}\right) l_y^2}{\left(\frac{q_e V_{Dy}}{m_e D_y v_x^2}\right) l_y L_Y} = \frac{l_y}{2}. \tag{7.12}$$

Hence, we see that point O is at the middle of the y-deflection plate. Substituting the value of v_z^2 from Eq. (7.1) into Eq. (7.11), we get

$$Y = \left(\frac{q_e m_e V_{Dy}}{m_e 2D_y q_e V_A}\right) l_y L_Y = \left(\frac{l_y L_Y V_{Dy}}{2D_y V_A}\right). \tag{7.13}$$

From Eq. (7.13) we can derive

$$\frac{V_{Dy}}{Y} = \left(\frac{2D_y V_A}{l_y L_Y}\right). \tag{7.14}$$

The quantity $\frac{V_{Dy}}{Y}$ in Eq. (7.14) is known as the deflection sensitivity of the oscilloscope and has the dimension of V/cm. We can get a desired deflection sensitivity $\frac{v_y}{Y}$ by simply amplifying the y-input v_y with an amplifier of gain A_y. With such an amplifier the deflection sensitivity $\frac{V_{Dy}}{Y}$ becomes

$$\frac{V_y}{Y} = \left(\frac{2D_y V_A}{l_y L_Y}\right) A_y. \tag{7.15}$$

By choosing suitable values of A_y, say $A_{y1}, A_{y2}, A_{y3} \ldots$, we can get the deflection sensitivities as, say

$$\left.\frac{V_y}{Y}\right|_{A_{y1}} = \left(\frac{2\,\mathrm{D_y V_A}}{l_y L_Y}\right) A_{y1} = 1\,\mathrm{mV/cm},$$

$$\left.\frac{V_y}{Y}\right|_{A_{y2}} = \left(\frac{2\mathrm{D_y V_A}}{l_y L_Y}\right) A_{y2} = 200\,\mathrm{mV/cm},$$

$$\left.\frac{V_y}{Y}\right|_{A_{y3}} = \left(\frac{2\,\mathrm{D_y V_A}}{l_y L_Y}\right) A_{y3} = 5\,\mathrm{V/cm}.$$

If we had set the y-sensitivity, say, at 1 V/div, the dot would move up for a y-input voltage of +0.5 V by 0.5 division. If the applied voltage is −4 V, then the dot will move down by 4 divisions. For this setting (1 V/div) any voltage more than 4 V of either polarity, will move the dot out of the viewing window (assuming that the screen size is 10 divisions in the x-axis and 8 divisions in the y-axis). Hence, if the dot has moved by say 1.2 divisions in the upward direction and the y-sensitivity is 20 mV/div then the y-input is $20^*1.2 = +24\,\mathrm{mV}$. Thus, by measuring the displacement on the screen in the y-direction and with the knowledge of y-sensitivity we can measure the amplitude of the y-input.

7.2.2 The x-Deflection

The set of equations derived for the y-deflection above is also applicable for the x-deflection plates. We will get a set of similar equations if we simply replace the variable y and the subscripts of variables with x. Hence deflection X for an applied voltage of V_{Dx} to the x-deflection plates will be

$$X = \left(\frac{l_x L_x V_{Dx}}{2D_x V_A}\right). \tag{7.16}$$

And the x-deflection sensitivity is

$$\frac{V_{Dx}}{X} = \left(\frac{2D_x V_A}{l_x L_x}\right). \tag{7.17}$$

Here X is the deflection of the light spot in the x-axis on the screen when the x-deflection plates of length l_x, separated by a distance of D_x, placed at a distance L_x from the screen is applied with a voltage V_{Dx}.

7.2.3 Extinction Frequency of a CRT

By design, it is ensured that the velocity of the electrons that travel from the cathode to the screen attain a steady-state value before they reach the x- and y-deflection plates. With v_z as the steady-state velocity of the electrons in the z-direction, an electron will take time t_d to cross across a deflection plate of length l_{d} ($l_d = l_x$ for x-deflection plates or l_y for the y-deflection plates). Here $t_d = l_d/v_z$ and any variations with a frequency $f_e = 1/t_d$ will have a net zero deflection of the electron beam at that frequency. During its transit under the deflecting plates, the electron beam will be accelerated upwards for the positive half cycle of the deflecting voltage and undergo a symmetric downward acceleration during the negative half cycle, thus emerging from the deflection zone with no change in the direction of its velocity with which it entered the zone. Hence the light spot will not move on the screen. Thus, t_d is known as the extinction time once a CRT is designed and fabricated. The maximum frequency of signals applied to either of the x- or y-deflection plates must be well below this extinction frequency f_e. Thus, the extinction frequency limits the "*bandwidth of operation*" of a CRT. The extinction frequency f_e of a CRT is proportional to the velocity v_z. The velocity v_z which is proportional to the anode voltage V_A. Hence to obtain a high extinction frequency, the anode voltage must be high. For this reason, the anode voltage of a CRT in a high bandwidth oscilloscope can be in the range of 100–150 kV.

7.3 The Synchronous Cathode Ray Oscilloscope (CRO)

From the foregone discussions, we see that the light spot is displayed at the position (x, y) depending on the value of V_{Dx} and V_{Dy}. To facilitate its observation, the light spot should not instantaneously shift from this location, once V_{Dx} and V_{Dy} change. To increase the duration of its appearance on the screen, screen phosphorescent material that has "*persistence*" of a few tens of milliseconds is used. This ensures that the light spot in its old location fades slowly. However, even with this increased persistence of the screen, the display on the screen will disappear before and one can register the path that the light spot had taken. Like the movie, to view the display as a stable picture, the display must be repeatedly refreshed (at least 25 times a second). For now, we will discuss the use of the CRO for display of waveforms of periodic voltages. Thus, for the y-input, say, $V_y\text{Sin}(\omega_y t)$ as shown in Fig. 7.7 to show one cycle of the waveform on the screen, the one cycle of the sine wave must be repeatedly "*written*" on the screen by suitably moving the spot in the x and y directions. To accomplish this feat, $V_y\text{Sin}(\omega_y t)$ (Marked (a) in Fig. 7.7) is given as the y-input to the oscilloscope and the waveform marked (b) in Fig. 7.7 is applied to the x-input. Here, the voltages $-V_{Xm}$ and $+V_{Xm}$ are the magnitudes of voltages required to deflect the light spot by 5 divisions toward the left and 5 divisions to the right respectively on the x-axis. Since the waveform looks like the teeth of a saw, it is popularly called the *sawtooth waveform*. During the period the sawtooth transitions from $+V_{Xm}$ to $-V_{Xm}$, the spot will travel from the leftmost point on the display to the rightmost point thus leaving a "f*aint line*" as shown in Fig. 7.7d. This horizontal line (which can also be an inclined line) is not part of the input.

7.3.1 The Blanking Pulse

To eliminate this intruding line drawn by the light spot during the period that the sawtooth wave transitions from $+V_{Xm}$ to $-V_{Xm}$, a pulse of magnitude $-V_B$ (V_B is the blanking voltage level required to switch off the light spot) is applied to the control grid to "*switch-off*" the light spot by not allowing any electrons to leave the cathode. Since during this transition time, electrons from the cathode are not allowed to reach the screen, the light spot is extinguished and hence the screen becomes "*blank*". Due to this reason, the pulse waveform marked (c) in Fig. 7.7 is popularly called the blanking pulse. Whenever the blanking pulse is applied to the control grid, it overrides the brightness setting. Once we apply the blanking pulse to the control grid, the return line in Fig. 7.7d will vanish and we will get the screen as shown in Fig. 7.7e. If we make the period of the sawtooth waveform $T_x = 2T_y$ as in Fig. 7.7f, with its corresponding blanking pulse as in Fig. 7.7g, then we will get two cycles of the y-input displayed on the screen as shown in Fig. 7.7h. It should be noted that the blanking pulse which is generated during the transition of the sawtooth from $+V_{Xm}$ to $-V_{Xm}$, shown in Fig. 7.7g, also will have a period $2T_y$. Initially, the x-input

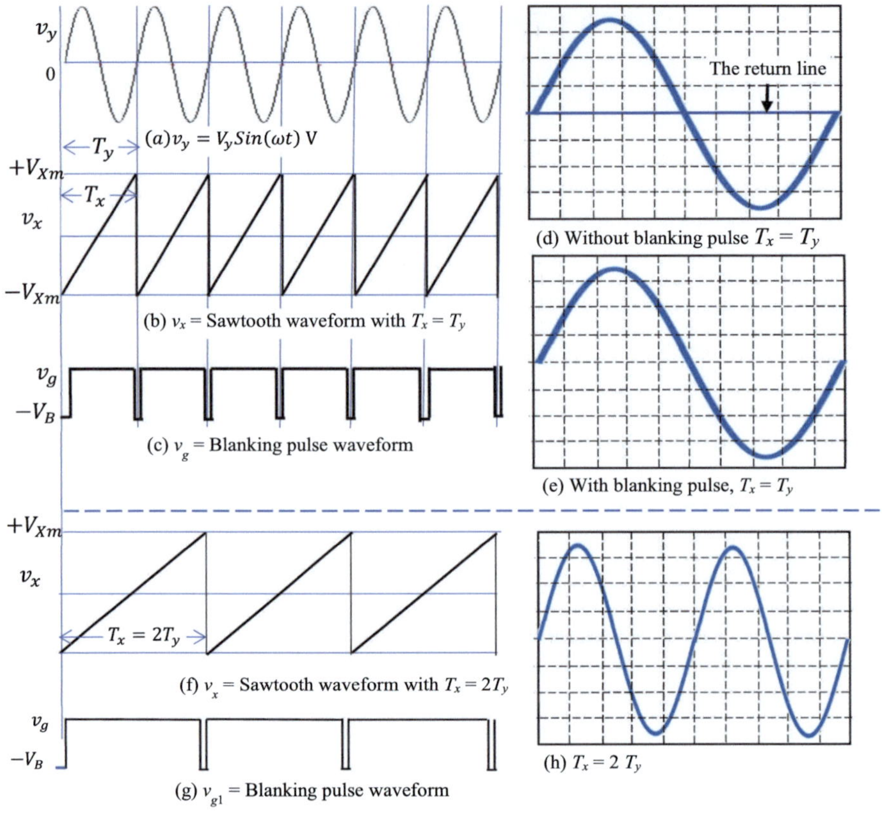

Fig. 7.7 Synchronous cathode ray oscilloscope (CRO)

sawtooth waveform as shown in Fig. 7.7b was generated using the y-input and will have the same frequency (and hence period) or integral multiples of the period T_y of the y-input. If we make the sawtooth period $= T_y$, then the display will show one full cycle as in Fig. 7.7e. If we make the sawtooth period $= 2\ T_y$, then the display will show two full cycles of the y-input as shown in Fig. 7.7h. If we make the period of the x-input depend on the y-input, then we cannot make any measurement of time on the x-axis without the knowledge of the frequency f_y (or time-period T_y) of the y-input. Such oscilloscopes wherein the x-axis period (time) is not prefixed but varies with the y-input are called *synchronous oscilloscopes* (the x-axis signal is synchronized to the y-input). Since measurement on the x-axis is not easy, the synchronous oscilloscopes were not popular and were soon replaced with the "*Triggered mode CRO*" or also called the "*time-base CRO*". Today, we simply call them ***analog oscilloscopes***. Though the general usage of synchronous oscilloscopes stopped once the triggered mode oscilloscopes were invented, the synchronous oscilloscopes are still employed as null detectors in AC potentiometers and AC bridges.

7.4 The Time-Base (Triggered) Oscilloscope

If we want to make measurements of time on the x-axis, independent of the y-input we can do this simply by generating the sawtooth waveform of Fig. 7.7, with a set value of period. If, say each period of the sawtooth is 10.0 ms, then each division on the x-axis will be exactly 1 ms. In such a case, to get a stable display, we need to ensure that the identical section of the y-input waveform is repeatedly written on the screen. Otherwise, the screen will not show any meaningful display. This is achieved by making the sawtooth generator designed as a start-stop generator, that is: it generates only one period of the sawtooth with a prefixed time once a start command is given and then stops. The next period of the sawtooth starts only when the next identical starting point on the y-input waveform is detected. From Fig. 7.8 we see that points P_1 and P_2 have the same trigger voltage level, but the slope at point P_1 is positive and at P_2 the slope is negative. Thus, a distinctive point on the y-input waveform is identified with a combination of two parameters shown below:

(1) The voltage level of the y-input.
(2) The sign (positive or negative) of the slope $\left(\frac{dy}{dt}\right)$ at that level.

The principle of operation of a triggered mode CRO is explained in Fig. 7.9. In Fig. 7.9, the y-input is a sinewave with a period of 11 ms. The sawtooth generator is set to provide one sawtooth waveform having a period of 10 ms (so that each division on the screen is 1 ms). The starting point (trigger) for the sawtooth is taken as: Voltage level zero and slope positive as indicated in Fig. 7.9a. As soon as the y-input reaches zero level, and the slope of change is positive the trigger circuit will generate a *trigger pulse*. Once the trigger pulse happens, the sawtooth generator will start from $-V_{Xm}$ increase linearly, reach $+V_{Xm}$ after 10 ms, drop back to $-V_{Xm,}$ and stop as shown in Fig. 7.9c. The sawtooth generator will not start till the next start signal (the *trigger*) gets generated. In the present example, the sawtooth generator will be idle for 1 ms and start again as shown in Fig. 7.9. Once the next trigger is generated (zero voltage and positive slope on the y-input), the sawtooth generator will generate another ramp exactly for 10 ms and stop. During the period that the sawtooth has stopped (that is between the end of the sawtooth and the beginning of the trigger, in the present example 10 ms to 11 ms of the y-input) the light spot will be at the extreme left-hand side of the x-axis and will create a display with an initial x-portion that is not of the input waveform, we wish to see. Hence, in a triggered

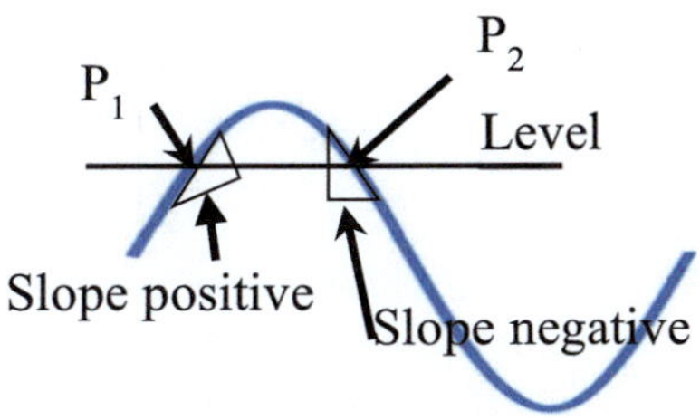

Fig. 7.8 Trigger point

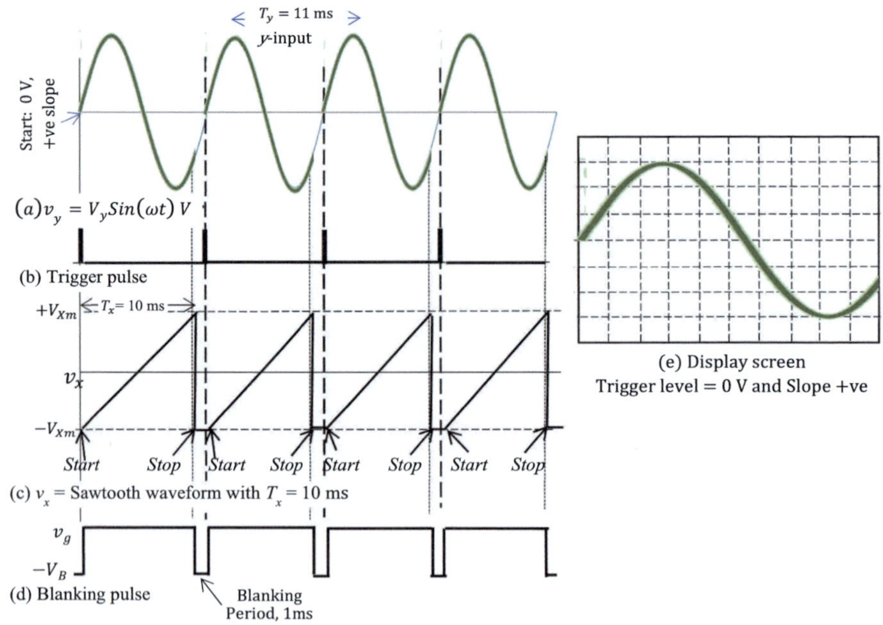

Fig. 7.9 Waveforms in time base (triggered) cathode ray oscilloscope (CRO)

mode CRO, blanking must be applied not only when the transition from $+V_{Xm}$ to $-V_{Xm}$ occurs on the sawtooth but also during the period the sawtooth had stopped (that is its output is $-V_{Xm}$) as shown in Fig. 7.9d. This process of trigger getting generated when the *y*-input level is zero and the slope is positive, and immediately the sawtooth generator outputting one sawtooth of period 10 ms continues and hence the first 10 ms of the sine wave will be repeatedly displayed on the screen as shown in Fig. 7.9e. Whenever the sawtooth transitions from $+V_{Xm}$ to $-V_{Xm}$ as well as during the time the sawtooth generator had stopped (that is its output is at $-V_{Xm}$) the control grid is applied the blanking voltage ($-V_B$.) as shown in Fig. 7.9c. With all the other settings unchanged and we change only the slope from positive to negative, the oscilloscope waveforms will change as shown in Fig. 7.10. Here we see that the trigger gets generated when the level is zero and the slope is negative, thus starting the sawtooth at that point. We also see that the blanking waveform is as shown in Fig. 7.10d, resulting in the display as in Fig. 7.10e.

7.4.1 The Trigger

The start signals for the start-stop sweep generator require the checking of whether the *y*-input has reached the selected voltage and whether the slope at that point is positive or negative. This is achieved by a circuit called the trigger generator circuit.

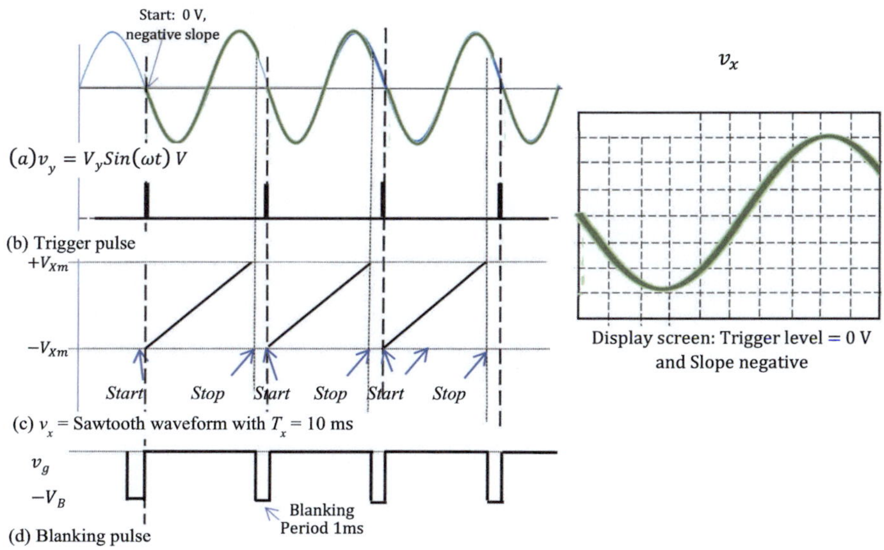

Fig. 7.10 Waveforms for trigger level 0 V and slope negative

The trigger generator circuit takes an input and provides the start signal when the input reaches the set trigger level and slope for the sawtooth generator to generate one period of the sawtooth waveform. A very simple trigger generator and start-stop sawtooth generator circuits are shown in Fig. 7.11a and b respectively.

Trigger level setting: In Fig. 6.11a whenever the input to the trigger circuit crosses the level V_L set on the potentiometer, the comparator OC1 output v_{c1} will change from "0" to 1 and as soon as the trigger input drops below the set level V_L, the comparator output v_{c1} will become "0" as indicated in Fig. 6.11c. Thus, according to the trigger level set on the potentiometer, the rising and falling edges are created at that level-crossing on the trigger input. The rising edge occurs when the input to the trigger generator has a positive slope, and the falling edge happens at the trigger level but when the slope is negative. Thus, to get a trigger pulse at the set trigger level and slope we need to generate a start pulse either at the rising edge (set level with positive slope) or at the falling edge as explained next.

Trigger slope setting: The monostable multivibrator, which is configured as a positive edge triggered device provides a trigger pulse v_{to} of pre-set width once the trigger input attains the level with a positive slope, if the input to the single pole double throw (SPDT) switch S_G is "0". If the input to the single pole double throw (SPDT) switch S_G is "1", then the output of the comparator OC1 is inverted as shown in Fig. 6.11d, and fed as input v_m to the monostable multivibrator. In this case, the trigger output v_{to} is generated by the monostable multivibrator whenever the input to the trigger circuit attains the level V_L with a negative slope as shown in Fig. 6.11d.

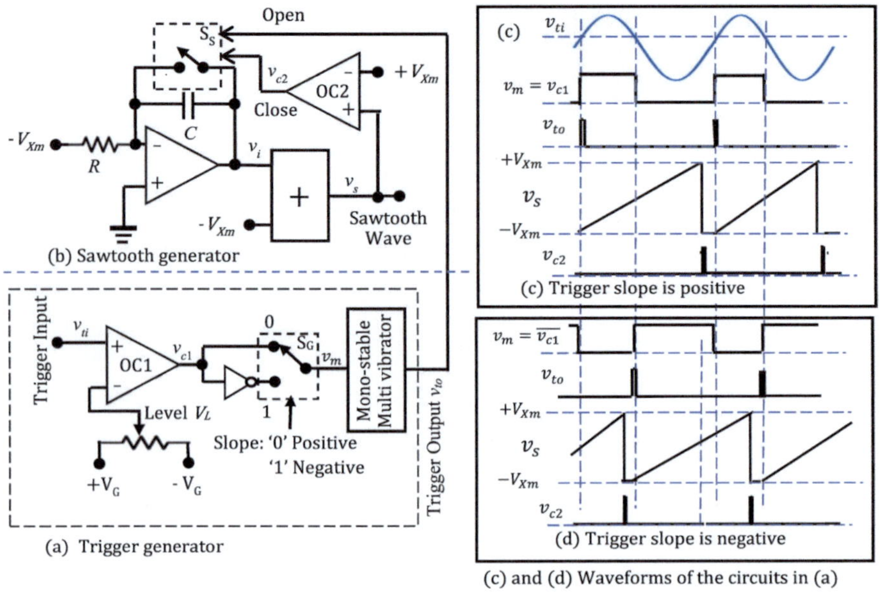

Fig. 7.11 Trigger and sawtooth generators

7.4.2 The Sawtooth Generator

The Fig. 7.11b shows, once again, a very simple sawtooth generator circuit. Here the single pole single throw (SPST) switch S_s is of latching type. That is: a pulse on the "open" signal will open the SPST switch S_s and keep it open till a pulse on the "close" input is applied. Once a "close" pulse is applied, switch S_s will close and will remain in the closed position till a new "open" pulse is applied. Hence once a trigger pulse is generated, switch S_s gets opened, and the integrator output v_i will be

$$v_i = \int \frac{V_{Xm}}{RC} \mathrm{d}t = \frac{V_{Xm}}{RC} \int \mathrm{d}t = \frac{V_{Xm}}{RC} t. \tag{7.18}$$

Hence the output v_i of the integrator will be a ramp that linearly increases with time t, from zero, with a slope $\frac{V_{Xm}}{RC}$. The output v_s of the two-input summary that sums v_i and $-V_{Xm}$ will be

$$v_s = -V_{Xm} + \frac{V_{Xm}}{RC} t. \tag{7.19}$$

Equation (7.19) indicates that once a trigger is generated, the output v_s of the summer will rise linearly from $-V_{Xm}$ with a slope $\frac{V_{Xm}}{RC}$. Once v_s reaches $+V_{Xm}$, the output of the comparator OC2 will become "1" and thus closing SPST switch S_s. Once S_s is closed, capacitor C is shorted, and the output of the integrator will drop to zero. Thus, the output v_s of the summer will become $-V_{Xm}$ and stay at $-V_{Xm}$ till

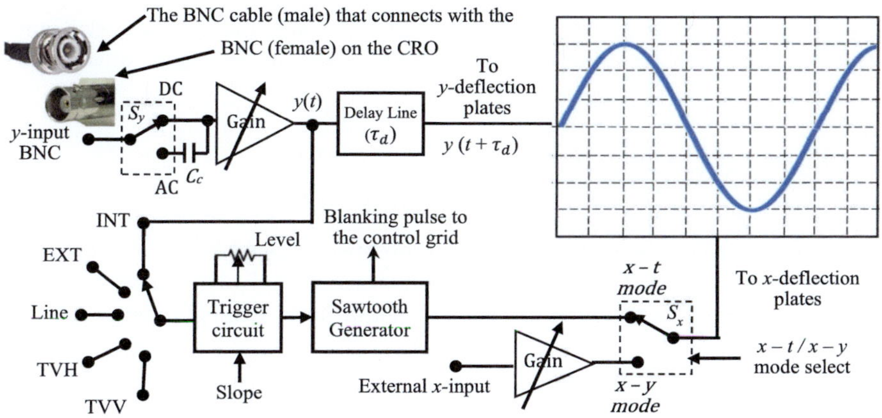

Fig. 7.12 The cathode ray oscilloscope (CRO)

the next trigger occurs. It is easily seen that the output v_s of the summer generates one cycle of a sawtooth once a trigger is generated by the trigger circuit. The period of the sawtooth v_s which travels from $-V_{Xm}$ to $+V_{Xm}$ with a slope $\frac{V_{Xm}}{RC}$ will be

$$v_s = \frac{2V_{Xm}}{\frac{V_{Xm}}{RC}} = 2RC. \tag{7.20}$$

Thus, the period of the sawtooth generator output will be $2RC$. By suitable selection of sets of RC values different time-base settings (sawtooth periods) can be obtained. If the slope of the trigger is set as negative, then the input v_m to the monostable will be an inverted version of v_{c1} as indicated in Fig. 7.11d. The sawtooth now will start when the trigger happens at the level V_L that is set on the potentiometer and the slope of the trigger input is negative. Figure 7.12 shows the block diagram of a typical triggered mode cathode ray oscilloscope (CRO). The first thing we note is that there is a delay line between the y-input amplifier and the y-deflection plates.

7.4.3 The Delay Line

When the level and slope of the trigger circuit are set, the trigger is generated once the level and slope are detected on the input to the trigger circuit, a start signal is issued to the sawtooth generator circuit. The sawtooth generator then starts the sawtooth waveform. All these serial actions take time and hence there will be a time delay τ_{tg}. Hence by the time the sawtooth starts, the y-input would have become $y(t + \tau_{tg})$ instead of $y(t)$. By introducing a delay line in the y-circuit with a delay of τ_d, the input to the y-deflection plates will be $y(t + \tau_d)$. If we ensure at the manufacturing stage itself $\tau_d = \tau_{tg}$, then when the sawtooth starts it will display the correct y-value. Hence all CROs will have a delay line inserted between the y-amplifier and the y-deflection plates to compensate for the delay in the trigger and the sawtooth generator circuits.

7.4.4 The Y-Amplifier

As discussed earlier, the y-input of the oscilloscope is standardized by all the oscilloscope manufacturers. That is the y-deflection sensitivity always follows the 1-2-5-10-20-50-100 sequence irrespective of the manufacturer. Similarly, the y-input can be

(i) Directly coupled (DC coupled and is marked as DC) or
(ii) Connected through a capacitor using the single pole double throw switch S_y as shown in Fig. 7.12. In the AC coupling, the capacitor comes between the input and the y-amplifier, and thus any DC voltage in the y-input will be blocked and only the AC (time varying) part of the y-input will go to the y-deflection plates and get displayed. Hence the introduction of a coupling capacitance at the input of the y-amplifier is called AC coupling. Switch S_y in Fig. 7.12 allows a user to select DC or AC coupling.

The input connector is also standardized and is popularly called the "*bayonet connector*" or simply BNC. The name BNC came about because the BNC connector uses the same mechanical latch-unlatch mechanism that was developed for connecting a *bayonet* to a rifle's end! Fig. 7.12 shows the photograph of a BNC fixed on the CRO, which is the female part of a two-part connector. The BNC cable end, which is the male part is also shown in Fig. 7.12.

7.4.5 The Trigger Selection

We see in Fig. 7.12, that apart from the level and slope settings on the trigger circuit, we can also select the input itself to the trigger circuit. The standard available choices are:

(i) **Internal (INT)**: If this choice is set, then the trigger circuit gets the input from the output of the y-amplifier.
(ii) **External (EXT)**: If this choice is set, then the trigger circuit gets its input from an external input. This facility enables the display of a signal (y-input) with reference to another signal.

Examples:

(1) With reference to the R-phase voltage other voltage or current can be viewed, of course, one at a time.

(2) With reference to a clock signal, other signals in a digital circuit can be viewed, one at a time.

(iii) **Line**: If this choice is set, then the trigger circuit gets its input from the power supply of the oscilloscope. This facility enables the display of a signal (y-input) with reference to the supply (line) frequency.

(iv) **TVH**: TV horizontal. This facility enables the display of one line of a raster-scanned TV signal.

(v) **TVV**: TV vertical. This facility enables the display of the y-input in synchronization with the vertical sync of a raster-scanned TV signal.

Over the years the oscilloscope manufacturers have also standardized several of the oscilloscope parameters and available settings. These are given below.

(i) The screen is always 8 divisions vertical and 10 divisions in the horizontal axes.

(ii) In most oscilloscopes, one division is $\approx$1 cm, but may not be exactly 1 cm. Hence sensitivity is given as 1 V/division instead of 1 V/cm. In an oscilloscope, the deflection sensitivity of the y-deflection can be set as

1 mV/div, 2 mV/div, 5 mV/div,

10 mV/div, 20 mV/div, 50 mV/div,

100 mV/div, 200 mV/div, 500 mV/div,
1 V/div, 2 V/div and 5 V/div.

(iii) The 1-2-5 sequence is maintained both on the x- and y-deflection sensitivities on all oscilloscopes and hence the x-sensitivities will be

1 μs/div, 2 μs/div, 5 μs/div,

10 μs/div, 20 μs/div, 50μs/div,

100 μs/div, 200 μs/div, 500 μs/div,

1 ms/div, 2 ms/div and 5 ms/div.

100 ms/div, 200 ms/div, 500 ms/div,
1 S/div, 2 s/div, 5 s/dv.

(iv) Continuously variable trigger level of the voltage and the choice of its slope.

(v) All the inputs are normally through the BNC type connectors. Also, all the inputs are designed to have an input impedance of 1 MΩ in parallel with $\approx$47 pF.

(vi) A standard 1 V peak-to-peak symmetrical square wave output of frequency 1 kHz is provided for calibrating the input, especially while using the divide by ten or divide by 100 probes (explained in the sequel).

(vii) The oscilloscope can be operated either in the y–t mode or x–y mode.

(viii) Facility to add a DC offset to the *y*-input (called the position adjust). Using this, the displayed waveform can be moved up or down on the screen.
(ix) Facility to give an external input to the control grid. This input is popularly called the *z*-modulation input.

7.4.6 *Measuring with an Oscilloscope*

In an oscilloscope, let the *y*-sensitivity is adjusted as 500 mV/div and the *x*-sensitivity is set as 5 ms/div. The trigger setting is 0 V (level) and slope positive. With these settings, the oscilloscope displays a waveform as shown in Fig. 7.12. From Fig. 7.12, we see that the input is a sinewave of magnitude (peak) of 3 divisions $*500\,\text{mV} = 1.5\,\text{V}$ and a period of 8 divisions $*5\,\text{ms/div} = 40\,\text{ms}$. Hence the frequency is 25 Hz. Thus, the input is

$$v_{\text{in}} = 1.5\,\text{Sin}\,(2 * \pi * 25\,\text{t} + \theta)\text{V} = 1.5\,\text{Sin}(50\pi\text{t} + \theta)\text{V}.$$

It may be noted here that as we trigger the input at an arbitrary level and slope, we may not be able to measure the phase angle θ. However, if we trigger the oscilloscope with an external reference, then we can determine θ as well.

7.4.7 *The* x–y *Mode of Operation*

If this choice is selected, then using the single pole double throw switch S_x, the *x*-deflection plates are disconnected from the output of the sawtooth generator and connected to the output of the *x*-amplifier as shown in Fig. 7.12. In the *x–y* mode of operation, the user must supply both the *x* and *y*-input signals. The *x*-input is given through the "external *x*-input" terminal through the *x*-amplifier as shown in Fig. 7.12. Provision is given to vary the *x*-amplifier gain to obtain the same 1-2-5 sequence of deflection sensitivities as available for the *y*-input. The *x–y* mode of operation enables the oscilloscope to be used for a variety of applications as given below:

(i) Measurement of phase,
(ii) Measurement of frequency, and
(iii) Any other parameter that can be measured using an *x–y* graph.

7.4.7.1 Measurement of Phase

To measure phase difference Ø between two sinusoids, say, $V_x\text{Sin}(\omega t)$ V and $V_y\text{Sin}(\omega t + \text{Ø})$ V, the oscilloscope is set in the *x–y* mode of operation and $V_x\text{Sin}(\omega t)$ V is given to the *x*-input and the *y*-input is connected to $V_y\text{Sin}(\omega t + \text{Ø})$ V. It can be shown that the resulting display will in general be an ellipse. In particular cases, it may be straight line, which is a degenerate case of an ellipse with a minor axis of zero length). Figures 7.13, 7.14, and 7.15 illustrate the displays that would be obtained for different cases. In the following, we take S_x and S_y to be the x- and y-axes sensitivities (volts/division) set on the oscilloscope.

(a) $\emptyset = 0^{\circ}$: The display will be an inclined line as seen in Fig. 7.13a, with a slope equal to:

$$= \left(\frac{V_y}{S_y}\right) / \left(\frac{V_x}{S_x}\right).$$

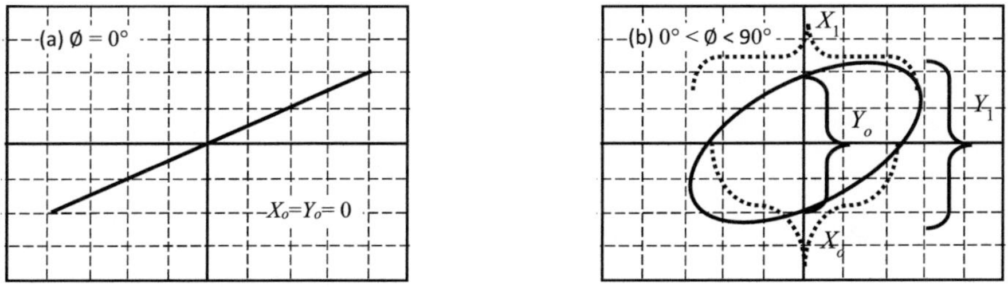

$x\ input = V_x Sin(\omega t);\ \ y\ input = V_y Sin(\omega t + \emptyset)$

Fig. 7.13 Phase measurement using oscilloscope in x–y mode

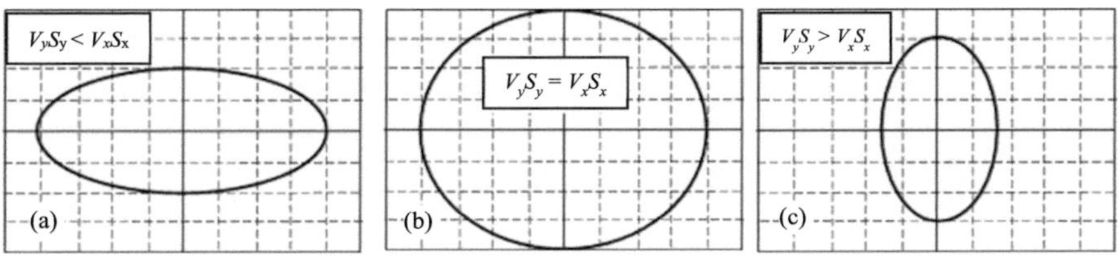

Fig. 7.14 Phase measurement with CRO ($\emptyset = 90°$)

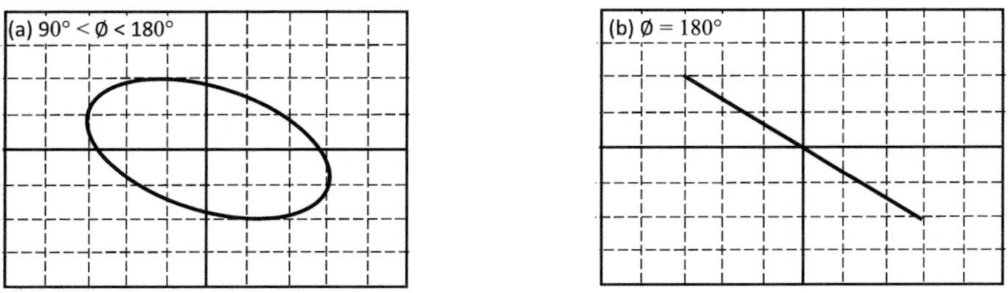

Fig. 7.15 Phase measurement using oscilloscope in X–Y mode

(b) $0^0 < \emptyset < 90^0$: We now get an inclined ellipse as shown in Fig. 7.13b. We note the following from it.

$$\text{Peak Vertical deflection} = \frac{V_y}{S_Y} = \frac{Y_1}{2}. \tag{7.21}$$

A zero-x-axis deflection occurs whenever $\omega t = 0$ or π. At this point, the vertical deflection should correspond to a vertical deflecting voltage of V_y sin Ø. From Fig. 7.13, we note that this deflection is $+\frac{Y_0}{2}$. Therefore, $\frac{V_y}{S_Y} = \frac{Y_0}{2}$. We thus have

$$\emptyset = \pm \text{Sin}^{-1} \frac{Y_0}{Y_1}. \tag{7.22}$$

Similarly, the x-intercept X_o is

$$X_0 = 2V_x \text{Sin}(\emptyset) X_s. \tag{7.23}$$

From Eq. (7.18), we can derive

$$\emptyset = +\text{Sin}^{-1}\left(\frac{X_o}{X_1}\right). \tag{7.24}$$

We thus note that we can deduce only the magnitude of the phase difference between v_y and v_x and cannot identify which one leads the other. This difficulty arises essentially because we cannot observe in slow motion how the ellipse is traced out on the screen, whether in the clockwise or the anticlockwise direction. A careful examination of Fig. 7.13 would reveal that the positive peak of v_y occurs before the positive peak of v_x (i.e., v_y leads v_x and Ø is positive) if it is in the clockwise direction and vice versa. The remarks in the above para apply equally well to the other cases that follow.

(c) If Ø = 90° and

   (i) $V_x S_x > V_y S_y$, the pattern displayed on the screen will be as shown in Fig. 7.14a.
   (ii) $V_x S_x = V_y S_y$, then the display will show a circle as in Fig. 7.14b and
   (iii) $V_x S_x < V_y S_y$, then the shape displayed will be as in Fig. 7.14c.

(d) For Ø in the range >90° but <180° we will get an inclined ellipse pattern as shown in Fig. 7.15a on the screen.
(e) When Ø = 180°, the oscilloscope will display the inclined line as illustrated in Fig. 7.15b.

For Ø in the range 180° to 360° or equivalently in the range $-180° < \emptyset < 0°$, the patterns will repeat as in Figs. 7.13, 7.14 and 7.15. For example, whether Ø is +90° or −90° the oscilloscope screen will show the patterns as in Fig. 7.13. Thus, using the x–y mode we cannot distinguish if the phase angle Ø is in the 1st or 4th quadrant and whether it is in the 2nd or 3rd quadrant.

7.4.7.2 VGK Murti's Method of Phase Measurement

With

$$v_x = V_x \text{Sin}(\omega t) \text{ V and}$$

$$v_y = V_y \text{Sin}(\omega t + \emptyset) \text{ V},$$

to measure $\emptyset$, we give the v_x signal as input to the non-inverting input of an opamp in the open circuit as shown in Fig. 7.16a. The opamp in Fig. 7.16 works as a zero-crossing detector and for an input of sine wave $V_x\text{Sin}(\omega t)$ as shown in Fig. 7.16b its output v_o will be a symmetrical square wave, as shown. Whenever the input to the opamp in Fig. 7.16a is positive, its output will rise and saturate at the positive supply voltage $+V_C$. Similarly, whenever the input to the opamp becomes negative, then its output v_o will attain the negative supply voltage $-V_C$. The corresponding waveforms of v_x, v_o and v_y are shown in Fig. 7.16b. To measure the phase $\emptyset$, here too, the oscilloscope is set in the x–y mode, and the output v_o of the opamp is connected to the x-input and the signal v_y is given to the y-input of the oscilloscope. Figure 7.17 illustrates the pattern that will be displayed on the oscilloscope screen when the phase angle $\emptyset$ is

(a) $\emptyset = 0$
(b) $\emptyset$ is in the range: $0^\circ < \emptyset < 90^\circ$
(c) $\emptyset = 90^\circ$
(d) $\emptyset$ is in the range: $90^\circ < \emptyset < 180^\circ$
(e) $\emptyset = 180^\circ$.

From Figs. 7.17b and d we see that

$$\emptyset = \text{Sin}^{-1}\left(\frac{Y_o}{Y_1}\right). \tag{7.25}$$

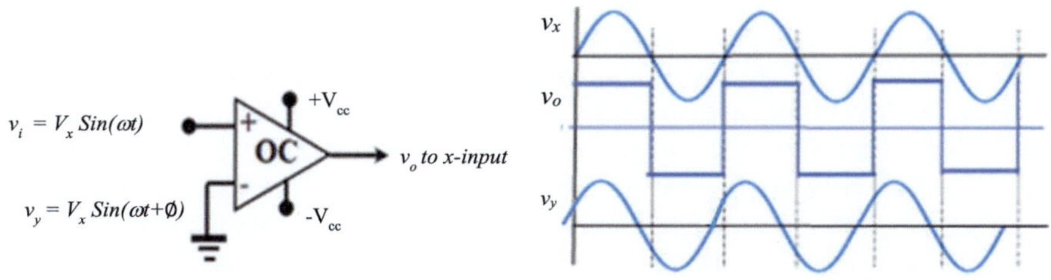

(a) VGK Murti's opamp phase measurement scheme (b) Waveforms of the circuit in Figure (a)

Fig. 7.16 Phase measurement using a comparator

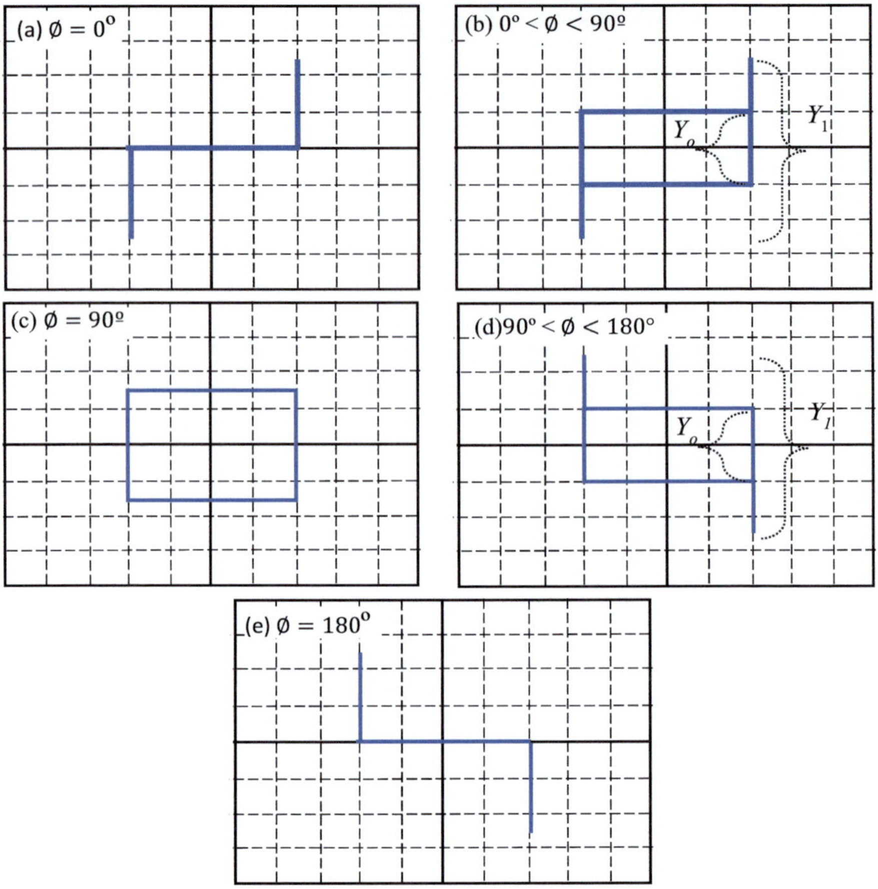

Fig. 7.17 Opamp comparator for phase measurement

The advantage of this method of phase measurement compared to the one illustrated earlier (Figs. 7.13, 7.14 and 7.15) is that the measurements Y_o and Y_1 can be easily accomplished as the distance between the end points of Y_o and Y_1, each pair being on the same vertical line, can be judged without any ambiguity. Here again, for the phase angle Ø in the range from 180º to 360º, the patterns repeat as indicated in Fig. 7.17a–e. That is whether we get Ø = 0° or 360º, the display will be as shown in Fig. 7.17a. Similarly, for Ø = 90° or 270º (that is Ø = ±90° we will get the display as in Fig. 7.17c. Thus, at the first glance it looks like it is not possible to resolve +Ø or −Ø, using this method. However, a practical reality, namely, the *overshoot* and finite slew-rate characteristics of a practical opamp comparator helps us to resolve the difference as illustrated in Fig. 7.18a and b. We see that by looking at the way the vertical portion of the pattern changes into horizontal is displayed on the screen

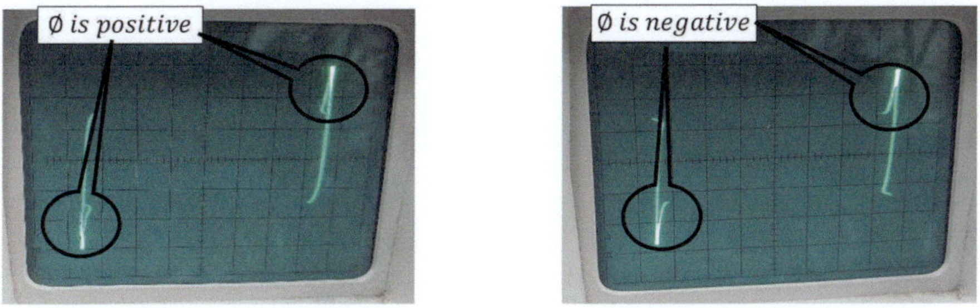

Fig. 7.18 Comparator for phase measurement: practical waveforms

we can easily identify the polarity of the phase Ø as positive or negative. Whenever Ø is negative, the vertical line overshoots at the inflexion point (vertical to horizontal change over).

7.4.8 Frequency Measurement Using the x–y *Mode*

If we set an oscilloscope in the x–y mode and feed the x-input with $v_x = V_x\text{Sin}(\omega t)$ V and the y-input with $v_y = V_y\text{Sin}(2\omega t)$ V, then we get the waveform as shown in Fig. 7.19a. On the other hand, if we feed the x-input with $V_x\text{Sin}(2\omega t)$ V and the y-input with $V_y\text{Sin}(3\omega t + \emptyset)$ V, then we get the display as in Fig. 7.19b. From Fig. 6.19, we can obtain the ratio $\frac{f_x}{f_y}$ of the x and y frequencies f_x and f_y respectively by simply drawing or imagining a horizontal line and a vertical line as indicated by the dotted lines and finding the maximum number of crossings with the horizontal line N_x and that with vertical line N_y. Once the values of N_x and N_y are known, then we can compute the ratio of the x and y frequencies as

$$\frac{f_x}{f_y} = \frac{N_y}{N_x}. \tag{7.26}$$

In Fig. 7.19a we get $N_x = 2$ and $N_y = 4$ therefore we get

$$\frac{f_x}{f_y} = \frac{N_y}{N_x} = \frac{4}{2} = 2.$$

Similarly in Fig. 7.19b we have $N_x = 6$ and $N_y = 4$ therefore we get

$$\frac{f_x}{f_y} = \frac{N_y}{N_x} = \frac{4}{6} = \frac{2}{3}.$$

Such figures were first studied by Jules Antoine Lissajous in 1857 almost a century before the oscilloscope was invented, hence we call these ***Lissajous figures*** in his

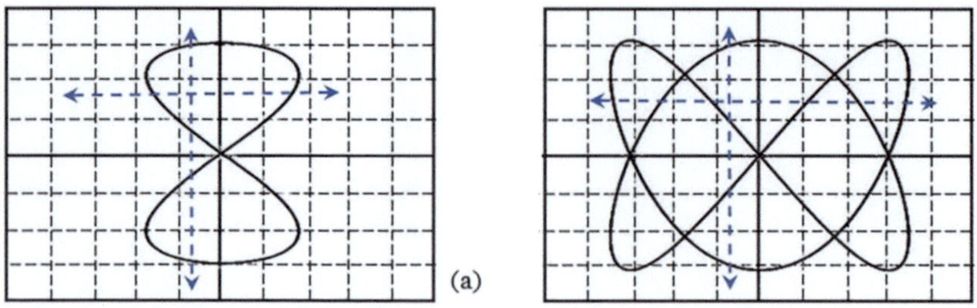

$x\ input = V_x Sin(\omega t)\ and\ y\ input = V_y Sin(2\omega t)$ (a) $x\ input = V_x Sin(2\omega t)\ and\ y\ input = V_y Sin(3\omega t)$ (b)

Fig. 7.19 Frequency measurement using oscilloscope in *x–y* mode

honor (More details at https://www.historyofinformation.com/detail.php?id=3263). The use of Lissajous figures for the measurement of frequency has limited application and is convenient only when an unknown frequency is related to a known reference frequency by a ratio of two small integers.

7.4.9 Hysteresis Characteristics and Iron Loss Measurement

Soft ferromagnetic cores used in solenoids, coils, and transformers are subject to hysteresis and eddy-current phenomena and suffer from the corresponding losses. Using an oscilloscope, it is possible to display the hysteresis characteristics (hysteresis loop at a particular AC excitation frequency and at a set peak flux density $\hat{B}$) and measure the iron losses of a soft ferromagnetic core. For this the soft ferromagnetic core under test is wound with the primary of N_1 turns and a secondary of N_2 turns as shown in Fig. 7.20a. The primary is excited with a sinusoidal voltage source at the required frequency and a magnitude that would yield the specified peak flux density. The output of the secondary is fed as input to an integrator circuit as shown in Fig. 7.20a. The exciting current $\underline{I}_1$ drawn by the primary is converted to a voltage $\underline{V}_{1i}$ by the opamp *i-to-v* converter shown in Fig. 7.20a and this voltage is fed as *x*-input to an oscilloscope, whose *y*-input is tied to the output $\underline{V}_{2i}$ of the integrator. The oscilloscope is set in the *x–y* mode of operation. The opamp will display the hysteresis loop as shown in Fig. 7.20b. Since the magnetomotive force $H(t) = \frac{N_1 i1}{l_m}$, where l_m is the mean magnetic path length of the flux φ in the core, and the voltage v_{Ii} represents $H(t)$ to a scale factor. The secondary induced emf $e_2 = N_2 \frac{\mathrm{d}\varphi}{\mathrm{d}t}$. The output of the integrator is $v_{2i} = N_2 \int \frac{\mathrm{d}\varphi}{\mathrm{d}t} = N_2\varphi = A_c N_2 B(t)$. Here, A_c is the area of cross section of the core and B is the flux density $\left(\frac{\varphi}{A_c}\right)$ in the core. Thus, on the oscilloscope display, the *x*-axis is the magnetic field intensity H applied to the core and the *y*-axis is flux density B in the core. Thus, the pattern displayed is the hysteresis (B–H) loop. The area of the hysteresis loop is the energy lost per cycle (iron loss) per

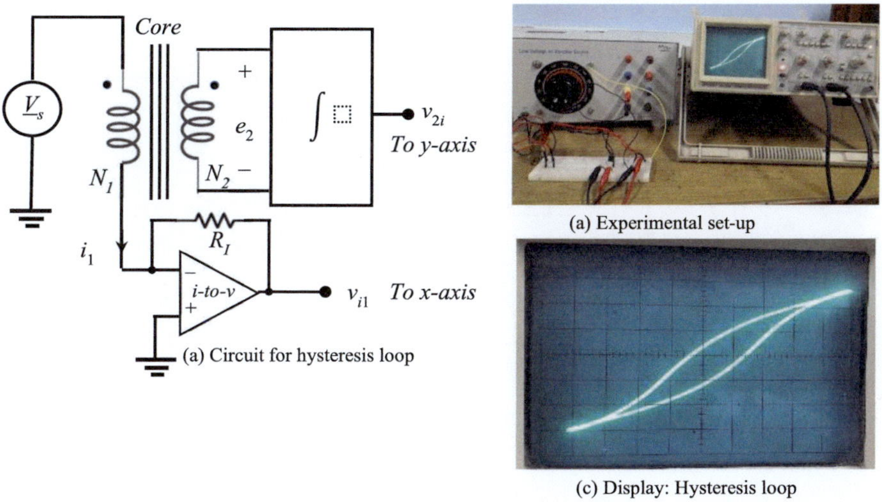

(a) Circuit for hysteresis loop
(a) Experimental set-up
(c) Display: Hysteresis loop

Fig. 7.20 Hysteresis loop display using Oscilloscope in *x–y* mode

unit volume. This value multiplied by the frequency of excitation gives the specific iron losses of the core.

7.5 The Dual Trace Oscilloscope

Once the cathode ray oscilloscope technology matured, the question that led to the next development was: Why not see two waveforms (channels y_1 and y_2) simultaneously? The initial solution for this question was the development of the *dual beam CRT* and the consequent *dual beam oscilloscope.* The dual beam CRT housed two identical sets comprising a cathode, a control grid, focusing anodes, and *x*- and *y*-deflection plates for each channel (channels y_1 and y_2) with only the screen along with the anode being common to both the channels y_1 and y_2. This solution was so complex that very few dual beam oscilloscopes were manufactured. Then came the development of the dual trace oscilloscope. As the name indicates, the dual trace oscilloscope has two traces of the same electron beam. The CRT of a single channel scope shown in Fig. 7.6 and that of a dual trace oscilloscopes are identical. The two traces are obtained by simply multiplexing the two *y*-input signals (channels y_1 and y_2). Figure 7.21 shows the block diagram of a dual trace oscilloscope. We see that there are now two *y*-inputs, namely, y_1 and y_2 with individual amplifiers. The outputs of these two *y*-amplifiers are multiplexed by a single pole double throw (SPDT) analog switch S_1 and the output of the analog switch S_1 is fed to the *y*-deflection plates through the delay line. By controlling the switch S_1, we can display either y_1 or y_2 channel on the screen at a given time. Using one of the two possible modes of

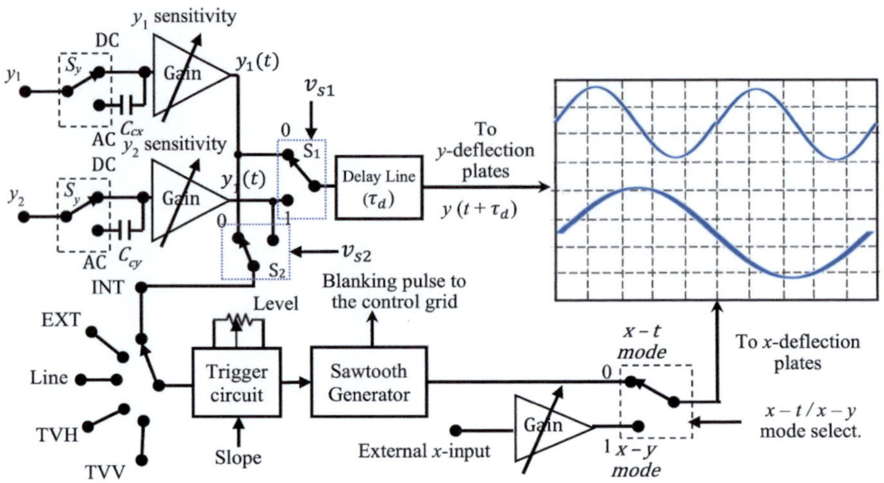

Fig. 7.21 The dual trace oscilloscope—block schematic

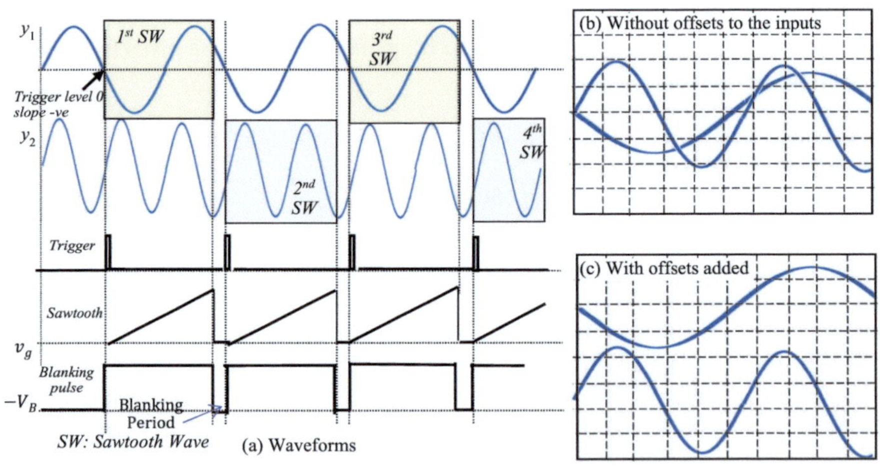

Fig. 7.22 Waveforms of dual trace oscilloscope in alternate mode

switching S_1, namely, the *alternate mode* (Fig. 7.22) or the *chopped mode* (Fig. 7.23) as illustrated next, we can show both channels on the oscilloscope screen.

7.5.1 The Alternate (ALT) Mode of Switching S_1

In the alternate mode, during one horizontal sweep we connect channel y_1 by setting switch S_1 in position 0 and display signal y_1. In the very next horizontal sweep, we

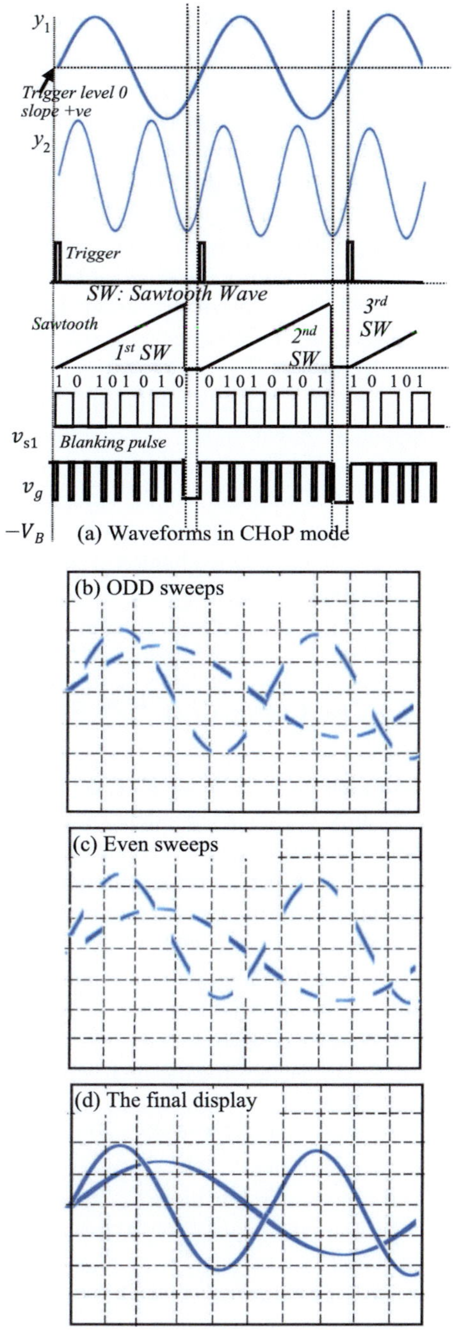

Fig. 7.23 Waveforms of dual trace oscilloscope in chop mode

set switch S_1 in position 1 and thus display signal y_2. In the very next sweep period, we select channel y_1 and during the next sweep select signal y_2, thus alternately displaying channels y_1 and y_2. Appropriately, we call this mode of switching y_1 and y_2 as an *Alternate* mode, having the acronym ALT. In the alternate mode the control signal v_{s1} to the SPDT switch S_1 will have a frequency half (or less than half if the stop period between trigger pulses is considerable) that of the sawtooth. Of course, we must generate a trigger to start the sweep, and in the internal trigger mode switch S_2 selects either y_1 or y_2 as input to the trigger circuit.

Figure 7.22a shows the relevant waveforms of a dual trace oscilloscope set in an alternate (ALT) mode of operation. Here channel y_1 is chosen for triggering with trigger level zero and slope negative. The x-axis sensitivity is chosen so that the x-axis (10 divisions) covers less than the period of y_1 (*vide* highlighted portions of input y_1 in Fig. 7.22a). The highlighted portions in these waveforms are displayed on the screen as shown in Fig. 7.22b. When we use alternate mode, the display shows channels y_1 and y_2 at different times and hence we cannot and should not make correlated measurements on the two channels. It is possible to add DC offsets to the channels y_1 and y_2 and thus separate the relative displays in the y-direction. Now if we add positive offset to channel y_1 and negative offset to channel y_2, the screen will display channels y_1 and y_2 as in Fig. 7.22c.

7.5.2 The Chopped (CHP) Mode of Switching S_1

In this mode, switch S_1 is clocked (operated) using a high-frequency clock signal of 50% duty cycle, independent of any other signal. Thus, we connect channel y_1 for one-half period of the high-frequency clock and then y_2 channel for the other half of the clock period to the y-deflection plates. Hence parts of channels y_1 and y_2 are displayed within one horizontal sweep as indicated in Fig. 7.23. To illustrate the chop mode, let us assume that the inputs to the channels y_1 and y_2 are as shown in Fig. 7.23. We take the clock frequency is approximately 10 times the frequency of channel y_1. We also assume that the trigger source is y_1, the trigger level is zero and trigger slope is positive and the resulting pulses as in the trigger waveform shown in Fig. 7.23a. Hence for the first half of this clock (when clock is "0") channel y_1 will be selected by switch S_1 and displayed on the screen and in the next half of the clock (when clock is "1") channel y_2 will be selected by switch S_1 and shown on the screen as in Fig. 7.23b. The missing portions can be "filled in" if we switch the channels in reverse during the next sweep. That is, in the *odd sweep periods* (1, 3, 5…), we switch S_1 with [0, 1, 0, 1, 0, 1, …) and thus we display parts of channels in the sequence $[y_1, y_2, y_1, y_2, y_1, \ldots]$ as indicated in Fig. 7.23b. Similarly in all the *even sweep periods* (2, 4, 6, 8 …) switch is controlled with [1, 0, 1, 0, 1, 0, 1, …) and thus the parts of channels $[y_2, y_1, y_2, y_1, \ldots]$ will be displayed as shown in Fig. 7.23c. The switching wave form v_{s1} illustrates this inverted switching in the odd and even sweep periods. Of course, the alternating displays will show the waveforms as in Fig. 7.23b and c, resulting in the "*filling up of voids*" that were created due to chopping the channels. In effect, we will see the waveforms as shown in Fig. 7.23d. During the channel transitions, the light spot will jump from one channel to the other and if we do not blank the screen during these transition periods, we will see lines during the transition periods that are not part of the inputs. Hence the blanking pulses must be applied during the channel transitions (y_1 to y_2 or y_2 to y_1) too as illustrated by the blanking pulse waveform shown in Fig. 7.23a. Here too, as in the ALT mode, it is possible to add different DC offsets to channels y_1 (positive offset) and y_2 (negative offset) and separate the two waveforms on the screen. It may be noted that on the oscilloscope the modes are marked as "ALT" for the alternate mode and "CHP" for the chop mode.

7.6 The Dual Time-Base CRO

In the dual trace oscilloscope, we had two y-channels. It is also possible to have two x-inputs, namely x-sweeps. Such an oscilloscope is called the dual time-base oscilloscope. Figure 7.24 shows the block schematic of a dual trace and dual time-base oscilloscope. Here we see that there are two sweep generators called the "*Main Sweep*" and "*Delayed Sweep*". The time period of the main sweep is independently set as T_M and the time period of the delayed sweep is also independently set as T_D, but with the constraint $T_D < T_M$. The start of the delayed sweep is controlled by a settable delay of Δt. The utility of the dual time base (also called the delayed time base) is illustrated with an input as shown in Fig. 7.25a. Here we see that there is a small glitch on the input waveform. If we expand the time base to view the glitch alone, then we will lose track of the rest of the waveform. With the dual time base, we can see both the full cycle of the input and the glitch portion as well. In Fig. 7.25, the waveform marked (a) is the input displayed with the main sawtooth, and the glitch portion alone as in Fig. 7.25b is displayed using the delayed sawtooth. When the first trigger occurs, as usual, the main sawtooth with time-period T_M is generated and the y-input is displayed as shown by the waveform of Fig. 7.25a. In the next trigger, after the delay Δt, the delayed sawtooth is generated with a time-period T_D and applied to the x-deflection plates. The portion of the input waveform for a period T_D from starting the trigger point plus Δt is displayed as in Fig. 7.25b. Thus, during the second sweep period, the input in the interval Δt to $(\Delta t + T_D)$ is displayed on the screen as the waveform in Fig. 7.25b.

This process of displaying the input for the main time base and then the delayed time base is repeated to obtain both waveforms. The selection of the main sweep or delayed sweep to be connected to the x-deflection plates is accomplished by SPDT switch S4 as seen in Fig. 7.24. Unlike the dual trace oscilloscope, the dual time-base oscilloscopes are not that popular.

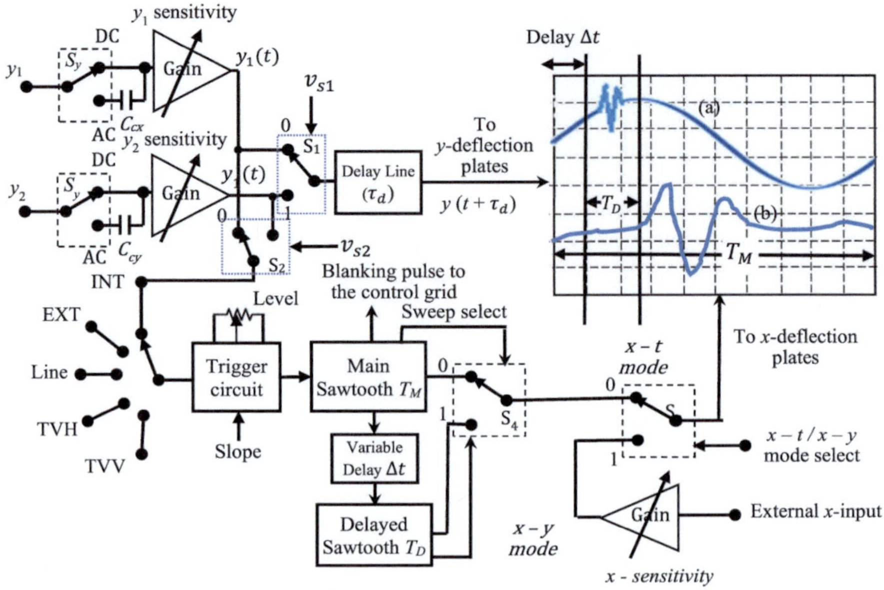

Fig. 7.24 The dual trace and dual time base oscilloscope

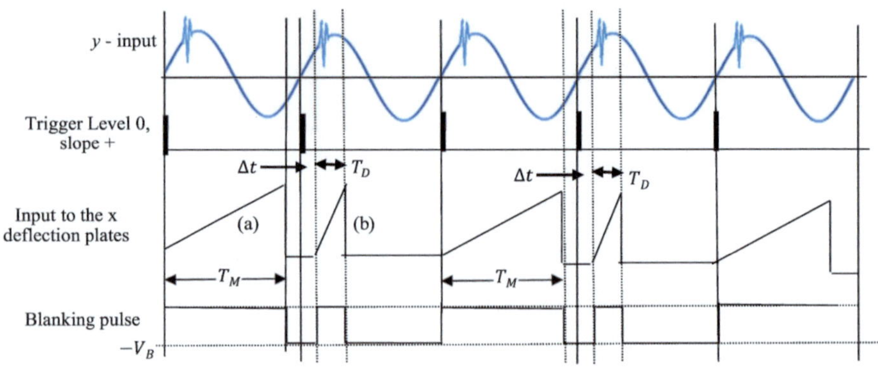

Fig. 7.25 Waveforms of dual time base oscilloscope

7.7 The Oscilloscope Probes

Oscilloscopes are built to have a standard value of 1 MΩ of input impedance and a lowest sensitivity of 5 V/div. Hence, we can apply input voltages of only up to 20 V if we feed the input directly. If we want to view the waveforms of higher voltages, then we need to step down that voltage and give the stepped-down voltage to the *y*-input of the oscilloscope. Oscilloscope manufacturers came up with "oscilloscope probes" to address this issue. Divided by 10 and divided by 100 are standard voltage attenuations

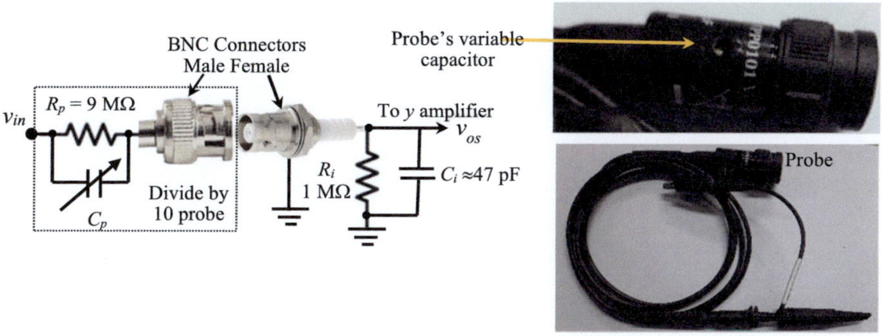

Fig. 7.26 Adjustment of divide by 10 oscilloscope probe

that are readily available from the oscilloscope manufacturers. Figure 7.26 shows a typical divide by ten oscilloscope probe. At DC the capacitors will be on an open circuit and hence the voltage v_{os} is

$$\underline{V}_{os} = \underline{V}_{\text{in}}\left(\frac{R_i}{R_p + R_i}\right) = \frac{\underline{V}_{\text{in}}}{10}. \tag{7.27}$$

However, for AC input we get

$$\underline{V}_{os} = \underline{V}_{\text{in}}\left(\frac{\frac{R_i}{1+j\omega C_i R_i}}{\frac{R_i}{1+j\omega C_i R_i} + \frac{R_P}{1+j\omega C_P R_P}}\right) = \underline{V}_{\text{in}}\left(\frac{R_i}{R_i + R_P\left(\frac{1+j\omega C_i R_i}{1+j\omega C_P R_P}\right)}\right). \tag{7.28}$$

To ensure that the reduced voltage fed to the oscilloscope preserves the waveform of the primary input to the probe, it is necessary to have the bracketed expression in the above equation to have the same value for all frequencies. This goal is achieved if the two-time constants $C_p R_p$ and $C_i R_i$ are equal. That is: $C_p = (R_i/R_p)C_i$. With this condition, we will have

$$\left(\frac{1 + \omega C_i R_i}{1 + \omega C_p R_p}\right) = 1$$

Hence the input voltage is now attenuated by a factor equal to

$$\left(\frac{R_i}{R_p + R_i}\right).$$

Now the attenuation is independent of the input frequency and hence the input waveform. For an attenuation factor of 10, we thus need $R_p = 9R_i$ and $C_p = (C_i/9)$. For an attenuation of 100, we need $R_p = 99R_i$ and $C_p = (C_i/9)$. While it is possible for an oscilloscope manufacturer to ensure that the input resistance R_i is exactly 1.000 MΩ, it may not be possible to ensure that C_i is exactly 47 pF. Thus, the value of $C_i \approx 47$ pF and may vary slightly from oscilloscope to oscilloscope. Hence a probe

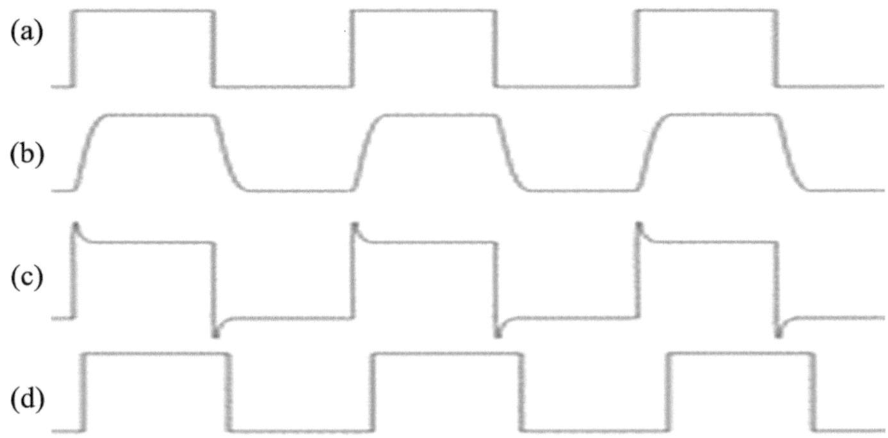

Fig. 7.27 The divide by 10 oscilloscope probe

manufacturer ensures that $R_p = 9R_i$ and provides a variable capacitor for C_p. Hence, a user must adjust such that $C_p = \frac{C_i}{9}$. In order to make the adjustment by the user to be simple all oscilloscopes will have an output square wave signal of frequency 1 kHz as shown in Fig. 7.27a. User, after installing the probe to the oscilloscope BNC, connects the probe end to the 1 kHz square wave output on the oscilloscope. If $9C_p < C_i$ then the high frequencies will be attenuated by a factor >10 and we will get the display as shown in Fig. 7.27b. On the other hand, if we have $9C_p > C_i$ then we will have the higher frequencies attenuated by a factor <10. If we have $9C_p = C_i$, then we will get the display as shown in Fig. 7.27c. Thus, after connecting the probe to the 1 kHz square wave, the user must adjust the value of C_p such that a correct square wave as given in Fig. 7.27d is seen on the oscilloscope so that $9C_p = C_i$. If we make, $R_p = 99R_i$ then we will get the attenuation as 1/100 and we get the divide-by-hundred probe. In such a case we need to ensure that $99C_p = C_i$. To do this we follow the same procedure as given for the divide-by-ten probe and adjust C_p to get a proper square wave display for a square wave input.

7.8 Problems and Exercise

E7.1 A voltage of 4 Cos $(200\pi t + {}^{\pi}/_{6})$ mV is applied to an oscilloscope whose X and Y sensitivities are set as 1 ms/cm and 1 mV/cm, respectively. If the trigger is set at 0 V and positive slope, draw the waveform traced on the screen. If the trigger slope is now changed to negative, draw the waveform seen on the screen.

E7.2 Draw the waveform traced in the above oscilloscope, if the input is 4 Cos $(2\pi\ 10^6\ t)$ mV, the X sensitivity is 100 ns/cm, the delay in the X deflection circuitry is 100 ns and the delay in the Y-channel circuitry is only 80 ns.

E7.3 A certain CRO with a screen size of 10 cm by 8 cm is operated in the x–y mode. If a signal 4 Cos 314t V is applied to the x-plates having a sensitivity of 500 mV/cm and 4 Sin 628t V is applied to the CRO's y-plates possessing a sensitivity of 1 V/cm, what image will be seen on the CRO's screen?

E7.4 A signal $(2 + 2 \sin 2\pi\, 3000t + 0.5 \sin 2\pi\, 9000t)$ V is to be viewed on a CRO. Determine the settings (x, y and trigger) you need to choose so that at least one full cycle of the waveform is seen on the screen of a CRO.

E7.5 List the constraints that determine the bandwidth of an oscilloscope in the order of significance of contribution.

E7.6 What does the input: Z-modulation provided on a CRO achieve? Give an application that would require this facility on the CRO.

E7.7 A voltage signal 4 Cos $(1000\pi t + 30°)$ V is given as input to a CRO. The screen displays one cycle of an inverted sine wave with peak-to-peak excursion occupying 4 divisions in the y-axis. Determine the x- and y-axis sensitivity settings and trigger settings (value and slope) of the CRO.

E7.8 If another signal 2 Cos $(1000\pi t)$ is available, show how you will use this signal and display the signal 4 Cos $(1000\pi t + 30°)$ V to be seen as a cosine wave with a 30° phase shift on the CRO screen.

E7.9 A signal 2 Cos $(1000\pi t)$ V is fed as input (non-inverting) to an opamp operated in open-loop mode (the opamp is powered from a 15 V symmetrical supply). The input and the opamp output are given as x- and y-inputs to an oscilloscope. Draw the waveform seen on the scope.

E7.10 A voltage signal 4 Sin $(1000\pi t + 30°)$ V is fed as input to a CRO. The x-axis sensitivity is set as 20 μs/division, and the y-sensitivity is set as 1 V/div. The trigger is selected as external and a id tied to a voltage of 3 Sin $(1000\pi t)$ V. Draw the waveform seen on the screen, if the trigger level is 0 V and the slope is positive. What will be the value of y displayed at $x = 0$?

Chapter 8
The Counter Timer

The digital revolution is far more significant than the invention of writing or even printing.
—Douglas Engelbart

8.1 The Number Systems

The term "*Digital*" means dealing with digits or in other words "*Numbers*". We are quite adapted to the decimal number system that has become universal now, but a few centuries back, the number systems around the world were distinctly different. For example, the Romans used the roman numerals (I, II, III, IV …). In the decimal system we have the numbers 0, 1, 2, 3, 4, 5, 6, 7, 8, and 9 (called digits) to indicate the value. After 9, we go with two digits and have 10, 11, 12, …19. After 19, we have 20, 21, 22, … 29. Once we come to 99, we need to add one more digit and start all over as 100, 101, … 999 … and so on. Apart from the integers, we also have decimal fractions and thus a general decimal number may be something like 9876.54321. Though the decimal system is quite convenient and easy for humans to deal with and comprehend, it is very difficult to implement the decimal number

V. Jagadeesh Kumar, *Principles of Electrical Measurement and Measuring Instruments*,
https://doi.org/10.1007/978-981-96-5475-8_8

system *directly* using electronics. On the other hand, the *binary number system,* and its corresponding Boolean algebra are easy to implement using electronics. Thus, today all digital systems use the binary to represent not just numbers (integers as well as fractions) but also pictures (diagrams, images, videos in multitude of colors), sound (speech and music) and even alphabets (and writings) of all the languages in the world!

8.2 The Binary Number

In the binary system we use only two distinct symbols, namely, "0" and "1". Thus, after writing 0, 1 we must add a second *bit*. We do not call the numeral in a binary number as a digit, because calling the positions as "*bits*" clearly indicates that we are dealing with the binary system. Thus, we have, in the binary number, 0, 1, 10, 11. These binary numbers indicate decimal numbers: (0, 1, 2, and 3) and must be read as zero; one; (one zero); (one, one). To represent decimal 4 in binary, we need to add a third bit and get 100, 101, 110, 111 … (the decimal 4, 5, 6 and 7 and to be read as (one, zero, zero); (one zero one) and (one, one, one). Once again, to represent decimal 8 we need four binary bits. Table 8.1 illustrates all possible 4-bit binary numbers ($b_3b_2b_1b_0$). and their decimal equivalent.

Table 8.1 The 4-bit binary number

Decimal	*Binary*			
	b_3	b_2	b_1	b_o
0	0	0	0	0
1	0	0	0	1
2	0	0	1	0
3	0	0	1	1
4	0	1	0	0
5	0	1	0	1
6	0	1	1	0
7	0	1	1	1
8	1	0	0	0
9	1	0	0	1
10	1	0	1	0
11	1	0	1	1
12	1	1	0	0
13	1	1	0	1
14	1	1	1	0
15	1	1	1	1

8.2.1 The Bit Position and Value

Like in the decimal system the number 1234 has “4” in the unit position (10^0), the number “3” occupies the ten’s position (10^1), the number “2” is in the 100th (10^2) position and the “1” occupies the thousandth (10^3) position. Similarly in a binary number, the right most bit position occupies the 2^0 (=1) position and is called the *least significant* bit (LSB). The next bit (left of the LSB) occupies the 2^1 position and the next bit occupies the 2^2 (=4) position and so on. Thus, a binary number 0101 is equal to

	b_3	b_2	b_1	b_0
Binary number	0	1	0	1
Positional value	2^3	2^2	2^1	2^0
Decimal value	$(8 * 0 + 4 * 1 + 2 * 0 + 1 * 1) = 5$			

advantage of the binary system is that it is easy to implement using electronics. For example, in a switch we can say: If the switch is open then it is binary “0” and if the switch is closed it is a binary “1”. Similarly, presence of a voltage (or current) can be made to represent a binary “1” and its absence a binary “0”. In the binary system each “*bit*” can either be “0” or “1”. We can also use the term “High” or “True” for binary “1” and “Low” or “false” for binary “0” to avoid confusion with decimal one and decimal zero. Each bit can also be thought of as a binary variable whose value can either be “1” or “0”. Needless to say, that an *n*-bit binary number (integer) can cover from $000 \cdots 0$ to $111 \cdots 1$ and thus can represent the maximum value of $(2^n - 1)$.

8.2.2 The Boolean Algebra

Independent of the development of the binary system, George Boole invented methods for handling logical sets in 1857, especially sets having binary elements in them. Henry Maurice Sheffer applied and expanded the work of Boole and coined the term “*Boolean Algebra*”. This elegant method helped the building of “*combinatorial logic*” and “*sequential logic*” that was the starting point of today’s digital explosion. Boolean algebra is also known as *Boolean logic* that defines functions that can operate on a Boolean variable or variables. The Boolean function or Boolean operator is also called “*Logical operator*” or “*Logic function*” or “*Logic gate*”. The simplest of the Boolean function is the “NOT” function that negates (or inverts) a Boolean variable.

Example 1 $B = \text{NOT}(A)$, where A and B are Boolean variables.

Here, the variable B becomes “0” if variable A is “1” and becomes a “1” if A is “0” as indicated in Table 8.2. It is customary to indicate the NOT function as: $B = \text{NOT}(A)$ as $B = \overline{A}$ or $B = A'$. Read as B equals A bar. We also call that variable “B” is a *complement* of variable "*A*". When we have $B = \overline{A}$, we can also say that $A = \overline{B}$ or $A = B'$ and hence "*A*" is a *complement* of variable "*B*". The logic function "NOT" (popularly called the NOT gate) is indicated with a symbol as shown in Table 8.2. The input–output relationship of a Boolean function when written in a tabular form is called the *truth table* and the truth table for a NOT gate is also shown in Table 8.2.

While the NOT function (also called Boolean operator) deals with just one Boolean variable, Boolean operators (also called basic Boolean functions or basic logic gates) "AND", "OR", and "XOR" deal with more than one binary variable. Table 8.3 indicates these for two input variables.

Table 8.2 The Boolean logic gates

No	Function	Description	Usage	Graphics	Truth Table		
1	NOT	$B = \text{NOT}(A)$	$B = \overline{A}$	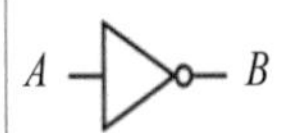	*A*		*B*
					0		1
					1		0
2	OR	$C = A$ OR B	$C = A + B$	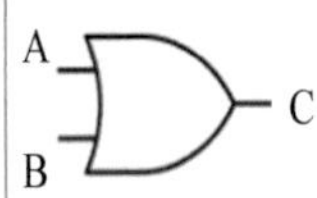	*B*	*A*	*C*
					0	0	0
					0	1	1
					1	0	1
					1	1	1
3	AND	$C = A$ AND B	$C = A.B$	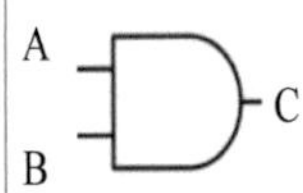	*B*	*A*	*C*
					0	0	0
					0	1	0
					1	0	1
					1	1	1
4	NOR	$C =$ NOT (A OR B)	$C = \overline{A + B}$	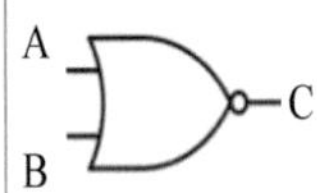	*B*	*A*	*C*
					0	0	1
					0	1	0
					1	0	0
					1	1	0
4	NAND	$C =$ NOT (A AND B)	$C = \overline{A.B}$	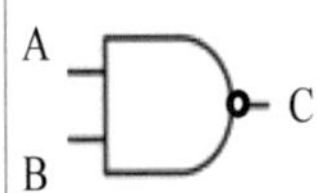	*B*	*A*	*C*
					0	0	1
					0	1	1
					1	0	1
					1	1	0
5	XOR	$C = A$ Exclusive OR B	$C = A \oplus B$	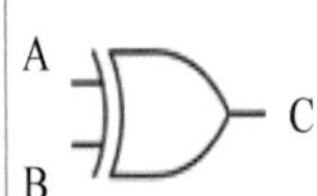	*B*	*A*	*C*
					0	0	0
					0	1	1
					1	0	1
					1	1	0
6	XNOR	$C = A$ Exclusive NOR B	$C = \overline{A \oplus B}$	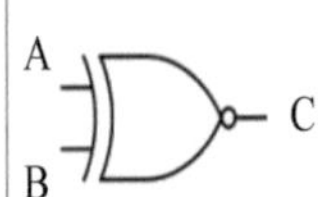	*B*	*A*	*C*
					0	0	1
					0	1	0
					1	0	0
					1	1	1

Table 8.3 Truth Table of 4-bit binary counter

Clock	Input	RST	Carry	b_3	b_2	b_1	b_o
X	X	1	0	0	0	0	0
1	↑	0	0	0	0	0	1
2	↑	0	0	0	0	1	0
3	↑	0	0	0	0	1	1
4	↑	0	0	0	1	0	0
5	↑	0	0	0	1	0	1
6	↑	0	0	0	1	1	0
7	↑	0	0	0	1	1	1
8	↑	0	0	1	0	0	0
9	↑	0	0	1	0	0	1
10	↑	0	0	1	0	1	0
11	↑	0	0	1	0	1	1
12	↑	0	0	1	1	0	0
13	↑	0	0	1	1	0	1
14	↑	0	0	1	1	1	0
15	↑	0	0	1	1	1	1
16	↑	0	1*	0	0	0	0
17	↑	0	1	0	0	0	1
18	↑	0	1	0	0	1	0
19	↑	0	1	0	0	1	1
20	↑	0	1	0	1	0	0
21	↑	0	1	0	1	0	1
...							

Example 2 The "OR" function: $C = (A + B)$, where A, B, and C are Boolean variables.

In the above example we should read as C equals A OR B and not A plus B! Another example is as follows.

Example 3 The "AND" function: $C = (A \cdot B)$, where A, B, and C are Boolean variables.

In the above example we should read as C equals A AND B. Reading this as C equals A dot B is also permitted. For the XOR Boolean operator we write $C = (A \oplus B)$ and to be read as C equals A XOR B. We can attach a NOT Boolean operator to the "AND", "OR", and "XOR" logic functions and thus get another set of Boolean functions: NAND, NOR, and XNOR. Here "NAND" is (Not AND), "NOR" is (Not OR) and "XNOR" is (Not XOR) as in Table 8.2.

All these Boolean functions are explained in Table 8.2 for two binary input variables. In general, we can write a Boolean function of several variables using the logic operators (OR, AND, NOR, NAND $\cdots$). Apart from the well-defined logic operators, it is also possible to define generic Boolean or logic functions using binary variables. The function:

$$f = A.B.C + \overline{A}.B.D + \overline{B}.C.D + \overline{A}.B.C.D$$

illustrates a typical logic function f using four binary (Boolean) variables A, B, C, and D, in the sum of products form. On the other hand,

$$Y = (A + BC).(\overline{A} + B + D).(\overline{B} + C + D).(\overline{A} + B + C + D)$$

is a Boolean function of four variables in the product of sum form. As is seen in Table 8.2, we also use distinct symbols (or pictures/graphics) to illustrate different Boolean operators. In this form these are called *logic gates* (more correctly 2-input logic gates). Thus, we need not specify the logic operation once we use the standard logic gate symbol. There is another set of standard symbols for logic gates proposed by the IEEE Standards as indicated in Fig. 8.1. When we combine several of these logic gates and realize a "*logic circuit*", we get either a combinatorial logic circuit or a sequential logic circuit.

8.3 Combinatorial Logic

Figure 8.2 shows the combinatorial logic circuit of a "*half adder*" that adds two binary bits A and B. It may be noted that adding two binary bits is different from ORing the bits. If we OR two binary bits A and B we get only a third binary bit C. On the other hand, adding two binary bits A and B we get two binary output bits *Sum* and *Carry*. The truth table of the addition of two binary bits A and B is shown in Fig. 8.2. The difference between the OR function of two binary variable and addition of 2 binary variables is easily seen if we look at the truth table of the OR function in Table 8.2 and the truth table of half adder shown in Fig. 8.2. As seen in Fig. 8.2, the half adder accepts the two binary bits A and B as input and provides the *Sum* and *Carry* binary outputs. If we compare the *Sum* bit in the truth table of Fig. 8.2 and the truth table of the XOR gate in Table 8.2, we see that *Sum* = (A XOR B). Similarly, the truth table entry of the AND gate in Table 8.2 matches with the Carry bit of the half adder in Fig. 8.2. Thus, we arrive at

$$Sum = (A\,XOR\,B)\ and\ Carry = (A\,AND\,B).$$

Once the integrated circuit technology was developed, the very first integrated circuits that were commercially available used the Transistor-Transistor Logic or simply TTL to realize the logic gates and logic circuits. In TTL itself we have four variations, namely,

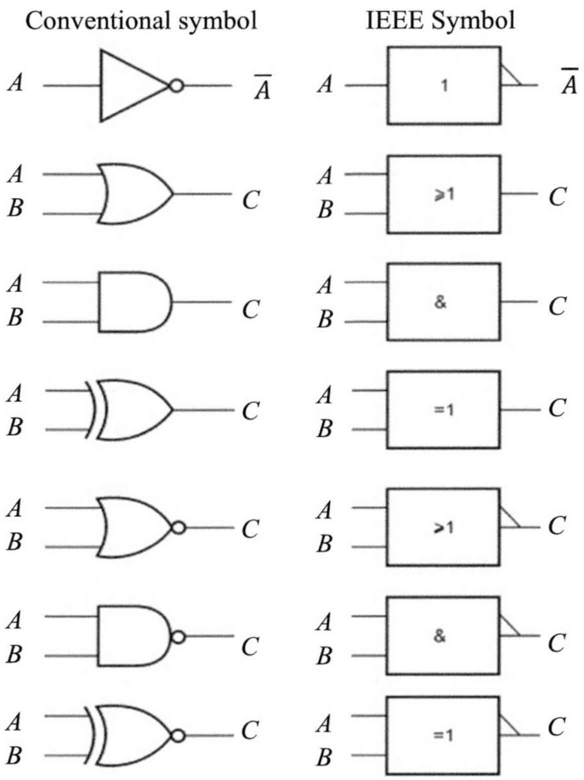

Fig. 8.1 IEEE Standard symbols for logic gates

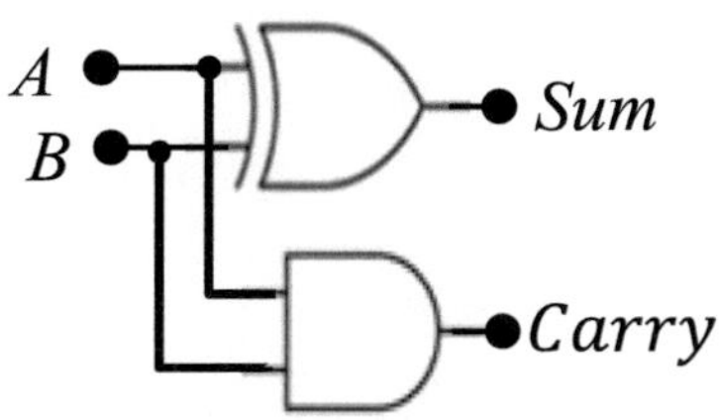

Fig. 8.2 The Half adder

Half adder truth table			
A	B	*Sum* (A⊕B)	*Carry* (A·B)
0	0	0	0
0	1	1	0
1	0	1	0
1	1	0	1

(i) TTL (ICs with numbers 54xx and 74xx),
(ii) Low power TTL (Numbers 54Lxx and 74Lxx),
(iii) Schottky TTL (Numbers 54Sxx and 74Sxx), and
(iv) Low power Schottky TTL (54LSxx and 74LSxx).

Further developments resulted in

(i) Emitter Coupled Logic (ECL),
(ii) Metal Oxide Semiconductor (MOS) logic, and
(iii) Complementary MOS (CMOS) logic.

These *logic* families differ in certain characteristics. For example, the output of a gate (which can be of any type like NOT/OR/AND …) when implemented with TTL, will settle after a delay of 10 ns from the time the input(s) had changed ("0" to "1" or "1" to "0"). On the other hand, in Schottky TTL logic gates the delay is only 2 ns. Initially these logic gates were realized as "Small scale integration (SSI) technology". Apart from the delay, the logic families also differ in terms of power consumption, logic levels and number of inputs that can be connected to an output (called Fanout). Using the logic gates, we can build *logic circuits* that perform certain functions such as half adder (shown in Fig. 8.2 and so named because it has no provision for a third input as *carry* from the addition in the previous lower bit position, as would be the case for a full adder), full adder, encoder, and decoder to name a few. In a combinatorial logic circuit, the output will be available after a small delay once the inputs of all the logic gates in the circuit have attained a steady state.

8.4 Sequential Logic

We can also build a logic circuit using feedback technique (that is an output or outputs in a logic circuit is brought back and connected as an input(s) in the same logic circuit). Hence in a sequential logic the output will be available only after the inputs to all the gates and the outputs that are fed back attain steady state. When a change occurs at the input in a sequential logic circuit, the output(s) then depends on the previous (state) outputs as well. The most popular and simplest logic circuit that uses feedback is Flipflop. There are different types of flipflops. The simplest form of the flipflop is the Set-Reset flipflop or popularly called the SR flipflop shown in Fig. 8.3. The logic of the SR flipflop in terms of logic-gate connection and its input–output relationship as a *truth table* are given in Fig. 8.3. Figure 8.4 shows the logic of a "Clocked J-K flipflop", which ensures that all changes in a logic circuit using several flipflops occur *synchronously* with reference to a clock signal. Such clocked flipflops can be made to respond to one of the parameters of the clock. The parameters of the clock signal that can be chosen for the status change of the flipflop are

Fig. 8.3 S-R Flipflop with NAND gates

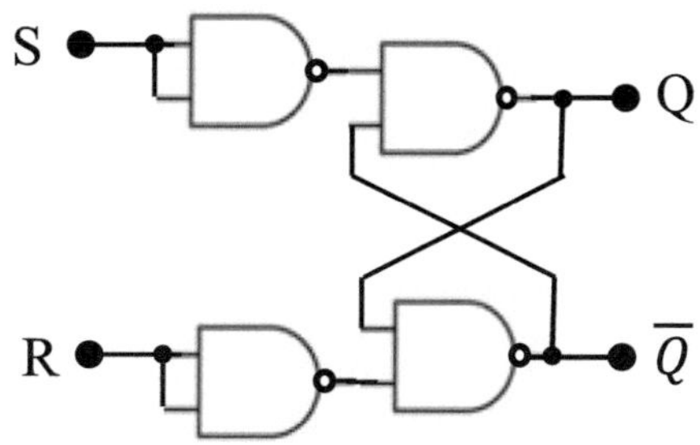

SR Flipflop truth table			
S	R	Q	$\overline{Q}$
1	0	1	0
0	1	0	1
1	1	No Change	
0	0	Not valid	

Fig. 8.4 Clocked JK Flipflop with NAND gates

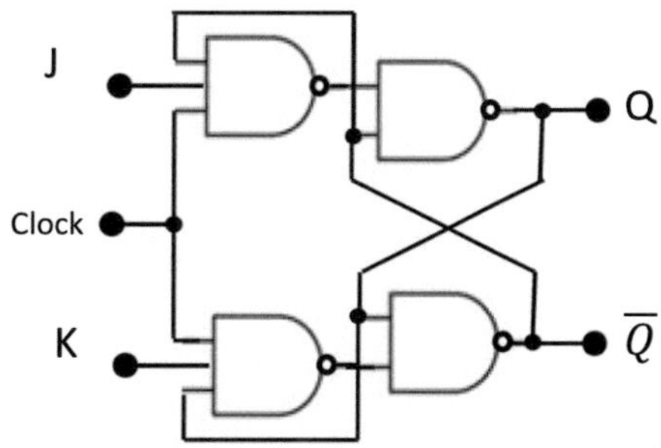

JK Flipflop - Truth table				
Clock	J	K	Q	$\overline{Q}$
↑	1	0	1	0
↑	0	1	0	1
↑	1	1	Toggle	
↑	0	0	No Change	

(i) Level (“0” or “1”) or
(ii) Rising or falling edge (Change from “0” to “1” is called the *rising edge or positive edge* and that from “1” to “0” is called the *falling edge or negative edge*).

Hence, we have either the “*level triggered*” or “*edge triggered*” logic. The J-K flipflop shown in Fig. 8.5 is a *positive edge* triggered flipflop. It is also possible to have

Fig. 8.5 Clocked JK Flipflop with RESET

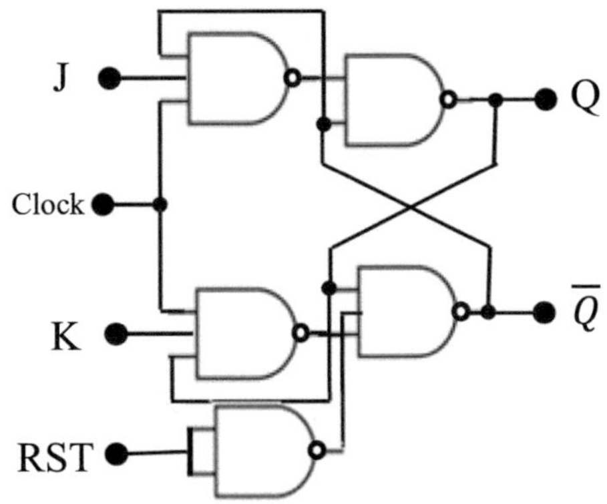

JK Flipflop with RST - Truth table					
RST	Clock	J	K	Q	$\overline{Q}$
1	x	x	x	0	1
0	↑	1	0	1	0
0	↑	0	1	0	1
0	↑	1	1	Toggle	
0	↑	0	0	No Change	

x: don't care can be '0' or '1'

another input in a flipflop that *resets* the output Q of the flipflop to "0" irrespective of other inputs, including the clock, as shown in Fig. 8.5. The *reset* input is referred to in Fig. 8.5 as "RST".

In the clocked J-K flipflop in Fig. 8.5, the output Q will not change as soon as the J and K inputs change. Instead, the output of a clocked flipflop remains unchanged till the succeeding clock arrives and only at that time, depending on the values of J and K, the output Q will change. Hence, we call the flipflop a one-bit memory. Once we add a clock input to a flipflop that flipflop becomes a "*memory*" that retains the value of one bit (as "0" or "1") for at least one clock period. For example, the output of the rising (positive) edge triggered J-K flipflop will change ("0" to "1" or "1" to "0") only when the clock signal rises from level "0" to "1". Once we add a memory element in a logic circuit, we call the logic circuit as the "*sequential logic*" circuit. It may be noted that most of the present-day digital devices starting from digital meters to digital computers (popularly called personal computers), and mobile phones are made with sequential logic circuits. A very simple example of a sequential circuit is the counter. The name "counter" is given to the sequential logic circuit because the circuit can count the number of clock inputs given to the counter.

8.4.1 The Binary Counter

Figure 8.6 shows four J-K flipflops cascaded to obtain a 4-*bit binary counter*. That is, the outputs b_3, b_2, b_1, b_0 of the flipflops will form a 4-bit binary number that provides (counts) the number of input pulses given to the first flipflop, after a reset pulse is applied for a short duration. Table 8.3 illustrates the operation of the 4-bit binary counter shown in Fig. 8.6. The waveforms of the inputs RST, clock, and the outputs b_3, b_2, b_1, b_0, shown in Fig. 8.7 illustrate the changes in the bits b_3, b_2, b_1, b_0 with respect to successive clock cycles. The Table 8.3 and the waveforms in Fig. 8.7 illustrate that the four flipflop sequential logic circuit of Fig. 8.7 functions as a 4-bit *binary counter*, becoming "0000" after reset, "0001" after getting the first clock pulse after reset, becoming "0010" after the second clock pulse, $\cdots$ "1110" after the 14th clock pulse and, "1111" after the 15th clock pulse (in decimal equivalent the counter counts: 0, 1, 2, 3 ... 14, 15). After "1111", the counter *rolls over* to "0000" and starts counting again as "0001", "0010", ... 1110, 1111. The roll over from 1111 to 0000 can be captured as an additional binary output, and if it is done, then the additional bit that provides the information that roll over has occurred, is called the *carry*. The output b_o of the first flipflop FF1 is called the *least significant bit* (LSB) of the 4-bit binary counter and the output b_3 of the 4th flipflop FF4 is called the most significant bit (MSB) of the 4-bit binary counter. Thus, whenever the MSB toggles (changes) from "1" to "0" the *carry bit* is toggled.

It is possible to cascade as many flipflops as we need and realize a general N-bit binary counter. The 4-bit binary counter structure shown in Fig. 8.7 is called the "*asynchronous*" 4-bit binary counter (also called the "*ripple* counter"), as the changes in the outputs b_3, b_2, b_1, b_0 will not happen at the same instant. That is only after the change in b_0 occurs, b_1 will change after a small delay followed by another time delay b_2 will change and after another delay finally b_3 will change. Normally the delay will be small, in nanoseconds. It is also possible to have a "synchronous" 4-bit binary counter using J-K flipflops as shown in Fig. 8.8. In the counter of Fig. 8.8, the changeover in flipflop outputs b_3, b_2, b_1, b_0 will occur synchronously at the rising edge of the clock.

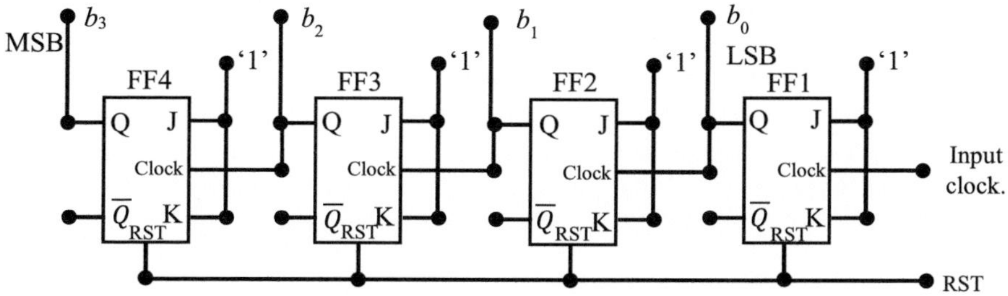

Fig. 8.6 A 4-bit binary ripple counter using clocked JK Flipflop with RESET

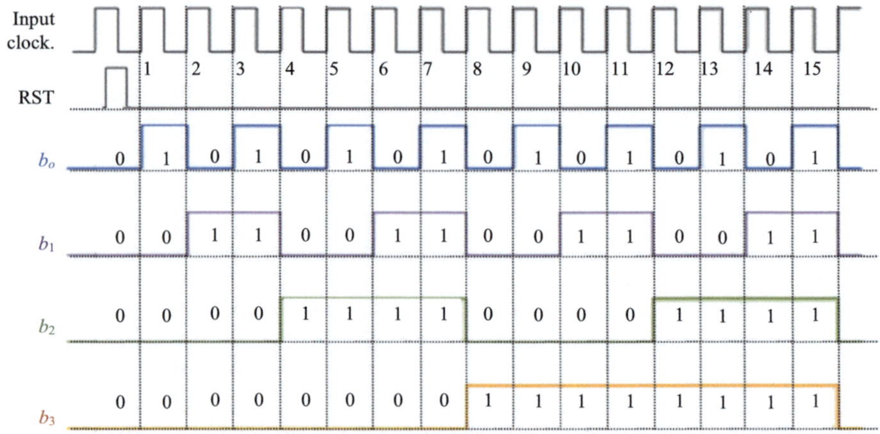

Fig. 8.7 Waveforms of the 4-bit binary counter of Fig. 8.6

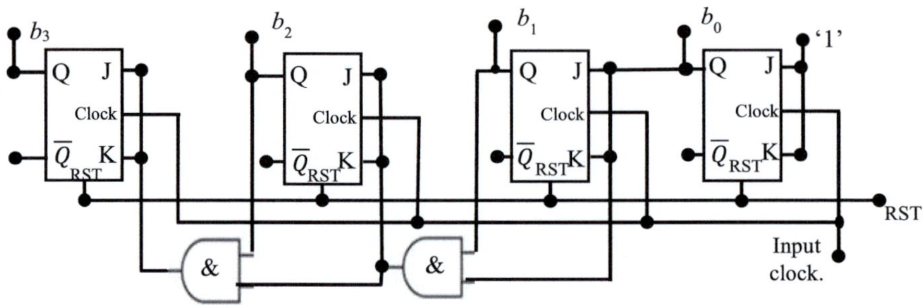

Fig. 8.8 Synchronous 4-bit binary counter using JK Flipflops with Reset

8.4.2 *The BCD Counter*

Since we are used to the decimal systems, it is useful to have a "*decimal counter*" that counts 1, 2, 3, 4 … 9, 10, 11, 12, … To realize this, a "*binary coded decimal counter*" or simply known as the BCD counter, shown in Fig. 8.9, was developed. We clearly see from this figure that the BCD counter uses a 4-bit binary counter but counts as though the counter is a decimal one, that is, after reaching decimal 9 (binary 1001), the next clock (10th clock after reset) resets the 4-bit counter to "0000". Thus, the BCD counter uses the 4-bit binary counter but "*rolls over*" to "0000" after "1001" (Decimal 9) as indicated in Table 8.4. The working of the BCD counter is detailed next.

Once the "Reset" is set to "1", bits b_3, b_2, b_1, b_0 and the carry bit CY are reset to "0". The counter then starts counting "0001", "0010" "0011", … "1001" for successive nine rising edges of the clock. At the tenth rising edge, the output b_3, b_2, b_1, b_0 becomes "1010" (decimal 10), the output of the AND gate becomes "1" and hence

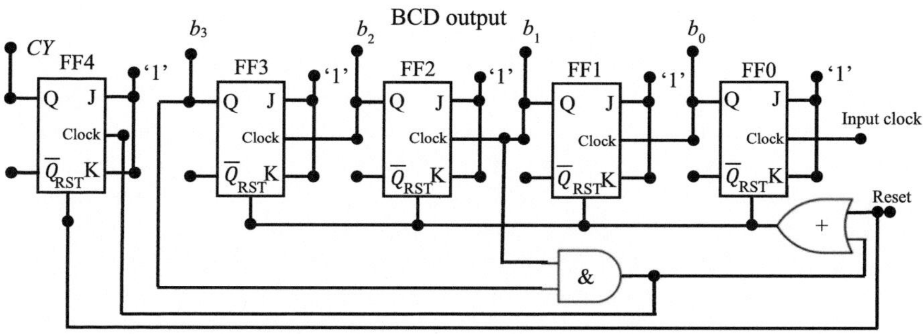

Fig. 8.9 One digit BCD counter with Reset and Carry

Table 8.4 The 4-bit BCD counter

Clock	Input	RST	Carry	b_3	b_2	b_1	b_o
X	X	1	0	0	0	0	0
1	←	0	0	0	0	0	1
2	←	0	0	0	0	1	0
3	←	0	0	0	0	1	1
4	←	0	0	0	1	0	0
5	←	0	0	0	1	0	1
6	←	0	0	0	1	1	0
7	←	0	0	0	1	1	1
8	←	0	0	1	0	0	0
9	←	0	0	1	0	0	1
10	←	0	1*	0	0	0	0
11	←	0	1	0	0	0	1
12	←	0	1	0	0	1	0
13	←	0	1	0	0	1	1
14	←	0	1	0	1	0	0
15	←	0	1	0	1	0	1
16	←	0	1	0	1	1	0
17	←	0	1	0	1	1	1
18	←	0	1	1	0	0	0
19	←	0	1	1	0	0	1
20	←	0	0*	0	0	0	0
...							

the output of the OR gate also raises to "1". Since the output of the OR gate is tied to all the RST inputs of the flipflops FF0 to FF3, their outputs, namely, bits b_3, b_2, b_1, b_0 are reset to "0". The output of the AND gate is also given as input to FF4 and hence carry bit CY toggles and becomes "1". Once the bits b_3, b_2, b_1, b_0 are reset to "0000", the output of the AND gate becomes "0" and the counter is put in the *counting mode* and will start counting "001", "00,002", ... It should be noted that after reaching "1000", the counter output will momentarily become "1010" and then "0000". The period for which the output of the counter stays temporarily at "1010" is determined by the sum of the propagation delays of the AND gate and OR gate. Table 8.4 illustrates the states of the BCD counter of Fig. 8.9. After b_3, b_2, b_1, b_0 reaching 1001 then the next clock pulse toggles the carry (CY) bit and resets b_3, b_2, b_1, b_0 to "0000". Thus, if we include the CY bit also then the tenth clock pulse will give an output CY $b_3, b_2, b_1, b_0 =$ "1" "0000". If we consider CY also as a decimal digit, then the output can be viewed as decimal 10. If we now give the 11th clock pulse, the output will become decimal 11. This will continue and the counter will count 12, 13, ... till we give the 19th clock pulse for which the output CY b_3, b_2, b_1, b_0 will be "1" "1001" (decimal 19). The next clock pulse (20th clock pulse) of course results in b_3, b_2, b_1, b_0 becoming "0000" (roll over) but will also toggle CY hence the CY b_3, b_2, b_1, b_0 will become "0 0000" (decimal 00). From Table 8.4, we see that the 4-bit binary b_3, b_2, b_1, b_0 counts from 0 to 9 and rolls over to 0 (binary "0000") after 9 and the roll over sets the carry. If we include the carry, then we see that the BCD counter in Fig. 8.9 can count from 00 to 19 and then rolls over to 00. Thus, without the carry bit, the BCD counter of Fig. 8.9 counts one decimal digit (0, 1, 2, ... 8, 9) and with the carry bit included, counts from 00 to 19. Since the second decimal digit is not full range (0 to 9), but only 0 or 1, the term 1 1/2 digit is coined to indicate that the count is more than one digit, but less than 2-digits! The BCD counter in Fig. 8.9 is a "ripple type BCD counter". As in the case of the 4-bit binary counter of Fig. 8.8, we can also design a synchronous BCD conter. We can cascade several, say N, BCD counters by connecting the carry of the preceding BCD counter to the input of the succeeding BCD counter and realize an N-digit (if we include the carry of the most significant digit, then N1/2 digit) BCD counter.

8.5 The Counter Timer

Figure 8.10 shows such an arrangement where four BCD counters are cascaded to realize a 4 1/2 digit binary coded *decimal counter*. This 4 1/2 digit BCD counter (4-digit BCD with a carry from the most significant digit) can count the number of clock cycles to its input *IN*0 from 00,000 to 19,999. Of course, a 0 or 1 for the fifth digit cannot be logically called 1/2 a digit, but this term was coined in the past and is in vogue for a long time and is accepted as such. On the other hand, if we have a counter which counts from 00,000 39,999 (that is the fifth digit has only two binary bits and hence can indicate the decimal digits 0, 1, 2, or 3), then that counter is called the 4¾th digit counter. In effect, we say that the 4 1/2 digit counter handles 20,000

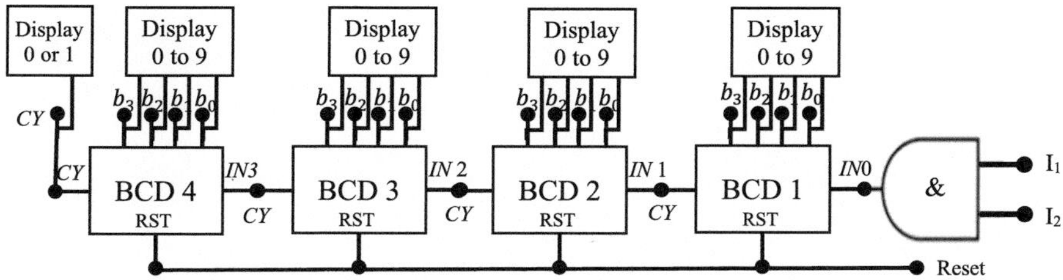

Fig. 8.10 A 4 1/2 digit BCD counter with Reset

counts (0000 is also a count). Similarly, the 4¾ digit counter handles 40,000 counts. The 4 1/2 digit BCD counter shown in Fig. 8.10 is "ripple" type and once again we can design a 4 1/2 digit synchronous type BCD counter. For a human to see the output in terms of decimal digits of the BCD counter, the 4-binary bits of each BCD counter must be shown as the corresponding decimal digit on a suitable display. That is if the output of a BCD digit (b_3, b_2, b_1, b_0) is "0000" then digit 0 must be displayed, if $b_3, b_2, b_1, b_0 ='' 0001''$ digit 1 must be displayed and so on. Initially this requirement was met with a BCD to decimal decoder, driving the appropriate decimal digit in the display array. Afterwards the *seven-segment display* was invented to display decimal numbers, the BCD output of each decimal digit of a $N1/2$ digit counter is connected to an individual BCD to 7-segment decoder and the outputs (a, b, c, d, e, f, and g) of the decoders were connected to individual segments of the display. Later the *dot-matrix display* was invented, and the BCD output was connected to a BCD to dot matrix decoder and the decoder connected to the dot-matrix display.

8.5.1 The Display

The display must show the decimal digit "0" if the BCD output b_3, b_2, b_1, b_0 is "0000" and display "3" if the output b_3, b_2, b_1, b_0 of that digit's BCD counter is "0011" and so on. Thus, the display should accept the 4-bit binary outputs of the BCD counter (one digit) as input and display the corresponding decimal digit.

8.5.1.1 The BCD to Decimal Decoder and Display

Initially, the 4-bit BCD was decoded in to 10 outputs one each for a digit from 0 to 9 using a BCD to decimal decoder. Each of these outputs individually indicated a decimal digit as shown in Fig. 8.11a. For example, if the input b_3, b_2, b_1, b_0 is "1000", then the output pin Y8 will be activated, and the corresponding decimal number (numeral 8) will be illuminated (or indicated) as illustrated in Fig. 8.12a. When the BCD number is "0000" the output 0 is selected and 0 is displayed. Similarly, when BCD is "0001" the output 1 is selected and digit 1 is displayed $\cdots$ and the output 9 is

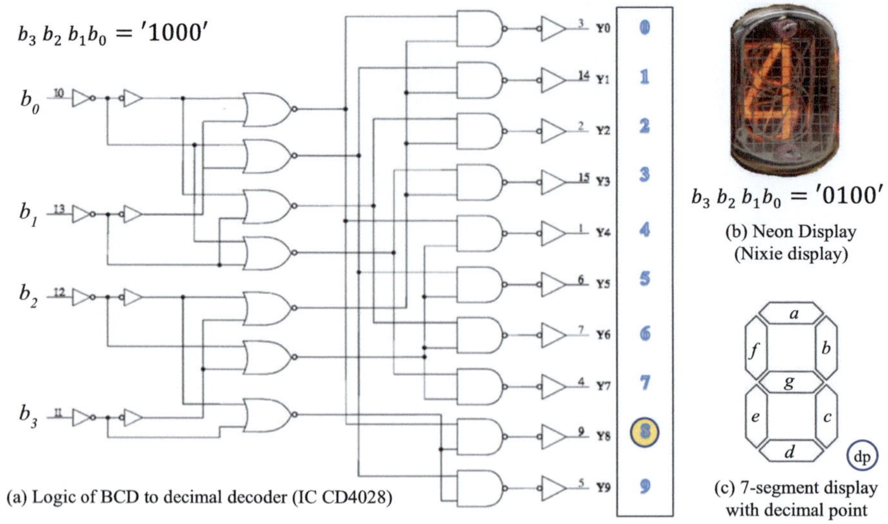

Fig. 8.11 BCD to decimal decoder (CD4028) and displays

selected when the BCD input is "1001" and the display will show the digit 9. In the beginning, the displays were specially fabricated incandescent lamps in the shape of the digits. Later, neon gas displays were developed and used. This type of display is popularly called "*Nixie Tube display*". The nixie displays used neon gas filled at very low-pressure glass tubes with one anode and eleven cathodes. Each of the ten cathodes is attached to a specific numeral (0 to 9) made of conductive electrode and the eleventh is attached to a decimal point (dot), all coated with phosphorus layers. By exciting the anode with a positive voltage and grounding the corresponding cathode, the appropriate numeral (plate) or the decimal dot is made to glow. Figure 8.11b shows a nixie tube display indicating the numeral "4". These displays required excitation voltages of few tens of volts and hence where replaced with the seven segment displays.

8.5.1.2 The 7-Segment Display

Today, we universally use the "seven segment display" shown in Fig. 8.11c to display. If the requirement is to display only numbers, then invariably seven segment displays are used. As the name indicates, the display for a digit is made of seven individual segments, named, *a*, *b*, *c*, *d*, *e*, *f*, and *g*. The seven segments together are arranged to form the numeral 8 as shown in Fig. 8.11c. By selectively illuminating some or all these segments we can display any numerals from 0 to 9. For example, if we illuminate all the seven segments, we will display numeral 8. On the other hand, if we illuminate all segments except segment *g* (that is illuminate only segments *a* to *f* but not illuminate segment *g*), then we will show the numeral 0. Apart from segments

(a) 7-segment vacuum fluorescent display

(c) 7-segment Liquid Crystal Display (LCD)

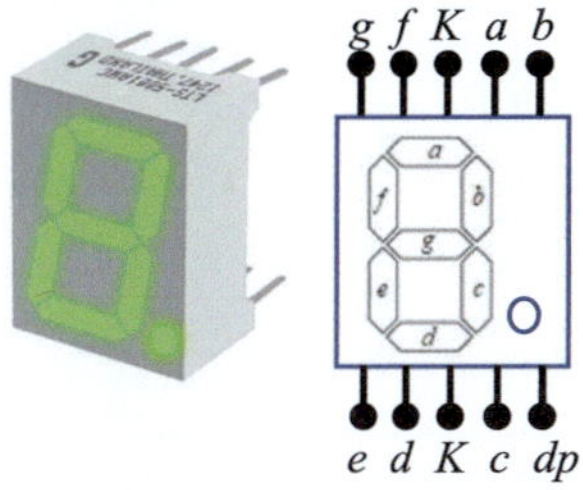

(b) 7-segment LED display pin connection K can be common anode or cathode

Fig. 8.12 Different 7-segment displays

a to *g*, normally a decimal point segment, marked as *dp* in Fig. 8.11c is also included in this display unit to show a decimal point.

To start with, the segments were realized using fluorescent strips as shown in Fig. 8.12a encapsulated in a vacuum glass enclosure and a segment is lit by bombarding it with electrons generated by a tungsten filament. These displays emitted bright light with a greenish blue hue. Today we either use the 7-segment displays with light emitting diodes (LED) as segments or Liquid crystal display (LCD) segments. The seven-segment LED display shown in Fig. 8.12b is quite popular as the LED displays have very high durability and are manufactured in different sizes and colors. Either all the cathodes of the eight LEDs (one for each segment plus one for the decimal point) are tied together to obtain a common cathode 7-segment LED display or all the anodes of the eight LEDs are connected to obtain a common anode 7-segment LED display. This common terminal (cathode or anode) is marked as *K* in Fig. 8.12b. The only drawback of LED displays is that they consume considerable power and thus make them unviable for battery operated devices. Once liquid crystal display technology was invented, the liquid crystal displays (popularly called LCDs) shown in Fig. 8.12c became the norm and are used extensively, especially in battery operated instruments/devices. LCDs require very low power to operate and hence are quite popular in such devices. Moreover, LCD displays can be illuminated from behind, using a *back panel* thus allowing the displays to be used even in very low light conditions, extending their use in other applications.

8.5.1.3 The Dot Matrix Display

As the name indicates, the display is made of "*dots*" arranged in a "*matrix*" form. Initially the 5 by 7 dot matrix (5 dots horizontally in a row and 7 such rows in the vertical direction) as shown in Fig. 8.13a was used and even today, this format is quite popular. Later, 7 by 9 and 8 by 8 formats were also used, especially in dot-matrix printers. Using the dot matrix, we can also display alphabets and even special characters as shown in Fig. 8.13b. Though these displays were first developed using

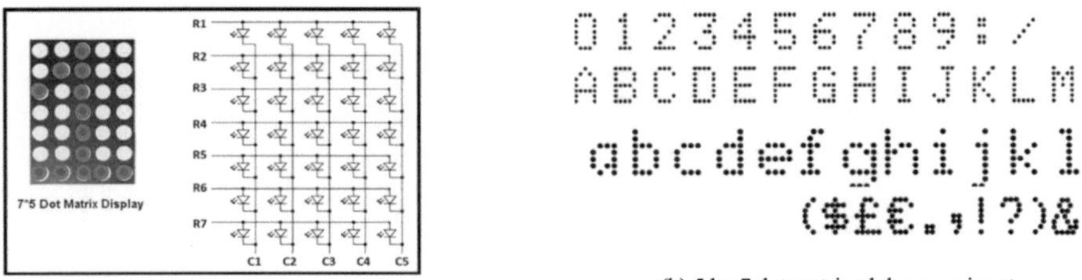

(a) 5 by 7 dot matrix LED one digit

(b) 5 by 7 dot matrix alphanumeric set

Fig. 8.13 The Dot Matrix display

LEDs, fluorescent and LCD dot matrix displays were also developed later. Dot matrix displays that can display pictures (called graphical displays) and movies (video) were also developed. The LCD dot matrix technology has now grown. Color LCD video displays in large-format 1920 by 1080 dot matrix (each dot is also called a pixel) are employed in computer monitor and color television. The most recent dot matrix color LCD technology is called the 4 k display. This uses 4096 horizontal dots (pixels) with 2160 pixels (2160 lines of 4096 horizontal dots per line) in the vertical direction. Today LCD dot matrix display technology is used in almost all devices that require the output to be shown on a color display and finds applications in digital storage oscilloscopes, color televisions, mobile phones, computer displays, household devices such as washing machines, microwave oven and even in smart watches.

8.5.2 *Measurement Using a Counter*

Using a multi-digit BCD counter connected to a suitable display, we can measure

(i) The frequency of an unknown signal, or
(ii) The period of an unknown signal, or
(iii) The ratio of the frequencies or the ratio of the periods of two unknown signals, or
(iv) Time between two events, or
(v) Elapsed time (as in a stopwatch).

8.5.2.1 The Counter for the Measurement of Frequency

To measure an unknown frequency f_x of a signal, the input *clock* to the counter is fed through a two input AND gate as shown in Fig. 8.14. The pertinent signal (input signal whose frequency needs to be measured) is fed to one of the inputs, say I_1, of the AND gate. Since the AND gate can only accept “0” or “1”, the unknown signal f_x must also be a pulse train as indicated in Fig. 8.15. If the input signal is not of binary form, then we must convert the input into binary pulse train form as detailed in the sequel. The other input I_2 of the AND gate is fed with a pulse of width exactly one second. A short pulse is applied to the reset input before applying the one second pulse as shown in Fig. 8.15. Once the reset pulse is applied, the counter will be reset to "000,000". After the reset period is over, the counter will count the number of pulses applied to the clock input during the one second period. Figure 8.15 shows the waveforms at different points of the configuration in Fig. 8.14. From Fig. 8.15 we see that the counter will count and display a count value N_{xf}, where N_{xf} is the number of cycles (of period T_x) of the input signal of unknown frequency f_x in one second. Hence the count value N_{xf} displayed by the counter is the unknown frequency f_x in cycles per seconds (Hz). We can express N_{xf} mathematically as

$$N_{xf} = \frac{T_s}{T_x}. \tag{8.1}$$

with $T_s = 1$,

$$N_{xf} = \frac{1}{T_x} = f_x \tag{8.2}$$

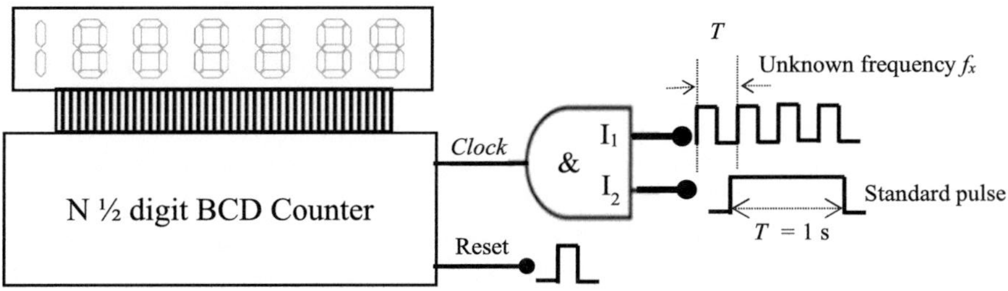

Fig. 8.14 The 6 1/2 digit BCD counter for frequency measurement

Fig. 8.15 Waveforms of Fig. 8.14

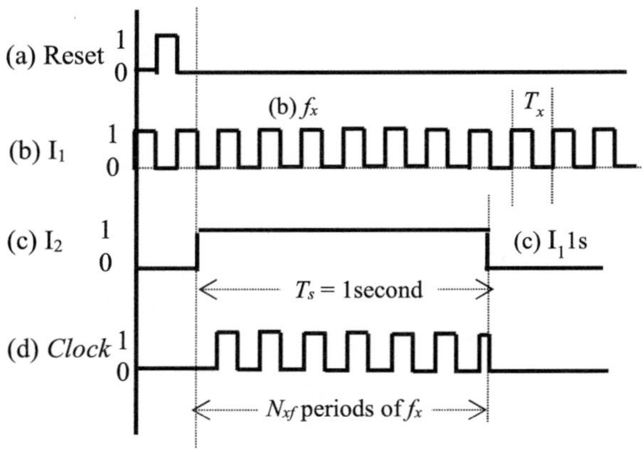

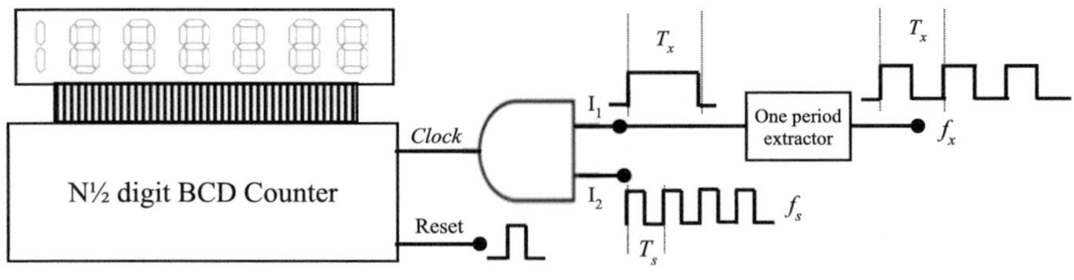

Fig. 8.16 BCD counter for period measurement

Now if we make $T_s = 1ms$ then the count N_{xf} will indicate the unknown frequency f_x in kHz. On the other hand, if we make $T_s = 1\mu s$ then the count N_{xf} will indicate the unknown frequency f_x in megahertz.

8.5.2.2 The Counter for the Measurement of Period

To measure the unknown period of a signal (of frequency f_x), first a pulse of width equal to the period T_x of the signal must be extracted from the pulse train f_x and fed to input I_1 of the AND gate as shown in Fig. 8.16. The other input I_2 of the AND gate is fed with a standard frequency f_s, say 1 Hz. Once again, a short pulse is applied to the Reset input before applying the T_x pulse as shown in Fig. 8.17. The counter will then be reset to "000…000" and will proceed to count the number of pulses of period T_s of the standard frequency f_s within one period T_x of the input of unknown frequency f_x. Figure 8.17 illustrates the waveforms at cardinal points of Fig. 8.16. From Fig. 8.17, we can deduce that the counter will count and display a count value N_{xt} where N_{xt} is the number of cycles (of period T_s) of the standard frequency f_s in one period of the unknown signal. Hence the count value N_{xt} displayed will indicate the period of the signal f_x in terms of T_s as shown in Eq. (8.3).

$$N_{xt} = \frac{T_x}{T_s} = \frac{T_x}{1} = T_x in\, seconds. \tag{8.3}$$

In Eq. (8.3) we see that with T_s as 1 s ($f_s = 1HZ$), the count N_{xt} directly indicates the period T_x in seconds. On the other hand, if $f_s = 1$ kHz (that is, the value of $T_s = 1ms$) then the count value N_{xt} will indicate T_x in milliseconds. Similarly, if f_s = 1 MHz ($T_s = 1\mu s$) then the count value N_{xt} will indicate T_x in microseconds.

8.5.2.3 The Counter for the Measurement of Ratio of Frequencies/ Periods

We can also get the ratio of two unknown frequencies f_a and f_b of two inputs or get the ratio of their periods T_a and T_b. To this end, the counter and its inputs are configured

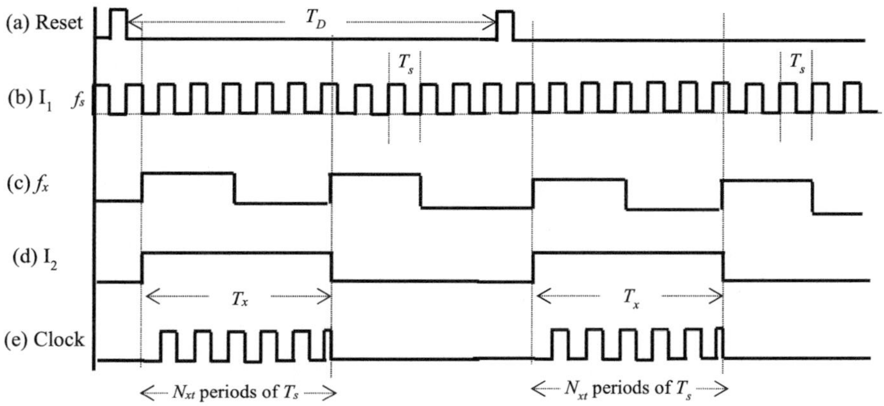

Fig. 8.17 Waveforms of Fig. 8.16

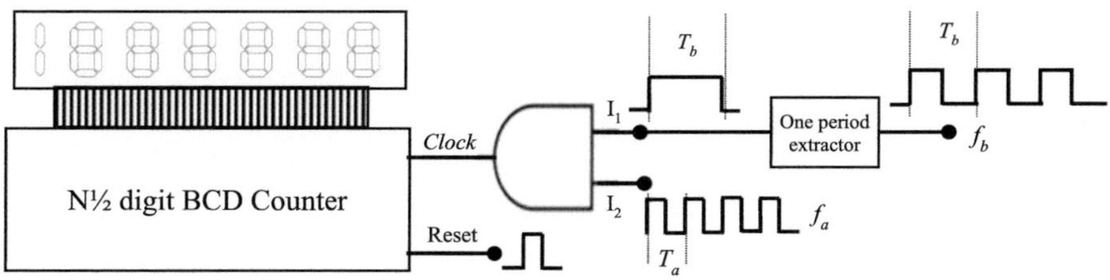

Fig. 8.18 BCD counter for ratio of frequencies/ratio of periods measurement

as shown in Fig. 8.18. Here, one period T_b of the input f_b is extracted and given as input I_1 of the AND gate. The other input I_2 of the AND gate is tied to the second unknown frequency f_a. Then the count value N_{xr} will be

$$N_{xr} = \frac{T_b}{T_a} = \frac{f_a}{f_b}. \tag{8.4}$$

Equation (8.4) indicates that the count value N_{xr} will indicate the ratio of the two frequencies f_a and f_b. Since the period T is the inverse of the frequency f, the ratio of frequencies $\frac{f_a}{f_b}$ will also be equal to the ratio of the periods $\frac{T_b}{T_a}$. Thus, it is possible to measure the unknown ratio of the periods or of the frequencies of two signals. This mode of operation of the counter is popularly called the ratio mode.

8.5.2.4 The Counter for the Measurement of Time Intervals

Using the counter, we can also measure the time interval (in seconds, milliseconds, or microseconds) between two *events*. This option is termed the $T_a \rightarrow T_b$ mode of operation of the counter. In this mode, the input I_2 of the AND gate is fed with a

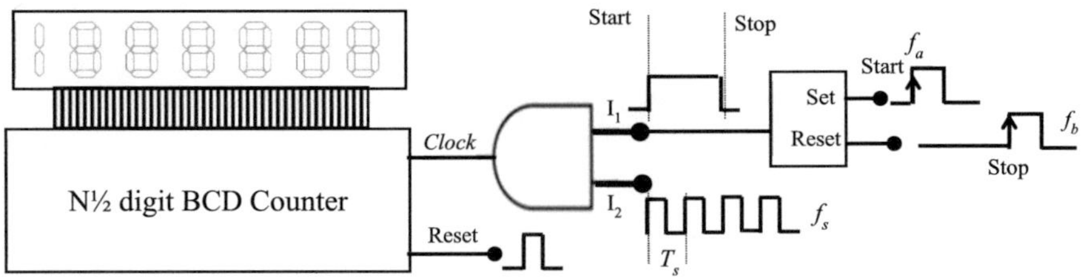

Fig. 8.19 BCD counter for elapsed time ($T_a \to T_b$) measurement

standard frequency (1 Hz or 1 kHz or 1 MHz) so that the time elapsed between any two events can be measured in terms of seconds or milliseconds or microseconds. The other input I_1 of the AND gate is set to "1" on the occurrence of the first event and reset to "0" on the occurrence of the second event. Thus, the time elapsed between the two events is measured. Figure 8.19 illustrates the $T_a \to T_b$ mode in determining the time elapsed between the rising edge of signal f_a (this sets input I_1 to "1") and the rising edge of signal f_b (which resets the input I_1 to "0"). In the $T_a \to T_b$ mode of operation of the counter, setting I_1 to "1" starts the counting of the standard frequency pulses and resetting I_1 to "0" stops the counter. Hence this mode is also called the "start-stop" mode or "stopwatch" mode or "Totalize" mode. In the foregone discussions we have understood that using a simple counter, we can measure the following:

(i) Frequency f_x of an unknown input in Hz or kHz or MHz.
(ii) Time period T_x of an unknown input in seconds, milliseconds, or microseconds.
(iii) The ratio $\frac{f_a}{f_b}$ of the frequencies f_a and f_b OR the ratio of their periods $\frac{T_b}{f_a}$.
(iv) The interval of (elapsed) time between two events.

8.5.2.5 Possible errors in the Measurement Using a Counter

Irrespective of whichever parameter is measured using the counter, we see that we simply count the number of pulses of one input within a period of the other input. A look at Eqs. (8.1), (8.3), and (8.4) shows that the left-hand side (LHS) is a count and hence can only be an integer as 1, 2, 3, … and the right-hand side is a ratio of two numbers and hence in most practical scenarios will be a fraction (proper fraction or improper fraction or composite fraction). Thus, any measurement made with a counter has an inherent error since only integer part of a fraction is measured. Thus, Eqs. (8.1), (8.3), and (8.4) must be modified as given below.

$$N_{xf} = INT\left(\frac{T_s}{T_x}\right) = INT\left(\frac{1}{T_x}\right) = INT(f_x) \tag{8.5}$$

$$N_{xt} = INT\left(\frac{T_x}{T_s}\right) = INT\left(\frac{T_x}{1}\right) = INT(T_x) \tag{8.6}$$

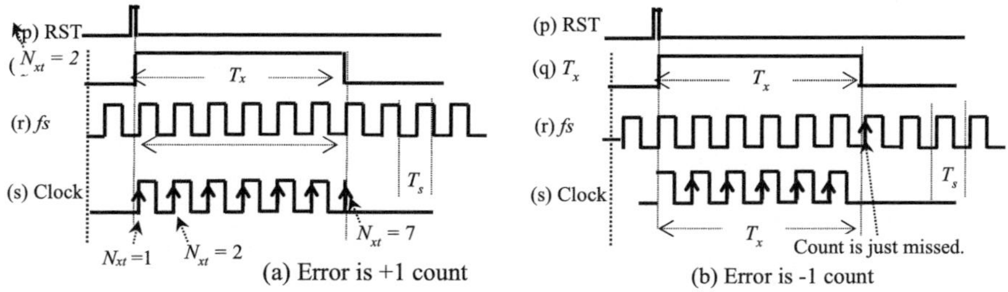

Fig. 8.20 Possible ±1 count error that can occur in a counter

$$N_{xr} = INT\left(\frac{T_b}{T_a}\right) = INT\left(\frac{f_a}{f_b}\right). \tag{8.7}$$

Apart from the "rounding off error" as indicated in Eqs. (8.5), (8.6), and (8.7), the way the counter works also introduces an error of ± 1 in the total count. Figure 8.8 pulseslustrates how the counting process for a given ratio of $\left(\frac{T_x}{T_s}\right)$ can lead to two different values differing by 1, depending on the relative positions of the first rising edge of clock pulse train after the start of the gate pulse. In Fig. 8.20a, we see that a rising edge occurs just after the rising edge of I_2 and the counter immediately counts 1 (remember we are using rising edge triggered logic). At the next rising edge, though only one full cycle (period) had elapsed, the count N_{xt} becomes 2. This additional one count will result in the higher count of 6, instead of 5 for the example shown in Fig. 8.20a. In Fig. 8.20b, the last pulse in the clock fails to be counted, since the timing (falling of 1 to 0 of I_1) just misses the succeeding rising edge. The count will be 5 instead of 6. Thus, we may (not necessarily true always) have an error of ±1 in the count value. If we analyze a similar situation for a ratio of T_x/T_s of say 5.X, we can easily show that the count value once again can be 6, 5, or 4 = (5 ± 1). Thus, whenever we use a counter we can anticipate, in a worst-case scenario, an error of ±1 count. Though Fig. 8.20 illustrates the period mode of operation, the situation is applicable to all the other modes (frequency measurement mode, ratio measurement mode and time interval measurement mode) of the counter. It should be noted here that the usual rounding-off error Of 0.5 of least significant digit is engulfed with the ± 1 count error.

8.6 The Universal Timer Counter

To harness the versatility of the counter in measuring different parameters, an instrument called "Universal Counter/Timer" was developed and was commercialized nearly seven decades ago. Figure 8.21 shows the block diagram of a typical Universal Counter/Timer. The universal timer counter contains:

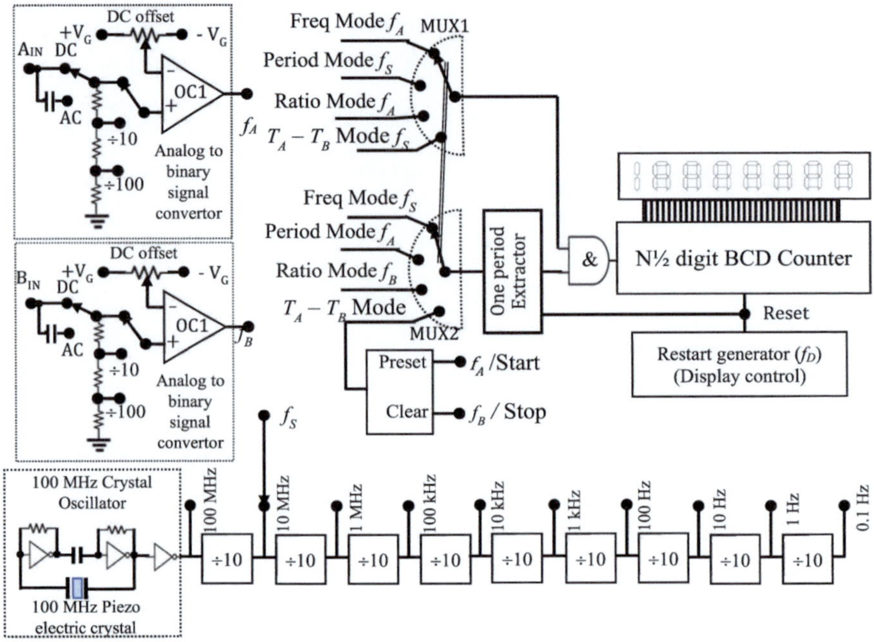

Fig. 8.21 Block diagram of a Universal counter/timer

(i) **A N1/2 Digit Counter** with a two-input AND gate (as in the Fig. 8.14) connected to its clock input.

(ii) **Mode-Select Multiplexers**

Each input of the AND gate is tied to the output of a 4 by 1 multiplexer. When we set the timer counter in the required mode of operation, the corresponding inputs to the AND gate are selected through these two multiplexers. For example, if we select the frequency mode, one input to the AND gate is set as f_A through multiplexer MUX1 and the input to the one period extractor (whose output serves as the second input I_2 to the AND gate) is selected as f_S through the multiplexer MUX2. It should be noted that MUX1 and MUX2 are operated in tandem as shown in Fig. 8.21.

(iii) **A "One Period" Extractor**

As we have seen, irrespective of the mode selected, the counter always counts the number of cycles (periods) of one of the selected bistate signal within a period of the other selected bistate signal. Hence, we need to extract the period of one of the selected inputs. Figure 8.22 illustrates a simple one period extractor from a single input f_{in}. $f_{in} = f_x$ in frequency mode or $f_{in} = f_s$ in period mode or $f_{in} = f_b$ in the ratio mode. High on the "Reset" signal latches a "1" to Q_1 and resets Q_2. The very next rising edge on the input f_{in} sets Q_2 to "1"

Fig. 8.22 A one period extractor

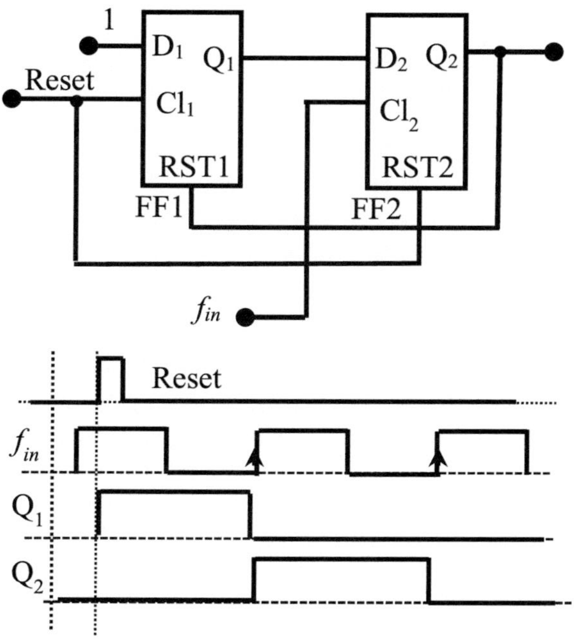

and resets Q_1. The succeeding rising edge on f_{in} latches a "0" to Q_2. Thus, the output Q_2 will be a one pulse whose width is one period of f_{in}.

(iv) **Analog to Binary (Bistate) Signal Converter**

Even though it was not mentioned, it would have been clear, at least from the waveforms that were shown, that the counter/timer we had learnt so far operates only on bistate or binary signals (that is these signals have only two states: either "0" or "1"). However, real-world signals are rarely of the bistate type and in general are analog signals. Thus, it is essential that real world analog signals are converted into bistate signals for doing any measurement using a counter/timer. Thus, the input stage of a universal counter/timer will have a couple of analog-to-bistate signal converters as shown in Fig. 8.21. Here the two inputs A_{IN} and B_{IN}, after attenuation or amplification (only an attenuator is shown in Fig. 8.21, but we can also add an amplifier), are given to individual comparators OC1 and OC2 that convert the real-world analog signals into bistate digital waveforms f_A and f_B. The crossover level (above which the input is taken as "1", otherwise "0") can be set at any voltage level in the range $+V_G$ to $-V_G$. This DC offset adjustment provides a mechanism to remove or add DC offset. Removing DC offset ensures that the crossover from "0" to "1" or "1" to "0" occurs at zero. Adding DC offset enables the "1" to "0" or "0" to "1" crossover to happen at a level other than zero. This feature helps to obtain stable readings even when the input signals are corrupted with offset and or noise.

To accommodate different input voltage magnitudes individual attenuators/amplifiers are employed. Generally, an attenuator with an attenuation factor settable as $\div 1$ or $\div 10$ or $\div 100$ is provided. Some manufacturers also give an amplification by a factor of 10 and or 100 for the inputs A_{IN} and B_{IN}. The coupling to the internal analog to binary signal converters (A_{IN} and B_{IN}) can be either direct coupled (marked as DC in Fig. 8.21) or coupled through a coupling capacitor (marked AC). Any DC present in A_{IN} and B_{IN} can be blocked using the AC input option. It should be noted that the offset adjustment described earlier is independent of this AC or DC coupling.

(v) **Standard Frequency Generator**.

From Eqs. (8.5), (8.6), and (8.7) we see that apart from the already discussed error in the counting process, the output also will have an error if f_S (or T_S) has an error. Thus, any error in f_S (or T_S) will directly affect the count. Hence f_S (or T_S) must be generated with very little error. Conventional oscillators (LC or RC) suffer from errors as the component values (especially L and or C) may change due to environmental factors such as temperature, humidity, and aging. Hence all universal counter/timers use a "crystal oscillator" that provides a stable output frequency with very little error and has negligibly small variation in the frequency due to environmental factors such as temperature and humidity.

Crystal Oscillator

A simple crystal oscillator, using a *piezoelectric crystal* and three inverters, is shown in Fig. 8.21. The piezoelectric property is exhibited in natural crystals such as "*quartz*" that possesses an anisotropic crystal structure. Quartz is a naturally occurring material with the chemical composition of SiO_2 (silicon dioxide). The crystal structure of quartz is as shown in Fig. 8.23. When a force F is applied, the crystal deforms by x and due to the pattern of the positive (Si) and negative (O_2) ions, an electric charge q is generated on the edge that is perpendicular to the force. The charge generated is directional in the sense that if we change the direction of the force, the polarity of the generated charge will also be reversed. The piezoelectric effect is reversible, that is, if we apply a charge between the surfaces of the crystal, the crystal will deform and thus can generate a force. Since the piezoelectric crystal is an electro-mechanical device, to analyze its behavior we need to address both the electrical and mechanical properties of the piezoelectric crystal. On the electric side, shown in Fig. 8.23b, the two plates A and B possess a capacitance C_e with the leakage resistance R_e in parallel.

On the mechanical side, we have the mass M of the crystal, its damping constant D, and the elastic constant K_s. We can relate the mechanical response of the piezoelectric crystal to the applied force F as

$$F = M\frac{d^2x}{dt^2} + D\frac{dx}{dt} + K_s x. \tag{8.8}$$

In Eq. (8.8), x is the deformation length of the element. To understand the behavior of the piezo electric crystal, we transfer the mechanical property of the crystal, using

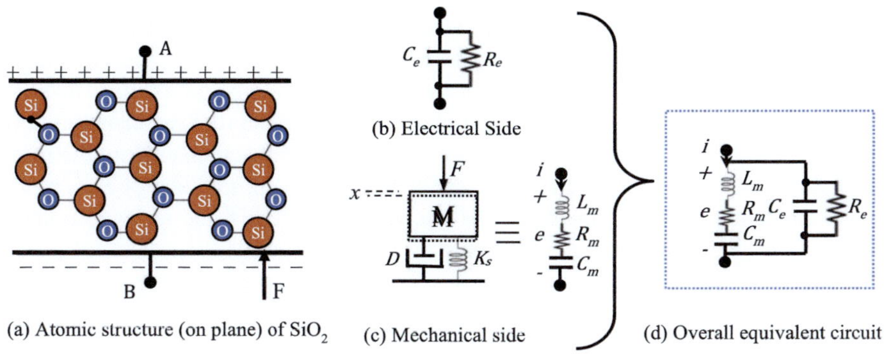

Fig. 8.23 Quartz (SiO_2) piezo electric crystal and its equivalent circuit

an appropriate equivalent to the electrical side and then analyze its characteristic using the total electrical representation. For doing this, we consider a series L-R–C circuit shown in Fig. 8.24c. The voltage e and current i relationship for the L_m-R_m-C_m circuit shown alongside the mechanical mass damper spring diagram is

$$e = L_m \frac{di}{dt} + R_m i + \frac{1}{C_m} \int idt. \tag{8.9}$$

Substituting $i = \frac{dq}{dt}$ (q is the electric charge) in Eq. (8.8) we get

$$e = L_m \frac{d^2 q}{dt^2} + R_m \frac{dq}{dt} + \frac{1}{C_m} q. \tag{8.10}$$

Comparing Eqs. (8.8) and (8.10), we realize that the mechanical properties can be analyzed using the series the L_m-R_m-C_m circuit shown in Fig. 8.24c. The facts that the voltage e generated across the piezoelectric element is proportional to the force F applied on the element and that the generated electric charge q is proportional to the deformation x of the element permit the insertion of this analogous electric equivalent with appropriate scaled element values into the existing electrical circuit

Fig. 8.24 Universal counter/ timer from Keysight

as a substitute for the actual mechanical part of the system. Thus, the overall performance of the piezoelectric crystal can be represented by the electrical equivalent circuit shown in Fig. 8.23d. The circuit indicates that the crystal exhibits a series resonance due to L_m and C_m (mass and elasticity) and a parallel resonance due to the mechanical equivalent inductance and the electrical capacitance. Here the mass M of the crystal is represented by the inductor L_m, the compliance of the crystal K_s is represented by C_m in the electrical equivalent. With this, the series resonance frequency of the piezoelectric crystal is

$$f_{sm} = \frac{1}{2\pi\sqrt{L_m C_m}}. \tag{8.11}$$

Once we manufacture a crystal, its mass M becomes a constant (resulting in L_m to be a constant) and if we maintain a constant temperature, its elasticity K_s (resulting in C_m) also becomes a constant. Thus, when we use the mechanical series resonance f_{sm} as the frequency for oscillation, we get a very stable frequency of oscillation. Though initially "quartz" crystals were employed and hence these oscillators were called quartz oscillators. Today, we use synthetic crystals made of ceramics like $BaTiO_3$ instead of quartz crystals. However, the name quarts oscillator is still used even when the piezoelectric crystal is a synthetic one and made of quartz. These oscillators can be built to have very small variation in the frequency of the generated wave. For example, the 100 MHz crystal oscillator shown in Fig. 8.21 can be practically built to have a frequency of

$$(100,000,000 \pm 10)\text{Hz}.$$

The low effective resistance at series resonance and the small difference in the values of the series and parallel resonance frequencies in the oscillator circuit lead to Q (selectivity factor) values of the order of 25,000, which are unattainable in conventional LC oscillators.

(vi) **Variable "Display Update" Generator**

Whenever we need to make a measurement with a universal counter/timer, we must reset the counter and the one period extractor. To do periodical measurements, say, every one second the reset signal must be applied every one second. To achieve this a simple oscillator whose frequency f_D can be varied is employed as Reset Generator in a universal counter/timer as shown in Fig. 8.21. Whenever the counter/timer is reset, a measurement is made, and the display is updated with the new measured value. After a time $T_D = 1/f_D$, the next reset pulse is applied, and the succeeding measurement cycle starts. Hence this oscillator (Reset Generator) is also called the *display update rate* controller.

Figure 8.24 shows a typical Universal Counter/Timer from Keysight. The universal counter/timer is the very first commercial digital instrument, manufactured and sold from the 1950s.

8.7 Problems and Exercise

E8.1 What are the inputs to the AND gate in a timer counter in the period mode when the range is set to 10 ns/count?

E8.2 A time counter is operated in the ratio mode (T_B/T_A). If the frequency given to input A is 1 kHz and the frequency of B input is 2 kHz, what value will be shown on the display?

E8.3 A timer counter uses a basic clock of 100 MHz. Find the input frequency that will have the same error whether we measure its frequency or period with this timer counter.

E8.4 A timer counter display 2000 in frequency mode with 1 s gating time. If the timer counter is now changed to period mode and the range is selected as microseconds/count, what value the timer counter will display?

E8.5 An 8 1/2 digit counter timer is to be used for the measurement of the phase difference of two 50 Hz sinusoidal signals. Describe the procedure for correctly measuring the phase.

E8.6 A timer counter is used for the measurement of the period of an unknown signal. The display shows two values (one small and the other large). List the reasons for this behavior. Show how you will eliminate this problem and get a stable display showing the correct value.

E8.7 The clock in a timer counter is derived using a crystal oscillator with the piezo-electric crystal maintained at a slightly higher temperature than the typical room temperature, why?

Chapter 9
Analog to Digital Converters

Digital technology has several features that can make it much easier for teachers to pay special attention to all their students.
—Bill Gates

9.1 Analog, Discrete, and Digital

Though at the fundamental level, our universe is based on "*quanta*", we perceive it as a continuum. Especially the *signals* that we encounter on a day-to-day basis are continuous. A signal is a quantity that we perceive of some entity. For example, the temperature variation in a day at a given geographical location is a signal. An example of such a signal is shown in Fig. 9.1. We see that this signal is indicated by a continuous curve. That is, the temperature can take every possible value both in time and magnitude, of course within the time limits T_1 and T_2 and magnitude between θ_1 and θ_2. We call this an *analog signal*. Mathematically, we represent the temperature variation as $\theta(t)$ using either of the two units Celsius or kelvin. *Note*: The unit of Fahrenheit being no longer a valid unit as per international standard (IS or SI) which

V. Jagadeesh Kumar, *Principles of Electrical Measurement and Measuring Instruments*,
https://doi.org/10.1007/978-981-96-5475-8_9

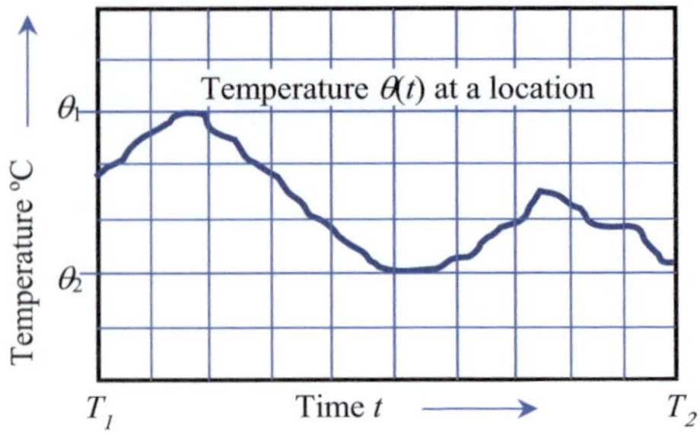

Fig. 9.1 Analog signal

is legal in more than 150 countries, including India. Thus, we can define an analog signal as:

> *Analog signal*: A quantity that varies continuously both with reference to the dependent and independent variables (parameters).

Here, in the example shown in Fig. 9.1, temperature θ is the dependent parameter (also popularly called the *dependent variable*) and time t is an *independent parameter* or *independent variable*. Another example of a signal is the elevation $h(x)$ along a road. Here the elevation $h(x)$ is the dependent variable, and the distance x on the road with respect to some starting point is the independent variable. $\theta(t)$ and $h(x)$ are called "*unidimensional*" or "*one dimensional*" signals since they have only one independent variable. A dependent variable can depend on more than one independent variable. An example of a signal that has more than one independent variable is the intensity of light $I(x, y)$ at a point (x, y) falling on a flat surface, say a wall. Here the dependent variable (signal), intensity of light, depends on two independent variables, namely, horizontal distance x and vertical height y. The signal $I(x, y)$ is called the *two-dimensional signal*. We can also think of the intensity of light falling on a flat surface and varying with respect to time as $I(x, y, t)$ as another signal. Of course, $I(x, y, t)$ is a three-dimensional signal as it has three independent variables.

Now if we decide to "look at" the temperature $\theta(t)$ in Fig. 9.1 at every hour and plot, then we get the plot shown in Fig. 9.2. We call this a *discrete signal* and can be defined as:

> *Discrete signal*: A signal that is defined or indicated at discrete intervals of the independent variable(s).

In the example shown in Fig. 9.2, the discrete signal is obtained by *sampling* an analog signal at discrete intervals of the independent variable. It may also happen that the dependent variable itself varies in discrete steps. A simple example is the height of steps in a staircase. It is also possible that we have the independent variable changing in discrete steps. For example, the heights of students in a class. It is also possible that both dependent and independent variables change in discrete steps. Again, a simple example is the number of students sitting in different rows in a classroom. We can also think of the height of a staircase as independent and dependent variables varying in discrete steps if we indicate the independent variable as: step 1, step 2,

Fig. 9.2 Discrete signal

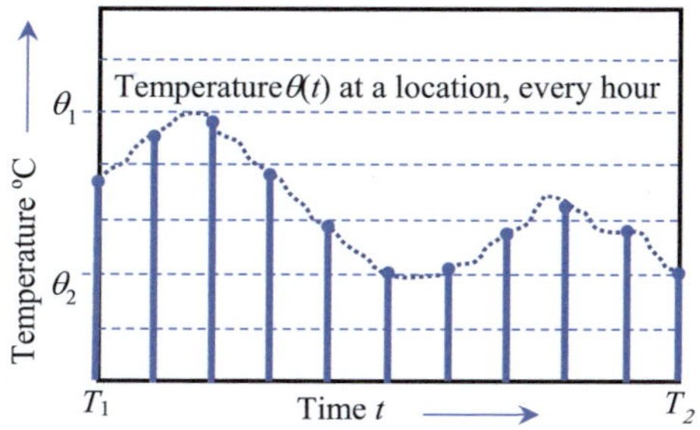

Table 9.1 Temperature measured at a location

Time (hours, n)	0	1	2	3	4	5	6	7	8	9
$\theta(n)$ in °C	24	24	24	25	25	25	26	27	28	28

step 3, ... The discrete signal (temperature at a point) in Fig. 9.2 can be represented mathematically as: $\theta(nT_s)|_{n=1,2,3,...}$ OR simply $\theta(n)$, if T_s is a known constant. If the interval T_s between two successive samples (readings) is a constant, then we term it as a *uniformly sampled* signal. It is also possible to sample a signal with varying T_s. If we do that, then we get a discrete signal that is non-uniformly sampled. Non-uniformly sampled signals can also be termed as "*randomly sampled*" signals. Now, if we make the measurements of $\theta(n)$ every hour with a thermometer that has a resolution of 1° then we get the data as shown in Table 9.1. We call this as the *digital signal* and a digital signal is defined as

> *Digital signal*: Is the signal (dependent variable) whose value is expressed in digits (number), at discrete values of the independent variable.

We can also say that a digital signal has discrete intervals both for the dependent and independent variables. Thus, whenever we make measurements in terms of digits (numbers) the result is called *digital*.

9.2 The Analog to Digital Conversion

While the timer counter described in Chap. 8 enables the measurement of frequency, period, phase, and time interval, we need devices to convert other analog signals into digital so that they can be read as a "*digital value*". Though such *digital reading* or *digital measurement* techniques in measurement were in vogue for several decades, the proliferation and explosion in their (digital measuring instruments) use for almost all measurement needs, thereby relegating analog indicating instruments to history, came about during the 1970s with the proliferation of:

(i) The dual slope analog to digital conversion (DSADC) technique and

(ii) The delta-sigma analog to digital conversion (SDADC) technique.

Humans started using the analog to digital conversion techniques in the later part of the eighteenth century (though not as ADC technique, but in an indirect manner). The following provides a rough timeline of the developments that can be thought of as the precursor to today's ADCs.

(i) The water usage measurement method was implemented in the Mahmud II dam in Turkey in the early nineteenth century.
(ii) The Morse code for representing alphabets, numerals, and punctuation marks in the 1830s.
(iii) The 5-bit Baudot code was developed for "teleprinter systems" in 1860.
(iv) The pulse code modulation (PCM) technique for facsimile was invented by Western Electric in 1921.
(v) The PCM and counter-based method of an analog to digital converter was patented by A.H. Reeves in 1939.
(vi) The initial analog to digital converters (ADCs) are called *Data converters*. They were huge in size and consumed hundreds of watts of power. One such data converter "DATRAC" was an 11-bit ADC with 50 kilo samples per second conversion speed (ksa/s), consumed 500 W, and was commercialized by EPSCO in 1954. The DATRAC was the first commercial successive approximation ADC.
(vii) The dual slope ADC method was published in 1954.
(viii) The dual slope ADC was manufactured in integrated circuit form in 1970s.
(ix) The delta sigma ADC technique was first known in 1962.
(x) The delta sigma ADC was commercialized, especially in the form of *codecs*, in the early 1980s.

9.2.1 Quantization in Analog to Digital Conversion

When we convert an analog signal to its digital value, we express the value of the analog signal in numbers at a particular time instant called the "*sampling time* T_s". It may be noted here that in a practical system we use a device called "*sample and hold*" (described in the sequel) to sample a signal before we convert the sampled value into digital. Except for the dual slope type ADC where we use the BCD number, in all other ADCs, we use the binary number for the digital value after this conversion. Once we decide on the digital representation, the step size gets decided. For example, if we use a 4-digit decimal number to represent the digital value, then the value can be from 0000 to 9999, in steps of 1. We generally call this step size as the "*resolution*", but in digital parlance we use a special term, namely, the "*quantization level*" or "*Least significant digit*" for decimal system/*least significant bit* for the binary system. Thus, for the 4-digit digital representation, if we have an analog signal whose value is 1234.5, we can represent the value in the 4-digit decimal number as 1234 only. Now the question arises as to what value we should assign for an input of 1234.9.

Table 9.2 Quantization levels in a 3-bit AD

Analog value	<0.5	0.5 to <1.5	1.5 to <2.5	2.5 to <3.5	3.5 to <4.5	4.5 to <5.5	5.5 to <6.5	>6.5
Binary number	000	001	010	011	100	101	110	111

Obviously, it must be 1235 as 1234.9 is much closer to 1235 than to 1234. Thus, whenever we convert an analog signal into a digital value, the stepwise digital value we choose must have the nearest least significant digit for decimal or least significant bit for binary as possible. This we do by:

(i) If the input minus the assigned digital value is less than half the least significant digit (or least significant bit for binary numbers), then we neglect the difference.
(ii) If the difference between the analog value and the assigned digital value is equal to or more than half the least significant digit (or least significant bit for binary output) then we step up the digital value by one least significant digit (or bit).

Since the digital value can only be in steps of the least significant digit (or bit), we call the process of assigning a value in steps to an analog input "*quantization*". To understand quantization, let us assume that the digital number is a 3-bit binary, having levels of 000, 001, 0010, 011, 100, 101, 110 and 111 (decimal 0, 1, 2, 3, 4, 5, 6 and 7). Here we see that the least significant bit (LSB) has a value of unity ($= 2^0$).

Example Q1: To represent an analog value from 0 to 7 using this 3-bit binary number, we will have to assign the binary numbers to corresponding analog values as shown in Table 9.2.

Quantization Error: In Table 9.2 we see that if our input is 0.5, then we read it as "001" or 1 in decimal which leads to an error of +0.5. Similarly, if our input is 6.4999, we read that as "binary 110 (one, one, zero)", or "decimal 6" resulting in an error of −0.4999. Thus, whenever we convert an along value into digital, we incur a worst-case error of ±0.5 least significant bit (LSB) if we use the binary number representation as shown in Fig. 9.3.

On the other hand, if we use a decimal number then the quantization error will be ±0.5 least significant digit (LSD). This ±0.5 LSB or ±0.5 LSD error is known as the "*quantization error*" and is inherent in all analog to digital conversions and thus in all analog to digital converters.

9.3 Analog to Digital Converters (ADC)

We will now learn about the following types of analog to digital converters (ADC) which are extensively used today.

(i) The FLASH or direct conversion ADC.

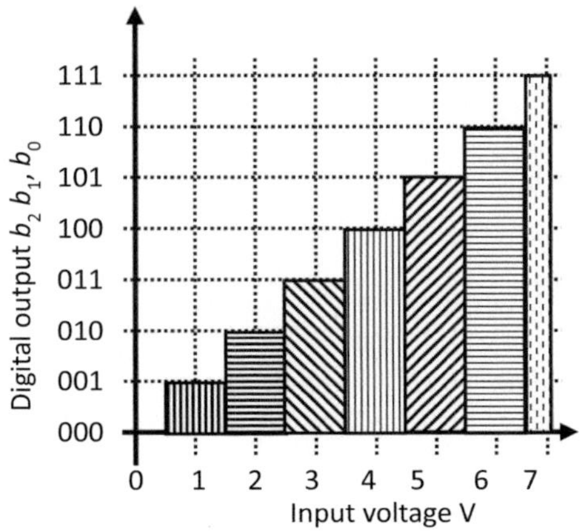

Fig. 9.3 Input–output characteristics of a 3-bit, 7 V full-scale ADC

(ii) The Successive Approximation Register (SAR) type ADC.
(iii) Successive subtraction ADC.
(iv) The dual slope ADC.
(v) The sigma-delta type ADC.

9.3.1 The FLASH ADC

To understand the principle of operation of a FLASH type ADC (also called the direct analog to digital conversion technique), let us take the example of a 3-bit, 7 V full-scale flash type ADC. The analog input to the digital output relationship, as illustrated in Table 9.2 for this ADC, is shown in graphical form in Fig. 9.3. From Fig. 9.3 and Table 9.2, we see that if we compare the input voltage V_{in} with corresponding voltage levels of 0.5, 1.5, 2.5, 3.5, 4.5, 5.5, and 6.5 V, we can determine the binary equivalent of the input analog voltage V_{in}. The FLASH ADC uses this principle and implements the comparison of the analog input voltage with different set voltage levels using comparators as shown in Fig. 9.4. In Fig. 9.4 we see that the full-scale voltage of 7 V is divided into seven levels, namely, 0.5, 1.5, 2.5, 3.5, 4.5, 5.5 V, and 6.5 V by the simple resistive divider circuit. The input voltage V_{in} is compared with each one of these levels by individual comparators marked C_1, C_2, C_3, C_4, C_5, C_6 and C_7. Comparator C_1 compares the input V_{in} to 0.5 V and hence its output O_1 will be "1" if the input V_{in} is $\geq$0.5 V. Similarly, Comparator C_2 compares the input V_{in} to 1.5 V and hence its output O_2 will be "1" if the input V_{iv} is $\geq$1.5 V, otherwise its output will be "0". Similarly, comparators C_3 to C_7 perform similarly and compare the input V_{in} to 2.5 V, 3.5 V, 4.5 V, 5.5 V and 6.5 V respectively. Table 9.3 shows the *truth table* of the outputs O_1–O_7 of the seven comparators C_1, C_2, C_3, C_4, C_5, C_6, and C_7 and the required binary bits b_2, b_1 and b_0 to realize the FLASH type 3-bit ADC. From the truth Table 9.3, we see that we need to decode the outputs O_1–O_7 of the seven comparators using a 7-input-to-3-output encoder as shown in Fig. 9.4 to extract the bits b_2, b_1 and b_0 from comparator outputs O_1–O_7. A simple look at Table 9.3 makes it very clear that the bit b_2 is "1" whenever comparator output O_4 is "1". Hence:

$$b_2 = O_4, \tag{9.1}$$

Similarly, we can derive from the entries highlighted with yellow in Table 9.3:

$$b_1 = O_6 + \overline{O_4}O_2 \text{ and} \tag{9.2}$$

Likewise, we can derive:

$$b_0 = O_7O_6 + \overline{O_6}O_5 + \overline{O_4}O_3 + \overline{O_2}O_1. \tag{9.3}$$

The logical implementation of the 7 bits to 3 bits encoder shown in Fig. 9.4 is for illustration only and has a one-to-one correspondence with Eqs. (9.1) to (9.3). In practice, the 7-bit to 3-bit encoder (in general a ($2^N - 1$) bits to N-bits encoder for an N-bit FLASH ADC) will be designed with all NAND gates or all NOR gates.

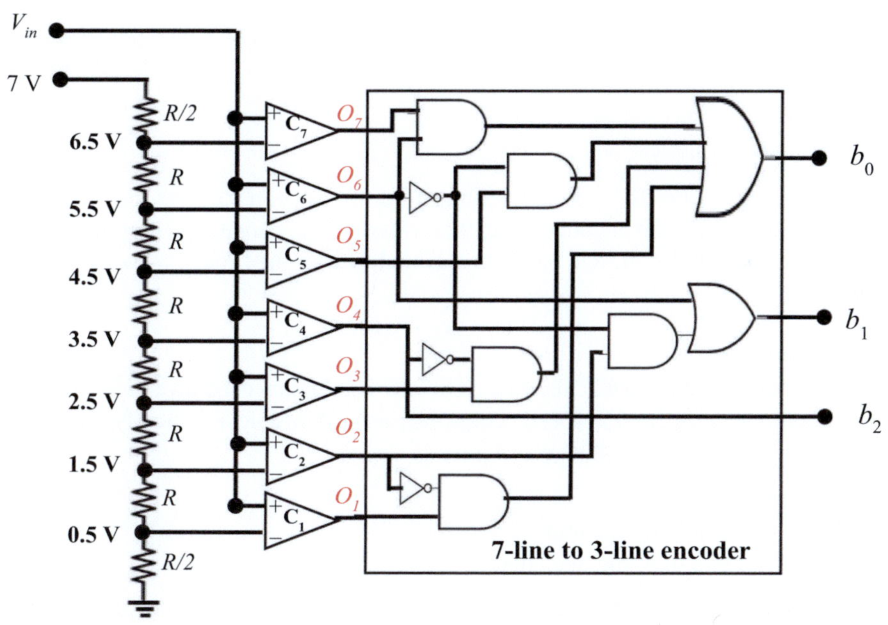

Fig. 9.4 Block diagram of a 3-bit FLASH ADC

Table 9.3 Truth table of 3-bit flash ADC

V_i V	O_7	O_6	O_5	O_4	O_3	O_2	O_1	b_2	b_1	b_0
< 0.5	0	0	0	0	0	0	0	0	0	0
≥ 0.5	0	0	0	0	0	0	1	0	0	1
≥ 1.5	0	0	0	0	0	1	1	0	1	0
≥ 2.5	0	0	0	0	1	1	1	0	1	1
≥ 3.5	0	0	0	1	1	1	1	1	0	0
≥ 4.5	0	0	1	1	1	1	1	1	0	1
≥ 5.5	0	1	1	1	1	1	1	1	1	0
≥ 6.5	1	1	1	1	1	1	1	1	1	1

9.3.1.1 Conversion Time of the FLASH ADC

The minimum time required for an ADC to convert an analog input to its corresponding digital output is called the conversion time T_{con}. If an ADC requires a conversion time of T_{con} then we can get maximum sampling rate $f_{sa} = \frac{1}{T_{\text{con}}}$ conversions per second (samples per second). f_{sa} is called the conversion speed (or conversion rate or simply sampling rate) and the unit used is sa/s (samples per second). For the flash ADC of Fig. 9.4, when the input V_{in} changes, then the output binary bits b_2, b_1 and b_0 do not change immediately. The bits b_2, b_1 and b_0 will settle only after the outputs of:

(1) The comparators and
(2) The encoder logic.

attain steady state. If τ_C is the settling time of the comparators and τ_E is the propagation delay of the encoder logic, then whenever a change in the input happens then the output bits b_2, b_1, and b_0 will be available after the time T_{con}, where:

$$T_{\text{con}} = (\tau_C + \tau_E). \tag{9.4}$$

Of all the available ADCs, the FLASH ADC has the lowest conversion time and hence the highest conversion speed (sampling rate) achievable, practically. However, to build an N-bit FLASH ADC we require

(i) 2^N Resistors,
(ii) $(2^N - 1)$ comparators, and
(iii) $(2^N - 1)$ bits to N-bits encoder.

Thus, most FLASH ADCs have limited resolution (number of bits N) of 8 bits. Recently, 10 bits or 12 bits resolution FALSH ADCs are becoming available. While the conversion speed of an 8-bit FLASH ADC can be as high as 2 Gsa/s, the conversion speed of a 10-bit or 12-bit FLASH ADC will be much lower than the 2 Gsa/s achievable for 8-bit FLASH ADC.

9.3.2 *The Successive Approximation Register (SAR) Type ADC*

Figure 9.5 illustrates the block diagram of an N-bit SAR type ADC. The digital part of the ADC is called the successive-approximation-register (SAR) that is implemented as a state machine, run using a clock v_{cl}. In Fig. 9.5, we see that the digital outputs (b_{N-1}, b_{N-2}, … b_1, b_0) of the SAR are given as input to an N-bit digital to analog converter (DAC). The DAC converts the digital input (binary number) to an equivalent analog voltage v_{dac}. A voltage equal to 0.5 LSB is subtracted from the output v_{dac} of the DAC using an opamp subtractor (unity gain difference amplifier) OS. The output v_{sub} of the opamp subtractor OS is tied to the inverting input of the comparator OC. The input voltage V_{in} is tied to the non-inverting input of the comparator OC, whose output v_c serves as the control input D to the SAR. The SAR logic tests each bit (by sensing the comparator voltage v_c), starting from the MSB (b_{N-1}) and determining whether that bit is a “1” or “0”.

An analog to digital conversion is started by eternally asserting the “*Start*” signal of the SAR and, at the end, a *conversion complete* signal called End of Conversion (EOC) is generated by the SAR. Once *Start signal* is initiated, all the bits (b_{N-1}, b_{N-2}, … b_1, b_0) are reset to “0” and the MSB (b_{N-1}) bit alone is set to “1” by the SAR logic shown in the form of a flowchart in Fig. 9.6. The DAC converts the digital output “100…00” of the SAR to equivalent voltage V_{dac}. The output of the subtractor V_{sub} which is (V_{dac^c} – 0.5 V_{LSB}) is compared with the input voltage V_{in} by the comparator OC.

(i) If $\{V_{\text{in}} \geq (V_{\text{dac}} - 0.5V_{\text{LSB}})\}$, then the output v_c of the comparator OC will be “1”. Since v_c is connected to the input D of the SAR, the SAR sensing D to be “1” retains b_{N-1} as “1” and moves to the next bit, that is b_{N-2}.
(ii) On the other hand:

If $\{V_{\text{in}} < (V_{\text{dac}} - 0.5\,V_{\text{LSB}})\}$, then the output V_c of the comparator OC will be “0” and hence input D of the SAR is “0”. Sensing D to be “0” the SAR resets b_{N-1} to “0” and moves to the next bit, that is b_{N-2}.

This process of setting the bit being tested to “1” first and either retaining that bit as “1” if D = “1” or resetting the bit as “0” if D = “0” is done for all the bits, one after another. The conversion becomes complete once the LSB is tested and its value

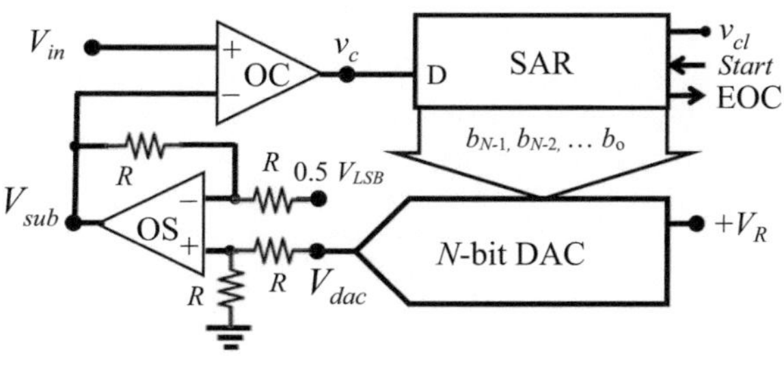

Fig. 9.5 Block diagram of the SAR type ADC

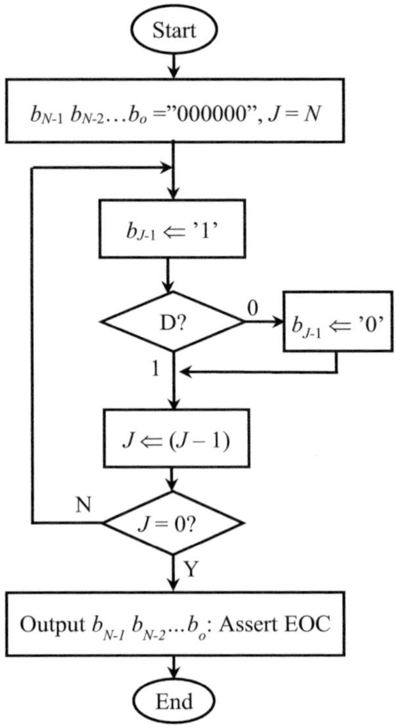

Fig. 9.6 Flowchart of the SAR logic

("1" or "0") is determined. Once all the bits are determined, the SAR logic indicates that the analog to digital conversion is completed by asserting the *end of conversion* signal (EOC).

Though the SAR logic explained above is clear and the flowchart of Fig. 9.6 illustrates this flow logic, let us understand it more by examining a 3-bit SAR ADC of full-scale 7 V. Obviously to build a 3-bit SAR ADC, we need a 3-bit SAR DAC. The input–output relationship of a 3-bit DAC of full-scale 7 V is shown in Table 9.4. It is seen that a binary input "001" to the 3-bit DAC results in an output voltage of 1 V, binary input "010" provides 2 V at the output of the DAC, and so on. Details and design steps of a DAC are illustrated in the sequel. The state diagram of the SAR logic of the 3-bit SAR ADC (structure shown in Fig. 9.5) is shown in Fig. 9.7. Here the voltages 0.5, 1.5, 2.5, 3.5, 4.5, 5.5, and 6.5 V are the *bit reference voltages* obtained by subtracting 0.5 LSB voltage from the output of the DAC for the corresponding 3-bit binary output of the SAR. To understand the operation of the 3-bit SAR type ADC, let us look at examples.

Table 9.4 Input output characteristics of A 3-bit 7 V Fs DAC

Input binary	000	001	010	011	100	101	110	111
Analog output (V)	0	1	2	3	4	5	6	7

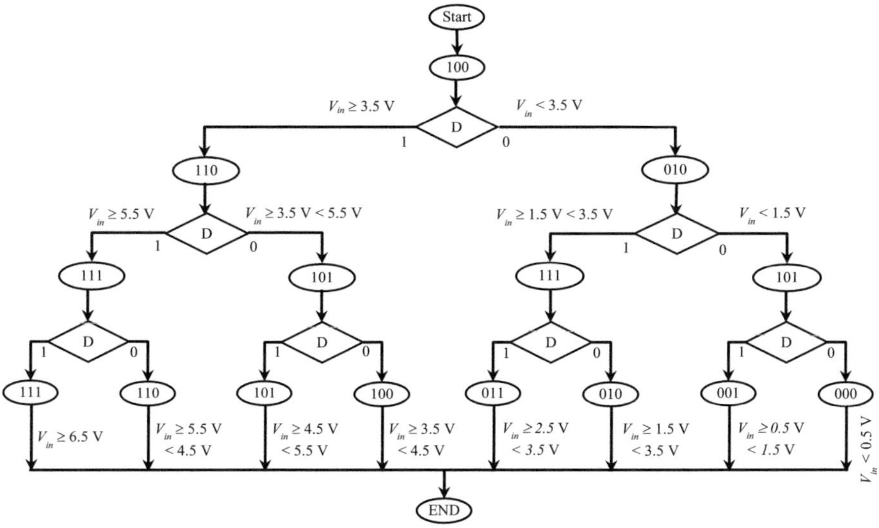

Fig. 9.7 State diagram of the 3-bit, 7 V Full-scale SAR ADC logic

Example 9.1 The steps the 3-bit SAR will take for an input V_{in} of 0 V.

Step 1: Reset bits b_1, b_0 and set bit b_2 to "1".
Step 2: Since $V_{\text{in}} < 3.5$ V, hence $D =$ "0": Reset b_2 to "0". Set b_1 to "1".
Step 3: Since $V_{\text{in}} < 2.5$ V, hence $D =$ "0": Reset b_1 to "0". Set b_0 to "1".
Step 4: Since $V_{\text{in}} < 0.5$ V, hence $D =$ "0": Reset b_0 to "0".
Step 5: Output $b_2, b_1, b_0. =$ "000" Assert EOC.

Example 9.2 The steps the SAR will take for an input of 7 V:

Step 1: Reset bits b_1, b_0 and set bit b_2 to "1".
Step 2: Since $V_{\text{in}} > 3.5$V, D = "1": St b_2 to "1". Set b_1 to "1". >>
Step 3: Since $V_{\text{in}} > 4.5$V, D = "1": Reset b_1 to "1". Set b_0 to "1".
Step 4: Since $V_{\text{in}} > 6.5$V, D = "1": Set b_0 to "1".
Step 5: Output $b_2, b_1, b_0. =$ "111" Assert EOC.

9.3.2.1 Conversion Time of an SAR ADC

From the above discussions, we see that the SAR needs: A start state, one state for testing and setting/resetting each bit, and an end state. Thus, an N-bit SAR ADC will require $(N + 2)$ states or $(N + 2)$ clock cycles to complete a conversion. If T_c is the period of the clock, then the conversion time T_{con} of the SAR type ADC will be

$$T_{\text{con}} = (N + 2)T_c$$

The minimum clock period T_c is decided by:

(i) The minimum time τ_{DAC} required by the DAC to attain steady state,
(ii) The time τ_{OS} required for the 0.5 LSB subtractor OS to settle,
(iii) The settling time τ_{OC} of comparator OC, and
(iv) The propagation delay τ_{SAR} of the SAR.

Thus, we get the minimum clock for each state will be

$$t_{\text{clock}}|_{\text{min}} = \tau_{\text{DAC}} + \tau_{\text{OS}} + \tau_{\text{OC}} + \tau_{\text{SAR}} = \tau_{\text{DAC}} + \tau_k \tag{9.5}$$

In Eq. (9.5), $\tau_k = (\tau_{\text{OS}} + \tau_{\text{OC}} + \tau_{\text{SAR}})$, which becomes a constant once we implement the SAR ADC with practical devices. Hence, the minimum conversion time of an N-bit SAR type ADC operated with $t_{\text{clock}}|_{\text{min}}$ is

$$T_{\text{conv}}|_{\text{min}} = (\tau_{\text{DAC}} + \tau_k)(N + 2) \tag{9.6}$$

To re-iterate, in Eqs. (9.5) and (9.6), τ_{DAC}, τ_{OS}, and τ_{OC} are the settling times of the DAC, opamp subtractor, and opamp comparator, while τ_{SAR} is the propagation delay of SAR logic.

9.3.2.2 Progressive Clock SAR ADC

While τ_{OS}, τ_{OC}, and τ_{SAR} are device specific and are decided once a device is chosen, τ_{DAC}, however, is not only device specific but also varies depending on the change in the DAC's output. For the 7 V full-scale 3-bit SAR ADC we have chosen as the example, when the first bit is decided, the output V_{dac} must change in its value by 4 V. If the DAC chosen has a slew rate of 1 V/μs then $\tau_{\text{DAC}} = 4$ μs for the MSB bit. For the (MSB-1) bit, the change required will be ±2 V and hence the value of τ_{DAC} for the (MSB-2) bit is 2 μs. For the LSB, the value of τ_{DAC} is 1 μs as shown in Fig. 9.8. Most modern SAR type ADCs exploit this reducing value of τ_{DAC} and thus operate the SAR with a clock that has progressively reduced periods. The optimum design is to design such a clock with successive periods as given in Table 9.5.

The first state (start state) we need to ensure that the digital part of the SAR has attained steady state hence the first state clock need to be $T_1 \geq \tau_{\text{SAR}}$. For the next N states that determine the bit values of the N-bits of the ADC, we must ensure that all the blocks have attained steady state. Thus, for the MSB we need the clock period T_2 to be $T_2 \geq (\tau_K + \tau_{\text{DAC}})$, where $\tau_K = (\tau_{\text{OS}} + \tau_{\text{OC}} + \tau_{\text{SAR}})$. For the (MSB-1) bit we must have $T_c \geq \left(\tau_K + \frac{\tau_{\text{DAC}}}{2}\right)$. Next for the (MSB-2) bit we need $T_c \geq \left(\tau_K + \frac{T_{\text{DAC}}}{4}\right)$ and so on. For the end state $\{(N + 2)\text{th state}\}$, once again we must have $T_c \geq \tau_{\text{SAR}}$. With such a "progressively reducing time period" clock the SAR ADC is called the *progressive clock* SAR type ADC. If we employ this optimal progressive clock, then the conversion time of optimal progressive clock SAR ADC is

$$T_{\text{con}} = (2\tau_{\text{DAC}} + N\tau_K + 2\tau_{\text{SAR}}). \tag{9.7}$$

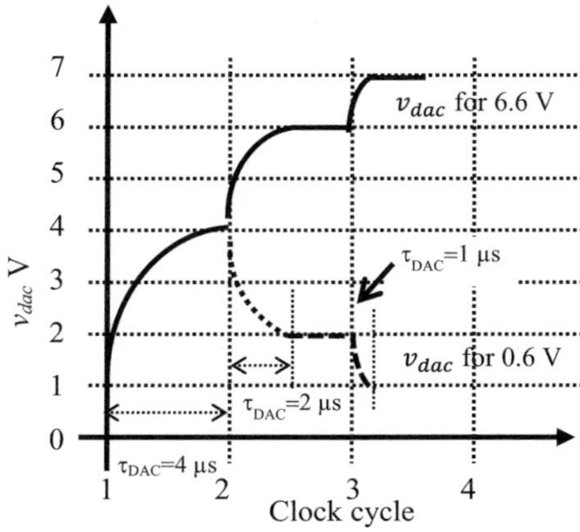

Fig. 9.8 Output of the 3-bit DAC

Table 9.5 Clock periods in a progressive clock SAR N-bit AD

Clock T_1 (start state)	T_2 (MSB)	T_3 (MSB-1)	T_4 (MSB-2)	…	T_N (LSB+1)	T_{N+1} (LSB)	T_{N+2} (end state)
τ_{SAR}	$\tau_K + \tau_{DAC}$*	$\tau_K + \frac{\tau_{DAC}}{2}$	$\tau_K + \frac{\tau_{DAC}}{4}$	…	$\tau_K + \frac{\tau_{DAC}}{2^{N-2}}$	$\tau_K + \frac{\tau_{DAC}}{2^{N-1}}$	τ_{SAR}

* $\tau_K = (\tau_{OS} + \tau_{OC} + \tau_{SAR})$

While using a progressive clock is advantageous, the generation of such a progressive clock is quite complex. Due to this, some manufacturers do not do such a fine, bit by bit clock period variation, but use two or three clock periods. For example, for a 12-bit SAR ADC, the first 4 bits are determined using clock with a period T_C, next 4 bits are decided with a clock $T_C/2$ and the last 4 bits are run with $T_C/4$. Thus, the conversion time instead of 14 T_C will become 7 T_C.

The name given to this type of ADC is Successive Approximation Register type ADC, because a locally generated voltage is *successively approximated* to become equal to the input voltage. When we say "equal to" what practically happens is that the locally generated DAC voltage v_{dac} is brought to be within $(V_{in} \pm 0.5V_{LSB})$. Figure 9.8 illustrates this aspect for the 3-bit SAR ADC chosen as an example. The solid line represents the changes in v_{dac} for V_{in} is 6.6 V: In the second clock cycle the output v_{dac} of the DAC will be 4 V. In the third clock cycle v_{dac} will become 6 V and in the fourth clock cycle v_{dac} will become 7 V. The dashed line represents the changes in v_{dac} for V_{in} is 0.6 V: In the second clock cycle the output v_{dac} of the DAC will change to 4 V. In the third clock cycle v_{dac} will become (4 V ± 2 V) and in the fourth clock cycle v_{dac} will become (2 V ± 1 V).

9.3.3 *The Successive Subtraction ADC*

The successive subtraction type ADC (SSADC) is not as popular as the SAR type ADC. However, the technique is quite simple and is worth learning. Like the SAR type ADC, the SSADC also determines each bit to be "0" or "1", starting from the MSB. If a bit is "1" then the corresponding voltage (for that bit level) is subtracted from the input and the modified voltage is chosen to decide the next bit. There are three methods to implement an *N*-bit SSADC, namely,

(i) *N*-reference voltage Successive Subtraction ADC (NR SSADC).
(ii) One reference voltage Successive Subtraction ADC (OR SSADC).
(iii) Optimal Successive Subtraction ADC (SSADC), that combines multiple low-resolution NR SSADCs to realize an SSADC. The optimal SSADC will have a higher resolution than either the NR SSADC or OR SSADC.

9.3.3.1 *N*-Reference Successive Subtraction ADC (NR SSADC)

The *N*-reference SSADC, as the name indicates, requires *N*-reference voltages (one for each bit) for an *N*-bit NR SSADC. These reference voltages can be derived either using a simple resistive potential divider or an *N*-bit *R*-2*R* ladder network (detailed in the sequel). To illustrate the principle of operation of an NR SSADC, let us study a simple 4-bit, 15 V full-scale NR SSADC shown in Fig. 9.9. Here, we have LSB = 1 V, MSB = 8 V. If V_D, V_C, V_B, and V_A are the voltages corresponding to bits *D*, *C*, *B*, and *A* then we have: $V_D = 8$ V, $V_C = 4$ V, $V_B = 2$ V and $V_A = 1$ V. Here, we see that we need to generate the four-bit voltages V_D, V_C, V_B, and V_A and four reference voltages V_D', V_C', V_B', V_A' corresponding to the voltages of bits *D*, *C*, *B*, and *A* reduced by the quantization bias of $V_L = \frac{V_{LSB}}{2} = 0.5$ V to implement the 4-bit NR SSADC. A simple $R - 2R$ resistive ladder and three unity gain subtractors can derive these voltages as shown in Fig. 9.9. With this, the logic that the 4-bit NR SSADC implements can be described as

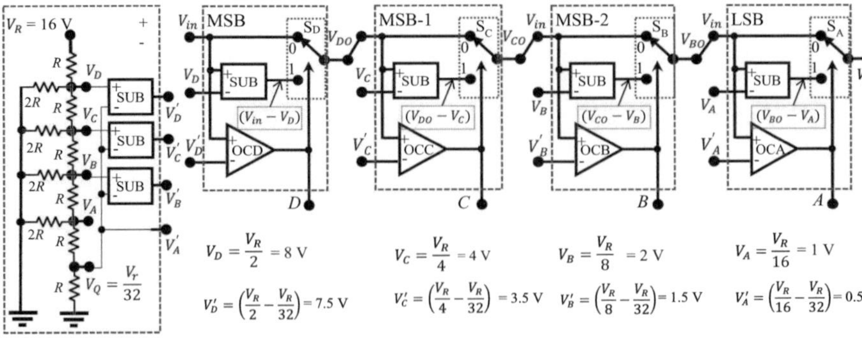

Fig. 9.9 Schematic of a 4-bit SSADC with 4-reference voltages

Stage 1:

If [$V_{\text{in}} \geq V_D'$ (7.5 V) Then {$D =$ '1', $V_{\text{DO}} \Leftarrow (V_{\text{in}} - V_D)$}]. Else {$D =$ '0', $V_{\text{DO}} \Leftarrow V_{\text{in}}$}.

Stage 2:

If [$V_{\text{DO}} \geq V_C'$ (3.5 V) Then {$C =$ '1', $V_{\text{CO}} \Leftarrow (V_{\text{DO}} - V_C)$}]. Else {$C =$ '0', $V_{\text{CO}} \Leftarrow V_{\text{DO}}$}.

Stage 3:

If [$V_{\text{CO}} \geq V_B'$ (1.5 V) Then {$B =$ '1', $V_{\text{BO}} \Leftarrow (V_{\text{CO}} - V_B)$}]. Else {$B =$ '0', $V_{\text{BO}} \Leftarrow V_{\text{CO}}$}.

Stage 4:

If [$V_{\text{BO}} \geq V_A'$ (0.5 V) Then {$A =$ '1'}]. Else {$A =$ '0'}.

From Fig. 9.9, we see that starting from the MSB, the logic for all the bits is the same. Each bit stage is identical wherein the input to that stage is compared with that bit's reference (that is bit voltage minus half LSB voltage). If the input to that stage is more than or equal to that bit's reference voltage, then that bit is set to "1" and the output is set to the input to that stage minus the bit's voltage. Otherwise, the bit of that stage is set to "0" and the input is passed on unaltered to the output. Of course, in the LSB stage, we only need the comparator OCA and can remove the switch, and the subtractor. The implementation of the NR SSADC has one practical problem, namely, as we go through from MSB to LSB, the inputs to the comparator become equal to or less than the non-idealities of the comparator such as offset and noise.

For example, if we choose a 10-bit NR SSADC of 1.023 V full scale then the comparator of the LSB will be comparing the $\frac{V_{\text{LSB}}}{2}$ voltage level of 0.5 mV. It should be remembered that the non-inverting input for the LSB is obtained through 9 stages of subtractors and 9 SPDT switches, thus the offset, offset drift, and noise of these 9 stages will affect the LSB stage. Hence beyond 8 bits, realizing an NR SSADC is fraught with practical problems.

Conversion time of NR SSADC: With τ_{co} as the settling time of the comparator, τ_{sw} being the settling time of the SPDT switch and τ_{su} is the settling time of the subtractor, the conversion time T_{con} of an N-bit NR SSADC will lie between

$$T_{\text{con}}|_{\text{min}} = (N\tau_{co} + (N-1)\tau_{sw}), \tag{9.8}$$

occurring when the input is zero, and the maximum conversion time of

$$T_{\text{con}}|_{\text{max}} = [N\tau_{co} + (N-1)(\tau_{su} + \tau_{sw})], \tag{9.9}$$

which happens when the input is full-scale. The logic of NR SSADC can be understood better by looking at examples.

Examples: NR(i) Input is 15 V

Stage 1: Since [$V_{\text{in}} \geq (7.5\text{ V})$, {$D =$ '1', $V_{\text{DO}} \Leftarrow (15 - 8) = 7.0$ V}].
Stage 2: Since [$V_{\text{DO}} \geq (3.5\text{ V})$, {$C =$ '1', $V_{\text{CO}} \Leftarrow (7.0 - 4) = 3$ V}].
Stage 3: Since [$V_{\text{CO}} \geq (1.5\text{ V})$, {$B =$ '1', $V_{\text{BO}} \Leftarrow (3.5 - 2) = 1$ V}].
Stage 4: Since [$V_{\text{BO}} \geq (0.5\text{ V})$, {$D =$ '1'}].
Hence "DCBA" = "1111"

NR(ii) Input is 8 V

Stage 1: Since [$V_{\text{in}} \geq (7.5\text{ V})$, {$D =$ '1', $V_{\text{DO}} \Leftarrow (8 - 8) = 0.0$ V}].
Stage 2: Since [$V_{\text{DO}} < (3.5\text{ V})$, {$C =$ '0', $V_{\text{CO}} \Leftarrow 0.0$ V}].
Stage 3: Since [$V_{\text{CO}} < (1.5\text{ V})$, {$B =$ '0', $V_{\text{BO}} = 0.0$ V}].
Stage 4: Since [$V_{\text{BO}} < (0.5\text{ V})$, {$A =$ '0'}].
Hence "DCBA" = "1000".

9.3.3.2 One Reference Successive Subtraction ADC (OR SSADC)

The OR SSADC, as the name indicates, requires just one reference voltage $V_1 = \left(\frac{V_R}{2} - \frac{V_{\text{LSB}}}{2}\right)$ and the value of just the MSB bit voltage viz. $\left(\frac{V_R}{2}\right)$ to determine all the N-bits in the result. Here too, the scheme contains N identical stages with one stage per bit. Each stage contains a comparator, a single pole double throw (SPDT) switch, a subtractor, and an amplifier with a gain of 2 as shown in Fig. 9.10. To illustrate the principle of operation of an OR SSADC, let us study a simple 4-bit, 15 V full-scale OR SSADC shown in Fig. 9.10. Comparing Fig. 9.10 with Fig. 9.9, we see that each bit stage apart from one comparator, one single pole double throw switch (SPDT), and one subtractor has an additional amplifier of gain 2. The introduction of the amplifier of gain 2 enables that all the bits are determined with the same set of one bit voltage (8 V) and one bit reference voltage (7.5 V). Again, here too the LSB stage needs to have only a comparator, and the rest such as the subtractor, the switch, and the amplifier of gain 2 can be dispensed with. The logic implemented by the OR SSADC scheme of Fig. 9.10 for each bit is identical starting from (MSB to LSB-1) bit and LSB just needs a comparison as given below:

Stage 1:

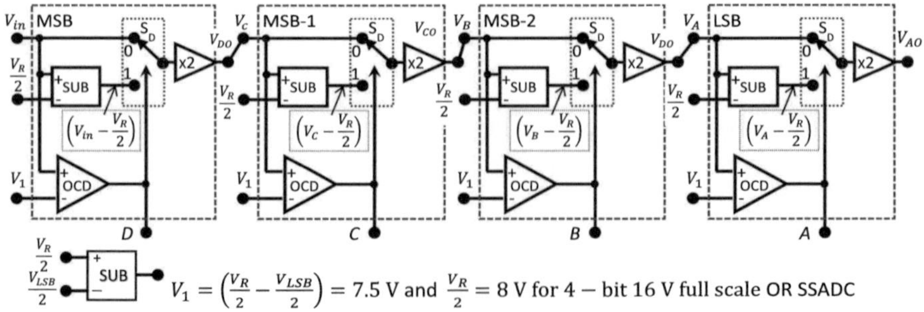

Fig. 9.10 Schematic of a four-bit OR SSADC

If[($V_{\text{in}} \geq V_1$) then {$D =$ '1', $V_{\text{DO}} \Leftarrow (V_{\text{in}} - V_R/2)$} Else {$D =$ '0', $V_{\text{DO}} \Leftarrow V_{\text{in}}$}], $V_{\text{DO}} \Leftarrow 2V_{\text{DO}}$.

Stage 2:

If[($V_{\text{DO}} \geq V_1$) then {$C =$ '1', $V_{\text{CO}} \Leftarrow (V_{\text{DO}} - V_R/2)$} Else {$C =$ '0', $V_{\text{CO}} \Leftarrow V_{\text{DO}}$}], $V_{\text{CO}} \Leftarrow 2V_{\text{CO}}$.

Stage 3:

If[($V_{\text{DO}} \geq V_1$) then {$C =$ '1', $V_{\text{CO}} \Leftarrow (V_{\text{DO}} - V_R/2)$} Else {$C =$ '0', $V_{\text{CO}} \Leftarrow V_{\text{DO}}$}], $V_{\text{CO}} \Leftarrow 2V_{\text{CO}}$.

Stage 4:

If[($V_{\text{BO}} \geq V_1$) then $A =$ '1' Else $A =$ '0'].

The implementation of the OR SSADC has one practical problem. As we pass from the MSB to the LSB stage, the input to the comparator goes through (N − 1) amplifiers of gain 2, and thus any noise or offset or drift at the MSB level gets amplified by a factor of 2^{N-1}. Similarly, the input to the (LSB+1) bit has the (noise + offset + drift) amplified by a factor of 2^{N-2} and so on. For example, if we choose a 10-bit OR SSADC, the input to the LSB (b_0) will have

$$2^9(V_{\text{OS}} + V_N)_{b_9} + 2^8(V_{\text{OS}} + V_N)_{b_8} + \cdots 2^2(V_{\text{OS}} + V_N)_{b_2} + 2(V_{\text{OS}} + V_N)_{b_1}. \tag{9.10}$$

Here, $(V_{\text{OS}} + V_N)_{b_9}$ is the offset and noise voltage level of the MSB (bit b_9), $(V_{\text{OS}} + V_N)_{b_8}$ is the offset and noise voltage level of the (MSB-1) bit (bit b_8) and so on. Hence, beyond 8 bits, realizing an OR SSADC is fraught with practical problems. Once again, the logic of OR SSADC can be understood by looking at examples, given below.

Examples: OR(i) Input is 12 V

Stage 1: Since [($V_{\text{in}} \geq 7.5$ V), {$D =$ '1', $V_{\text{DO}} \Leftarrow 2*(12 - 8) = 8$ V}].
Stage 2: Since [($V_{\text{DO}} \geq 7.5$ V), {$C =$ '1', $V_{\text{CO}} \Leftarrow 2*(8 - 8) = 0$ V}].
Stage 3: Since [($V_{\text{CO}} < 7.5$ V), {$B =$ '0', $V_{\text{BO}} \Leftarrow 2*0 = 0$ V}].
Stage 4: Since [($V_{\text{CO}} < 7.5$ V), {$A =$ '0'}].
Hence "DCBA" = "1100".

OR(ii) Input is 3 V

Stage 1: Since [($V_{\text{in}} < 7.5$ V), {$D =$ '0', $V_{\text{DO}} \Leftarrow 2*3 = 6$ V}].
Stage 2: Since [($V_{\text{DO}} < 7.5$ V), {$C =$ '0', $V_{\text{CO}} \Leftarrow 2*6 = 12$ V}].
Stage 3: Since [($V_{\text{CO}} < 7.5$ V), {$B =$ '1', $V_{\text{BO}} \Leftarrow 2*(12 - 8) = 8$ V}].
Stage 4: Since [($V_{\text{BO}} \geq 7.5$ V), {$A =$ '1'}].
Hence "DCBA" = "0011".

Conversion time of the OR SSADC: With τ_{co} as the settling time of the comparator, τ_{sw} being the settling time of the SPDT switch, τ_{su} is the settling time of the subtractor and τ_A is the settling time of the gain 2 amplifier, the conversion time T_{con} of an N-bit OR SSADC will lie between minimum conversion time, (when the input is zero and none of the stages require subtraction):

$$T_{\text{con}}|_{\text{min}} = (N\tau_{co} + (N-1)(\tau_{sw} + \tau_A), \tag{9.11}$$

and the maximum conversion time (when the input is full-scale, and every bit requires subtraction):

$$T_{\text{con}}|_{\text{max}} = (N\tau_{co} + (N-1)(\tau_{su} + \tau_{sw} + \tau_A)). \tag{9.12}$$

9.3.3.3 Hybrid Successive Subtraction ADC (HSS ADC)

The diminishing reference voltages of the NR SSADC and the amplification by 2 in every stage of the OR SSADC pose practical problems to build SSADC of either kind beyond 8 bits. However, by combining both methods resulting in a hybrid SSADC, we can build successive subtraction type ADC for 12 or even 16-bit resolution. The hybrid structure uses the required number of 3-bit or 4-bit (preferred) NR SSADC concatenated using the OR SSADC (multiplication by 2 for every bit) principle.

Figure 9.11 shows three 4-bit NR SSADC concatenated to obtain a 12-bit hybrid successive subtraction ADC. Here each block, marked as X, Y, or Z is a 4-bit NR SSADC. The four bits of the Z block serve as the most significant 4 bits of the 12-bit SSADC realized. The four bits of the Y block serve as the next most significant 4 bits of the 12-bit SSADC, and the four bits of the X block serve as the least significant 4 bits of the 12-bit SSADC thus realized. As is seen in Fig. 9.11, we interconnect blocks Z and Y with an amplifier of gain 16 and similarly, we amplify the output of block Y by a factor of sixteen and then connect the amplified signal to the input of block X. If we had implemented using the OR SSADC then the LSB stage will have the multiplication factor of 2048 of the input. With the hybrid structure, the input is multiplied by a factor of 256.

We can also implement the 12-bit hybrid SSADC with four 3-bit SSADC as shown in Fig. 9.12. Here the blocks marked P, Q, R, and S are identical and each block implements a 3-bit NR SSADC. The outputs V_{OP}, V_{OQ}, V_{OR} of blocks P, Q, and R, are multiplied by 8 and passed on to the succeeding blocks Q, R, and S, respectively. But compared to the three 4-bit schemes of Fig. 9.11, the scheme in Fig. 9.12 has a multiplication of the input voltage by 512 when it reaches the 4th stage. Thus, the noise and offset voltage will also be amplified by the factor of 512.

It is possible to cascade two 6-bit NR SSADCs with just one amplifier of 64, but the comparator of the 6th bit in the 6-bit NR SSADC will have to handle a voltage of $\frac{V_R}{128}$. Thus, the hybrid SSADC optimal implementation requires a comparison of input noise and offset voltage amplification with the comparators ability to handle low input voltage levels. However, while OR SSADC and NR SSADC can be implemented only for 8-bit resolution, the hybrid SSADC can be designed for 12-bit or even 16-bit

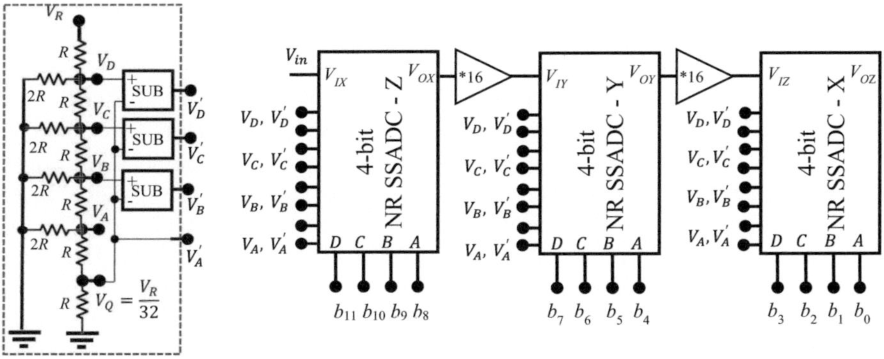

Fig. 9.11 Schematic of a 12-bit hybrid SSADC with three 4-bit NR SSADC

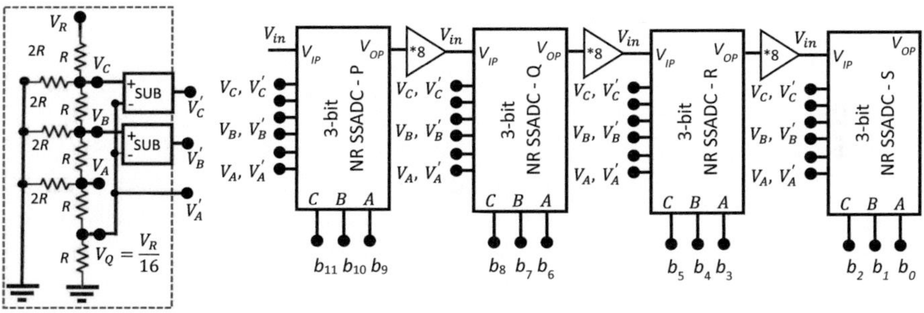

Fig. 9.12 Schematic of a 12-bit hybrid SSADC with four 3-bit NR SSADC

resolution. While all other types of ADCs are commercially exploited, the SSADC remains an academic curiosity.

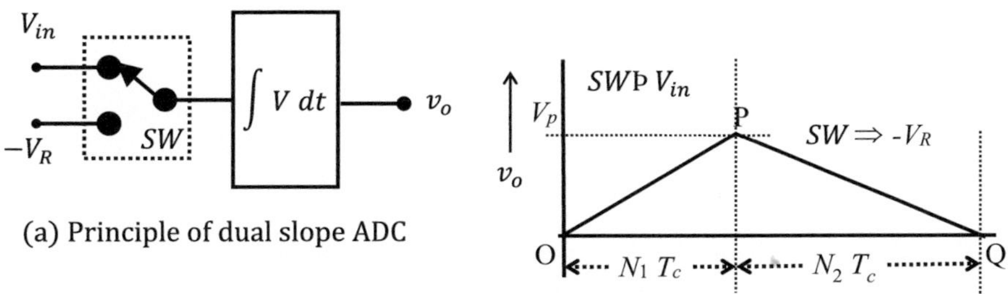

Fig. 9.13 Principle of operation of dual slope ADC

9.3.4 The Dual Slope ADC

Though the principle of operation of the dual slope type ADC was published in the 1950s, this type of ADC became a reality only after the integrated circuit technology advanced in the late 1970s. Today, any meter that uses the digital measurement of voltage or current to be read by a user (not a machine) invariably uses the dual slope ADC. All digital multimeters in use or being sold are based on this technique. Prior to learning the details of its related electronic circuitry, it is useful to first understand the essential principle of this method. Let an analog input V_{in} (assumed to be constant) be fed as input through a single pole double throw switch *SW* to an integrator as shown in Fig. 9.13a for a prefixed number, say, N_1 clock cycles of period T_c. Consequently, the output v_{oi} of the integrator will vary as shown by the straight-line OP in Fig. 9.13b, the integrator output v_{oi} rises at a constant rate kV_{in} per second, where k is a constant involved with the integration hardware. Taking the integrator output to be zero to start with, it reaches a value V_p at the end of the N_1 clock cycles, where

$$V_p = kV_{\text{in}}t\big|_{t=0}^{t=N_1T_c} = kV_{\text{in}}N_1T_c \tag{9.13}$$

At this point, the input to the integrator is switched, using *SW*, to a known reference voltage $-V_R$. The voltage at the integrator output v_{oi} now linearly decreases (at a rate of kV_R per second), as shown by the line PQ the Fig. 9.13b. The number of clock cycles N_2 it takes for v_{oi} to fall from V_p to zero is counted and is the digital measure of the analog input V_{in}, as the following analysis shows. When the integrator output reaches zero, we have

$$V_p = kV_RN_2T_c \tag{9.14}$$

Combining Eqs. (9.13) and (9.14), we get:

$$N_2 = \left(\frac{N_1}{V_R}\right)V_{\text{in}} \tag{9.15}$$

With a suitable choice of V_R and N_1, the digital number N_2 can be made to read the input voltage V_{in} with the appropriate units. The supreme merit of this method is that all consistent errors in the values of the integrator circuit elements (which determine k) and of the clock period T_c cancel out in the two integrations. The accuracy of the final result depends only on the values of V_R and the two counts N_1 and N_2, all three of which can be obtained with great accuracy. Figure 9.14 illustrates the block schematic of a typical dual slope ADC. The digital part is made of a state machine called the control and logical unit (CLU) that houses a N_d ½ digit BCD counter for BCD output or N_b bit binary counter for binary output dual slope ADC, clocked by a signal with period T_c to count N_1 and N_2, implement the control logic of starting the analog to digital conversion, as well as switching the two SPDT analog switches S_1 and S_2, and providing the final digital output. The analog part of the dual slope ADC is made of:

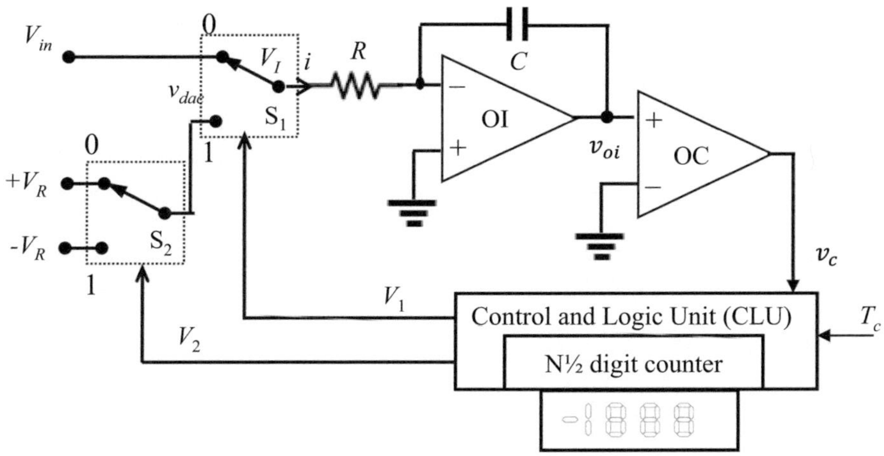

Fig. 9.14 Block Schematic of The dual Slope ADC

(i) Two single pole double throw (SPDT) analog switches S_1 and S_2,
(ii) An opamp RC integrator OI, and
(iii) An opamp comparator OC that functions as a zero-crossing detector.

The opamp integrator OI has: (i) its non-inverting input terminal grounded and (ii) a capacitor C connected between its inverting input terminal and its output, with the inverting input also connected to one terminal of a resistor R. The other terminal of R is connected to the pole of the first SPDT switch S_1. One input terminal (position 0) of S_1 is tied to the input V_{in} and the other input terminal of S_1 is connected to the pole of SPDT switch S_2. SPDT switch S_2 selects either $+V_R$ or $-V_R$ depending on its control signal V_2. Switches S_1 and S_2 are individually controlled (set in position "0" or "1") by the CLU through the control lines V_1 and V_2 respectively. Whenever a control signal (either V_1 or V_2) is "0" then the corresponding switch (S_1 or S_2) is set in position "0". On the other hand, if the control signal (either V_1 or V_2) is "1" then the corresponding switch (S_1 or S_2) is set in position "1". From Fig. 9.14, we see that the input V_I to the RC integrator can be V_{in} or $+V_R$ or $-V_R$ depending on the way the switches S_1 and S_2 are set. Since the end of resistance R that is connected to the inverting terminal of OI will be at virtual ground potential, the current

$$i = \frac{V_I}{R}, \tag{9.16}$$

and hence the output v_{oi} of the integrator will be

$$v_{oi} = -\frac{1}{C}\int i\mathrm{d}t = -\frac{1}{C}\int \frac{V_I}{R}\mathrm{d}t = -\frac{V_I}{RC}t. \tag{9.17}$$

Here we have assumed that R, C and V_I are constants and do not change with respect to time, at least within the period of interest. The analog output v_{oi} of OI is converted to a binary "0" or "1" by the opamp comparator OC which serves as the interface between the analog part and the digital part of the dual slope ADC. The output v_c of OC will be "1" if the output v_{oi} of the integrator OI is ≥ 0 V, otherwise, V_c will be "0". A "0" to "1" transition on v_c indicates that v_{oi} is crossing zero in the positive direction and a "1" to "0" change in v_c indicates that v_{oi} is crossing zero in the negative direction. The output v_c of the comparator is given as input to the CLU. By sensing v_c, the CLU determines whether the output v_{oi} of the integrator is negative, positive, or transitioning zero in the positive direction or in the negative direction. The CLU controls the switches S_1 and S_2 appropriately to alter the inputs to the RC integrator OI and accomplishes an analog to digital conversion of input V_{in}. To accomplish an analog to digital conversion, the CLU performs two successive periods of integrations (T_1 and T_2), as explained earlier, and obtain a digital count that directly indicates the input voltage V_{in}. The CLU sets the first integration period T_1 and measures the second integration period T_2 using the N½ digit BCD counter (or N_b bit binary counter). Before the CLU performs the two integrations required for an analog to digital conversion, first the CLU must control switches S_1 and S_2 appropriately and ensure that the output v_{oi} of the integrator is zero so that the first integration can start. The process of ensuring the output v_{oi} of the integrator to be zero before starting the first integration is called the "*Auto-Zero Phase*" of the

9.3.4.1 3.4.1. The Auto-Zero Phase

In the auto-zero phase, the CLU checks the output v_c of the comparator.

If v_c is "0" (indicating to the CLU that the output v_{oi} is negative):

Then:
{The CLU sets the switch S_1 to position 1 and switch S_2 to position 1. Thus connecting $-V_R$ to the resistor R (input of the RC integrator). The current i will be $-\frac{V_R}{R}$ A that would charge the capacitor C with a positive polarity at the output of OI. Thus, the negative charge in the capacitor C will be removed and the voltage across the capacitor will decrease hence the magnitude of the output v_{oi} of OI will also decrease and reach zero as indicated by the dotted line in Fig. 9.15. Once v_{oi} reaches zero, v_c will rise from "0" to "1". Sensing this change (that is v_{oi} is crossing zero), the CLU terminates the auto-zero phase and enters the conversion phase}.

Else: (that is at the start v_c is "1" indicating to the CLU that the output v_{oi} is positive):
{The CLU sets the switch S_1 to position 1 and switch S_2 to position 0. Thus connecting $+V_R$ to the resistor R (input of the RC integrator). The current i will be $+\frac{V_R}{R}$ A that would charge the capacitor C with a negative polarity at the output of OI. Thus, the positive charge in the capacitor C will be removed and the voltage across the capacitor will decrease hence the output v_{oi} of OI will also decrease and reach zero as shown by the dashed line in Fig. 9.15. Once v_{oi} reaches zero, comparator output v_c will change from "1" to "0". Sensing this change (that is v_{oi} is

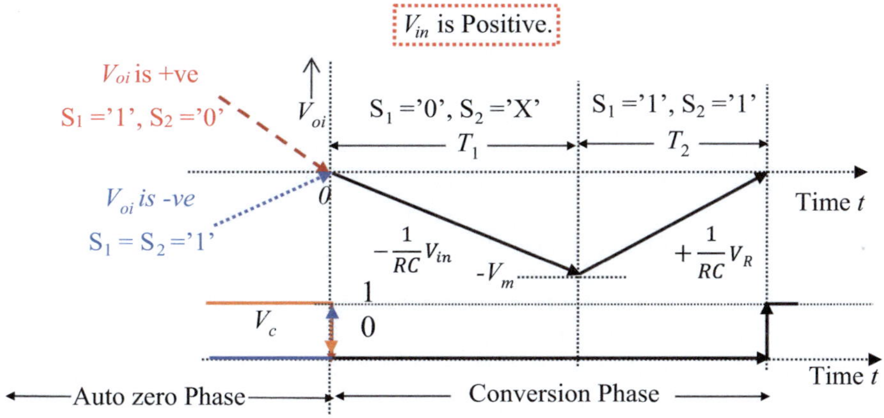

Fig. 9.15 Waveforms at cardinal points on the Dual Slope ADC input is positive

crossing zero), the CLU terminates the auto-zero phase and enters the conversion phase}.

In the above explanation, we know that the terminal of capacitor C connected to the inverting input will be at the ground potential (virtual ground) and hence the charge in the capacitor is taken as positive if v_{oi} is positive, otherwise the charge in the capacitor is taken to be negative. Here, the opamp is assumed to be ideal with zero bias current, zero offset voltage, and a forward gain of infinity.

9.3.4.2 The Conversion (Logic) Phase of a Dual Slope ADC

Once the output of the comparator v_c transits from "0" to "1" or "1" to "0" (indicating that the output v_{oi} of the integrator just crossed zero) the CLU enters the conversion phase. First, switch S_1 is set in position 0, thus connecting V_{in} to the resistor R (input of the RC integrator). The CLU resets the counter and starts counting and waits for a pre-set number of counts, say N_1 (N_1 can be BCD or binary). There are two possibilities, namely, the input V_{in} is either positive or negative. First, let us assume that the input V_{in} is positive.

(i) **When V_{in} is positive**:

During the period of N_1 counts, the current i will be $+\frac{V_{\text{in}}}{R}$ A that would charge the capacitor C with a negative polarity at the output of OI. Thus, the voltage across the capacitor v_{oi} will increase with negative polarity (with a slope $-\frac{V_{\text{in}}}{RC}t$) as shown in Fig. 9.15. The comparator output v_c will be "0" during the period T_1 of first integration, where T_1 is set as

$$T_1 = N_1 T_c. \tag{9.18}$$

Here T_c is the clock given to the counter (BCD or binary). At the end of the first integration period T_1 (that is after N_1 counts), v_{oi} will reach a voltage $-V_m \left(= -\frac{V_{\text{in}}}{RC}T_1\right)$ and the comparator output $v_c =$ '0'. The CLU senses the output v_c of the comparator OC to be "0" and hence sets the switch S_1 to position 1, switch S_2 to position 1, and records the polarity of V_{in} as +ve. The CLU also resets the counter to zero and starts a second count N_2 and thus the second integration. With these switch settings, the RC integrator OI is connected to $-V_R$. The current i will be $-\frac{V_R}{R}$ A and would charge the capacitor C tending to promote a positive polarity at the output of OI. Thus, the charge in the capacitor C (stored in the opposite direction during T_1) will decrease and hence output v_{oi} of OI will rise and reach zero as shown in Fig. 9.15 after a time, say, T_2 seconds. At the end of T_2, v_{oi} reaches zero and hence v_c will rise from "0" to "1". Sensing this transition, the CLU stops the counter and outputs the count value N_2 at that time with the polarity as +ve as the digital equivalent of V_{in}. Since T_2 is measured as N_2 counts, we have

$$T_2 = N_2 T_c. \tag{9.19}$$

Since the charge acquired by the capacitor C during the first integration period T_1 is discharged completely during the second integration period T_2, we get

$$\frac{V_{\text{in}}}{RC}T_1 = \frac{V_R}{RC}T_2. \tag{9.20}$$

Substituting the values of T_1 and T_2 from Eqs. (9.19) and (9.19) into Eq. (9.20) we get

$$\frac{V_{\text{in}}}{RC}N_1 T_c = \frac{V_R}{RC}N_2 T_c. \tag{9.21}$$

Simplifying equation (9.21), we get

$$N_2 = \left(\frac{N_1}{V_R}\right)V_{\text{in}} \tag{9.22}$$

In Eq. (9.22), if we make $\left(\frac{N_1}{V_R}\right)$ a constant, say $\left(\frac{N_1}{V_R}\right) = 1$, then the output count N_2 directly indicates the input V_{in}. The discussion so far assumed that input V_{in} is positive.

(ii) **When V_{in} is negative**:

During the period of N_1 counts, the current i will be $-\frac{V_{\text{in}}}{R}$ A that would charge the capacitor C with a positive polarity at the output of OI. Thus, charge in the capacitor C will increase from zero and the voltage v_{oi} will increase with positive polarity (with a slope $+\frac{V_{\text{in}}}{RC}t$) as shown in Fig. 9.16. The comparator output v_c will be "1" during T_1. At the end of first integration, v_{oi} will reach a voltage $+V_p \left(= +\frac{V_{\text{in}}}{RC}T_1\right)$. At the end of time T_1 (that is after N_1 counts), the CLU senses the output v_c of the comparator OC to be "1" and hence sets the switch S_1 to position 1, switch S_2 to position 0 and records the polarity of V_{in} as negative. The CLU also resets the counter to zero and

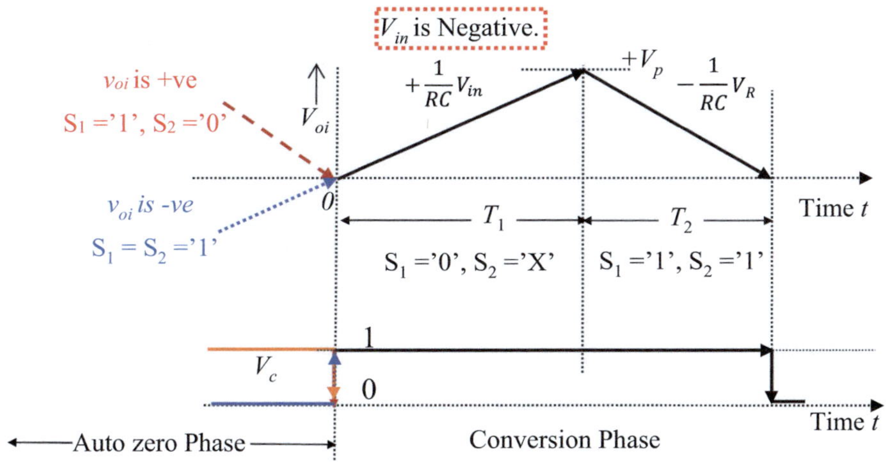

Fig. 9.16 Waveforms at cardinal points on the Dual Slope ADC, input is negative

starts a second count N_2 and thus the second integration. With these switch settings, the RC integrator OI is connected to $+V_R$. The current i will be $+\frac{V_R}{R}$ A that would discharge the capacitor C (which has a positive polarity at the output of OI). Thus, charge in the capacitor C (stored in the opposite direction during T_1) will decrease and hence output V_{oi} of OI will decrease and reach zero as shown in Fig. 9.16 after a time, say, T_2. At the end of T_2, v_{oi} reaches zero and hence v_c will transit from "1" to "0". Sensing this transition, the CLU stops the counter and outputs the count value N_2 at that time with the polarity as −ve as the digital equivalent of V_{in}. Here too the charge acquired by C during T_{c1} is removed from C in T_2. Hence Eqs. (9.20), (9.21), and (9.22) hold good.

If we make N_1 to be 1000 and V_R to be 1000 mV then $\frac{N_1}{V_R} = 1$ mV and hence every count in N_2 as per Eq. (9.22) will be 1 mV. The flowchart shown in Fig. 9.17 illustrates the logic of the CLU of a dual slope ADC. We see that at the end of a conversion, the output of the integrator is zero and hence a new conversion can begin immediately. Thus, most dual slope ADCs have two modes of operation, namely,

- **Start-stop mode**: In this mode, an auto-zero phase is initiated before a conversion. Once a conversion is over, the ADC stops and waits for an instruction to start the next conversion. Hence an auto-zero phase precedes every conversion phase.
- **Continuous conversion mode:** In this mode, an auto-zero phase is initiated before the first conversion and once a conversion is over, the output is updated, and the succeeding (next) conversion is initiated without an auto-zero phase (because we know that at the end of a conversion, the output of the integrator OI is zero).

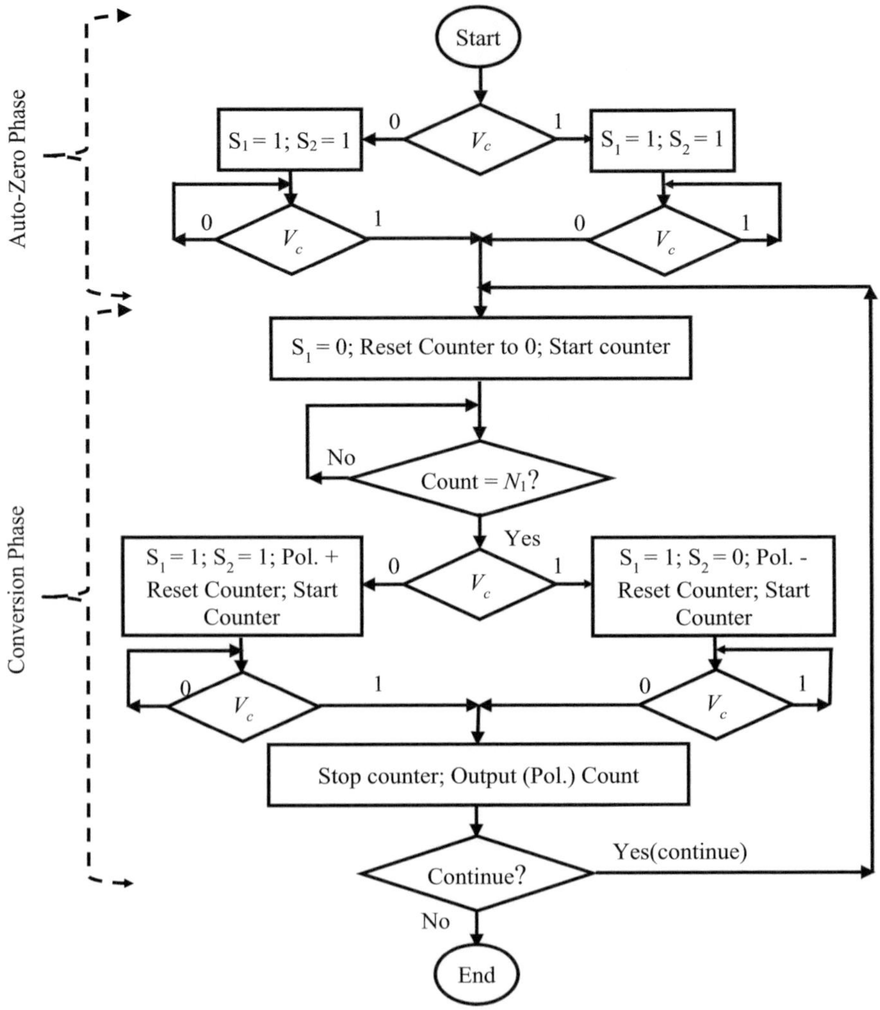

Fig. 9.17 Flowchart of the Logic of a Dual Slope ADC

9.3.4.3 The Conversion Time of a Dual Slope ADC

The conversion time required for a dual slope ADC in the "*Start-Stop mode*" is

$$T_{\text{con}} = T_{\text{AZ}} + T_1 + T_2. \tag{9.23}$$

In Eq. (9.23), T_{AZ} is the time required for the auto-zero phase, T_1 is the set time of the first integration ($N_1 T_c$) and T_2 is the measured time ($N_2 T_c$).

The conversion time of a dual slope ADC in the "*continuous conversion mode*" is

1. For the first conversion the conversion will be as per Eq. (9.23).
2. From the second conversion onwards the conversion time is

$$T_{\text{con}} = T_1 + T_2. \tag{9.24}$$

In Eqs. (9.23) and (9.24), T_1 is constant but T_2 depends on the input V_{in}. If V_{in} is zero, then T_2 is also zero. The maximum value of T_2 is when V_{in} is full-scale and will have a value $2N_1T_c$. Thus, the conversion time T_{con} of a dual slope ADC varies with the input voltage V_{in} being converted as well as the time taken for the output of the integrator to be made zero before the start of a conversion.

9.3.4.4 Noise and Interference Rejection in Dual Slope ADC

Since the input V_{in} is getting integrated for the set time T_1, by choosing T_1 appropriately, the effect of noise or interference that may be present in V_{in} on the digital output can be reduced or even eliminated. However, to accomplish this, the integration time must be large enough so that the noise/interference can be minimized.

Reduction in the effect of noise on the digital output: Noise signal has "*zero mean*". Thus, if we average (integrate) the input V_{in} over a sufficiently large period T_1 then we can reduce or even eliminate the effect of noise present in V_{in}. Thus, in most dual slope ADC, T_1 is set in milliseconds or even hundreds on milliseconds.

Reduction in the effect of interference on the digital output: To reduce or eliminate the effect of interference we need to choose T_1 such that integral of the interference over the period T_1 is zero. When making voltage measurements, one interference that can affect the measurement (especially voltages in the range few millivolts) is interference due to the power line. By making T_1 to be an integral multiple of the power supply period, the power supply interference can be altogether eliminated. Since most countries use either 50 Hz or 60 Hz power frequency, most dual slope ADCs available use a first integration time T_1 of 100 ms that would eliminate both 50 Hz and or 60 Hz interference. With $T_1 = 100$ ms, then T_2 will vary from 0 (for $V_{\text{in}} = 0$) to $2T_1$ (200 ms) when V_{in} has full-scale value (200 mV for a 200 mV full-scale dual slope ADC). Hence conversion time in the continuous mode of operation will vary between 100 ms ($V_{\text{in}} = 0$) and 300 ms ($V_{\text{in}} = 199.9$ mV). Due to this reason, the conversion speed of a dual slope ADC is very low, in the range few conversions per second. However, we can obtain a few hundred conversions per second by choosing T_1 to be in the range of milliseconds but will lose the ability of the dual slope ADC to suppress noise and or interference.

In the foregone discussions, we assumed that the counter in the CLU of a dual slope ADC is a BCD counter. However, this is not mandatory, and we can use a binary counter instead of a BCD counter. The explanations and the derivations, Eqs. (9.13) to (9.24), will all be valid with just one change, the numbers represented in these equations will be binary. The dual slope ADCs in integrated circuit form, designed in the late 1970s are still in use with little modifications. Of these, the ICL 7106 and

ICL 7107 ICs implement a 3½ digit dual slope ADCs and the ICL7109 IC provides a sign plus 12-bit binary output dual slope ADC. The IC ICL7135, on the other hand is a 4½ digit dual slope ADC.

9.3.5 The Dual Slope ADC as a Digital Voltmeter (DVM)

We see that the digital output of a dual slope ADC indicates the input voltage V_{in} and hence is typically a Digital Voltmeter (DVM). The most popular DVM is the 3½ digit DVM with a full-scale of 200 mV, i.e., the meter can measure 000.0–199.9 mV. Popular integrated circuits that implement a 3½ digit DVM are ICL7106 and ICL 7107. While the ICL 7106 supports an LCD display, the ICL7107 can drive 7-segment LED displays. Figure 9.18 portrays the circuit schematic of a 200 mV full-scale DVM using the ICL7107 IC and external components. In this configuration, the input voltage is integrated for 100 ms, during this 100 ms period the input cannot change and must remain constant. Moreover, the conversion time for a full-scale of 200 mV will be 300 ms in continuous conversion mode, resulting in $\approx$3 conversions per second. Hence, in practice a dual slope ADC (especially the 3½ digit DVM shown in Fig. 9.18) is a DC voltmeter. In order to ensure that the input is filtered, the input is connected through a first-order RC low-pass filter using R_5 and C_5 in Fig. 9.18.

9.3.5.1 Multi Range DVM

Once we have a basic DC voltmeter of full-scale 200 mV, voltmeters of higher full-scale DC voltages can be easily built using a simple potential divider as shown

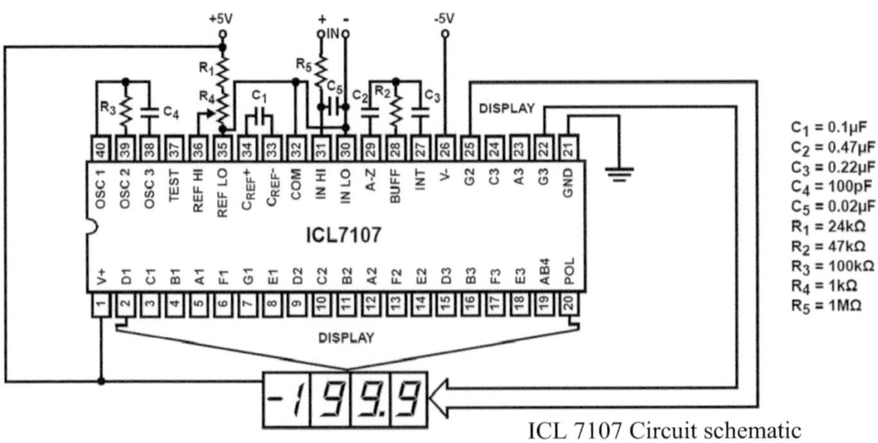

Fig. 9.18 The IC ICL7107: 3½ digit 200 millivolt DVM (dual slope ADC)

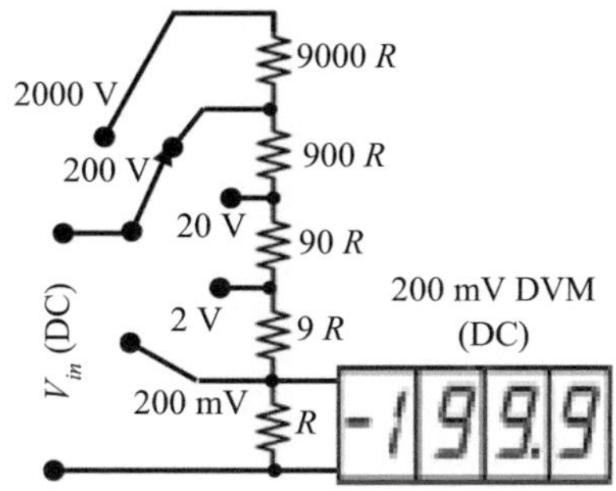

Fig. 9.19 Multi range DC DVM

in Fig. 9.19. Since the current drawn by the DVM (dual slope ADC) is small, the input resistance seen at the input terminals of a typical DVM will be in the order of megohms. The lowest looking in resistance of 1 MΩ will be applicable to the 200 mV full-scale DVM.

9.3.5.2 Measurement of DC Current Using the DVM

Measurement of current using a DVM is accomplished simply by converting the current to be measured into a voltage by employing the following schemes.

(i) A simple standard resistor R_s is shunted across the terminals of the DVM as illustrated in Fig. 9.20. If we have $R_s = 1\ \Omega$, then when the current $I = 1$ mA, the voltage developed across R_s will be 1 mV and hence the DVM will read 1.0. Similarly, a current I of 100 mA will develop a voltage of 100 mV across the resistance $R_s = 1\ \Omega$ and thus the DVM will read 100.0. Thus, with $R_s = 1\ \Omega$, the 200 mV full-scale DVM in Fig. 9.20 gets converted into a 200 mA full-scale digital DC ammeter. On the other hand, if we have $R_s = 100$ m Ω, then we get a 2 A full-scale DC ammeter. Hence, we can convert a DVM into different range ammeters easily as in Fig. 9.21.

(ii) **An active *i*-to-*v* converter instead of a simple standard resistance**: We can use an active *i*-to-*v* converter and measure the converter's output voltage to measure the current. Use of an active current to voltage converter reduces the internal resistance of the ammeter but the solution is quite complex and hence implemented very rarely, that too only in high-end digital multimeters.

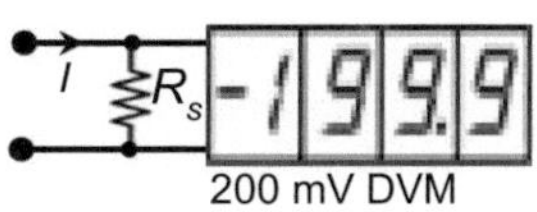

Fig. 9.20 Digital DC ammeter

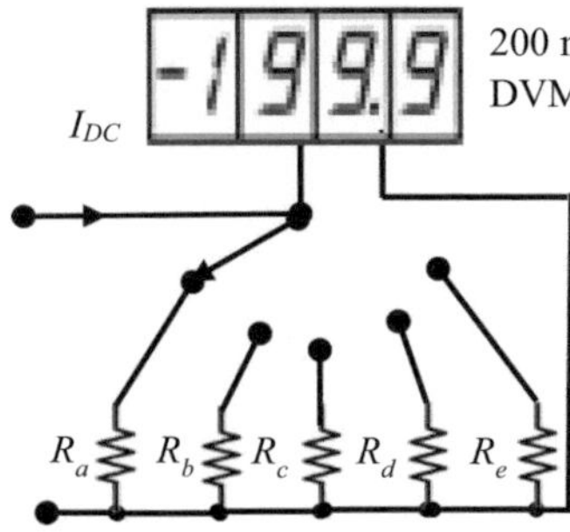

Fig. 9.21 Multi range digital DC ammeter

9.3.5.3 Measurement of AC Voltage Using the DVM

While the basic PMMC ammeter is a DC ammeter, the basic digital voltmeter is a DC voltmeter. We have already seen that the basic PMMC DC ammeter can be employed with additional circuitry and made to read AC voltages and currents. Similarly, the basic DVM (a DC voltmeter) can also be made to read AC voltages or AC currents as well using the schemes indicated below.

(i) **Rectifier type DVM** for AC measurement: Herae, the AC voltage is converted into DC using a full-bridge rectifier and RC flow-pass filter as shown in Fig. 9.22. The DC voltage is read by the DVM. The cut-in voltage V_γ of (0.7 V) of practical diodes makes this type of AC voltmeter suitable for a full-scale voltage of tens of volts or higher.

(ii) **Precision Rectifier type DVM**: To enable measurement of AC voltages starting from the range of 200 mV, instead of a diode bridge, a precision rectifier shown in Fig. 9.23a can be used. The first opamp OA1 half wave rectifies any AC input voltage: $V_{AC} > \frac{V_\gamma}{A_{OL}}$. The second opamp adds the missing half wave portions and thus realizes a full-wave rectifier. This fact is illustrated by the *v*-*i* characteristics of the precision rectifier shown in Fig. 9.23c. It should be noted here that the *v*-*i* characteristics shown in Fig. 9.23b and c are for both positive and negative input voltages (full-wave rectification). A comparison of the *v*-*i* characteristics of precision rectifier of Fig. 9.23c with the *v*-*i* characteristics of a bridge rectifier shown in Fig. 9.23b, we see that:

   (a) The cut-in voltage (below which the rectifier does not function) reduces from $2V_\gamma$ to $\frac{V_\gamma}{A_{OL}}$, where A_{OL} is the open-loop gain of the opamp OA1 and

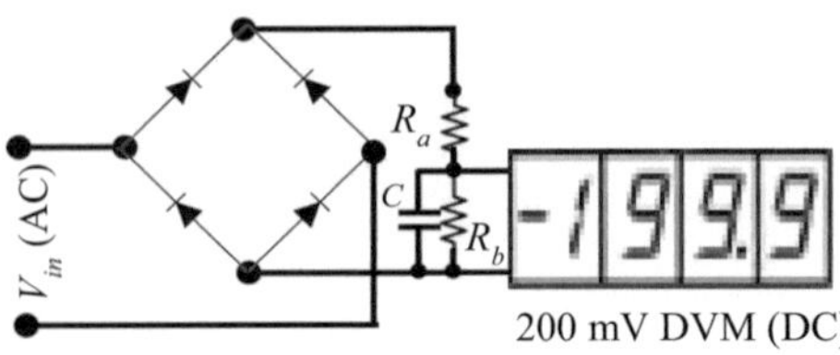

Fig. 9.22 Rectifier type AC DVM (Full scale >20 V)

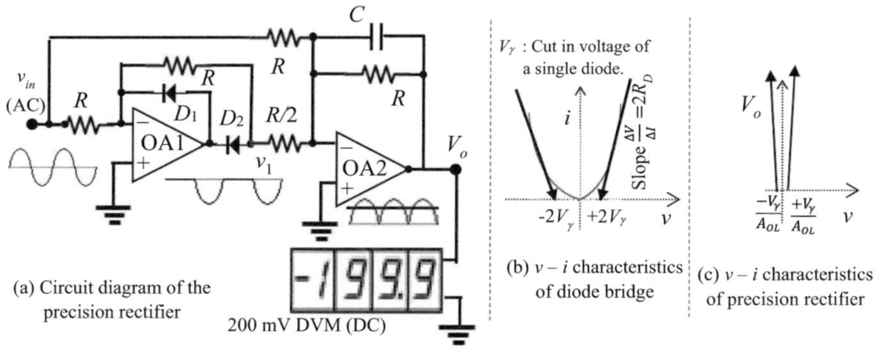

Fig. 9.23 AC voltage measurement using DVM and precision rectifier for

(b) The ON resistance of the precision rectifier is opamp OA2's output resistance R_o.

Figure 9.24 shows the circuit schematic of a precision rectifier type multi-range AC DVM. Like any rectifier type meter, the rectifier type DVM (including the precision rectifier type) responds to the DC value of the rectified input and is calibrated to read the RMS value only for sinusoidal AC voltages. If the AC input has a waveform different from a sinewave, the rectifier type (simple diode bridge or precision rectifier type) meter will read erroneously. It may be noted that there is a distinct difference between the rectifier type Permanent Magnet Moving Coil meter and the rectifier (normal or precision) type DVM. While the coil of the PMMC meter itself filters the harmonics after rectification, in rectifier type DVM, an R–C low-pass filter is implemented along with a rectifier to filter the harmonics.

(iii) **True RMS AC DVM**:

Using an analog multiplier and an integrator (a low-pass filter, in practice), it is possible to extract a DC voltage proportional to the root mean square value of an AC voltage. Figure 9.25 illustrates such a scheme. Here, the multiplier can be of transconductance, or log-antilog type described in Chap. 10. Both the *x*- and *y*-inputs

Fig. 9.24 Multi range precision rectifier type digital AC voltmeter

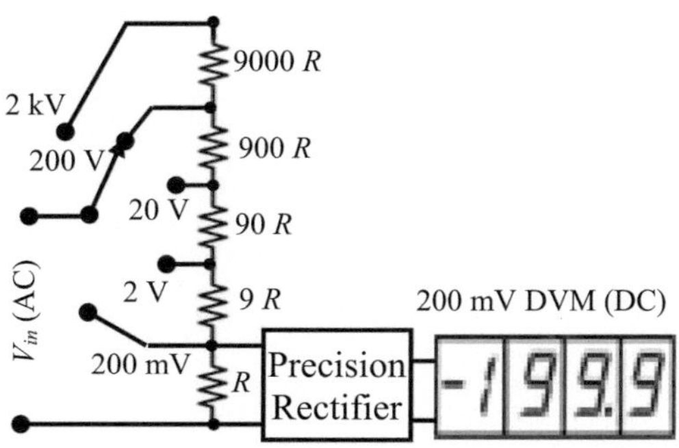

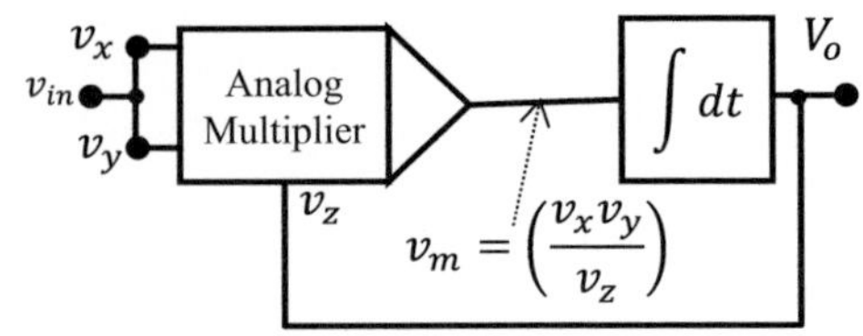

Fig. 9.25 AC RMS to DC extractor

of the analog multiplier are shorted and fed with the input AC voltage v_{in}. The multiplier output v_m is fed as input to an integrator as shown in Fig. 9.25 and the output of the integrator is fed back as the z-input to the multiplier. We now have

$$V_o = \int \frac{v_x v_y}{v_z} \mathrm{d}t = \frac{1}{V_o} \int v_{\text{in}}^2 \mathrm{d}tx \tag{9.25}$$

Rearranging Eq. (9.25) we get

$$V_o^2 = \int v_{\text{in}}^2 \mathrm{d}t \rightarrow V_o = \sqrt{\int v_{\text{in}}^2 \mathrm{d}t} \tag{9.26}$$

In practice, the integrator in Fig. 9.25 is replaced with a low-pass filter with a very low cut-off frequency. Most integrated circuit manufacturers offer ICs for AC RMS to DC conversion. Examples are

(a) LTC 1966 from Linear Technology.
(b) LH009: National semiconductors.
(c) MX636 Maxim Semiconductors.
(d) AD736: Analog Devices.

Figure 9.26 shows a block diagram of a multi-range true RMS AC voltmeter that uses an AC RMS to DC converter.

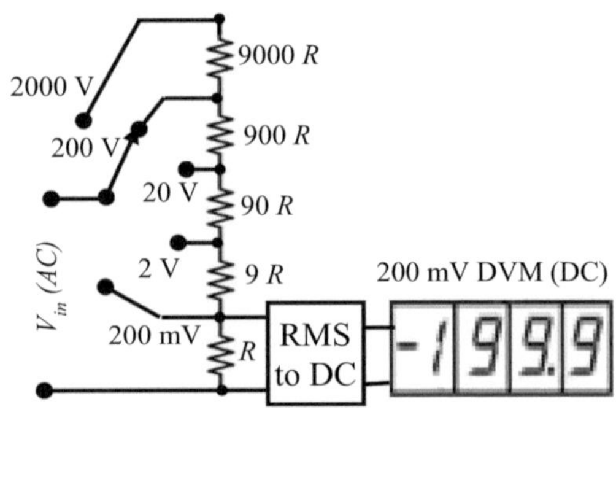

Fig. 9.26 Multi range true RMS digital AC voltmeter

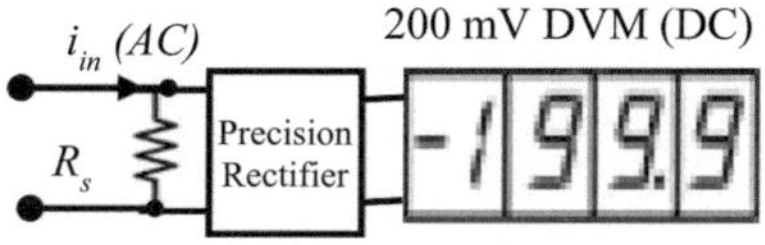

Fig. 9.27 Precision rectifier type AC ammeter

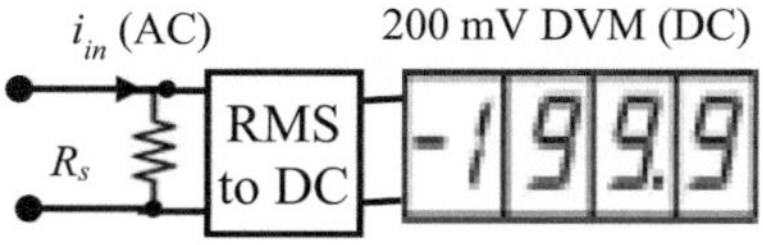

Fig. 9.28 True RMS digital AC ammeter

9.3.5.4 Measurement of AC Current Using the DVM

For measuring AC currents, we first convert the AC current into an AC voltage using a standard resistance R_s and then measure the AC voltage.

Precsion Rectifier type digital AC ammeter: Unlike the rectifier type multi-range AC voltmeter (for voltages >20 V RMS), decsign of a rectifier type AC ammeter using a DVM is practically not feasible. The reason is such an AC ammeter will have a voltage drop of a minimum of 1.4 V across the ammeter. However, it is possible to use a precision rectifier and measure AC current as shown in Fig. 9.27. As was the case of AC voltmeter, here too, the precision rectifier type digital ammeter will read the RMS value of only the sinusoidal currents and the true RMS type ammeters will read correctly even if the AC current is non-sinusoidal.

True RMS digital AC ammeter: As shown in Fig. 9.28, here too the AC current is converted into proportional AC voltage using a standard resistance, and the resulting voltage is converted into equivalent DC voltage and measured using a DVM. Though it is possible to measure both AC and DC voltages/currents once we use a RMS to DC converter, it is not practiced as such an option will result in lower accuracy in the readings obtained in DC. Thus, in a true RMS meter, when a DC voltage or current measurement function is selected, the RMS to DC converter is not employed (switched off in battery powered multimeters).

It is also possible to design a multi-range "precision rectifier type" or "true RMS reading AC" ammeter as was done for DC ammeters with different shunt resistors.

9.3.5.5 Measurement of Resistance Using the DVM

With the addition of a simple current source of value I_s, we can measure the value of an unknown resistance R_x (under DC excitation), using a DVM as shown in Fig. 9.29. If we keep $I_s = 1$ mA, then R_x of value 1 Ω will develop a 1 mV and will be read by the DVM as 1.0. Similarly, $R_x = 100$ Ω will develop 100 mV and hence read as 100.0. Thus, the simple addition of a current source with output values of:

1 mA, 100 μa, 10 μa, 1 μa, and 100 nA

We can convert the 200 mV full-scale DVM into a multi-range ohmmeter of full-scale values:

200 Ω, 2 k Ω, 20 k Ω, 200 k Ω, and 2 MΩ.

Howland Current Source:

Figure 9.30 shows the circuit schematic of a current source popularly known as the Howland current source. To analyze this circuit, let the voltage at the non-inverting input of opamp OA be V_x. Hence the inverting input also will have a potential V_x. Furthermore, there being no current input into the inverting terminal of OA1, $V_o = 2V_x$. Thus, we have

$$I_1 = \frac{(V_S - V_x)}{R_s} \text{ and } I_2 = \frac{(V_o - V_x)}{R_s} = \frac{(2V_x - V_x)}{R_s} = \frac{V_x}{R_s}. \tag{9.27}$$

Hence current I_s will be

$$I_s = I_1 + I_2 = \frac{(V_s - V_x)}{R_s} + \frac{V_x}{R_s} = \frac{V_s}{R_s} \tag{9.28}$$

Equation (9.28) indicates that the current I_s is independent of R_x and is set only by V_s and R_s. Today we have digital multimeters (DMM) that provide digital measurement of DC and AC voltages in the range 200 mV to 1000 V and currents in the range 2 mA to 2 A full-scale as well as resistance in the range 200 Ω full-scale to 2 M Ω full-scale. With the current source added in the digital multimeter, it is very simple to add the following features.

(i) **Continuity testing**: In its 200 Ω ohmmeter mode, if we connect a simple wire across its terminals, the meter will read near to 0 Ω. Thus, we can check whether the wire is intact or if there is a break in the wire. Most manufacturers provide

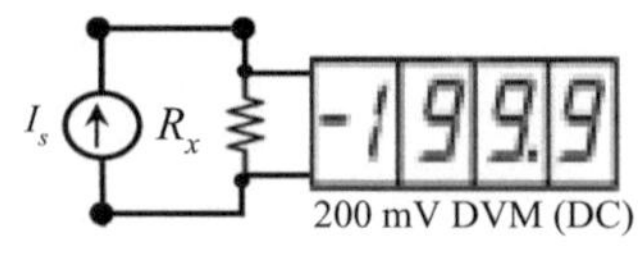

Fig. 9.29 Digital ohm meter

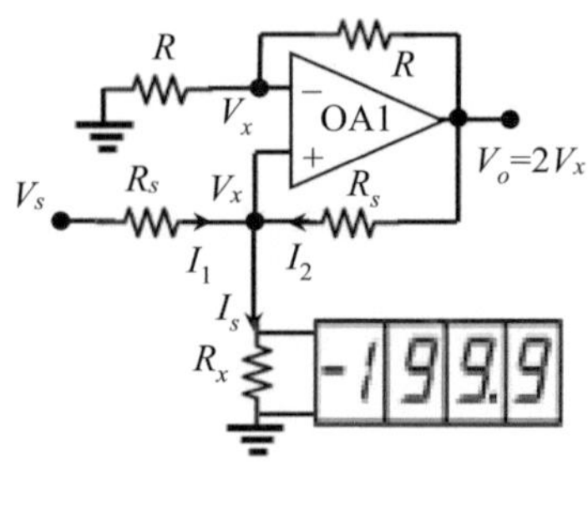

Fig. 9.30 Howland current source

a "buzzer" which will sound whenever there is continuity, and the user may not even have to look at the reading. The continuity of a wire or a track on a printed circuit board (PCB) can be easily ascertained in this manner. Almost all digital multimeters have this continuity testing feature.

9.3.5.6 Testing of Diode Using a DVM

The diode testing circuit is shown in Fig. 9.30, where a DVM (full scale of 2 V or preferably 20 V) is connected to a current source of value I_s. The diode is now connected across the DVM.

(i) **Diode functioning normally**: If the diode is working normally, then if it is connected as shown in Fig. 9.31a, then the diode will be forward biased and the DVM will read the forward voltage drop of the diode. Having the DVM with a full-scale 20 V in Fig. 9.31a facilitates testing of LEDs also as LEDs normally have a voltage drop between 1.8 and 3.3 V. When the diode connection is reversed as in Fig. 9.31b, the diode will be reverse biased, and the current source will get open circuited and the DVM will show "overload" (the current source will saturate).

(ii) **The diode is open**: With the diode open (damaged) then whichever way the diode is connected (Fig. 9.31a or b) the current source is open circuited and hence in both the connections the DVM will indicate "overload".

(iii) **The diode is short**: On the other hand, if the diode is a short circuit, then the DVM will show near zero values in both forward and reverse connections.

Thus, we can test a diode using a DVM and an additional current source. Once again, most digital multimeters have this facility too. The diode testing is normally integrated into the 2 kΩ full-scale ohmmeter setting. It may be noted that the display part of the DVM in all the figures was typically shown as -1999, which is the standard way of showing a 3½ digit seven-segment display. While negative values are shown with the—segment lit, positive values are shown with the—segment off. That is positive values do not have "+" symbol in front of the digits. Another standard form of display is to blank all the digits except the most significant digit (normally the "1") and maybe the negative sign as shown in Fig. 9.31b to indicate overload or open circuit (during resistance or diode measurement).

(a) Diode in forward bias

(b) Diode in reverse bias

Fig. 9.31 DVM for diode testing

9.3.5.7 Measurement of the Current Gain β of an NPN or PNP Transistor

Quite a few digital multimeters have this feature as the additional circuitry required is minimal. The configuration to convert a DVM for the measurement of the current gain β of an NPN transistor is shown in Fig. 9.32a. The current source of value I_b is now connected to the base of the NPN transistor while its emitter is grounded. Let the resistance connected to the collector be R_c. From Fig. 9.32a we see that, with the collector current being βI_b, the voltage read by the DVM would be $\beta I_b R_c$. If we make $I_b R_c = 10$ mV, then when $\beta = 1$, then the voltage read by the DVM is 10 mV. Similarly, if $\beta = 100$, then the voltage read by the DVM is 1000 mV. Thus, if we set the DVM in 20 V full-scale range, then the DVM will indicate β directly. Making $I_b R_c = 10$ mV can be accomplished with $I_b = 1$ μA and $R_c = 10$ kΩ. We can also make $I_b = 10$ μA and $R_c = 1000$ Ω and get $I_b R_c = 10$ mV. For example, if $\beta = 1$ then the voltage developed will be 10 mV, and the last digit of a 20 V full-scale 3½ digit DVM will indicate 1. If the β of the transistor in Fig. 9.32a is 234, then the voltage input to the DVM is 2340 mV and the DVM will indicate 234. The situation is the same for a PNP transistor, except for the reversal of the current source and the supply voltage as shown in Fig. 9.32b. Here, we assumed that the β of the PNP transistor is 123. It may be noted that the design illustrated can measure β up to 1999 (assuming a 3½ digit DVM) is not unique. We can also use a 2 V full-scale and generate an output of 1 mV when $\beta = 1$. In such a case we can measure β up to 1999. On the other hand, if we design such that when $\beta = 1$, we develop a voltage of 1 mV across the collector resistor and use a 200 mV full-scale DVM, we can measure β up to 199.9 only.

9.3.6 *The Delta Sigma ADC*

The delta sigma ADC gained prominence in the early 1970s and is extensively employed in *codecs* (audio coder/decoder) for voice communications. Figure 9.33 shows the block schematic of a delta sigma ADC. Here, we see that the input voltage is summed up with a reference voltage (either $+V_R$or $-\ V_R$) and then the sum is

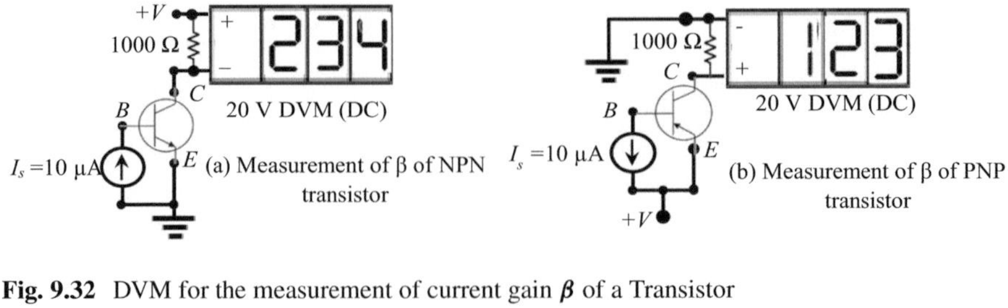

Fig. 9.32 DVM for the measurement of current gain $\boldsymbol{\beta}$ of a Transistor

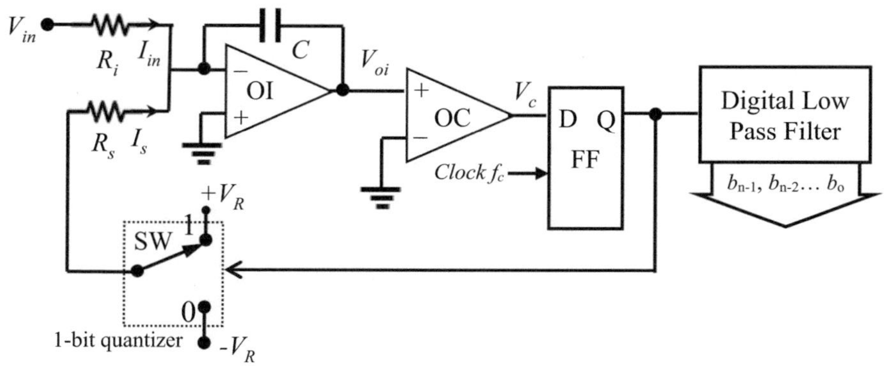

Fig. 9.33 Block schematic of delta sigma ADC

integrated. The output of the integrator is converted to binary by the comparator and the comparator output is synched with a clock using the D flipflop. The output of the D flipflop determines the input to the integrator to be either $+V_R$or $-$ V_R. The average value of the output bit stream of the D flipflop is determined using a digital low-pass filter, whose output is taken as the digital equivalent of the analog input. Since the circuit shown in Fig. 9.33 entwines both the analog delta modulator and part of the digital sigma operation into one integrated feedback loop, it is hard to explain and harder to understand the working of the circuit. We can easily understand the principle of operation of the delta-sigma ADC if we decouple the delta modulator and perform the sigma operation on the output of the decoupled delta modulator as explained next.

9.3.6.1 The Delta Sigma ADC with the Delta Modulator Delinked

The principle of operation of a delta sigma ADC after delinking the delta and sigma parts is illustrated in Fig. 9.34. The delta modulator (analog part) is made of a 3-input RC integrator OI, an opamp comparator OC, and a single pole double throw (SPDT) switch SW. One input to the integrator OI is the input voltage V_{in}. A low-frequency symmetrical square wave v_c with a frequency f_c (period T_c) as the second input to the integrator OI. The third input to the integrator is selected as $+V_R$ or $-V_R$ through the switch SW controlled by the output v_{oc} of the comparator OC. If v_{oc} is "1" then $-V_R$ is tied to the third input of OI. On the other hand, if v_{oc} is "0" then $-V_R$ is tied to the third input of OI. The digital (*sigma* = counting) part is simply an up/down counter whose up/down (U/D) control input is tied to the output v_{oc} of the comparator OC. If v_{oc} is "1" then the counter will count-up otherwise the counter will count down. The clock to the Up/Down counter is the high-frequency f_{hc}, where $Nf_{hc} = f_c$. Here, $f_{hc} = \frac{1}{T_{hc}}$ where $T_{hc} = \frac{1}{f_{hc}}$ is the period of the clock given to the up/down counter. The output of the up/down counter is updated after each cycle of the frequency f_c (that is the output of the up/down counter at the end of time T_c is latched to the output). For

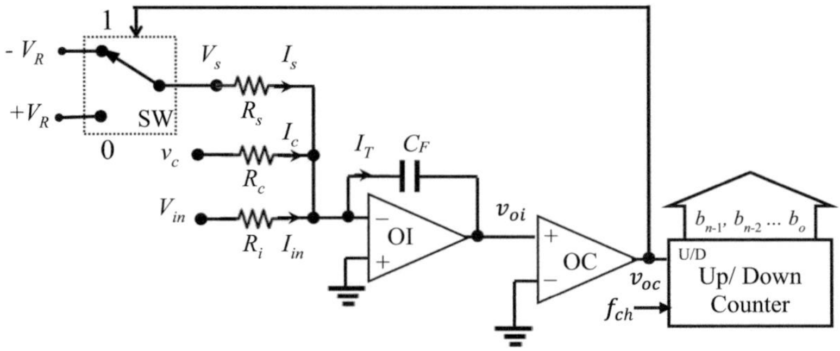

Fig. 9.34 Principle of operation of delta sigma ADC

the negative feedback scheme made of OI, OC, and SW to work, the output v_{oi} of the integrator OI must be bounded. That can happen if and only if the net charge Q_T accumulated by the capacitor C_F of integrator OI is zero for every cycle of f_c. The net charge Q_T accumulated in each cycle is

$$Q_T = \int_0^{T_c} (I_s + I_c + I_i)\mathrm{d}t = \int_0^{T_c} \left(\frac{V_s}{R_s} + \frac{\pm V_c}{R_c} + \frac{V_{\mathrm{in}}}{R_i}\right)\mathrm{d}t = \left(\frac{\pm V_R}{R_s} + \frac{\pm V_c}{R_c} + \frac{V_{\mathrm{in}}}{R_i}\right)t\Big|_{T_c} = 0 \tag{9.29}$$

In Eq. (9.29), V_c and V_R are constant, and we assumed that V_{in} is not changing within the period T_c of the clock. The clock f_c is a symmetrical square wave and at the start of the clock (rising edge), let us assume that $v_{oi} = +V_o$ as indicated in Fig. 9.35, and $V_c > (V_{\mathrm{in}} + V_R)$. Hence, the net current I_T is steady and positive, and hence the voltage v_{oi} reduces from $+V_o$ and reaches zero. During this period, say T_1, the charge Q_1 removed from the capacitor C_F Is:

$$Q_1 = \left(\frac{-V_R}{R_s} + \frac{+V_c}{R_c} + \frac{V_{\mathrm{in}}}{R_i}\right)T_1 \tag{9.30}$$

Once v_{oi} reaches zero, the comparator will change from “1” to “0” and thus changing the voltage to $+V_R$ at the output of the single pole double throw switch to $+V_R$. The additiona l charge Q_2 removed from the capacitor C_F after T_1 till the end of $T_c/2$, say T_2, is

$$Q_2 = \left(\frac{+V_R}{R_s} + \frac{+V_c}{R_c} + \frac{V_{\mathrm{in}}}{R_i}\right)T_2 \tag{9.31}$$

At $T_c/2$ the clock changes from $+V_c$ to $-V_c$ hence the current I_T will be reversed, the negative charge on the capacitor will start to reduce and v_{oi} will rise and reach zero after a time of say T_3. The charge Q_3 acquired by the capacitor C_F during T_3 is

$$Q_3 = \left(\frac{+V_R}{R_s} + \frac{-V_c}{R_c} + \frac{-V_{\mathrm{in}}}{R_i}\right)T_3 \tag{9.32}$$

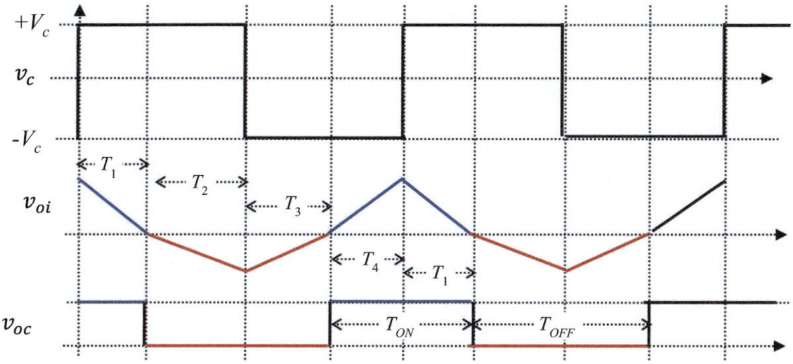

Fig. 9.35 Waveforms at cardinal points in the circuit of Fig. 9.34

Once v_{oi} reaches zero, the comparator will change from "0" to "1" and thus changing the output of the single pole double throw switch V_s as $-V_R$ and the charge Q_4 acquired by the capacitor C_F after the interval T_3 till the end of T_c, say during an interval T_4, is

$$Q_4 = \left(\frac{-V_R}{R_s} + \frac{-V_c}{R_c} + \frac{-V_{\text{in}}}{R_i}\right)T_4 \tag{9.33}$$

From Fig. 9.35 we see that whenever the comparator output V_c is "1", that is during the period marked as T_{ON} and is equal to $(T_1 + T_4)$, the capacitor retains a positive charge Q_P as

$$Q_P = \left(\frac{-V_R}{R_s} + \frac{+V_c}{R_c} + \frac{V_{\text{in}}}{R_i}\right)T_1 + \left(\frac{-V_R}{R_s} + \frac{-V_c}{R_c} + \frac{V_{\text{in}}}{R_i}\right)T_4 \tag{9.34}$$

Similarly, from Fig. 9.35 we see that whenever the comparator output V_c is "0", that is during the period marked as T_{OFF} and is equal to $(T_2 + T_3)$, the capacitor retains a negative charge Q_N as

$$Q_N = \left(\frac{+V_R}{R_s} + \frac{+V_c}{R_c} + \frac{V_{\text{in}}}{R_i}\right)T_2 + \left(\frac{+V_R}{R_s} + \frac{-V_c}{R_c} + \frac{V_{\text{in}}}{R_i}\right)T_3 \tag{9.35}$$

At this point, let us ask as to what level the capacitor voltage v_{oi} would rise at the end of the interval T_4. The answer to this question lies in the fact that the negative feedback around the integrator drives the effective DC input of the latter (that is: the average value of the input current I_T over a complete clock period T_c) toward zero. Therefore, the net change in the charge of the capacitor over a clock cycle is driven toward zero. Hence the value of v_{oi} at the end of the interval T_4 would be the same $(= +V_o)$ as it was at the beginning of the interval T_1. Viewed in this light, we conclude that the total amount of the two decreases of charge that occur in the intervals T_1 and T_2 would be equal to the total amount of increase of charge in the intervals T_3 and T_4. Thus, for the scheme to work, we must have $Q_P = Q_N$, hence we have

$$\left(\frac{-V_R}{R_s}+\frac{+V_c}{R_c}+\frac{V_{\text{in}}}{R_i}\right)T_1+\left(\frac{-V_R}{R_s}+\frac{-V_c}{R_c}+\frac{V_{\text{in}}}{R_i}\right)T_4$$
$$=\left(\frac{+V_R}{R_s}+\frac{+V_c}{R_c}+\frac{V_{\text{in}}}{R_i}\right)T_2+\left(\frac{+V_R}{R_s}+\frac{-V_c}{R_c}+\frac{V_{\text{in}}}{R_i}\right)T_3 \tag{9.36}$$

Rearranging Eq. (9.36) we get

$$\tfrac{V_R}{R_s}\{(T_2+T_3)-(T_1+T_4)\}+\tfrac{V_{\text{in}}}{R_i}\{(T_1+T_2)+(T_3+T_4)\}=0 \tag{9.37}$$

Equation (9.37) does not contain V_c as we see that in Eq. (9.36):

$$\frac{V_c}{R_c}\{(T_1+T_2)-(T_3+T_4)\}=\frac{V_c}{R_c}\left\{\frac{T}{2}-\frac{T}{2}\right\}=0.$$

We further note that from Fig. 9.35 that during the period (T_1+T_4), the integrator output voltage v_{oi} is positive and hence the comparator voltage v_{oc} is "1". Accordingly, this interval is marked T_{ON}. Similarly, during the complementary interval (T_2+T_3) of the clock period, the integrator output voltage is negative, the comparator output is "0" and hence this period indicated as T_{OFF}. Thus, we have

$$\tfrac{V_R}{R_s}(T_{\text{ON}}-T_{\text{OFF}})=\tfrac{V_{\text{in}}}{R_i}(T_{\text{ON}}+T_{\text{OFF}}), \tag{9.38}$$

resulting in:

$$\tfrac{(T_{\text{ON}}-T_{\text{OFF}})}{(T_{\text{ON}}+T_{\text{OFF}})}=\left(\tfrac{R_s}{V_R R_i}\right)V_{\text{in}} \tag{9.39}$$

The up/down counter in Fig. 9.34 will count-up say N_{ON} during T_{ON} and count down N_{OFF} during T_{OFF}, where $N_{\text{ON}}=\frac{T_{\text{ON}}}{T_{hc}}$ is the count during T_{ON} and $N_{\text{OFF}}=\frac{T_{\text{OFF}}}{T_{hc}}$ is the count during T_{OFF}. Here, $T_{hc}=\frac{1}{f_{hc}}$ is the period of the clock given to the up/down counter. At the end of every period T_c of the low-frequency clock f_c, the output $b_{n-1},b_{n-2},\ldots,b_1,b_0$ of the counter will be $(N_{\text{ON}}-N_{\text{OFF}})$, where

$$b_{n-1},b_{n-2},\ldots,b_1,b_0=(N_{\text{ON}}-N_{\text{OFF}})=\tfrac{(T_{\text{ON}}-T_{\text{OFF}})}{T_{ch}} \tag{9.40}$$

In the foregone discussion, we made

$$T_{ch}=\tfrac{T_c}{N}=\tfrac{(T_{\text{ON}}+T_{\text{OFF}})}{N}, \tag{9.41}$$

From Eqs. (9.39), (9.40), and (9.41), we get

$$b_{n-1},b_{n-2},\ldots,b_1,b_0=\tfrac{N(T_{\text{ON}}-T_{\text{OFF}})}{(T_{\text{ON}}+T_{\text{OFF}})}=N\left(\tfrac{R_s}{V_R R_i}\right)V_{\text{in}}. \tag{9.42}$$

Equation (9.42) indicates that the digital output $b_{n-1}, b_{n-2}, \ldots, b_1, b_0$ is proportional to the input V_{in}. If we make $\left(\frac{NR_s}{V_R R_i}\right)$ a round number, say 1000, then the output of the up/down counter $b_{n-1}, b_{n-1}, \ldots, b_1, b_0$ will directly indicate V_{in} in millivolt. It should be noted here that the average value of the output V_s of the switch SW is

$$\overline{V}_s = \frac{1}{T}\int_0^T V_s dt = \frac{1}{T}\left(\int_0^{T_{\text{ON}}} V_{\text{R}}\, dt - \int_{T_{\text{ON}}}^{T_{\text{OFF}}} V_R\, dt\right) = \frac{(T_{\text{ON}}-T_{\text{OFF}})}{(T_{\text{ON}}+T_{\text{OFF}})} V_R = \frac{R_s}{R_i} V_{\text{in}} \quad (9.43)$$

Thus, if we evaluate the average value of the delta modulated signal, we get an output which is proportional to the input V_{in}. The scheme of delta sigma ADC given in Fig. 9.32 achieves this with the digital low-pass filter extracting the average value of the delta modulated bit stream at the output Q of the D flipflop and hence implements Eq. (9.43).

9.3.6.2 Qualitative Analysis of the Delta Sigma ADC (Fig. 9.33)

In Fig. 9.33: **If we have $V_{\text{in}} = 0$**, then the input to the integrator will be a symmetrical square wave of $\pm V_R$ and that would be integrated by OI to give a triangular wave as shown in Fig. 9.36. Whenever the triangular wave voltage $V_{oi} \geq$ zero, then the output V_c of the comparator will be "1". These "1"s will be latched to the output Q at the next rising edge of the clock f_{ch}. Thus, the bit stream out of Q will have alternate "1" s and "0"s as shown in Fig. 9.36. Since the number of "0"s equals number of "1"s the digital low-pass filter output will be "half the full-scale value" (That is $b_{n-1}, b_{n-2}, \ldots, b_1, b_0 = 1000\cdots000$).

If we have $V_{\text{in}} > 0$, then the input to the integrator will be $(V_{\text{in}} + V_R)$ whenever $Q =$ "1" and whenever Q = "0" the input will be $(V_{\text{in}} - V_R)$. Thus, the output V_{oi} will transit to negative voltage faster that it transits to positive voltage as shown in Fig. 9.37. Here too, whenever the triangular wave voltage $V_{oi} \geq$ zero, then the output V_c of the comparator will be "1". These "1"s will be latched to the output Q at the next rising edge of the clock f_c. Thus, the bit stream out of Q will have a larger number of "1" s compared to number of "0"s as shown in Fig. 9.37. Since the number of "1"s

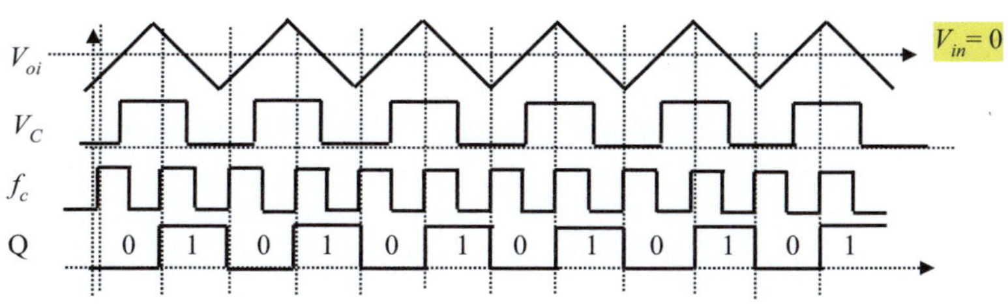

Fig. 9.36 Waveforms at cardinal points in Fig. 9.33 for zero input voltage

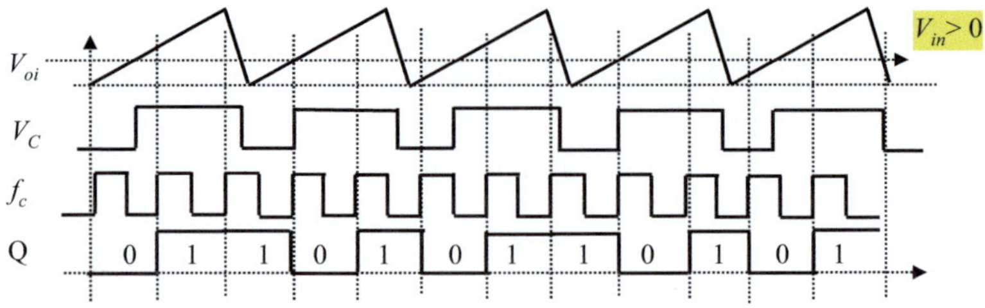

Fig. 9.37 Waveforms, for positive input voltage, at cardinal points in the circuit of Fig. 9.33

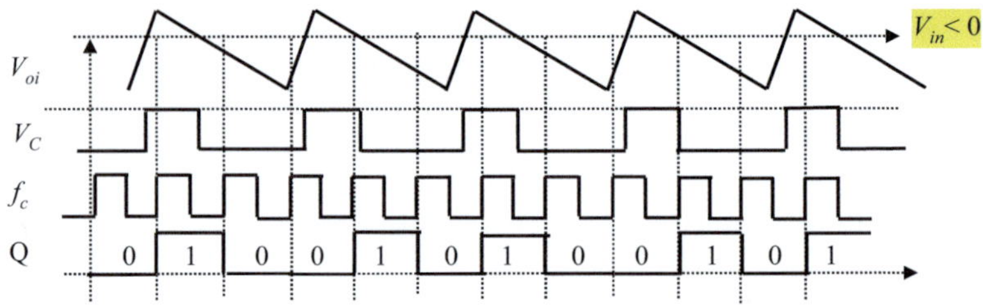

Fig. 9.38 Waveforms, for negative input voltage, at cardinal points in the circuit of Fig. 9.33

is more than the number of "0"s the digital low-pass filter output will be more than the "half the full-scale value" (That is $b_{n-1}, b_{n-2}, \ldots, b_1, b_0 > 1000\cdots000$).

On the other hand, if we have $V_{\mathbf{in}}$ < 0, then the input to the integrator will be $(-V_{in} + V_R)$ whenever $Q =$ "1" and whenever $Q =$ "0" the input will be $(-V_{in} - V_R)$. Thus, the output V_{oi} will transit to positive voltage faster that it transits to negative voltage as shown in Fig. 9.38. Here too, whenever the triangular wave voltage $V_{oi} \geq$ zero, then the output V_c of the comparator will be "1". These "1"s will be latched to the output Q at the next rising edge of the clock f_c. Thus, the bit stream out of Q will have a greater number of "0" s compared to number of "1"s as shown in Fig. 9.37. Since the number of "1"s is less than the number of "0"s the digital low-pass filter output will be less than the "half the full-scale value" (That is $b_{n-1}, b_{n-2}, \ldots, b_1, b_0 < 1000\cdots000$). Hence the output of the low-pass filter will indicate the digital value of the input voltage.

9.3.6.3 The Conversion Time and Latency (Delay) of Delta Sigma ADC

We saw that the conversion time for flash, SAR and dual slope ADCs are defined easily, and the digital (converted) data is available immediately after the conversion time for these ADCs. However, in a delta sigma ADC, a conversion time is not that easy to define. The main reason is that the output is computed by digital low-pass filtering the bit stream of the output Q of the D flipflop of the delta modulator as seen in Fig. 9.33. The accuracy of the digital output is determined by how closely the computed average value of the output bit stream matches the input voltage. When it comes to averaging a data set, we know that the higher the number of data (number of bits), the better resolution in the computed average value. Thus, higher the number of bits (periods of T_{hc}) we take from the output Q of the D flipflop for averaging, we would get better results. Typically, the bit length for averaging may be 512 or 1024 bits (clock cycles of f_{hc}). If we assume 512 bits then we need to wait 512 clock cycles to compute the average, that means the complete 512 bits of input data to the digital low-pass filter will be available only after 512 clock cycles. Let us say that the clock to the delta modulator is 1 MHz and the digital filter output is 20-bit wide. After collecting the 512 data points (that is bit one to bit five hundred and twelve), let us say the digital filter takes one clock cycle to compute the average. Hence, we will get the first digital output only after 513 clock cycles or with our example, after 513 μs. Thus, we may call the conversion time is 513 μs. However, this is not correct, because when the next bit arrives from Q (at the 513th microsecond), we can throw out the first bit (bit one) and take the new bit (that is bit 2 to bit 513 thus getting 512 bits to average) and compute the average within one clock cycle, which then gives us the next 20-bit digital data at the 514 μs instant. This we can continue then onwards and get the 20-bit digital output every 1 μs. This way of padding the new bit at the end of the bit stream and discarding the first bit in the 512-bit que is known as implementing the "*moving average*". Thus, the time we need to wait to get first data is 513 μs but only 1 μs for each of its subsequent updates. Now it is easy to understand with this example that when we start afresh, we need to wait for a certain time before we get a digital value. This initial waiting time is called the *latency* of the delta sigma ADC and the time between the digital output updates after the initial waiting time is called the *sampling time* or also as "*conversion time*" (not strictly correct!). The delta sigma ADC can be viewed as a one bit quantizer (quantization happens in the delta modulator) that is resolved into an N-bit digital output by the digital low-pass filter.

9.3.6.4 "*Noise Shaping*" in Delta Sigma ADC

If we look at the delta sigma ADC of Fig. 9.33, we see that it is a negative feedback system with an integrator in the forward path thus introducing a $\left(\frac{1}{s}\right)$ factor in the system transfer function. Thus, the feedback system is inherently of low pass structure. Let f_D be the cut-off frequency of the delta modulator. To obtain *as large a number of samples* as practically possible to average the bit stream using the digital low-pass filter, we will invariably select the clock frequency $f_c = Kf_s$, where $f_s > 2f_i$ and f_i is the input signal bandwidth. K is a constant and is chosen to be large. The digital low-pass filter reduces the quantization noise because of the fact $f_c = Kf_s$. This aspect is clearly brought out in Fig. 9.39a wherein the portion marked (hashed) is removed by the digital low-pass filter. This aspect is called the "*noise reduction due to oversampling* in delta sigma ADCs". Over and above this, the integrator in the delta modulator shapes the in-band noise as given in Fig. 9.39b. Thus, shifting the in-band noise out of the band of the digital low-pass filter. Thus, the noise in the input

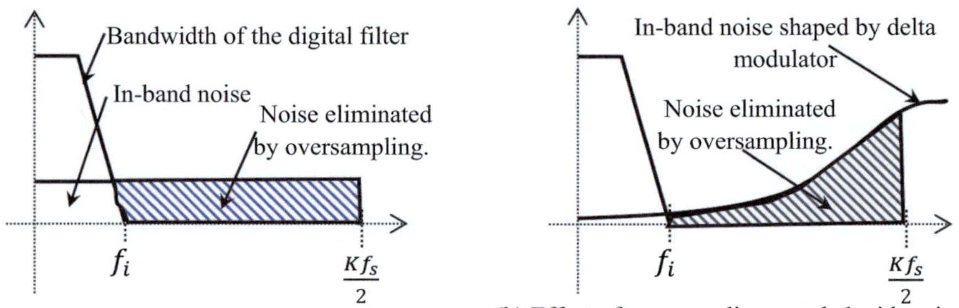

(a) Effect of oversampling in SD ADC

(b) Effect of oversampling coupled with noise shaping in SD ADC

Fig. 9.39 Noise in Delta Sigma ADC

is attenuated due to the cascaded low pass action of the digital filter combined with the shifting of the in-band noise by the integrator in the delta modulator and to that extent the digital output of the delta sigma ADC has reduced noise content. The noise shaping of the delta modulator combined with the noise elimination characteristic of the digital filter shown in Fig. 9.39b is popularly called the *noise shaping characteristic* of the delta sigma ADC. To improve the noise shaping effect, some delta sigma ADCs have a double integrator, thus introducing another $\left(\frac{1}{s}\right)$ factor resulting in a $\left(\frac{1}{s^2}\right)$ factor in the system transfer function of the delta modulator, but this is not very common in practice.

9.4 Characteristics of ADCs

So far, we have learnt different types of analog to digital converters. When we want to select an ADC for an application then we must look at the characteristics of the ADCs and choose. We now will look at different parameters that define the characteristics of an ADC.

9.4.1 Resolution of an ADC

Resolution of an ADC is specified by the number of bits (binary output ADC) or number of digits (for BCD output ADC). Example specifications: 12-bit or 24-bit or 3½ digit or 12½ digit. As we have seen earlier, resolution will be very high for the dual slope and Delta Sigma type ADCs. In terms of resolution, on the lower end stands the flash type ADC. The setting of the resolution automatically determines the quantization (QE) as (±½ LSB or ±½ LSD).

9.4.2 Full-Scale

Full-scale is the maximum value of the analog input that the ADC can convert to digital. Typically, the full-scale of an ADC can be 2.5, 5, 10, ±2.5, and ±5 V. While the full-scale will normally be a round number for BCD output ADCs (like 20 mV, 200 mV, or 2 V), for the binary output ADCs the full-scale is normally a number that is related to 2^N, where N is the number of bits of the ADC. It is common to see the full-scale binary output of a binary output ADC as: 10-bit, 1.023 V (resolution 1 mV/bit), or 12-bit 4.095 (resolution 1 mV/bit).

9.4.3 Conversion Time or Conversion Speed

We have already learned about the conversion time T_{conv} of an ADC. The conversion speed is obtained as:

$$f_{\text{conv}} = \frac{1}{T_{\text{conv}}}. \tag{9.44}$$

The unit used is samples per second (sa/s), or ksa/s or Msa/s. Some manufacturers give the conversion speed (sampling rate) in Hz, kHz, or MHz. For the sigma-delta ADC, apart from conversion rate, another important characteristic, namely, *delay* or *latency* should also be considered.

9.4.4 Signal to Noise Ratio (SNR)

We already know that every ADC suffers from quantization error (QE). We can also look at QE as a noise in the output and hence call it as a quantization noise. If we apply an input that linearly increases from zero to full-scale to an ADC and look at the quantization error, we will get the result as shown in Fig. 9.40a. While the ramp is the analog input, the digital output is a stepwise ramp. Thus, we can visualize the quantization error as a "*quantization noise*" introduced in the digital output by the ADC process. The quantization noise (also called the quantization noise of an ideal ADC) is a peak-to-peak value of 1 LSB sawtooth as indicated in Fig. 9.40b. The RMS value of the sawtooth wave of Fig. 9.40b is

$$\left(\frac{1}{\sqrt{12}}\right)\text{LSB}. \tag{9.45}$$

If we assume that the input is a sine wave of full-scale $\left(2^N\right)$LSB, then its RMS value is

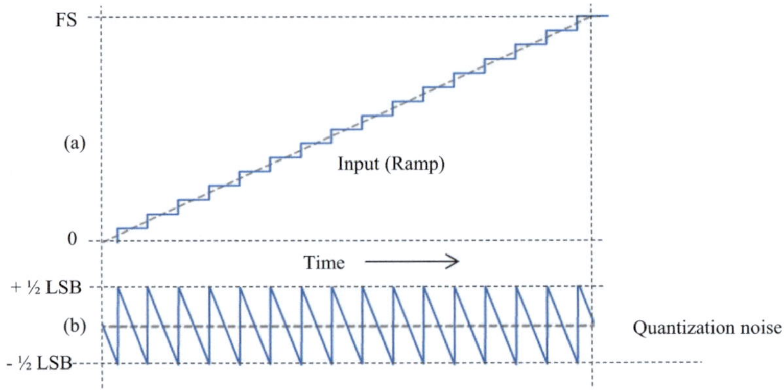

Fig. 9.40 Quantization Noise in ADC

$$\left(\frac{2^N}{2\sqrt{2}}\right)\text{LSB}. \tag{9.46}$$

With this, we compute the signal to noise ratio (SNR) expressed in dB at the full-scale level as

$$\text{SNR} = 20\log\left(\frac{\left(\frac{2^N}{2\sqrt{2}}\right)}{\left(\frac{1}{\sqrt{12}}\right)}\right) = 20\log(2^N) + 10\log\sqrt{\frac{3}{2}} = (6.02N + 1.76)\text{dB} \tag{9.47}$$

It may be noted that Eq. (9.45) is for an ideal N-bit ADC of any type. For a practical ADC the actual circuit noise will be more than the quantization noise and hence SNR of practical ADC will be $<(6.02N + 1.76)$dB.

Improvement in SNR by oversampling: If we sample the signal with a sampling frequency that is several times the Nyquist rate and then low pass or bandpass filter (of course using a digital filter) the resulting digital signal will have an enhanced signal and reduced noise. In such a scenario, the SNR can be computed as

$$\text{SNR} = (6.02N + 1.76)\text{dB} + 10\log\left(\frac{f_{sa}}{2*BW}\right) \tag{9.48}$$

In Eq. (9.48), BW represents the bandwidth of the band-limited signal. This fact is utilized in a delta sigma ADC to obtain a higher SNR than an ideal ADC!

9.4.5 *Effective Number of Bits (ENOB)*

When a practical ADC is designed and implemented (even in integrated circuit form), there will be additional noise added by various practical components in building the ADC. In addition to noise, a practical ADC may also distort the input. With the additional noise and the distortion added, a practical ADC is specified with SINAD and not SNR (Signal to noise and distortion ratio, again in the units of dB). In such a case we find an ideal N_{Ideal} bit ADC that will have the SNR equals the SINAD of the practical ADC. Then the practical ADC is equivalent to having an ideal ADC of N_{Ideal} bit resolution. The N_{Ideal} is then called the effective number of bits (ENOB) of the practical ADC. Once SINAD is known, we set:

$$[\text{SINAD} = (6.02 * \text{ENOB} + 1.76)]$$

and compute the ENOB as:

$$\text{ENOB} = \frac{(\text{SINAD} - 1.76)}{6.02} \tag{9.49}$$

It must be noted that the ENOB of a practical ADC will have the highest value for DC input and will gradually decrease as we increase the frequency components in the input.

9.5 The Digital to Analog Converter (DAC)

We saw earlier that to implement an N-bit SAR type analog to digital converter we require an N-bit digital to analog converter (DAC). Apart from the SAR ADC, we may also need a digital signal converted into analog signal using a DAC. Examples are: (i) In digital communication, analog signals such as speech or music are converted into bit streams and at the receiving end must be converted back to analog signal to hear them (ii) Situations arise wherein we may have to control (vary) the magnitude of an analog signal digitally. Hence let us learn about the available techniques of digital to analog converter (DAC), namely,

(i) Weighted R digital to analog converter (WR DAC).
(ii) (R – 2R) ladder (current mode) digital to analog converter (written as R2R DAC).

Today, the (R – 2R) ladder type DAC and its derivatives are used, and the weighted R DAC is not preferred.

9.5.1 *The Weighted* R *DAC*

Figure 9.41 illustrates the circuit schematic of a weighted R, 4-bit digital to analog converter with a full-scale voltage of 15 V. Switches S_3, S_2, S_1, and S_0 are single pole double throw (SPDT) switches controlled by the corresponding binary inputs (bits) b_3, b_2, b_1, and b_0. The poles of switches S_3, S_2, S_1, and S_0 are connected to resistors of value R, 2R, 4R, and 8R respectively. The free ends of resistors R, 2R, 4R, and 8R

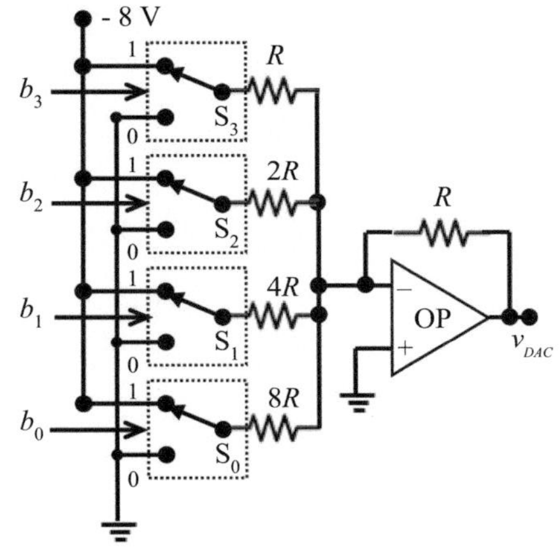

Fig. 9.41 Weighted R 4-bit DAC

are tied together and given to the inverting input of opamp OP with its non-inverting input grounded. A resistor of value R is connected between the output of OP and its inverting input. Thus, opamp OP in Fig. 9.41 becomes a 4-input inverting summer. Whenever a bit (say b_3) is "1" then the corresponding switch (S_3) is set in position 1 connecting—8 V to the b_3 resistor (for b_3 it is R) connected to its pole. If a bit (say b_2) is "0" then the corresponding switch (S_2) is set in position 0, grounding the b_2 resistor (2R) connected to that switch. When b_3, b_2, b_1, b_0 is "0001" (binary) then switch S_0 alone is set in position "1" and the switches S_3, S_2, and S_1 are set in position "0". Thus, whenever a switch is set to position "0", one end of the corresponding resistor (in this case "1000", R, 2R, and 4R) is at the ground and the other end is at virtual ground. Thus, the current through the resistors R, 2R, and 4R will be zero. Hence the output v_{dac} of the opamp OP will be

$$-\frac{R}{8R}(-8) = +1V. \tag{9.50}$$

On the other hand, if we have b_3, b_2, b_1, b_0 as "0011" (binary) then switches S_0 and S_1 are set in position "1" and the switches S_3 and S_2 are set in position "0". Thus, one end of resistors of value R and 2R is at ground and the other end is at virtual ground. Once again currents in R and 2R will be zero. Hence the output v_{dac} of the opamp OP will be

$$-\frac{R}{4R}(-8) - \frac{R}{8R}(-8) = +3V. \tag{9.51}$$

Thus, the output v_{dac} of the opamp OP will vary depending on b_3, b_2, b_1, b_0 as shown in Table 9.6. Table 9.6 illustrates the analog output v_{dac} of the DAC of Fig. 9.41 for the variations in its digital input (digital value) bits $b_3b_2b_1b_0$. Since the values of the resistors used in a weighted R DAC progress by a factor of 2, an N-bit weighted R DAC requires resistors in the range R to $R*2^{N-1}$. It is much more difficult to realize these many different valued resistors in the integrated circuit form. Thus,

Table 9.6 Output of a 15 V full-scale 4-bit DAC

Bit b_3	0	0	0	0	0	0	0	0	1	1	1	1	1	1	1	1
Bit b_2	0	0	0	0	1	1	1	1	0	0	0	0	1	1	1	1
Bit b_1	0	0	1	1	0	0	1	1	0	0	1	1	0	0	1	1
Bit b_0	0	1	0	1	0	1	0	1	0	1	0	1	0	1	0	1
Output of v_{DAC} V	0	1	2	3	4	5	6	7	8	9	10	11	12	13	14	15

it is practically very difficult to build weighted *R* DAC beyond 10-bit resolution and hence is not popular.

9.5.2 *The* R – 2R *Ladder DAC*

Figure 9.42 illustrates the circuit schematic of an $R - 2R$ ladder type, 4-bit digital to analog converter with a full-scale voltage of 15 V. Switches S_3, S_2, S_1, and S_0 are single pole double throw (SPDT) switches controlled by corresponding binary inputs (bits) b_3, b_2, b_1 and b_0 of the DAC as shown in Fig. 9.42. The poles of switches S_3, S_2, S_1, and S_0 are connected to resistors of value $2R$. Since each one of these poles is also either connected to ground (when the corresponding bit is "0") or virtual ground of opamp OP (when the switch control is "1"), the currents through the switches will be

$$I_3 = \frac{V_3}{2R}, I_2 = \frac{V_2}{2R}, I_1 = \frac{V_1}{2R} \text{ and } I_0 = \frac{V_0}{2R}. \tag{9.52}$$

These currents either go to ground (when the controlling bit is "0") or go to the inverting input of opamp OP (when the control bit of the corresponding switch is

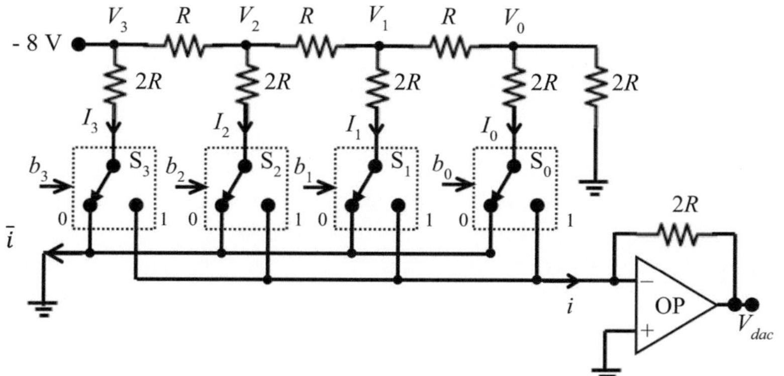

Fig. 9.42 THE $R - 2R$ Ladder DAC

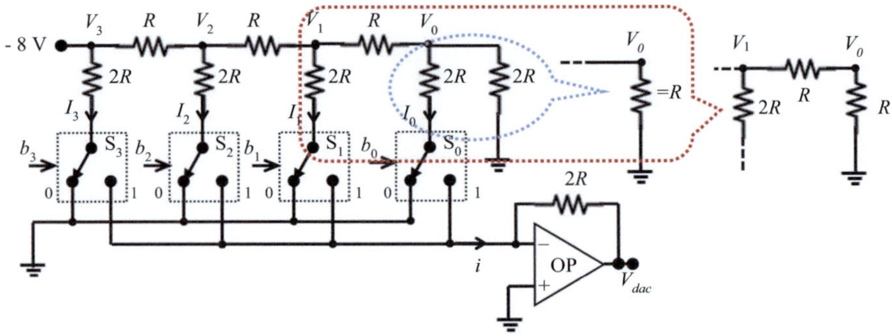

Fig. 9.43 The $R - 2R$ Ladder DAC—equivalent circuit

"1"). A resistor of value $2R$ is connected between the output of opamp OP and its inverting input. Thus, opamp OP in Fig. 9.42 becomes an *i-to-v* converter and its output is $V_{DAC} = -i(2R)$. We can easily ascertain that the equivalent resistance between node V_0 and ground is $2R$ in parallel with $2R$ and hence equal to R as shown in Fig. 9.43, resulting in:

$$V_0 = \frac{R}{2R} V_1 = \frac{1}{2} V_1. \tag{9.53}$$

Similarly, the equivalent resistance between node V_1 and ground is equal to R, resulting in:

$$V_1 = \frac{R}{2R} V_2 = \frac{1}{2} V_2. \tag{9.54}$$

In the same manner we get $V_2 = \frac{1}{2} V_3$. Since $V_3 = -8\text{V}$, we will have

$$V_2 = -4V,\ V_1 = -2V,\ \text{and}\ V_0 = -1V. \tag{9.55}$$

The currents I_3, I_2, I_1 and I_0 will be

$$I_3 = -\frac{8}{2R} A,\ I_2 = -\frac{4}{2R} A,\ I_1 = -\frac{2}{2R}\ \text{and}\ I_0 = -\frac{1}{2R}. \tag{9.56}$$

When b_3, b_2, b_1, b_0 are "0000" (binary) then all switches are set in position "0", and hence no current flows through the inverting input of opamp OP. Thus, the current $i = 0$, and hence output V_{dac} of opamp OP will be zero. When b_3, b_2, b_1, b_0 is "0001" (binary), switch S_0 alone is set in position "1" and the switches S_3, S_2, and S_1 are set in position "0". Thus, only current I_0 is sent into the inverting input of opamp OP and all other currents (I_3, I_2 and I_1) will directly flow into the ground. Hence the output V_{dac} of the opamp OP will be

$$V_{\text{dac}} = -\frac{2R}{2R}(-1) = +1V. \tag{9.57}$$

If we have b_3, b_2, b_1, b_0 is "0011" (binary) then switches S_0 and S_1 are set in position "1" and the switches S_3 and S_2 are set in position "0". Now the current $(i = I_1 + I_0) = \frac{3}{2R}$ A. Hence the output V_{dac} of the opamp OP will be:

$$V_{dac} = -\tfrac{3}{2R}(-2R) = +3V. \tag{9.58}$$

Thus, the output V_{DAC} of the opamp OP in Fig. 9.10 will vary depending on b_3, b_2, b_1, b_0 as shown in Table 9.6. Since the final output is obtained through the summing of the currents of individual binary stages, this type of DAC is also popularly called the current mode DAC. The output V_{dac} of an N-bit DAC be mathematically expressed as

$$V_{dac} = V_R\left(\tfrac{1}{2^0}b_{N-1} + \tfrac{1}{2^1}b_{N-2} + \tfrac{1}{2^2}b_{N-2} + \cdots + \tfrac{1}{2^{N-1}}b_1 + \tfrac{1}{2^N}b_0\right) \tag{9.59}$$

Equation (9.59) illustrates that the output V_{dac} of a DAC is the product of a reference voltage V_R and the combination of the binary number (b_{N-1}, b_{N-2}, ..., b_1, b_0). Hence the DAC is also called the multiplying DAC (or *m*DAC). In a practical sense, all DACs are multiplying DACs if V_R is a DC voltage. If V_R is a time varying AC voltage, then the DACs must be specially made such that the switches and the resistors are "AC compatible". We also see that when we vary the bits b_{N-1}, b_{N-2}, ..., b_1, b_0 from "00 ... 00" to "11...11", the output of the DAC varies from 0 to ≈2 V_R. Hence a multiplying DAC can also be used as a digitally controlled attenuator (or simply as a digital potentiometer). Instead of resistances, it is also possible to *switch in* capacitances and obtain a capacitance $2C - C$ ladder array based on current mode DAC. Such a capacitance-switched current mode DAC will consume much less power compared to a similar resistive type of current mode DAC, but these are not commercially available.

9.6 The Sample and Hold

In all the ADCs we see that the input to the ADC must remain constant and not change during the conversion time of the ADC. Instead of assuming that it will not change, it is better to ensure that the input to the ADC remains constant during the conversion time. The device (circuit) that ensures that the input remains constant during the conversion time of an ADC is called the *sample and hold* circuit. Figure 9.44a shows the block schematic of a sample and hold. We see that a sample and hold unit has an input v_i, an output v_o and a control input $S/\overline{H}$. Figure 9.44b shows the input–output characteristics of a sample and hold unit. We see from Fig. 9.44b that the output is equal to (follows) the input whenever the sample and hold command input $S/\overline{H}$ is in the sample mode ($S/\overline{H} =$ "1"). Whenever the control input $S/\overline{H}$ is in the hold mode ($S/\overline{H} =$ "0") then the output is held constant to the value of the input at the time the

$S/\overline{H}$ transitions from "1" to "0" as shown in Fig. 9.43b. Once the $S/\overline{H}$ signal is set to sample mode ($S/\overline{H}$ = "1") once again the output follows the input.

A very simple sample and hold circuit, shown in Fig. 9.45a, uses a single pole single throw (SPST) switch S and a capacitor C. When the control $S/\overline{H}$ to the switch S is "1" the switch S is closed and hence the capacitor voltage as well as the output voltage v_o will be equal to the input voltage v_i thus the circuit will be in the "sample" mode. As soon as $S/\overline{H}$ goes to "0", switch S gets open circuited, the circuit enters the "hold" mode, and the output voltage v_o will be the voltage across the capacitor at the time instant the switch S is opened. However, for the simple sample and hold circuit of Fig. 9.45a to work, the switch and the capacitor must be ideal, the input must supply the charging/discharging current of the capacitor during the "sample" mode and no current can be drawn from the capacitor in the hold mode.

The sample and hold circuit shown in Fig. 9.45b is quite popular and is available in integrated circuit form from several manufacturers (Example LF398 from National semiconductor corporation). The advantages of the circuit are:

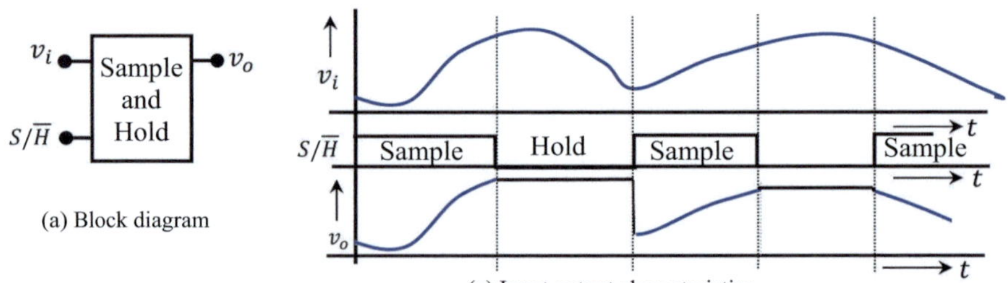

(a) Block diagram

(a) Input output characteristics

Fig. 9.44 Sample and hold characteristic

Fig. 9.45 Sample and hold circuit

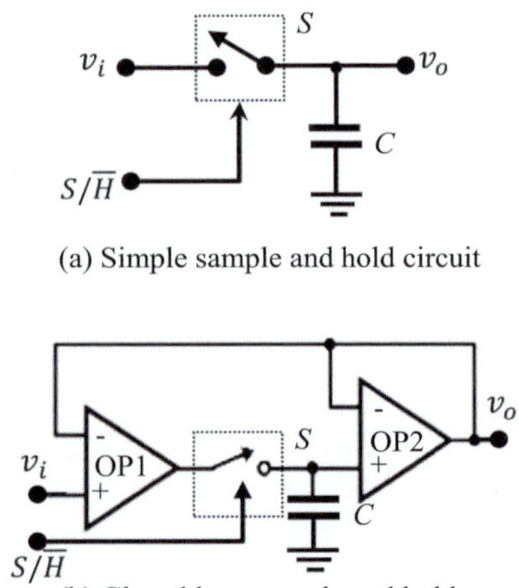

(a) Simple sample and hold circuit

(b) Closed loop sample and hold

(i) The capacitor is connected to the output of OP1 during sample mode thus the output v_o will follow the input v_i as the required charging current will be provided by OP1. The input v_i supplies only the input bias current of the non-inverting input of opamp OP1.

(ii) In the hold mode, the output current is supplied by the OP2 and hence the current demand on the capacitor during the hold mode is only the input bias current of the non-inverting input terminal of opamp OP2.

(iii) An added advantage is that as soon as the circuit enters the "hold" mode, the output of OP1 will be at the supply voltage (positive supply voltage if the input is positive otherwise negative supply voltage). Hence, when the state changes from hold mode to sample mode, capacitor C will be quickly charged to v_i.

9.6.1 Important Characteristics of a Sample and Hold are

(i) **The acquisition time**: This is the minimum time required for the output v_o of the sample and hold to be within $\pm 0.01\%$ of the input v_i when the mode of the sample and hold is changed from sample mode to hold mode. The time is 100 ns for the HA5351 and is 10 μs for the LF398.

(ii) **The droop rate**: This is the rate at which the output v_o of the sample and hold changes in the hold mode. The droop occurs mainly due to the input bias current and leakage currents. The droop rate is given in microvolts per microsecond. During design, it must be ensured that the droop rate multiplied by the conversion time of the ADC is less than half the LSB voltage.

(iii) **Input offset voltage**: The voltage at the input of the sample and hold (that is the voltage between the non-inverting input and inverting input of OP1 in Fig. 9.45b.

(iv) **Input bias current**: The current drawn by the input of the sample and hold (the bias current of the non-inverting input of OP1 in Fig. 9.45b.

(v) **Hold step**: When the mode is changed from sample to hold, the switch S is opened and that will introduce unwanted charge into the hold capacitor C, resulting in a small step in the output v_o. This is absent in modern sample and hold ICs that use zero charge injection switch configuration for switch S.

(vi) **Feedthrough**: Once again parasitic capacitances between the input and output terminals cause some portion of the input affecting the output v_o. The feedthrough describes the extent the input alters the output and is expressed in the units of dB.

(vii) **Gain error**: In most sample and hold, the gain $\frac{|v_o|}{|v_i|}$ is unity but in practicality the gain may not be 1.000 and can differ, leading to gain error. The gain error is expressed in percentage as

$$\left[\frac{\text{Actual gain} - \text{Design gain}}{\text{Design gain}} 100\%\right] \tag{9.60}$$

The typical value may be 0.001%. Sample and hold ICs also allow setting a gain >1.

9.7 Sampling

When a continuous time domain signal $x(t)$ is *sampled*, we get a series of values $x(n)$ at every time instant, say nT_s, where n is an integer 0, 1, 2 $\cdots$. Thus, $x(n)$ is the discrete form of $x(t)$. If the sampling interval T_s remains constant, then the samples $x(n)$ are called uniform samples and the process itself is termed as "*uniform sampling*". On the other hand, if T_s is not a constant then the sampling is called "*non-uniform*" or "*Random*" sampling. We assume the sampling to be uniform in all our further discussions. Furthermore, we use the term: *sampling frequency* $f_s = 1/T_s$ to indicate the sampling rate, again assuming uniform sampling.

Needless to say, we would like the discrete samples $x(n)$ to faithfully represent the characteristics of $x(t)$. One way to ensure this is that if $x(t)$ is represented by a *frequency spectrum*, say $F(\omega)$, then $x(n)$ should also have an identical spectrum possibly within a scale factor. Here, by the term "spectrum of $x(n)$" we mean the spectrum of the continuous signal obtained from the sample train by standard analog signal reconstruction techniques. This is strictly possible if $x(t)$ is band-limited, i.e., it does not contain any components with frequency higher than a specific value frequency $f_{\max}$.

The analog reconstruction process involves the use of an ideal low-pass filter with a passband extending from 0 to $f_{\max}$, thereby cutting out the extraneous higher frequencies that arise in the sample formation. Now, to satisfy the condition that the spectrum of $x(n)$ should be a replica of the spectrum $F(\omega)$ of $x(t)$, it is necessary that the digital sampling scheme is able to correctly represent a signal of the form $\sqrt{2}V_{\max}\mathrm{Sin}(2\pi f_{\max}t + \theta)$. A general form of a sinusoid like this has three defining parameters, viz. magnitude, frequency, and phase. With two samples per cycle, it will not be possible to reconstruct, in general, a sine wave that satisfies all these three specifications. Hence, we need to have a sampling frequency $f_s > 2f_{\max}$. This is explained further in the following paragraph.

Let us consider sampling a sinusoid of frequency f_{max} at a sampling frequency $f_s = 2f_{\max}$. We will get the samples as 0, 0, 0 … if we start sampling at the zero crossing as shown in Fig. 9.46a. However, if we start sampling not at the zero crossing of $f_{\max}$ then we will get two samples per cycle as shown in Fig. 9.46b. In either case, it is easy to see that more than one waveform (differing from the original one from which the samples are taken) can be reconstructed from this sample set. For instance, the first set may arise from a sine wave of different amplitudes or even from zero signal. The second set of sample points permits sine waves of the same frequency but with different phases.

Fig. 9.46 Sampling $f_{\max}$ at $2f_{\max}$

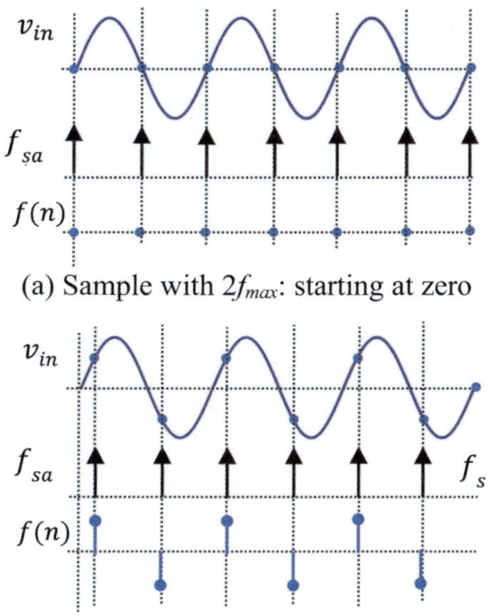

(a) Sample with $2f_{max}$: starting at zero

(b) Sample with $2f_{max}$: start not at zero

9.7.1 *Nyquist Sampling*

Such ambiguities, listed above, can be avoided if the sampling rate is more than $2f_{\max}$. This lower bound on the required sampling rate is called the *Nyquist sampling frequency*.

A few additional details of the Nyquist criterion will be presented later in Sect. 9.7: Digital Storage Oscilloscope. In a practical scenario, we set the sampling frequency to be much higher than the Nyquist frequency $2f_{\max}$.

9.7.2 *Aliasing due to Sampling*

What happens if we sample a signal $x(t)$ at a frequency f_{sa} which is below its *Nyquist* sampling frequency $2f_{\max}$?

1. The first thing that we can ascertain is that the spectrum of the resulting signal $x(n)$ will not contain frequencies $> f_{\max}$.
2. Apart from the above, the spectrum of $x(n)$ will be distorted because when a high-frequency f_h (the term high-frequency points to a frequency greater than $\frac{f_{sa}}{2}$, f_{sa} being the sampling frequency) folding down and portraying the samples that would have been obtained if a low-frequency sinusoid of frequency f_a was present in $x(t)$. We call this "*aliasing*" as the high-frequency sinusoid f_h shows itself (aliases) as samples of a low-frequency sinusoid f_a in $x(n)$. Figure 9.47 shows an example where the samples of a higher frequency (f_h) sinusoid aliases

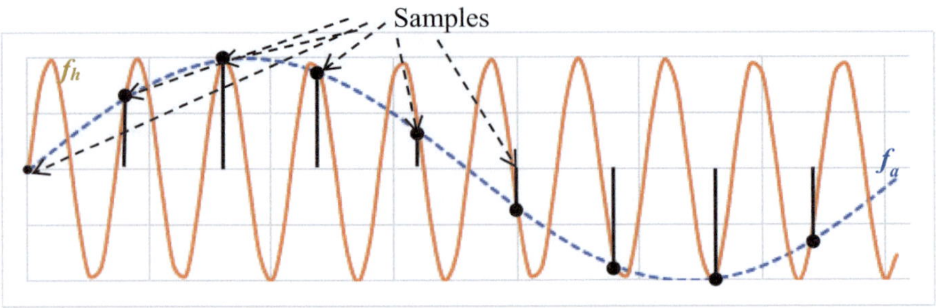

Fig. 9.47 Sub Nyquist sampling resulting in aliasing

as a lower frequency (f_a) sinusoid. We can compute the aliased frequency f_a due to the high-frequency f_h when sampled at sub-Nyquist frequency as

$$f_a = |\text{NCIM of} f_{sa} - f_h| \tag{9.61}$$

In Eq. (9.61), the acronym NCIM expands to: Nearest Common Integer Multiple.

Let us consider an example:

Let $x(t)$ contain: 50, 200, 400, 700, 950, 1100, and 5010 Hz as indicated in the spectrum shown in Fig. 9.47a, and sampled at the rate of 1000 sa/s. Then Nyquist frequency is 500 Hz. Hence $x(n)$ will contain 50, 200, and 400 Hz (as these are less than the Nyquist frequency) and will not have 700, 950, 1100, and 5010 Hz, but will contain aliases of these frequencies a:

(1) 700 Hz will alias to $|1000 - 700| = 300$ Hz,
(2) 950 Hz will alias to $|1000 - 950| = 50$ Hz,
(3) 1100 Hz will alias to $|1000 - 1100| = 100$ Hz,
(4) 5010 Hz will alias to $|5 * 1000 - 5010| = 10$ Hz.

If the aliased frequency f_a is not present in the spectrum of $x(t)$ then the spectrum of $x(n)$ will contain a new component. Thus, due to (1), (3) and (4) above $x(n)$ will contain the additional frequency components of: 10, 100, and 300 Hz. The aliasing of 950 Hz produces a sinusoid of 50 Hz. But there is already a 50 Hz present in $x(n)$ and the new 50 Hz samples will combine with the samples of the existing 50 Hz in $x(n)$ and the resulting 50 Hz component in $x(n)$ will be altered (in magnitude and phase). Figure 9.48a shows the spectrum (frequency components in) of $x(t)$ and Fig. 9.48b illustrates the frequency components of $x(n)$. Here we assumed that the magnitude of 50 Hz present in $x(t)$ and the magnitude of 50 Hz that came in due to the aliasing of 950 Hz constructively added in $x(n)$. In general, this may not be so, and the alias component may even completely cancel the original component present in $x(t)$! Thus, aliasing may give rise to:

(a) The amplitude and or phase of frequency component(s) of $x(t)$ may be altered in $x(n)$ and or

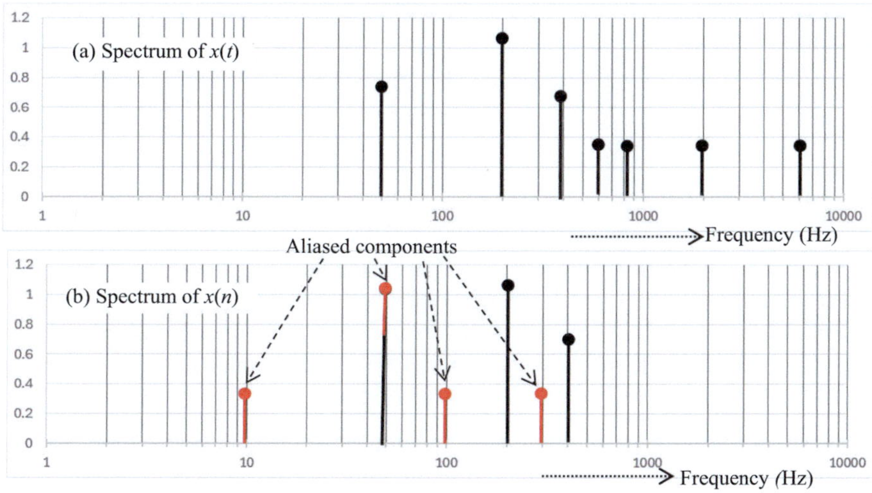

Fig. 9.48 Aliasing due to sub-Nyquist sampling

(b) New frequency components that are not present in $x(t)$ may be added in $x(n)$.

Such errors (sub-Nyquist frequency components introduced through aliasing due to sub-Nyquist sampling) are called *Aliasing errors*. This type of error occurs also while sampling non-band-limited waveforms and we take $f_{\max}$ to be the highest significant frequency component in it. In practice, we chose the sampling frequency to be much higher than the Nyquist frequency with a view to reduce the errors due to aliasing and the non-ideality of the low-pass filters in the analog reconstruction.

9.7.3 Anti-Aliasing Filter

To ensure that the sampled data does not suffer from aliasing errors, we introduce a low-pass filter between the input and the sample and hold as seen in the block diagram of a typical data acquisition system shown in Fig. 9.49. The low-pass filter is designed to ensure that its output does not contain frequencies $>(f_s/2)$. This low-pass filter is generally called an "*Anti-aliasing filter* (AAF)" as it ensures that the spectrum of the digitized samples $x(n)$ does not contain spurious aliased frequency components that are not present in the input. Of course, the AAF may remove components $> (f_s/2)$ from the input and thus the digital samples $x(n)$ will also not have these components. Thus, the AAF must be properly chosen so that a minimal difference exists between the spectrum of the input $x(t)$ and the spectrum of the digital counterpart $x(n)$.

In a general scenario, the sampling rate of an analog to digital converter is higher than the sampling rate required for a given signal. In such a case, we may use a channel multiplexer and use a single ADC to convert several analog signals, say,

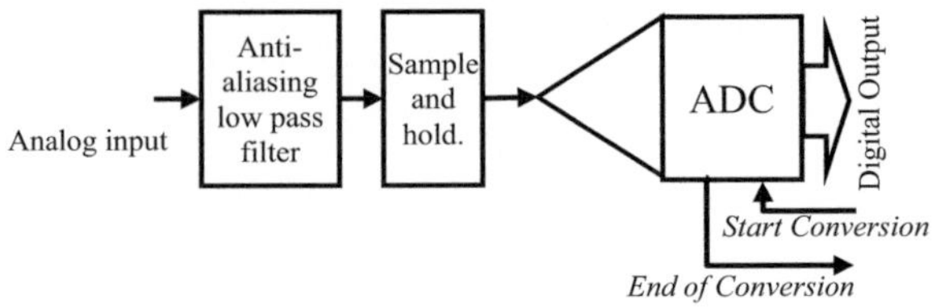

Fig. 9.49 Single channel data acquisition system

A_1, A_2, A_3 ..., as shown in Fig. 9.50. Such an arrangement is called "*multi-channel*" data acquisition system. Again, in general the number of analog signals multiplexed can be 2 or 4 or 8 or 16. Figure 9.50 shows the analog inputs are multiplexed and then the multiplexed signal is sampled. Obviously, the samples of different analog inputs A_1, A_2, A_3, ..., will be at different time points.

Simultaneous sampling: In some specific applications it may be required that the samples of A_1, A_2, A_3, ... are acquired at the same instant of time. In such a case, the anti-aliasing filter and the sample and hold must be placed in every channel as illustrated in Fig. 9.51. Such a method is popularly called "*Simultaneous sampling*".

Synchronous sampling: In certain applications, it may be required to sample a signal that contains fundamental + harmonics of a single frequency. An example is the power system currents and voltages. In such a case it is quite advantageous (makes further processing of the digital samples less complex and results in reduced errors) to set the sampling frequency to be an integral multiple of the fundamental frequency f_1, that is: $f_{sa} = Nf_1$.

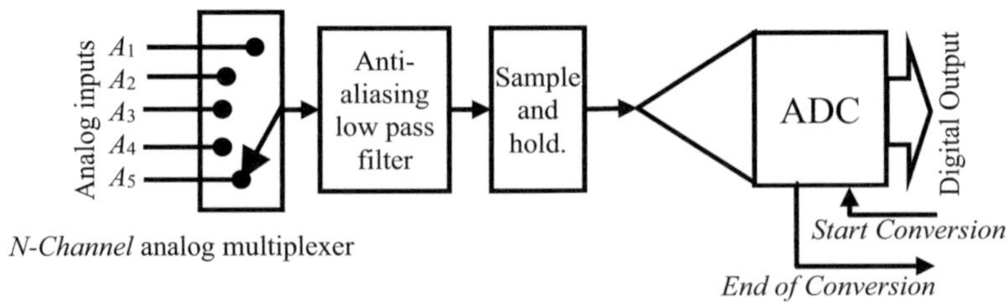

Fig. 9.50 Multi-channel data acquisition system

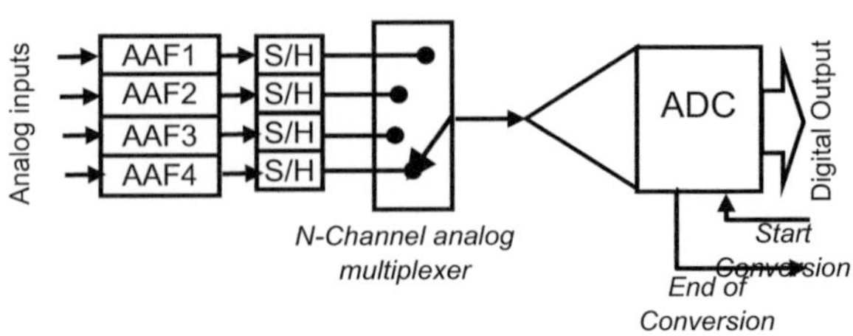

Fig. 9.51 Multi-channel data acquisition system simultaneous sampling

9.8 The Digital Storage Oscilloscope

We studied the principle of operation of the Cathode Ray Oscilloscope (CRO) in Chap. 7. Today CROs are no longer manufactured as they are now replaced with the digital storage oscilloscopes (DSOs). Accordingly, we nowadays call the latter simply an oscilloscope. Figure 9.52 illustrates the block diagram of a digital storage oscilloscope. The analog input is first signal conditioned (magnitude is increased or decreased so that the resulting magnitude is within the optimal range of the ADC) and low-pass filtered using an anti-aliasing filter. The ADC converts the analog signal into digital samples. The digital output of the ADC is stored directly into the memory of a digital processing unit (DPU) and display system. The DPU then displays this as a curve (mostly as a picture as seen in Fig. 9.52) on a suitable display. The individual blocks of a DSO are explained next.

The Y-amplifier: The Y-amplifier in a DSO is similar to the Y-amplifier in a CRO, with two differences. The first is that the output of the Y-amplifier in a DSO will be few volts in magnitude compared to few tens of volts in a CRO and, secondly, the Y-amplifier output will not go through a delay line as in a CRO but connected to an anti-aliasing filter. If the DSO has more than one channel (say as in a 4-channel DSO) then each channel will have an individual Y-amplifier such as Y_1 amplifier, Y_2 amplifier, Y_3 amplifier and Y_4 amplifier. Each Y-amplifier will have the provision of direct coupling (DC coupling) or AC coupling (connecting the input through a series (internal) capacitor). Like the CRO the inputs of a DSO are also standardized with a BNC connector and an input impedance of 1 MΩ in parallel with a 47 pF capacitance. Thus, the divide by 10 or divide by 100 probes of an analog CRO will work with the inputs of a DSO as well without any need for alterations.

The anti-aliasing filter: Let us now go through a summary of what we already learnt about the principles of sampling analog signals along with some interesting historical details. Whenever we convert an analog signal into digital, the digital samples too must contain all the information in the analog signal. One way to ensure is to look at the frequency domain components of an analog signal and ensure that all the frequency components in the analog signal are faithfully captured in the

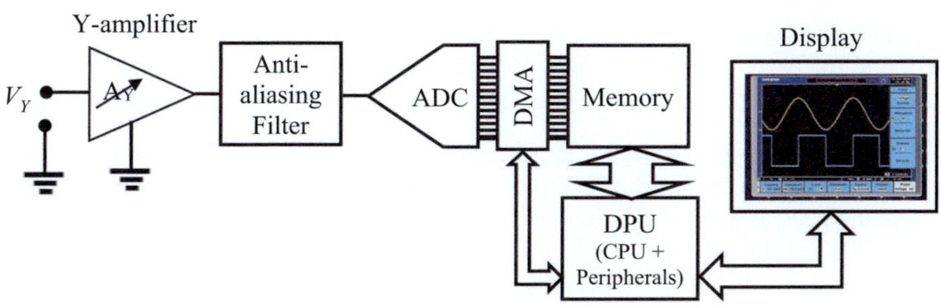

Fig. 9.52 Bock diagram of the digital storage oscilloscope

digital samples. The frequency domain components of repetitive analog signals can be obtained by using the Fourier series. The frequency spectrum of non-repetitive analog signals can be obtained by applying Fourier transform. Now, if we ensure that the maximum significant frequency component of the analog signal is faithfully represented in the digital samples, then all other lower frequency components will also be represented correctly in the digital samples. To represent a sinusoid in the digital sample, we need at least three samples per period (as each sinusoid is characterized by three variables, namely, magnitude, frequency, and phase) as already discussed. This aspect was first studied by E. T. Whittaker in 1915 but was not noticed at all. Later in 1924 Harry Nyquist elaborated this aspect, but again his work was not noticed. When the famous mathematician-electrical engineer Claude Shannon published the now popularly known "*Shannon sampling theorem*" in 1945, he received the entire credit for this result (though Shannon had cited Nyquist's paper in his paper). Shannon sampling theorem states that to represent an analog signal faithfully in digital samples, the signal must be sampled at a rate greater than twice the maximum frequency contained in the analog signal. Under certain conditions it is possible to sample just at twice the maximum frequency and represent the maximum frequency component in the digital samples, Because of this, Shannon used the phrase "greater than or equal to" in his theorem. However, as we may miss the maximum frequency signal completely at certain other conditions if we sample at the "twice the maximum frequency", it is better to specify, in general, that the required sampling frequency as greater than twice the maximum frequency contained in the signal. Later people realized that the credit should have gone to Nyquist and not to Shannon but by that time the name "*Shannon sampling theorem*" had become very popular. Hence, to give due recognition to Nyquist too, we call this criterion, as the "*Nyquist criterion*" and twice the maximum frequency in the analog signal as the *Nyquist frequency*. As already discussed, if we sample an analog signal at a rate less than the Nyquist sampling frequency, then the digital samples will contain an error called the "*aliasing error*" as the higher frequency components in the analog signal may manifest into lower frequency components (maybe this will add to the existing low-frequency components in the analog signal or simply create a new low-frequency signal that is not part of the analog signal) in the digital samples. Hence to avoid this aliasing error in the DSO, if the conversion rate is N sa/s in the ADC we must ensure that the analog input does not contain frequencies above $\left(\frac{N}{2}\right)$ Hz. As a sure way of ensuring that this is the case, a low-pass filter with a cut-off of $\leq \left(\frac{N}{2}\right)$ Hz is tied to the output of the Y-amplifier. Such a low-pass filter is popularly known as "*Anti-aliasing filter*" and always precedes the ADC in a digital storage oscilloscope.

The ADC: We see that the maximum frequency that can be handled by a DSO is half the sampling frequency of the ADC. Hence, to have a wide bandwidth, the sampling rate of the ADC is as high as possible. The result is that nearly all the DSOs available today use the FLASH type ADCs. Most DSOs are manufactured with an 8-bit flash type ADC, but some high-end DSOs use 10-bit or 12-bit flash ADC. Typically, the sampling rate of the ADC in a DSO can be in the range 50 Msa/s (low end DSO) to

4 Gsa/s (high-end DSO). Few special DSOs use 16-bit or 18-bit SAR type ADCs, but bandwidth of such high-resolution DSOs will be low ($\approx$10 MHz).

The DMA: The DMA stands for Direct Memory Access. In a normal data acquisition system using an ADCs of sampling rate in the range few sa/s to few hundred ksa/s, the output of the ADC is read by the processor (central processing unit, CPU) and stored in the memory. But when the sampling rate is high, then it will not be at all possible for the CPU to do this and the only solution is to directly store the sampled data from the ADC into the memory. This is achieved by the direct memory access (DMA) block. One popular method of implementing direct memory access is called "Cycle stealing". In this method, whenever the CPU is executing instructions that involve only internal registers, the DMA accesses the memory (as the CPU is not using the memory)!

The DPU: The digital processing unit is essentially a digital system that consists of a central processing unit with program and data memory and peripheral controllers to control the other devices, namely, the ADC, the DMA, input keys, and the display. Today the DPU is more or less a personal computer (PC) and may port an operating system as well. The DPU is programmed to set the ADC and DMA parameters for proper operation, acquire the required number of samples, process them, and present them on the display. The processing of the samples could be to make measurements as well as to obtain, say, the frequency spectrum of the input. Thus, the DPU may contain, other than a CPU, some dedicated processing units, such as FFT, peak, RMS, period, and frequency calculators. Several manufacturers provide these units as standard, but some manufacturers may give these units as add-on modules.

The Display: Initially the display was a cathode ray tube (CRT) and hence the digital data from the memory was converted back to analog using a digital to analog converter and then displayed on the screen of the cathode ray tube (CRT). Today, the display itself is a digital display (such as a liquid crystal display or simply the LCD). While the CRT display is controlled as several horizontal lines on the screen, the digital display is simply made of "dots" (called *pixels*) on the screen. Initially, with the VGA (acronym for versatile graphics adaptor), each screen is made of 640 pixels in the horizontal direction and 480 pixels in the vertical direction (a matrix of 640 columns and 480 rows). Later this became 1280 pixels in the horizontal direction and 720 pixels in the vertical direction (This is popularly known as 720p or high definition or standard HD or simply HD). Of course, today the VGA can support graphics display of 1920 pixels in the horizontal direction and 1080 pixels in the vertical direction (called full high definition or FHD or 1080p). The ultra-high definition or UHD display has 3840 pixels in the horizontal direction and 2160 pixels in the vertical direction (Movie UHD uses a slightly different specification of 4096 horizontal pixels by 2160 vertical pixels) is currently evolving and is popularly known as the 4 k display. In the LCD display the intensity and color of each pixel are represented by 24-bits. The display controller of the DPU simply writes the value (8-bit binary number) of R, G, and B intensities into the corresponding memory dedicated to a pixel. This pixel value determines the intensity and color of that pixel.

The control and setting up of the DSO: Since the DSO evolved from a CRO, the DSO replicates most of the controls of the CRO as seen in Sect. 7.3 and listed below once again.

Channel Mode Select: Like the CRO, in the DSO you can also choose Alternate (ALT) mode or Chopped (CHOP) mode to display the channels. It is also possible to have the $Y - t$ mode or $X - Y$ mode as in a CRO. Additionally, a channel can be switched ON or OFF as well.

Channel sensitivity select: This aspect is also identical to the CRO. The Y-channel sensitivity in mV/div or V/div can be adjusted in the 1-2-5-10-20 …steps and the X-channel sensitivity (in the $Y - t$ mode) can be selected again in μs/div or ms/div, once again in the 1-2-5-10-20 sequence.

The trigger: The trigger is again set with level and slope. The trigger modes are the same as in a CRO. However, there is one big difference: In the CRO, the trigger point is always on the left most point on the screen. In a DSO, since the samples are anyway available, the trigger point can be anywhere on the screen. That is: A CRO displays the input after the trigger had occurred and we cannot see the waveform before the trigger point. In a DSO we can set the trigger point to be at the middle of the screen, thus we can see the waveform(s) before and after the trigger has occurred. In a DSO, it is also possible to select just one trigger (not repeated trigger) and thus display events that occur only once and do not repeat. Thus, all DSOs have the trigger to be selected as (i) Continuous trigger (popularly called RUN) mode or (ii) One trigger (popularly called RUN-STOP) mode. The trigger input selection in a DSO is similar to the CRO, namely, the trigger can be selected from:

1. One of the y-inputs (marked INTernal).
2. External trigger input (marked EXTernal).
3. The power line (marked as LINe).
4. TVV or TVH.

Additional Control Parameters in a DSO

The Display Update Rate: Apart from the above there is another big difference between a CRO and a DSO. While in a CRO the display is always in real time, in a DSO there will always be a latency (delay). That is: what is displayed on the screen is the input captured sometime earlier. Thus, it is very important to consider an additional specification of the DSO (which is not present in a CRO), namely, the display update rate (DUR). Typically, the DUR can lie in the range of few hundred frames to few hundred thousand frames per second.

The Channel memory: Apart from the DUR, another specification that is particular to the DSO is the Channel memory size. Typically, the memory could be a few kilobytes to few megabytes per channel. The higher the memory, the larger the number of data points stored, and hence we will get better resolution not just in the x-axis but also in the y-axis, a valuable feature especially in measurements.

Figure 9.53 shows two different DSOs.

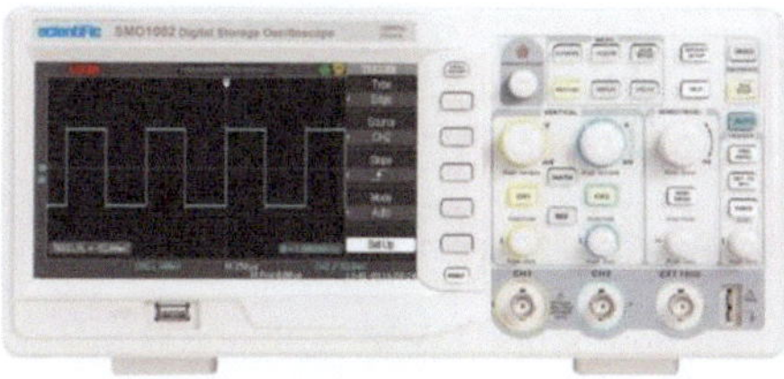

(a) A 2-channel, 100 MHz DSO with 1 G sa/s ADC with 2 M datapoints per channel

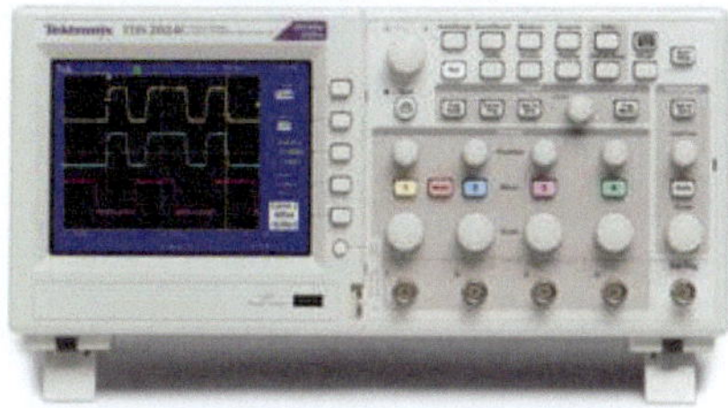

(b) A 4-channel, 200 MHz DSO with 2 G sa/s ADC, 2500 data points per channel

Fig. 9.53 Digital storage oscilloscopes

9.9 Problems and Exercise

E9.1 A DAC manufacturer claims that a particular DAC is monotonic. What do you understand by this term?

E9.2 A flash ADC uses comparators with a settling time of 40 ns and a 3-stage digital decoder made of gates possessing a delay of 2 ns. Determine the conversion time of the ADC.

E9.3 Design a weighted R 8-bit DAC for a full scale of 2.55 V.

E9.4 The reference to a 12-bit R-2R ladder type DAC is 2.048 V. Derive an expression for its output.

E9.5 Show how you will obtain a ±2.047 V from the above DAC for the range of input 000_h to FFF_h.

E9.6 Draw the state diagram for a 4-bit successive-approximation-register (SAR Type) ADC with a full scale of 15 V.

E9.7 The DAC of problem 8.4 is used to build a SAR type ADC using an SAR of delay 10 ns, a comparator with a settling time of 40 ns and a 0.5 LSB subtractor with a settling time of 10 ns. Derive the values of the characteristics of this ADC.

E9.8 In a 4-bit SAR Type ADC, the MSB is stuck at zero. If the full scale is 15 V, determine the output from the ADC for an input of (i) 5 V and (ii) 10 V. Draw the state diagram in each case.

E9.9 Repeat problem 8.1 if the MSB is stuck at 1.

E9.10 An n-bit SAR type ADC is fabricated with a SAR having a delay time of 10 ns, DAC with a full-scale settling time of 1 μs, a subtractor with a settling time of 50 ns and a comparator with a settling time of 40 ns. Determine the maximum clock frequency that can be applied to the SAR.

E9.11 If a progressive clock is to be applied to the above ADC determine the optimum time periods for first 8 clocks. Comment on the results.

E9.12 A progressive clock SAR type 16-bit ADC uses successive clock frequencies of f_c, $f_c/2$, $f_c/4$ and $f_c/8$. Determine the total conversion time in terms of T_c.

E9.13 Find the resolution and quantization errors of the following ADCs.

a. SAR-08, 8-bit, full scale = 5 V
b. SAR-12, 12-bit, full scale = 4.096 V
c. Dual slope (DS-12), 12 bits, full scale = 2.048 V
d. SAR-16, 16-bit, full scale = 5 V
e. DS-3.5, 3½ digit, full scale = 20 mV.

E9.14 Show how you will convert a 3½ digit dual slope DVM with a 200 mV full scale to: (a) an ohmmeter to measure 20 MΩ full scale. (b) Diode tester (C) Measure the current gain of a PNP or NPN transistor.

E9.15 A sample and hold uses a switch having a settling time of 40 ns, 1000 pF hold capacitor, opamps having a slew rate of 20 V/μs, bias current of 10 nA and 10 mA short circuit current capacity. Determine the minimum sampling time that can be obtained from this S/H for a 10.23 V FS ADC. Calculate the droop rate.

E9.16 An 8-bit flash ADC uses comparators with a settling time of 20 ns, 4-stage digital decoder using TTL gates possessing a propagation delay of 5 ns. Determine the conversion time.

E9.17 In a dual slope ADC, two capacitors C_S and C_x are connected in the feedback path of the opamp as indicated in Fig. 8E.1. If the controller keeps all the switches in position 1 for T_1 s and switches to position 2 at the end of T_1 and determines the time taken T_2 for the integrator output to reach zero, what value the period T_2 will indicate?

Chapter 10
Miscellaneous Measurement Methods

"One accurate measurement is worth more than one thousand experts' opinion."
—Grace Hopper

10.1 The Miscellaneous Measurement Methods

In Chaps. 1–9, we learnt different methods to measure electrical quantities such as voltage, current, power, energy, resistance, inductance, and capacitance; grouped together based on certain common "*principle of measurement*". In this chapter, we will continue to learn additional methods of measurement that do not come under any of those principles of measurement discussed in the earlier chapters. Of these, the ***Q meter*** explained next helps us make measurements on an inductor.

V. Jagadeesh Kumar, *Principles of Electrical Measurement and Measuring Instruments*,
https://doi.org/10.1007/978-981-96-5475-8_10

10.2 The *Q* Meter

The *Q* meter, shown in Fig. 10.1, is a simple device that provides measurement of parameters of an inductor. Here, the test coil is represented as a series of resistance R_X (that includes winding resistance and any power dissipation in the core of the inductor due to hysteresis and eddy current losses), a self-inductance L_X and a shunt capacitance C_X that represents the distributed inter-turn capacitance and any other capacitance that comes across the terminals as shown in Fig. 10.1a. In Fig. 10.1, C_S is a variable standard capacitor. Before learning how to make the measurements with a *Q* meter, let us analyze the *Q* meter circuit of Fig. 10.1a.

10.2.1 Analysis of the *Q* Meter

If we consider the Nortan equivalent across the terminals *P* and *Q* (left of capacitor C_S encircled by dashed box) in Fig. 10.1a, we get the circuit of Fig. 10.1b. Once again, we replace the circuit part consisting of the Norton current source $\underline{I}_N$ in parallel with the series circuit made of L_X and R_X (encircled by dotted box) in Fig. 10.1b, with its Thevenin equivalent, we get the circuit shown in Fig. 10.2. It should be noted here that the Thevenin source $\underline{V}_{Th}$ in Fig. 10.2 is not the excitation source $\underline{V}_S$ in Fig. 10.1a. However, we may note that the readings of the meters I_1, V_1, and V_2 are valid for the circuit shown in Fig. 10.2. In Fig. 10.2, the impedance across terminals *O* and *Q* is

$$Z = (R_X + j\omega L_X) + \frac{1}{j\omega(C_S + C_X)}. \tag{10.1}$$

When we get the reading in V_2 (or the reading in I_1) is maximum, then the series circuit L_X and $(C_S + C_X)$ is under resonance and the frequency of resonance is

$$\omega = \sqrt{\frac{1}{L_X(C_S+C_X)}} \rightarrow f = \frac{1}{2\pi\sqrt{L_X(C_S+C_X)}}. \tag{10.2}$$

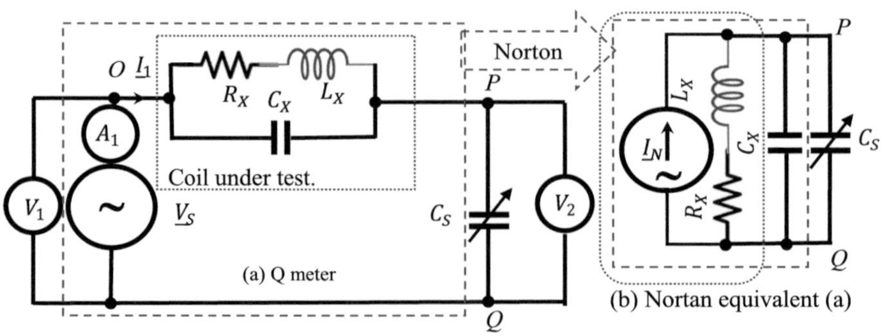

Fig. 10.1 The *Q* meter

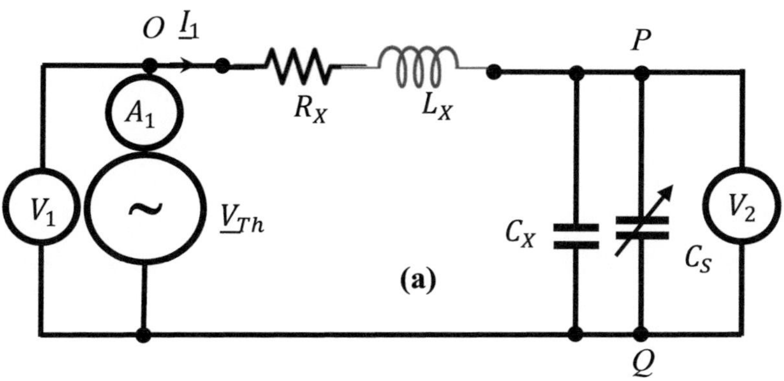

Fig. 10.2 Thevenin equivalent of Fig. 10.1b

10.2.2 Measurements with Q Meter

Measurement procedure for using the Q meter is:

Step 1: First, the measurement frequency f_1 is set on the AC source. Then capacitor C_S is varied till the AC voltmeter V_2 (or ammeter I_1) reads maximum. At that point, the reading of C_S say C_{S1} readings of the ammeter I_1 AC voltmeters V_1 and V_2 say I_{11} V_{11} and V_{21} are read.

Step 2: The measurement frequency f_2 is set as $2f_1$ on the AC source. Once again, capacitor C_S is varied till the AC voltmeter V_2 (or ammeter I_1) reads maximum. At that point, the reading of C_S say C_{S2} readings of the ammeter I_1 say, I_{12} AC voltmeters V_1 and V_2 say, V_{12} and V_{22} are read.

From these values, the value of inductance L_X, the series resistance R_X, the inter-winding (distributed) capacitance C_X, and the quality factor Q_X of the coil (inductor) under test can be determined, as explained next.

With measurements made using step 1, we get

$$f_1 = \frac{1}{2\pi\sqrt{L_X(C_{S1}+C_X)}} \to f_1^2 = \frac{1}{4\pi^2[L_X(C_{S1}+C_X)]}. \tag{10.3}$$

With measurements made using step 2, we get

$$f_2 = 2f_1 = \frac{1}{2\pi\sqrt{L_X(C_{S2}+C_X)}} \to 4f_1^2 = \frac{1}{4\pi^2[L_X(C_{S2}+C_X)]}. \tag{10.4}$$

If we divide Eq. (10.4) by Eq. (10.3) we get

$$\frac{4f_1^2}{f_1^2} = \frac{4\pi^2[L_X(C_{S1}+C_X)]}{4\pi^2[L_X(C_{S2}+C_X)]}. \tag{10.5}$$

Simplifying Eq. (10.5), we get

$$C_X = \frac{(C_{S1} - 4C_{S2})}{3}. \tag{10.6}$$

Thus, we can obtain the value of the distributed capacitance C_X. Once we get the value of C_X, we can determine the value of L_X using Eq. (10.2) as

$$L_X = \frac{1}{4\pi^2 f_1^2 (C_{S1} + C_X)}. \tag{10.7}$$

On similar lines, we can also get the value of L_X from the readings f_2 and C_{S2} as

$$L_X = \frac{1}{4\pi^2 f_2^2 (C_{S2} + C_X)}. \tag{10.8}$$

We obtain the quality factor Q_{meas} of the coil as

$$Q_{\text{meas}} = \frac{V_{21}}{V_{11}} = \frac{V_{22}}{V_{12}}. \tag{10.9}$$

On similar lines, we can also obtain the value of R_X as

$$R_X = \frac{V_{11}}{I_{11}} = \frac{V_{12}}{I_{12}}. \tag{10.10}$$

10.2.3 Corrections to be Applied to Q Meter Measurements

(i) *Correction for compensating the error introduced by the internal resistance of the excitation source.*

In deriving the above Eqs. (10.1)–(10.10), one of the assumptions we made (though not explicitly stated then) is that the source possesses zero internal resistance. In practice, an AC sinusoidal source may possess an internal resistance, say, R_s. In such case, the measured Q_{meas} will be that of the entire circuit shown in Fig. 10.3 and not that of the coil (inductor). In such a scenario, we have

$$Q_{\text{meas}} = \frac{\omega L_x}{(R_X + R_s)} = \frac{\omega L_x}{R_X\left(1 + \frac{R_s}{R_X}\right)}. \tag{10.11}$$

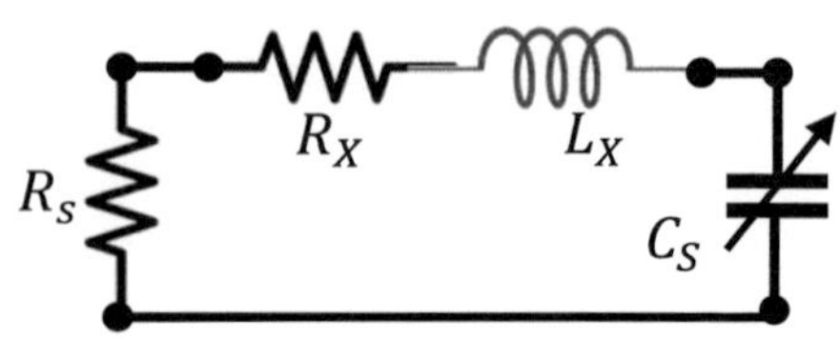

Fig. 10.3 Source-free equivalent circuit of a *Q* meter

Rearranging Eq. (10.11), we get

$$Q_{\text{coil}} = \frac{\omega L_x}{R_X} = \left(1 + \frac{R_s}{R_X}\right)Q_{\text{meas}} = \left(1 + \frac{R_s}{R_X}\right)\left(\frac{V_{21}}{V_{11}}\right). \tag{10.12}$$

We can use $\left(\frac{V_{22}}{V_{12}}\right)$ instead of $\left(\frac{V_{21}}{V_{11}}\right)$. in Eq. (10.12) and get

$$Q_{\text{coil}} = \frac{\omega L_x}{R_X} = \left(1 + \frac{R_s}{R_X}\right)Q_{\text{meas}} = \left(1 + \frac{R_s}{R_X}\right)\left(\frac{V_{22}}{V_{12}}\right). \tag{10.13}$$

Practical oscillators will have appreciable internal resistance (typically 50 Ω) and thus if we directly use an oscillator, we will incur large errors in the measurement of Q. To overcome this problem, a typical Q meter is constructed using the circuit shown in Fig. 10.4. The typical value of R_{sh} in Fig. 10.4 will be in the range of 10 s of milliohm (typically 0.02 Ω). Even with such a small internal resistance for the source, it is advisable to compute the Q of the coil as

$$Q_{\text{coil}} = Q_{\text{meas}}\left(1 + \frac{R_{sh}}{R_X}\right) = \left(1 + \frac{R_{sh}}{R_X}\right)\left(\frac{V_{21}}{V_{11}}\right) OR \left(1 + \frac{R_{sh}}{R_X}\right)\left(\frac{V_{22}}{V_{12}}\right). \tag{10.14}$$

(ii) *Correction for compensating the error introduced by the distributed capacitance C_X on the measured Q of the coil.*

When we compute the Q of the coil using Eq. (10.9), we measure

$$Q_{\text{meas}} = \frac{1}{\omega(C_S + C_X)R_X} = \frac{1}{\omega C_S R_X\left(1 + \frac{C_X}{C_S}\right)}. \tag{10.15}$$

Rearranging Eq. (10.15), we get

$$Q_{\text{coil}} = \frac{1}{\omega C_S R_X} = Q_{\text{meas}}\left(1 + \frac{C_X}{C_S}\right). \tag{10.16}$$

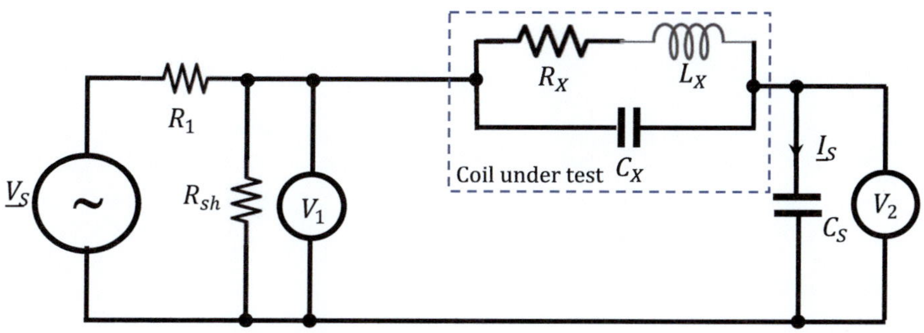

Fig. 10.4 The Q meter circuit using a practical oscillator

Thus, whenever we make measurements, we compute the Q of the coil as

$$Q_{\mathrm{coil}} = \left(\frac{V_{21}}{V_{11}}\right)\left(1 + \frac{R_{sh}}{R_X}\right)\left(1 + \frac{C_X}{C_{S1}}\right) \tag{10.17}$$

OR

$$Q_{\mathrm{coil}} = \left(\frac{V_{22}}{V_{12}}\right)\left(1 + \frac{R_{sh}}{R_X}\right)\left(1 + \frac{C_X}{C_{S2}}\right). \tag{10.18}$$

Though in step 2 we have set the excitation frequency to be twice that of the first step, we need not stick to the factor to be two. We can have any other factor other than one. For example, if we chose:

$$f_{2(3)} = 3f_1. \tag{10.19}$$

Then we have

$$f_{2(3)} = 3f_1 = \frac{1}{2\pi\sqrt{L_X\left(C_{S2(3)}+C_X\right)}} \rightarrow 9f_1^2 = \frac{1}{4\pi^2\left(L_X\left(C_{S2(3)}+C_X\right)\right)}. \tag{10.20}$$

Resulting in:

$$C_X = \frac{\left(C_{S1}-9C_{S2(3)}\right)}{8}. \tag{10.21}$$

The balancing of the Q meter (that is vary C_S and obtain maximum value on V_2 or I_1) and evaluating relevant equations to obtain Q_X, L_X, R_X, and C_X can be automated as in commercial Q meters shown in Fig. 10.5. From the foregone discussions, we see that the measurements using a Q meter involve errors in the measurements that are minimized with suitable corrections. Since the principle of operation of a Q meter itself is based on *estimation*, Q meters are not preferred for making accurate measurements. After the introduction of the impedance analyzer (described in the sequel), the popularity of Q meters vaned.

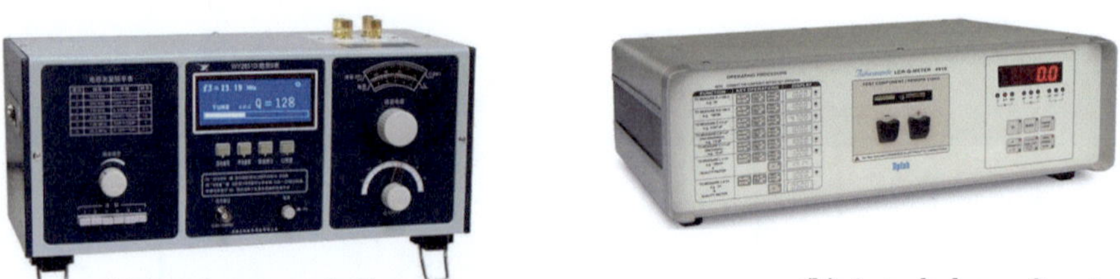

(a) Semi-automatic Q meter Model WY2851D

(b) Auto-balance Q meter Aplab model 4910

Fig. 10.5 Commercial Q meters

10.3 Electronic Wattmeters

Though the electrodynamic type of wattmeter continued to serve the purpose of measuring electrical power, especially at the power system frequency (50 Hz or 60 Hz) for more than a century, several electronic methods of measuring electrical power at power supply frequency (50 Hz or 60 Hz) were developed. Of these methods, except for one analog electronic technique of power measurement and the digital method of measurement of power popularly known as sampling wattmeter, the rest are no longer in vogue.

10.3.1 Power in AC System

While in DC system power can be easily obtained just by multiplying the current I_{DC} by the system voltage V_{DC}, in AC system we have different components of power as given below.

Instantaneous power $p(t)$: If we represent the voltage across a load as $v(t)$, the current is taken by the load as $i(t)$, then the instantaneous power $p(t)$ supplied to the load is

$$p(t) = v(t) \cdot i(t). \tag{10.22}$$

Average power P: Thus, to get instantaneous power we need to multiply the voltage and current signals. Then the average power P (work done per unit of time) can be obtained using Eq. (10.23).

$$P = \int v(t).i(t)\mathrm{d}t. \tag{10.23}$$

The average power P with the current and voltage as repetitive signals of frequency 50 Hz or 60 Hz as in a power system, Eq. (10.23) becomes

$$P = \frac{1}{T}\int_0^T v(t) \cdot i(t)\mathrm{d}t. \tag{10.24}$$

In Eq. (10.24), T is the period of the 50 Hz or 60 Hz as the case may be. Thus, to measure the average or active power in a circuit we need to obtain instantaneous power by multiplying the voltage across and current through the load and integrating the instantaneous power as indicated in the block diagram of Fig. 10.6. Different methods of analog electronic multipliers were developed, of which only the "Time division multiplier" is in use for power measurement. All the analog multipliers were designed to multiply two voltages, say v_1 and v_2. To use these multipliers in an electronic wattmeter to measure instantaneous or average power, the circuit current

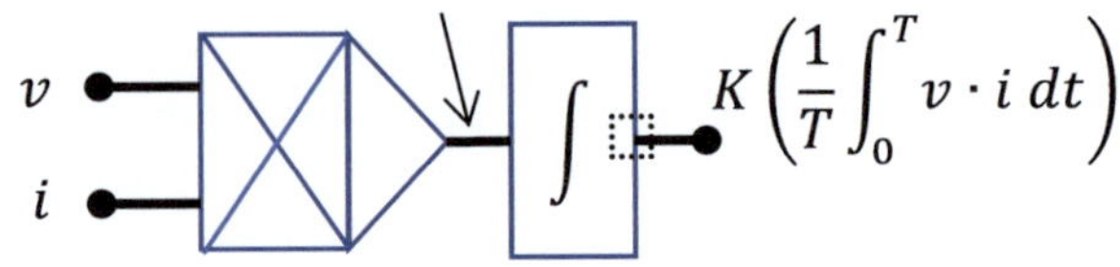

Fig. 10.6 Block diagram of an electronic wattmeter

i is converted into a voltage (say v_2) and the multiplier of two voltages is used. The integrator in Fig. 10.6 is normally realized with a low-pass filter with a cut-off frequency very close to dc, as pure integrating circuits suffer from instability due to offset drift.

10.3.2 The Log-Anti Log Multiplier

The very first analog multiplier developed is the log-antilog multiplier. If we apply logarithm to Eq. (10.22) we get

$$\mathrm{Log}_b p(t) = \mathrm{Log}_b v(t) + \mathrm{Log}_b i(t). \tag{10.25}$$

Thus, if we apply logarithm to base b, multiplication simply becomes the summation. Thus, we can take logarithm $\mathrm{Log}_b v(t)$ of voltage $v(t)$ and logarithm $\mathrm{Log}_b i(t)$ of the current $i(t)$ and add them. Then take the inverse logarithm of the resulting sum to get the multiplication of the voltage and current. In the circuit of Fig. 10.7a, if the diode current $i_D \gg I_o$, where I_o is the reverse saturation current of the diode, then we get

$$i_D = \frac{v_{i1}}{R} \approx I_0 e^{\frac{v_D}{\eta V_T}}. \tag{10.26}$$

In Eq. (10.26), η is called the ideality factor (lies between 1 and 2) and $V_T = \frac{K\theta}{q}$ is the thermal voltage with K being Boltzmann constant, θ is the diode temperature in kelvin and q is the charge of the electron. Rearranging Eq. (10.26) and applying

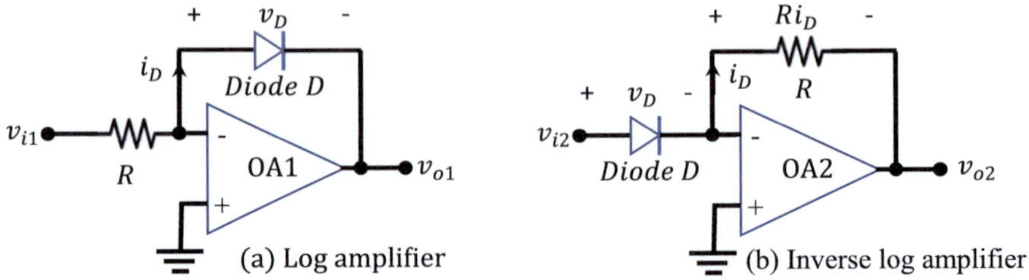

Fig. 10.7 Log-anti log amplifiers

logarithm to the base e, we get

$$v_D = -v_{o1} = \eta V_T \mathrm{Log}_e\left(\frac{v_{i1}}{RI_0}\right). \tag{10.27}$$

Equation (10.27) shows that the output voltage v_{o1} is the logarithm to the base e of the input voltage v_i. In Fig. 10.7b, the output v_{o2} of opamp OA2 is

$$v_{02} = -Ri_D = I_0 e^{\frac{v_{i2}}{\eta V_T}} = K_A \mathrm{Log}_e^{-1}(v_{i2}). \tag{10.28}$$

In Eq. (10.28), the constant $K_A = \left(I_0 e^{\frac{1}{\eta V_T}}\right)$. In a practical log or antilog amplifier we do not use a diode but use the base emitter junction of a bipolar junction transistor as shown in Fig. 10.8. Figure 10.8 shows the block diagram of the IC AD538 which is a log-antilog multiplier integrated circuit (IC) from the semiconductor manufacturer Analog Devices. The AD538 can implement complex functions apart from functioning as a multiplier. Today, log-antilog multipliers are no longer in use.

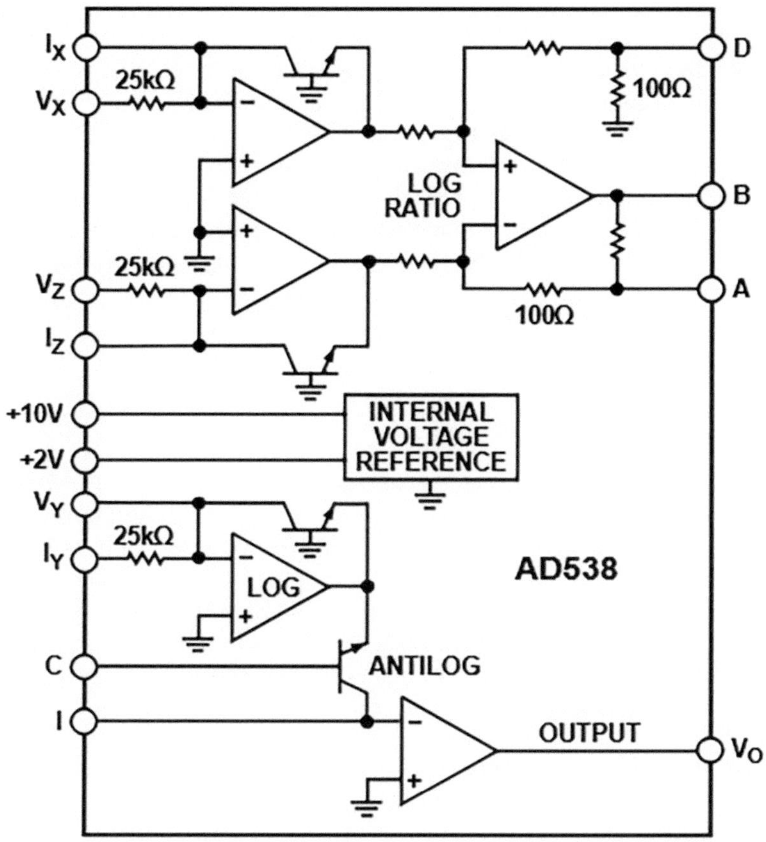

Fig. 10.8 The block diagrams of IC AD538 analog multiplier

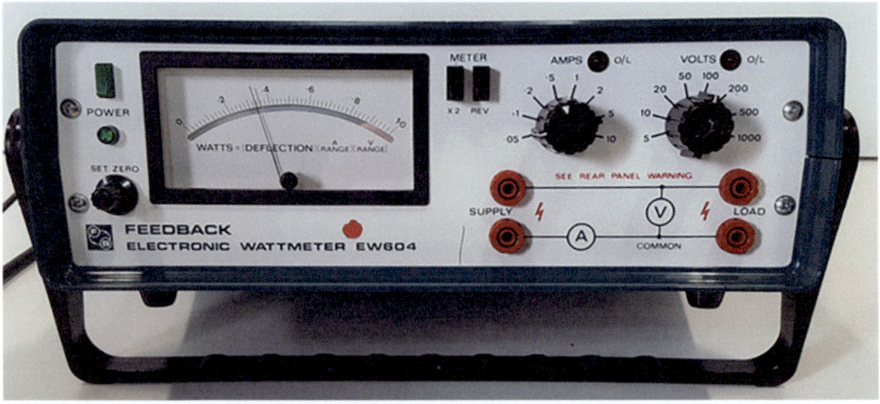

Fig. 10.9 Electronic wattmeter using analog multiplier

10.3.3 The Transconductance (Gilberts) Multiplier

Professor Barrie Gilbert in 1968 invented the now famous Gilbert's transconductance multiplier circuit. Today Gilbert's transconductance multiplier is available in integrated circuit form. While the AD532 IC manufactured by Analog Devices that uses the transconductance multiplier invented by Professor Barrie Gilbert provides an error of $\pm 1.0\%$ over the 75 kHz range, the upgraded AD632 (also uses the transconductance multiplication principle) provides 0.5% accuracy with reduced bandwidth of 10 kHz. While analog multiplier (such as AD532/AD632) based electronic wattmeters shown in Fig. 10.9 were manufactured and sold two decades ago, today these types of wattmeters are not manufactured.

10.3.4 The Time Division Multiplier

Of all the analog multipliers developed, the time division multiplier (TDM) invented by Miyaji Tomata, Takashi Sugiyama, and Keiki Yamaguchi is still in use and will be in use in the foreseeable future, especially for building standard wattmeters of high accuracy. Figure 10.10 illustrates the principle of operation of the time division multiplier. The opamp OA1 is configured as a three-input integrating, inverting summer with the capacitor C_F connected between its inverting input terminal and its output terminal. With the circuit as given in Fig. 10.10, the output v_{oi} of opamp OA1 will be

$$v_{oi} = -\frac{1}{C_F}\int i_T \mathrm{d}t = -\frac{1}{C_F}\int \left(\frac{v_c}{R_c} + \frac{v_X}{R_I} + \frac{\pm V_R}{R_S}\right)\mathrm{d}t. \tag{10.29}$$

Here the signal v_c is a symmetrical square wave clock signal of magnitude $\pm V_c$ and period T_c, v_X is one of the input signals to be multiplied (multiplier input) and $+V_R$ and $-V_R$ are standard DC reference voltages of equal magnitude V_R and

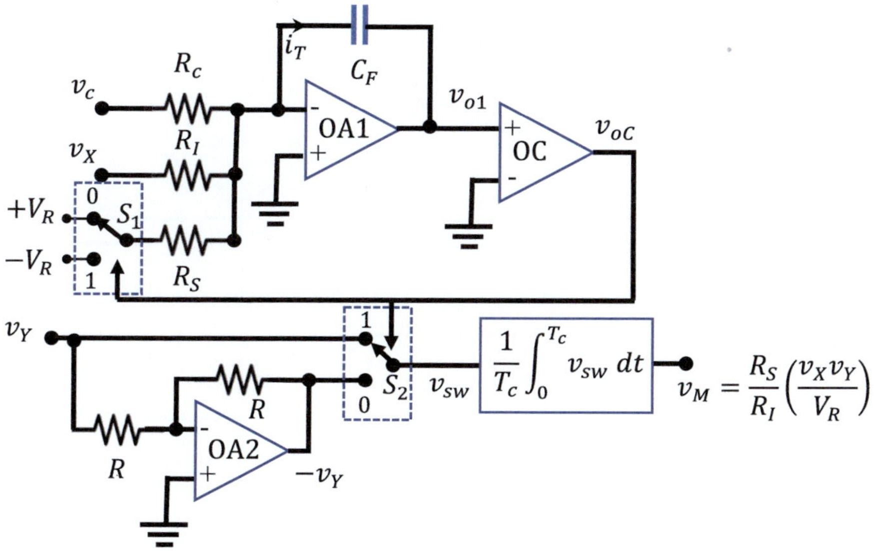

Fig. 10.10 The time division multiplier

of opposite polarity, and one of them is selected by the single pole double throw (SPDT) switch S_1. Input resistances R_c, R_I and R_S connect the inputs v_c, v_X and $\pm V_R$ to the non-inverting terminal of OA1. The output v_{oi} of OA1 is fed as input to the opamp comparator OC which works as a zero-crossing detector. If $v_{oi} \geq 0$ then the output v_{oc} of the comparator will be "1", otherwise v_{oc} will be "0". The output v_{oc} of the comparator OC controls the SPDT switches S_1 and S_2. Switch S_1 connects either $+\mathrm{V_R}$ (whenever v_{oc} = '0') or $-\mathrm{V_R}$ (whenever v_{oc} = '1') to the resistance R_S. Switch S_2 connects either $+v_Y$ (whenever v_{oc}= '1') or $-v_Y$ (whenever v_{oc} = '0') to the integrator, whose output v_M is the final output of the time division multiplier. If we assume that the inputs v_X and v_Y are constant during the clock period T_c, then Eq. (10.29) can be simplified, within a clock period, as

$$v_{oi} = -\frac{1}{C_F}\left(\frac{v_c}{R_c} + \frac{v_X}{R_I} + \frac{\pm V_R}{R_S}\right)t. \tag{10.30}$$

For the multiplier to work properly, the charge gained by C_F over a cycle $\left(\int_0^{T_c} i_T dt\right)$ must be zero. Otherwise, the remnant charge in the capacitor C_F in one clock cycle will accumulate over several clock cycles and hence the output v_{oi} will reach either positive or negative saturation. For $\int_0^T i_T dt = 0$, we must have $V_c > (V_R + |v_X|)$ and the clock must have 50% duty cycle. With these conditions satisfied, the waveforms of the voltages v_c, v_{oi}, v_{oc} and v_{sw} will be as shown in Fig. 10.11. In Fig. 10.11, we see that whenever the output v_{oi} of the three-input integrator OA1 crosses zero, the output v_{oc} of the comparator flips from "0" to "1" or "1" to "0". Thus, connecting either $-V_R$ or $+V_R$ to resistance R_S. In Fig. 10.11, we arbitrarily choose the voltage

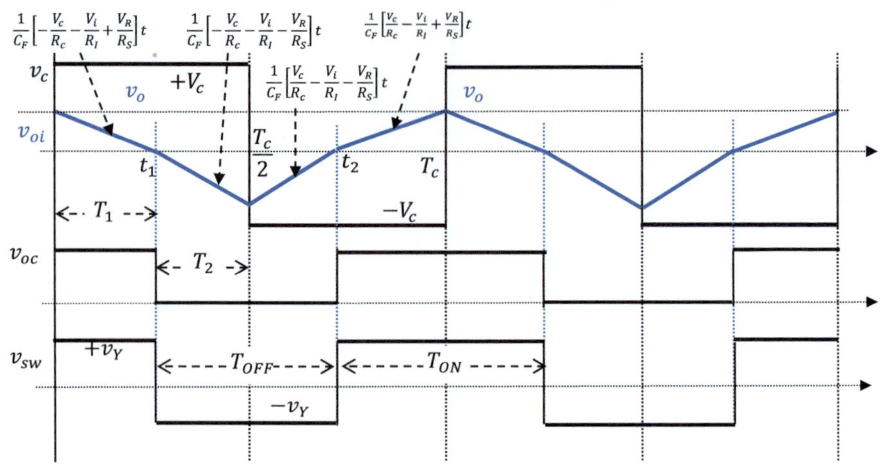

Fig. 10.11 Waveforms of the time division multiplier

v_{oi} to be at some positive value, say v_o. From Fig. 10.11, we see that during the period $t = 0$ to $t = t_1$ (say period T_1 seconds) the output v_{oi} of the integrator slopes down from v_o and reaches zero. Hence, the change in the charge of the capacitor C_F in the period T_1 will be

$$\int_0^{t_1} i_T \,\mathrm{d}t = -\left(\tfrac{+v_c}{R_c} + \tfrac{v_X}{R_I} + \tfrac{-V_R}{R_S}\right)T_1. \quad (10.31)$$

At time t_1, output v_{oc} of the comparator OC flips from “1” to “0” and thus switch S_1 connects $+V_R$ to resistance R_S. This condition is maintained till time t_2. From end of time t_1 till time $T_c/2$ (say period T_2 seconds), the charge in the capacitor C_F changes by

$$\int_{t_1}^{T_c/2} i_T \,\mathrm{d}t = -\left(\tfrac{+v_c}{R_c} + \tfrac{v_X}{R_I} + \tfrac{+V_R}{R_S}\right)T_2. \quad (10.32)$$

Since between time $t = 0$ to $t = T_c/2$, the output v_{oi} (and hence the voltage across the capacitor C_F) is changing with a negative slope, the change in the charge of the capacitor C_F is taken as negative. During the period $T_c/2$ to t_2 (say period T_3 seconds), the charge accumulated by the capacitor C_F is

$$\int_{T_c/2}^{t_2} i_T \mathrm{d}t = \left(\tfrac{-v_c}{R_c} + \tfrac{v_X}{R_I} + \tfrac{+V_R}{R_S}\right)T_3. \quad (10.33)$$

At time t_2, the output v_{oi} of the integrator turns positive and hence the output of the comparator v_{oc} flips from “0” to “1” and forces the switch S_1 to connect $-V_R$ to resistance R_S. Thus, during the period t_2 to T_c (say period T_4 seconds), the charge gained by the capacitor C_F is

$$\int_{t_2}^{T_c} i_T \mathrm{d}t = \left(\tfrac{-v_c}{R_c} + \tfrac{v_X}{R_I} + \tfrac{-V_R}{R_S}\right)T_4. \tag{10.34}$$

Since between time $t = T_c/2$ to $t = T_c$, the output v_{oi} (and hence the voltage across the capacitor) is changing with a positive slope, the charge accumulated in the capacitor is taken as positive. Combining Eqs. (10.31)–(10.34) and using the condition $\int_0^T i_T\,\mathrm{d}t = 0$, that is the net charge accumulated by the capacitor C_F over a period T_c must be zero, we get

$$\begin{aligned} &-\left(\frac{v_c}{R_c} + \frac{v_X}{R_I} - \frac{V_R}{R_S}\right)T_1 - \left(\frac{v_c}{R_c} + \frac{v_X}{R_I} + \frac{V_R}{R_S}\right)T_2 \\ &+\left(-\frac{v_c}{R_c} + \frac{v_X}{R_I} + \frac{V_R}{R_S}\right)T_3 + \left(-\frac{v_c}{R_c} + \frac{v_X}{R_I} - \frac{V_R}{R_S}\right)T_4 = 0. \end{aligned} \tag{10.35}$$

Rearranging Eq. (10.35) we get

$$\begin{aligned} &\frac{v_c}{R_c}[(T_1 + T_2) - (T_3 + T_4)] \\ &+\frac{v_X}{R_I}(T_1 + T_2 + T_3 + T_4) + \frac{V_R}{R_S}(-T_1 + T_2 + T_3 - T_4) = 0. \end{aligned} \tag{10.36}$$

Since $(T_1 + T_2) = (T_3 + T_4)$ we get

$$\tfrac{v_c}{R_c}[(T_1 + T_2) - (T_3 + T_4)] = 0. \tag{10.37}$$

Hence Eq. (10.36) becomes

$$\tfrac{v_X}{R_I}(T_1 + T_2 + T_3 + T_4) = \tfrac{V_R}{R_S}[(T_1 + T_4) - (T_2 + T_3)]. \tag{10.38}$$

If we take $(T_1 + T_4) = T_{\mathrm{ON}}$ (that is the period in which the output v_{oc} of the comparator is "1") and the period $(T_2 + T_3) = T_{\mathrm{OFF}}$ (that is the period in which the output of the comparator is "0") then Eq. (10.38) reduces to

$$\tfrac{(T_{\mathrm{ON}} - T_{\mathrm{OFF}})}{(T_{\mathrm{ON}} + T_{\mathrm{OFF}})} = \left(\tfrac{R_S}{R_I V_R}\right)v_X. \tag{10.39}$$

During the period T_{ON} (that is v_{oc} is "1") the output v_{sw} of switch S_2 is $+v_{\mathrm{Y}}$ and during the period T_{OFF} (that is v_{oc} is "0") the output v_{sw} of switch S_2 is $-v_{\mathrm{Y}}$. Hence, the output v_M of the integrator will be

$$v_M = \tfrac{1}{T_c}\int_0^{T_c} v_{sw} dt = \tfrac{1}{T_c}\left[+v_Y t|_{T_{\mathrm{ON}}} - v_Y t|_{T_{\mathrm{Off}}}\right] = \tfrac{(T_{\mathrm{ON}} - T_{\mathrm{OFF}})}{T_c} v_Y. \tag{10.40}$$

Simplifying Eq. (10.40) with $\mathrm{T_c} = (T_{\mathrm{ON}} + T_{\mathrm{OFF}})$, we get

$$v_M = \tfrac{(T_{\mathrm{ON}} - T_{\mathrm{OFF}})}{(T_{\mathrm{ON}} + T_{\mathrm{OFF}})} v_Y. \tag{10.41}$$

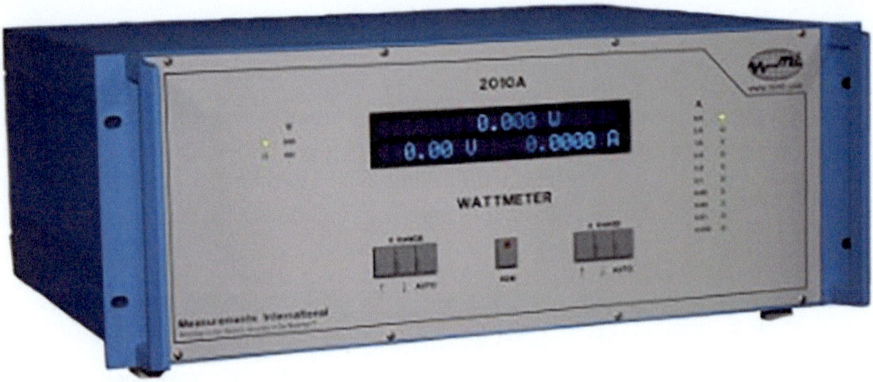

Fig. 10.12 The time division multiplier standard wattmeter model 2010

Substituting Eq. (10.39) in Eq. (10.41) we get

$$v_M = \left(\frac{R_S}{R_I}\right)\frac{v_X v_Y}{V_R}. \tag{10.42}$$

Equation (10.42) illustrates that the output is the product of the inputs v_X and v_Y. Since the other terms in Eq. (10.42), namely, R_S, R_I and V_R can be made to have very high accuracy and thus the time division multiplier can be made to possess very high accuracy. Hence almost all standard wattmeters of very high accuracy use the TDM principle. It should be noted that the diagram shown in Fig. 10.10 is a simple one just to illustrate the principle of operation of the TDM. A practical TDM internal circuit will be more complex than the one shown in Fig. 10.10. Figure 10.12 shows the model 2010 standard wattmeter of TDM type from Measurements International that has an accuracy of ± 50 ppm.

10.3.5 The Digital Sampling Wattmeter

With the advent of personal computers, digital technology is fast replacing analog measurement techniques. On these lines, digital sampling wattmeters are becoming more and more popular. In the digital sampling wattmeters, as the name indicates the voltage and current signals are sampled using a sample and hold circuit and converted to digital using analog to digital converter. The samples are then multiplied, and the product is digitally integrated to obtain power. Figure 10.13 illustrates the block schematic of a typical digital sampling wattmeter. The voltage and current signals are suitably scaled so that the magnitude of these signals becomes compatible with the analog to digital converters. The signals are then sampled simultaneously using the sample and hold circuit, and converted into digital samples, say, V_j and I_j by the ADCs. Initially, only one ADC with an analog multiplexer was employed. As the cost of ADCs reduced, most digital sampling wattmeters today use individual ADCs one each for the current and the voltage. Most digital wattmeters are designed to take N

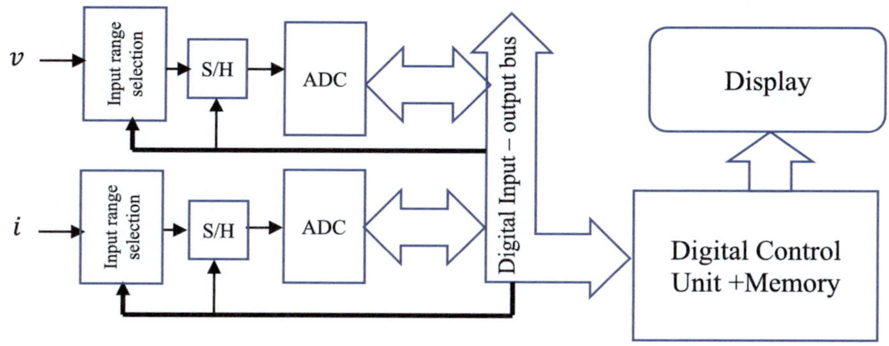

Fig. 10.13 Block diagram of a digital sampling wattmeter

equidistance samples (called synchronous sampling) within a cycle (normally 50 Hz or 60 Hz) of the voltage and current signals. The average power is then computed as

$$P_A = \frac{1}{N}\sum_{j=1}^{N} V_j I_j W. \tag{10.43}$$

Comparison of Eq. (10.43) with Eq. (10.24) illustrates that the integration in the analog domain is implemented using summation in the digital domain. The advantage of a digital wattmeter is that we can easily measure the reactive power Q_A by employing Eq. (10.44).

$$Q_A = \frac{1}{N}\sum_{j=1}^{N} V_j I_{(j+\frac{N}{4})} VAr. \tag{10.44}$$

Once we have the voltage and current samples V_j and I_j, the RMS values of the voltage V_{rms} and current I_{rms} can be obtained easily by evaluating Eqs. (10.45) and (10.46).

$$V_{\mathrm{rms}} = \sqrt{\frac{1}{N}\sum_{j=1}^{N} V_j^2}\, V \tag{10.45}$$

$$I_{\mathrm{rms}} = \sqrt{\frac{1}{N}\sum_{j=1}^{N} I_j^2}\, A. \tag{10.46}$$

The apparent power $\Re$ can be obtained by multiplying V_{rms} and I_{rms} as

$$\Re = V_{\mathrm{rms}} I_{\mathrm{rms}} VA \tag{10.47}$$

Apart from these, a digital sampling wattmeter can also obtain the fundamental component $V_F\angle\theta_{VF}$ and $I_{\mathrm{F}}\angle\theta_{IF}$ second harmonic component $V_2\angle\theta_{V2}$ and $I_2\angle\theta_{I2}$, third harmonic component $V_3\angle\theta_{V3}$ and $I_3\angle\theta_{I3}$ and so on of the voltage and current. To determine the harmonic components of, say the voltage, first, we find the in-phase component V_{FP} and the quadrature component V_{FQ} of the fundamental component V_F as

$$V_{FP} = \frac{1}{N}\left(\sum_{j=1}^{N} V_j \mathrm{Cos}\left(\frac{2\pi j}{N}\right)\right) \text{ and } V_{FQ} = \frac{1}{N}\left(\sum_{j=1}^{N} V_j \mathrm{Sin}\left(\frac{2\pi j}{N}\right)\right). \tag{10.48}$$

From V_{FP} and V_{FQ} we can obtain V_F and θ_F as

$$V_F = \sqrt{\left(V_{FP}^2 + V_{FQ}^2\right)} \text{ and } \theta_F = \mathrm{Tan}^{-1}\left(\frac{V_{FQ}}{V_{FP}}\right). \tag{10.49}$$

Similarly, the in-phase component V_{2P} and the quadrature component V_{2Q} of the second harmonic of the voltage is obtained as

$$V_{2P} = \frac{1}{2N}\left(\sum_{j=1}^{N} V_j \mathrm{Cos}\left(\frac{4\pi j}{N}\right)\right) \text{ and } V_{2Q} = \frac{1}{2N}\left(\sum_{j=1}^{N} V_j \mathrm{Sin}\left(\frac{4\pi j}{N}\right)\right). \tag{10.50}$$

From V_{2P} and V_{2Q} we can obtain the second harmonic magnitude V_2 and phase angle θ_2 as

$$V_2 = \sqrt{\left(V_{2P}^2 + V_{2Q}^2\right)} \text{ and } \theta_2 = \mathrm{Tan}^{-1}\left(\frac{V_{2Q}}{V_{2P}}\right). \tag{10.51}$$

In these lines, the in-phase component V_{nP} of the nth harmonic and the quadrature component V_{nQ} of the voltage is obtained as

$$V_{nP} = \frac{1}{nN}\left(\sum_{j=1}^{N/2} V_j \mathrm{Cos}\left(\frac{2n\pi j}{N}\right)\right) \text{ and } V_{nQ} = \frac{1}{nN}\left(\sum_{j=1}^{N} V_j \mathrm{Sin}\left(\frac{2n\pi j}{N}\right)\right). \tag{10.52}$$

From V_{nP} and V_{nQ} we can obtain the nth harmonic magnitude V_n and phase angle θ_n as

$$V_n = \sqrt{\left(V_{nP}^2 + V_{nQ}^2\right)} \text{ and } \theta_n = \mathrm{Tan}^{-1}\left(\frac{V_{nQ}}{V_{nP}}\right). \tag{10.53}$$

If we replace V by I in Eqs. (10.48)–(10.53) we will get the fundamental and harmonic components of the current as given in Eqs. (10.54)–(10.57).

$$I_{FP} = \frac{1}{N}\left(\sum_{j=1}^{N} I_j \mathrm{Cos}\left(\frac{2\pi j}{N}\right)\right) \text{ and } I_{FQ} = \frac{1}{N}\left(\sum_{j=1}^{N} I_j \mathrm{Sin}\left(\frac{2\pi j}{N}\right)\right) \tag{10.54}$$

$$I_F = \sqrt{\left(V_{FP}^2 + V_{FQ}^2\right)} \text{ and } \theta_F = \mathrm{Tan}^{-1}\left(\frac{I_{FQ}}{I_{FP}}\right). \tag{10.55}$$

The nth harmonic in-phase component I_{nP} and the quadrature component I_{nQ} of the current is obtained as

$$I_{nP} = \frac{1}{nN}\left(\sum_{j=1}^{N/2} I_j \,\mathrm{Cos}\left(\frac{2n\pi j}{N}\right)\right) \text{ and } I_{nQ} = \frac{1}{nN}\left(\sum_{j=1}^{N} I_j \,\mathrm{Sin}\left(\frac{2n\pi j}{N}\right)\right). \tag{10.56}$$

From I_{nP} and I_{nQ} we can obtain the nth harmonic magnitude I_n and phase angle θ_{In} as

$$I_n = \sqrt{\left(I_{nP}^2 + I_{nQ}^2\right)} \text{ and } \theta_{In} = \mathrm{Tan}^{-1}\left(\frac{I_{nQ}}{I_{nP}}\right). \tag{10.57}$$

Once we get the fundamental and harmonic components of the voltage $V_F\angle\theta_F$ and $V_n\angle\theta_n$ and current $I_F\angle\theta_F$ and $I_n\angle\theta_n$, the fundamental components of the average power P_F, reactive power Q_F and apparent power $\Re_F$ of the fundamental component can be obtained as

$$P_F = V_F I_F \mathrm{Cos}(\theta_F - \theta_{IF}),\ Q_F = V_F I_F \mathrm{Sin}(\theta_F - \theta_{IF}) \text{ and } \Re_F = V_F I_F. \tag{10.58}$$

Similarly, the average power P_n, reactive power Q_n and apparent power $\Re_n$ of the nth harmonic component can be computed using Eq. (10.46):

$$P_n = V_n I_n \mathrm{Cos}(\theta_n - \theta_{In}),\ Q_n = V_n I_n \mathrm{Sin}(\theta_n - \theta_{In}) \text{ and } \Re_n = V_n I_n. \tag{10.59}$$

Thus, a digital sampling wattmeter can not only measure the power, but also can provide harmonic components of the voltage, current, and power (active, reactive, and apparent). Such sampling type digital wattmeters that can measure harmonics are called the power quality analyzers. Figure 10.14 illustrates a typical power quality analyzer that can measure voltage, current, power, reactive power, and apparent power up to the 50th harmonics.

10.4 The Electronic Energy Meter

The induction type energy meter was in use for more than a century and is now replaced by the electronic energy meter. In general, energy meters (induction type or electronic type) are classified as:

(i) Single-phase energy meter.
(ii) Three-phase, three-wire (two-element) energy meter.
(iii) Three-phase, four-wire (three-element) energy meter.

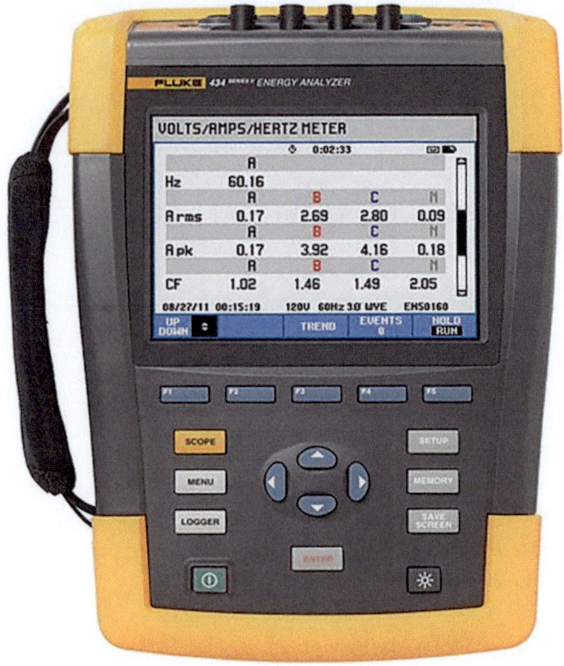

Fig. 10.14 Power quality analyzer

Apart from these classifications, the electronic energy meter is of two distinct types (based on how the power is first computed) as given below.

(i) Electronic energy meter that uses analog multiplier.
(ii) Digital sampling energy meter that uses samples of current(s) and voltage(s) and digital multiplier.

10.4.1 Electronic Energy Meter Using an Analog Multiplier

Figure 10.15 shows the block schematic of a low-cost electronic single-phase energy meter wherein the power is computed using an analog multiplier-integrator and the energy is obtained using a voltage to frequency converter cascaded to a counter. The integration of the power to obtain energy is implemented using the "*Integration by Summation*" principle, with the counter providing the *summation.* We see in Fig. 10.15 (and Fig. 10.16) that the voltage and current levels are scaled to levels compatible with the electronic circuitry of the energy meter using a voltage transformer and a current transformer respectively.

Figure 10.16 shows the block schematic of a two-element electronic energy meter utilizing analog multiplier and is suitable for three-phase three-wire systems. A three-phase, four-wire (three-element) energy meter will combine three single-phase elements shown in Fig. 10.15 and hence can provide individual (phase) energy as

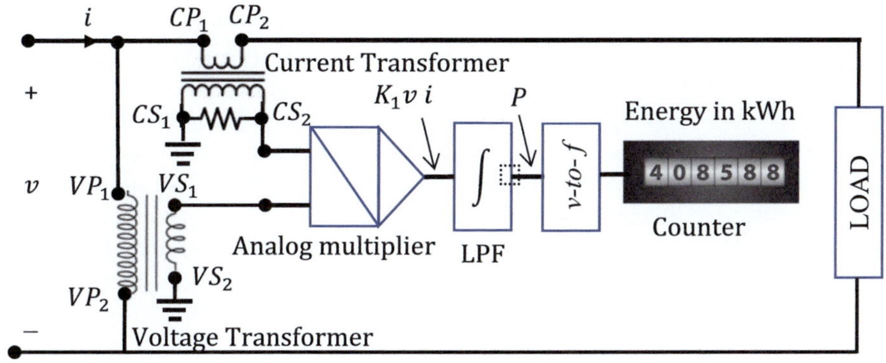

Fig. 10.15 Block diagram of a single-phase electronic energy meter

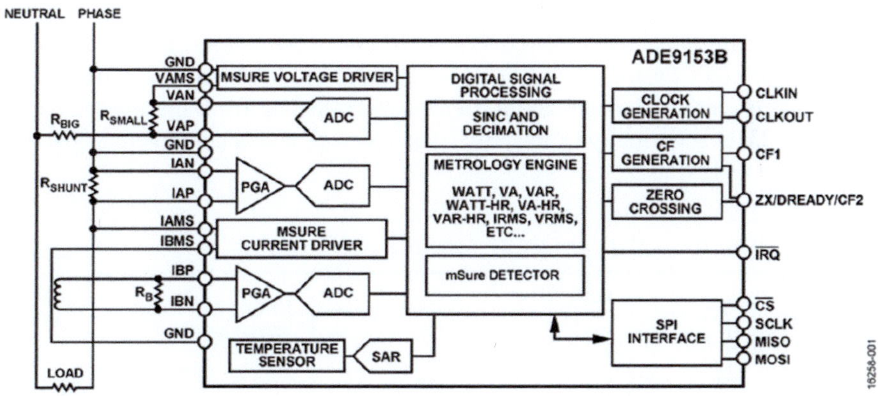

Fig. 10.16 Block diagram of a two-element three-phase electronic energy meter using analog multipliers

well as total energy. Since these energy meters provide only active power and active energy, this type of energy meter is not popular and is no longer manufactured. However, many such meters have been installed in the past and thus are currently in use. The digital energy meters, especially the digital sampling energy meters are replacing the analog ones. Once we get the samples of the voltage(s) and current(s) it is quite simple to compute all the parameters such as active power, reactive power, apparent power, and even harmonic distortion if present in the voltage(s) or current(s).

10.4.2 *Electronic Energy Meter Using Sampling Wattmeter*

Electronic energy meters that obtain the digital samples of the voltage(s) and current(s) and all required quantities are computed using these digital samples (as detailed in the sampling digital wattmeter of Fig. 10.13). Once the samples of voltage and current are obtained, the required quantities can be computed using Eqs. (10.43)–(10.59). The active power is simply accumulated to obtain active energy. Since the sampling type energy meter provides all required parameters, this type of energy meter is quite popular. Several integrated circuit manufacturers have designed application specific integrated circuit (ASIC) for building an electronic energy meter using this principle. For example, Fig. 10.17 shows the block diagram of IC ADE 9153 manufactured by Analog Devices to build a single-phase electronic energy meter based on sampling principle. This sampling energy meter is capable of not only measuring the active energy (kWh), but can also provide active power (kWh), apparent power (kVA), reactive power (kVAR), RMS voltage, RMS current, and power quality. Many of the digital energy meters also have a provision to remotely read them using one or more of the standard protocols. One such open standard protocol is called the Device Language Message Specification—Companion Specification for Energy Metering (DLMS-COSEM).

Apart from the parameters listed in Eqs. (10.43)–(10.59), another parameter called the Total Harmonic Distortion (THD) in voltage and in current provides information on to what extent the voltage and current signals deviate from the expected sine wave. The THD is normally expressed as a percentage and is computed as given in Eqs. (10.60) and (10.61).

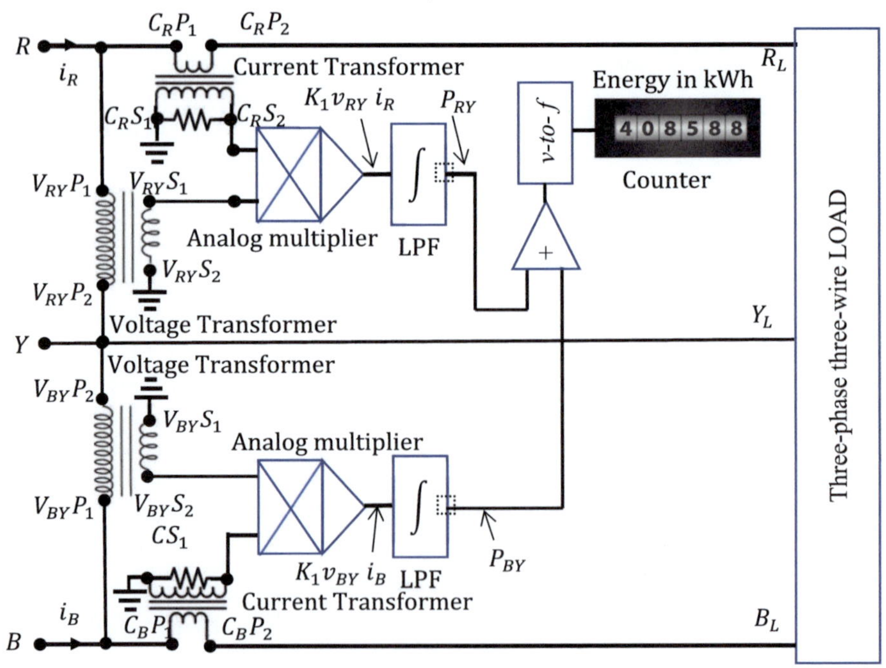

Fig. 10.17 Block diagram of a sampling type single-phase electronic energy meter IC ADE9153B

$$\%\text{THD in the current} = \tfrac{(I-I_F)}{I}100. \tag{10.60}$$

$$\%\text{THD in the voltage} = \tfrac{(V-V_F)}{V}100. \tag{10.61}$$

In Eq. (10.60) I is the RMS value of the line/phase current and I_F is the RMS value of the fundamental component of I. Similarly, V is the RMS value of the line/phase voltage and V_F is the RMS value of the fundamental component of V in Eq. (10.61). As in the sampling wattmeter, the sampling type energy meter can also extract harmonic components of voltage, current, power and energy.

10.5 The Lock-In Amplifier

Another, quite popular instrument for measuring signals, especially very low signals buried in noise, is the "lock-in amplifier". Figure 10.18 shows the block diagram of a typical lock-in amplifier. A typical lock-in amplifier contains a couple of phase sensitive detectors (PSD) one working as the in-phase PSD and the other quadrature PSD. The working principle of a PSD (multiplier type and synchronous type PSDs) was explained in Sect. 4.5.2.3. As seen in Fig. 10.18, the input signal is corrupted by the additive noise $\overline{v_n}$ as given in Eq. (10.62).

$$v_i = V_i\text{Sin}(\omega t + \theta) + \overline{v_n} \tag{10.62}$$

The PSD has a reference signal v_{RP} as

$$v_{RP} = V_R\text{Sin}(\omega t) \tag{10.63}$$

As we have already seen, the IPSD will provide an output V_{oP} as in Eq. (10.64).

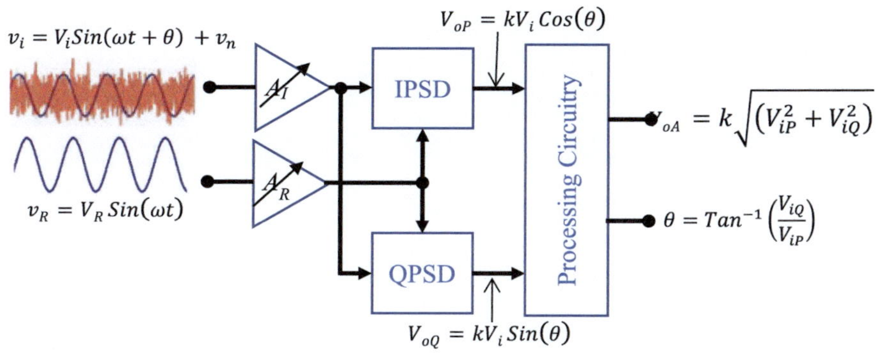

Fig. 10.18 The lock-in amplifier

$$V_{oP} = kV_i\text{Cos}(\theta) \tag{10.64}$$

The output V_{oQ} of the QPSD that has the reference $v_{RQ} = V_R Cos(\omega t)$

$$V_{oQ} = kV_i\text{Sin}(\theta) \tag{10.65}$$

The processing circuitry then derives the magnitude V_i and phase θ of the input signal. From the above discussions it is seen that the lock-in amplifier possesses the ability to reject the noise and selectively detect the input sinusoid. Some lock-in amplifier manufacturers provide a tunable bandpass filter at the input to enhance the signal to noise ratio. Most lock-in amplifiers are designed to detect input signals of magnitude at the nanovolt level. While the PSDs of yesteryear lock-in amplifiers are of the analog type, today most lock-in amplifiers are of the digital sampling kind. Here, the input is simply sampled at very high sampling rate (typically in hundreds of Msa/s) and the IPSD and QPSD are implemented in the digital domain as

$$V_{op} = \frac{1}{N}\left(\sum_{j=1}^{N} V_{in}\,\text{Cos}\left(\frac{2\pi j}{N}\right)\right) \text{ and } V_{oq} = \frac{1}{N}\left(\sum_{j=1}^{N} I_j\text{Sin}\left(\frac{2\pi j}{N}\right)\right). \tag{10.66}$$

In Eq. (10.66), N is the number of samples in one cycle of the input V_{in}, V_{op} is the output of IPSD and V_{oq} is the output of the QPSD. Figure 10.19 shows a typical commercial lock-in amplifier, model SR865A from Stanford Research Systems. Because of its ability to reject noise and detect the single frequency sinusoidal signal, the lock-in amplifier finds application in detecting signals buried in excessive additive noise.

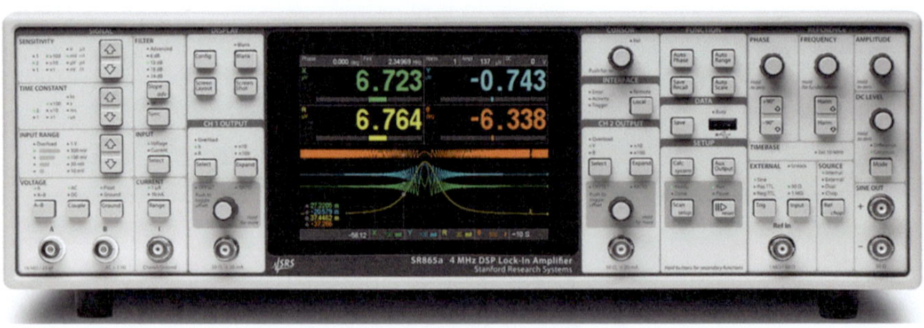

Fig. 10.19 The SARA 865A lock-in amplifier (stanford research)

10.6 The Impedance Analyzer

Like the lock-in amplifier, the impedance analyzer also uses in-phase (IPSD) and quadrature-phase sensitive detectors (QPSD) as seen in Fig. 10.20. The difference is that the impedance analyzer uses three PSD blocks (two IPSDs and one QPSD) and a variable frequency, variable voltage source $v_s = \sqrt{2}\,V_s\,\mathrm{Sin}(\omega t)$ V as shown in Fig. 10.18. The unknown element X is excited by the sinusoidal voltage v_s and the current i_X of the unknown element X is then converted to a voltage v_{iX} using an opamp current to voltage converter, whose output $v_{iX} = R_s i_X$. Here R_s is the resistance connected between the output and the inverting input of opamp OA as shown in Fig. 10.20. The impedance analyzer can be set in two modes of operation, namely,

(i) The impedance measurement mode and
(ii) The admittance measurement mode.

(1) **The impedance measurement mode**:

In the impedance measurement mode, the unknown element X is assumed to be a resistance R_X in series with a reactance X_X (can be inductive reactance and hence positive X_X or capacitive reactance and hence negative X_X) in series as indicated in Fig. 10.20. In such a case we have

$$\underline{V}_X = V_s\angle 0 \text{ and } \underline{I}_X = I_X\angle\theta = \frac{V_s\angle 0}{R_X + jX_X}. \tag{10.67}$$

In Eq. (10.67) $\underline{V}_X$ is the phasor of the voltage across the unknown element X and $\underline{I}_X$ is the phasor current flowing through X. Rearranging Eq. (10.58) we get

$$R_X + jX_X = \frac{V_s\angle 0}{I_X\angle\theta} = \frac{V_s\mathrm{Cos}\theta}{R_s I_X} - \frac{V_s\mathrm{Sin}\theta}{R_s I_X}. \tag{10.68}$$

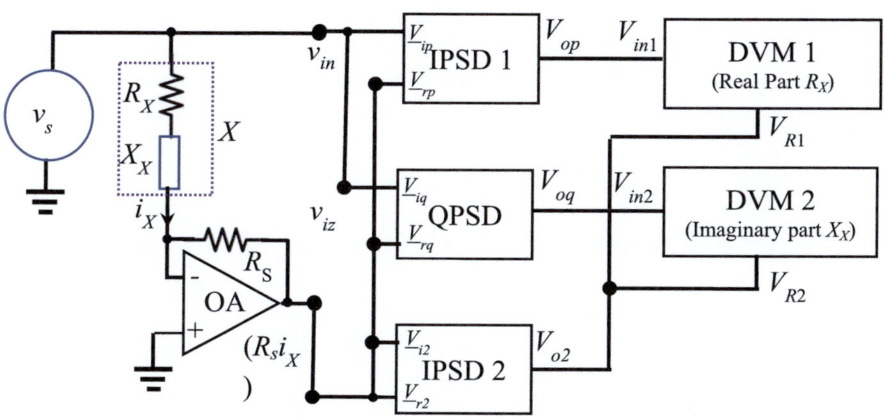

Fig. 10.20 Impedance analyzer in the impedance mode

Equation (10.68) illustrates that if we resolve the voltage phasor $\underline{V}_X$ into its in-phase component $V_s\text{Cos}\theta$ and quadrature component $V_s Sin\theta$ with respect to the current phasor $\underline{I}_X$, series equivalent R_X and reactance X_X can be determined. To obtain this, the inputs to the IPSD 1 and the QPSD are chosen as the excitation voltage v_s and v_{iz} is taken as the reference voltage v_R for both the PSDs as shown in Fig. 10.18. With this connection the output V_{op} of IPSD 1 and output V_{oq} of the QPSD will be

$$V_{op} = \frac{2\sqrt{2}V_s\text{Cos}\theta}{\pi} \text{ and } V_{oq} = \frac{2\sqrt{2}V_s\text{Sin}\theta}{\pi}. \tag{10.69}$$

The input v_{i2} and the reference v_{r2} of the IPSD 2 is $\underline{I}_X$ and hence its output V_{o2} is

$$V_{o2} = \frac{2\sqrt{2}R_sI_X}{\pi}. \tag{10.70}$$

The output V_{op} of the IPSD 1 is given as input $V_{\text{in}1}$ to the digital voltmeter DVM1 and its reference V_{R1} is tied to the output V_{o2} of IPSD 2. Hence DVM1 will read

$$\text{DVM1} = 1000\frac{V_{\text{in}1}}{V_{R1}} = 1000\frac{V_s\text{Cos}\theta}{R_sI_X} = \frac{1000}{R_s}R_X. \tag{10.71}$$

Similarly, DVM2 has its input $V_{\text{in}2} = V_{oq}$ and its reference $V_{R2} = V_{o2}$, hence DVM2 will read

$$\text{DVM2} = 1000\frac{V_{\text{in}2}}{V_{R2}} = 1000\frac{V_s\text{Sin}\theta}{R_sI_X} = \frac{1000}{R_s}X_X. \tag{10.72}$$

It is assumed that the DVMs are 3½ digit DVMs. If we use a 4½ digit DVM then the multiplication factor will be 10,000 instead of 1000.

(ii) **The admittance measurement mode**:

In the admittance measurement mode of operation of an impedance analyzer, the unknown element is considered as the parallel combination of conductance G_X and susceptance B_X as indicated in Fig. 10.21. Here the phasor of voltage v_X is taken as the reference sinusoid and the in-phase component I_{PX} and the quadrature component I_{QX} of the current i_X are obtained using the IPSD and QPSD respectively. From I_{PX} the I_{QX}, the conductance G_{XP} and the susceptance B_{XP} are computed as

$$G_{XP} = \frac{I_{XP}}{V_X} \text{ and } B_{XP} = \frac{I_{XQ}}{V_X}. \tag{10.73}$$

With these modifications, the 3½ digit DVMs, namely, DVM1 and DVM2 will read

$$\text{DVM1} = 1000R_s\,G_X \text{ and DVM2} = 1000R_sB_X. \tag{10.74}$$

Once again as in Eqs. (10.73) and (10.74) too it is assumed that the DVM1 and DVM2 are 3½ digit DVMs. Thus, in the admittance measurement mode, the voltage-current relationship provides a parallel equivalent circuit comprising a resistance

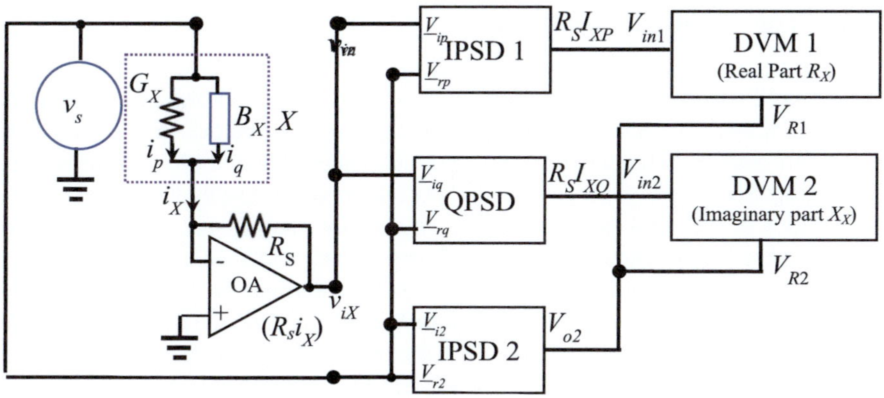

Fig. 10.21 Impedance analyzer in the admittance mode

G_{XP} in parallel with a capacitive or inductive susceptance B_{XP}. As in the lock-in amplifier, here too the PSDs can be of analog type or digital sampling type. Today impedance analyzers can obtain these values and plot the impedance or admittance as a function of frequency. Figure 10.22 shows the impedance analyzer model E4990A from Keysight that has a working bandwidth of 20 Hz to 120 MHz.

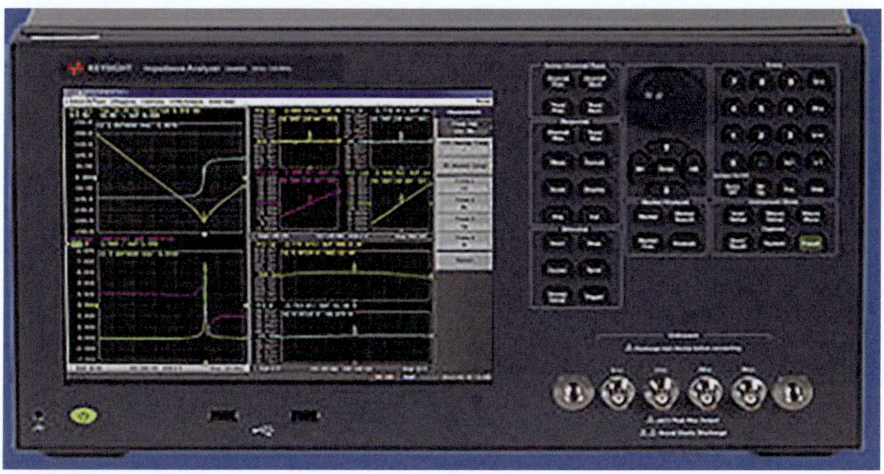

Fig. 10.22 Keysight model E4990A impedance analyzer

10.7 The Network Analyzer

The network analyzer can be a scalar network analyzer (SNA) or a vector network analyzer (VNA). The SNAs deal only with magnitudes of voltages and currents in a two-port network whereas the VNAs provide magnitude and phase information. Thus, today SNAs are extinct and only VNAs are in vogue. The underlying principle of operation of the VNA is the same as that of an impedance analyzer, namely, the voltage and or current phasors are resolved into in-phase and quadrature components. Once we have the voltage and current phasors in the coordinate (a + jb) form it is easy to convert them into polar form ($V\angle\theta$) format. While impedance analyzers handle one port element (two connection points), a VNA handles a two-port element as shown in Fig. 10.23. The input and output ports' voltages and currents are resolved into phasors and from the measurements, the network impedances (or s-parameters) of the two-port network are obtained. It is possible to completely automate the measurements and plotting of the characteristics using either an internal source or an external source. Examples of two-port elements are:

(i) Connectors,
(ii) cables,
(iii) attenuators, and
(iv) amplifiers.

Most VNAs can obtain these values in a range of frequency f. Here f is varied and at cardinal values of f, the measurements are made. The data is then processed and plotted as a graph as shown in Fig. 10.24. The commercial VNA from Rohde & Schwartz is shown in Fig. 10.24. While impedance analyzers work in the range of few kilohertz to few tens of megahertz, VNAs work in the RF range (few kilohertz to few gigahertz).

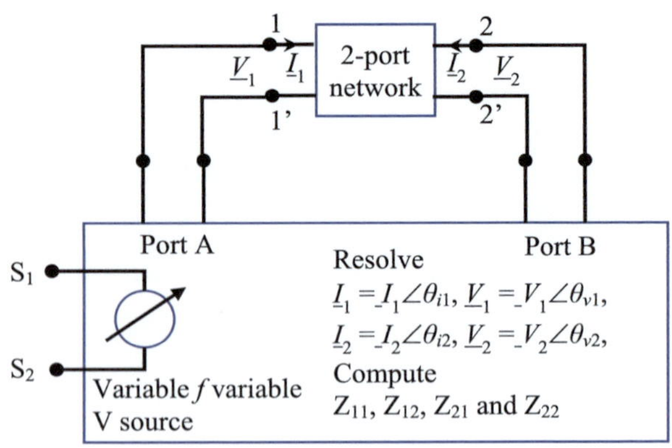

Fig. 10.23 Block diagram of the vector network analyzer

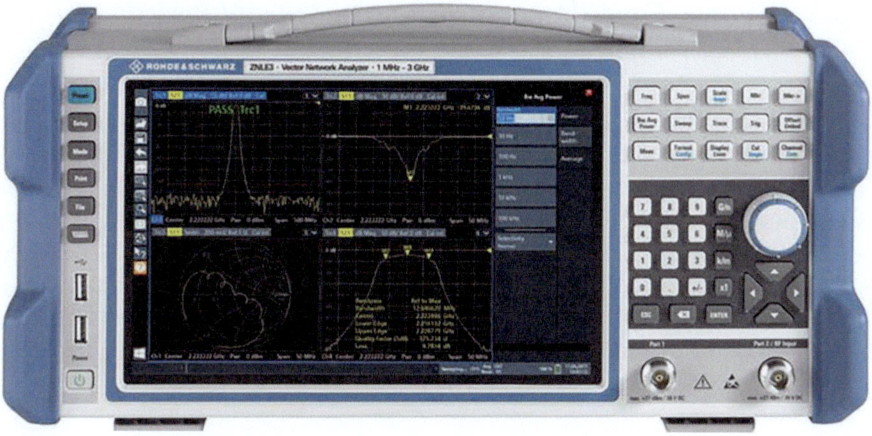

Fig. 10.24 Vector network analyzer model ZNLE3 from Rhode and Schwarz

10.8 Measurement of Earth Resistance

To ensure safety, every electrical installation (home, office, or industry) in a power system must be provided with an earth wire connected to an earth electrode. For example, a power distribution system is designed such that when an insulation failure occurs the resulting fault current is diverted by the earth wire to the earth electrode. The diversion of the fault current to earth prevents system voltages rising to dangerous levels. Thus, for electrical safety, the earth electrode must not only be properly created, but also maintained on a regular basis. The earth current can be sensed using appropriate circuitry and the power can be switched off if the earth current is more than a set threshold value. Such a device is called the earth leakage circuit breaker (ELCB). Two parameters, namely, the earth resistance and the earth resistivity indicate the status of the earth system. While earth resistivity is needed and hence measured for the design of the earthing electrode, the earth resistance is measured to ascertain the status of the earthing terminal. Thus, it is enough to measure the earth resistivity once, but the earth resistance must be periodically measured to ascertain the state of earthing. For measuring earth resistance, the following techniques are available.

(i) Wenner four-probe method.
(ii) Schlumberger four-probe method.
(iii) Three terminal method (falloff potential method/62% method/slope method).
(iv) Two-point method (dead earth method).
(v) Clamp-on test method.

10.8.1 Wenner Four-Probe Earth Resistance and Resistivity Measurement

Frank Wenner in 1915 devised a four-probe method for measuring the earth resistance (R_E) and earth resistivity (ρ_E). In this method, four electrodes (normally conductive rods) marked E_{c1}, E_{v1}, E_{v2} and E_{c2} separated from each other by equal distance l_p are inserted into the earth to a depth d_p as shown in Fig. 10.25. The outer electrodes E_{c1} and E_{c2} are called the current electrodes, and the two current electrodes are excited by the AC voltage source v_B at the power frequency and the current i_E in the circuit is measured with an ammeter as indicated in Fig. 10.25. The inner electrodes E_{v1} and E_{v2} are called the potential electrodes and the potential difference v_E between them is measured with a voltmeter. With these measurements the earth resistance R_E is calculated as

$$R_E = \frac{V_E}{I_E}. \tag{10.75}$$

In Eq. (10.75), V_E is the RMS value of the measured voltage v_E and I_E is the RMS value of the current i_E. Once the earth resistance is computed, the earth resistivity ρ_E is computed as

$$\rho_E = \frac{4\pi l_p R_E}{1+\frac{2l_p}{\sqrt{l_p^2+4d_p^2}}-\frac{l_p}{\sqrt{l_p^2+d_p^2}}}. \tag{10.76}$$

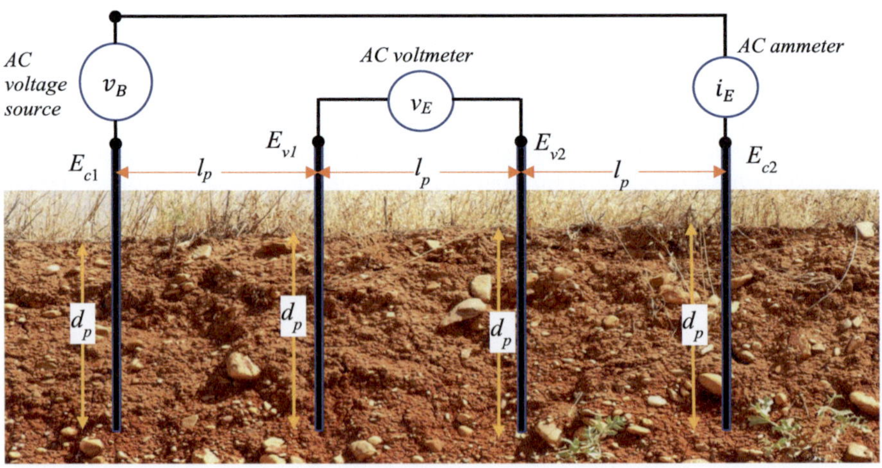

Fig. 10.25 Wenner method of earth resistance/resistivity measurement

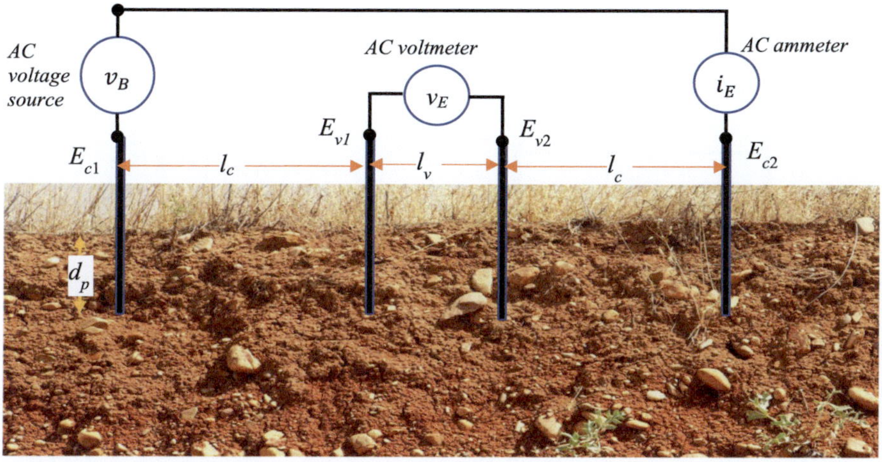

Fig. 10.26 Electrode placing in Schlumberger method of earth resistance measurement

10.8.2 Schlumberger Four-Probe Method of Measurement of Earth Resistance and Resistivity

In this method too, four conducting electrodes are driven to a depth d_p into the ground. However, the probes are separated with the distance between the inner electrodes E_{v1} and E_{v2} as l_v and the distance between the current electrodes (E_{c1} and E_{c2}) and potential electrodes (E_{v1} and E_{v2}) is set as l_c as shown in Fig. 10.26. The electrodes are placed such that $d_p < l_v$ and $l_v < l_c$ as shown in Fig. 10.26. Here too, the earth resistance R_E is computed using Eq. (10.75). The earth's resistivity is then calculated as given in Eq. (10.77).

$$\rho_E = \pi \frac{l_c(l_c + l_v)}{l_v} R_E \tag{10.77}$$

While Wenner and Schlumberger methods use four probes, there are methods available that use only three probes as given next.

10.8.3 Three-Probe Method of Measurement of Earth Resistance

The three-probe method, shown in Fig. 10.27, uses only three electrodes (probes). The outer electrodes E_{c1} and E_{c2} are the current probes that drive a current i_E through the ground. The voltage v_E between the current probe E_{c1} and the potential (inner) electrode E_v is measured. Two different measurement procedures can be adopted to compute the earth resistance using the three-probe method as listed below:

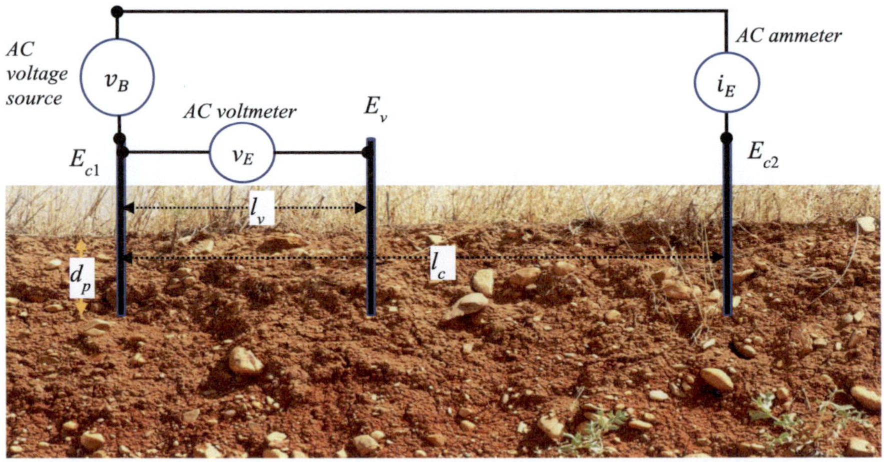

Fig. 10.27 Electrode placing in three-probe method of earth resistance measurement

(i) **The falloff potential method**: The distance between E_{c1} and E_{c2} is set to be >20 m. Then the distance between E_{c1} and E_v is varied in small steps and for each position of l_v the current I_E and the voltage V_E are measured. Then the resistance R_E is computed using Eq. (10.60) and plotted. A typical plot will look like the one portrayed in Fig. 10.28.

We see that in Fig. 10.28 the computed earth resistance R_E increases initially. Once electrodes E_{c1} and E_{c2} are set I_E remains constant. The value of R_E at the flat region is taken as the resistance of the earth. Initially, for the value of R_E to increase, potential V_E must increase as l_v is increased. Once l_v is >4 m, we see that R_E plateaus (at 12 Ω). That is between $l_v = 4$ m and $l_v = 12$ m, measured V_E remains constant. When $l_v > 12$ m, the measured V_E increases and hence computed R_E increases. Since the voltage-current ratio remains unaltered in the flat region, the method since the voltage remains constant (the voltage falls off), this method is called the *falloff potential method*.

Fig. 10.28 Earth resistance R_E as the distance l_v is varied

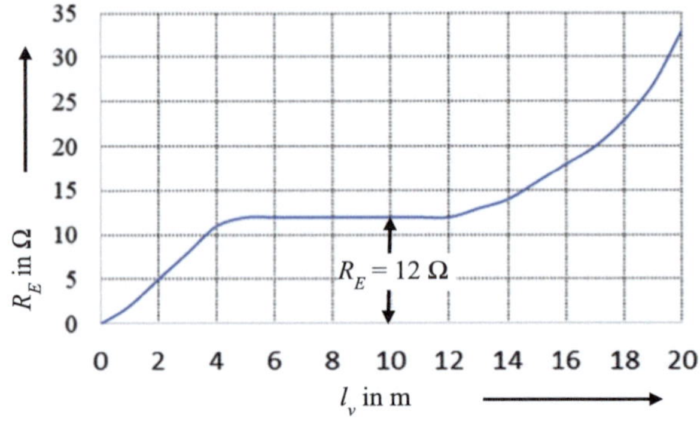

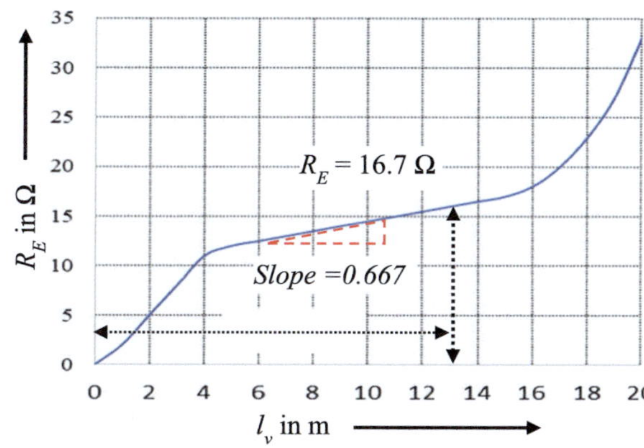

Fig. 10.29 Earth resistance R_E measurement—Slope method

(ii) **The "slope" method**: When we implement the "*falloff potential method*", in most practical cases, the flat region may not be horizontal, but with a very small slope as shown in Fig. 10.29. In such a case the slope m is determined and the value of R_E at the distance m l_c is taken as the earth resistance. As an example, the slope of the plot in Fig. 10.29 is 0.667 and hence the resistance at $0.667 * 20 = 13.34$ m is taken as earth resistance R_E.

(iii) **The 62% three-point method**: Here too, the distance between E_{c1} and E_{c2} is set to be >20 m. Then the distance $l_v = 0.618\, l_c$ is set and once again the current I_E and the voltage V_E are measured. Then the resistance R_E is computed using Eq. (10.75). Hence this method is called the 62% method.

10.8.4 The Two-Point Method of Measurement of Earth Resistance

It may happen that it may not be possible for driving ground rods at three or four points, especially in built-up urban areas. In such cases it is practical to use the two-point method. In this method, the earth electrode is taken as one terminal and another electrode not in the "field of influence" of the earth electrode is chosen as seen in Fig. 10.30. Normally the second electrode may be even a buried conductive pipe (water or gas). To each electrode two contact points, say $\mathbf{P_1}$ **and** $\mathbf{C_1}$ for the earth electrode and $\mathbf{P_2}$ **and** $\mathbf{C_2}$ for the second electrode are chosen. A sinusoidal voltage v_B at the supply frequency is then connected between C_1 and $\mathbf{C_2}$ and the current I_E is measured using the ammeter i_E as shown in Fig. 10.30. The voltage drop V_E between terminals P_1 and $\mathbf{P_2}$ is measured. The earth resistance R_E is then calculated using Eq. (10.75). The two-point method is also called the "dead earth method" and is the simplest technique to measure earth resistance reading but is not as accurate as the three-point method or four-point method. However, this method is most effective for quickly measuring the earth's resistance.

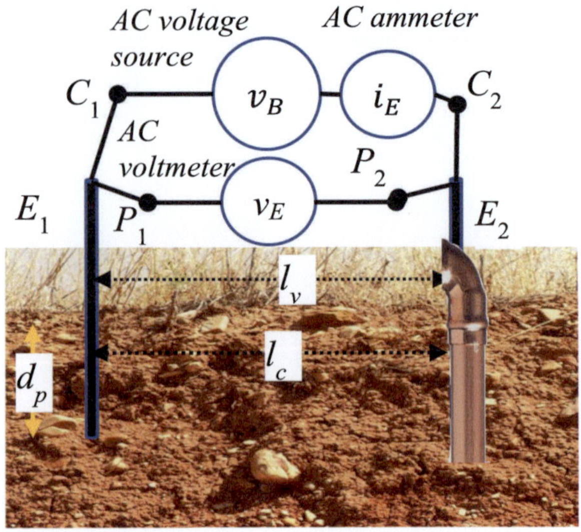

Fig. 10.30 Two-point earth resistance measurement

10.8.5 Clamp-On Method of Measurement of Earth Resistance

In several instances it may not even be possible to get a second terminal (electrode inside the earth that is deep enough). In such cases the clamp on earth resistance measurement can be made. Figure 10.31a illustrates the principle of the clamp on earth resistance method. The earth wire, close to the earth pit, is encircled by two toroidal cores. Of the two cores, the excitation core marked C_E is wound with a coil of N_E turns and the detector core C_D has a winding of N_D turns. The winding on the excitation core marked C_E is driven with a current i_s that drives a current i_E into the earth loop made of earth resistance R_E and earth wire resistance R_W. The current i_E induces a current i_D into the winding of N_D turns on the detector core C_D. The equivalent circuit of Fig. 10.31a is illustrated in Fig. 10.31b assuming negligible winding resistances for the excitation and detector windings. Using v_E, N_E, N_D and i_D, the earth (loop) resistance $(R_E + R_W)$ is computed as given below. From Fig. 10.31b we can deduce:

$$i_E = \frac{v_E}{N_E(R_E+R_W)}. \tag{10.78}$$

Hence the detector current i_D is

$$i_d = \frac{i_E}{N_D} = \frac{v_E}{N_E N_D(R_E+R_W)}. \tag{10.79}$$

Rearranging Eq. (10.79), we get

$$(R_E + R_W) = \frac{V_E}{I_d N_E N_D}. \tag{10.80}$$

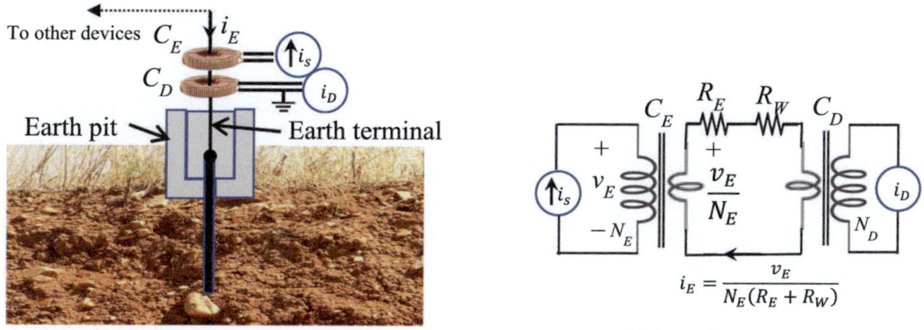

(a) Clamp on earth resistance measurement

(b) Equivalent Circuit

Fig. 10.31 Clamp-on earth resistance measurement

In Eq. (10.80), V_E is the RMS value of v_E and I_D is the RMS value of i_D. It should be noted here that this scheme measures the earth loop resistance (resistance of the earth electrode and the rest of the wire above the earth) and not the earth resistance. Normally $R_E > R_W$ and hence the error in determining R_E will not be high. Figure 10.32 shows a commercial clamp on earth resistance measuring unit Hioki model FT6380. In a real clamp on earth resistance tester the core is split in two halves and hinged at the bottom. A press to open lever opens the cores so that they can be positioned to encircle the earth wire as shown in Fig. 10.32b.

It should be noted that while the four-probe method conforms to measuring standards such as IEC 61557 series and IEEE 81 series, the other methods, though not part of any standard, are in use as well.

(a) Hioki model FT6380

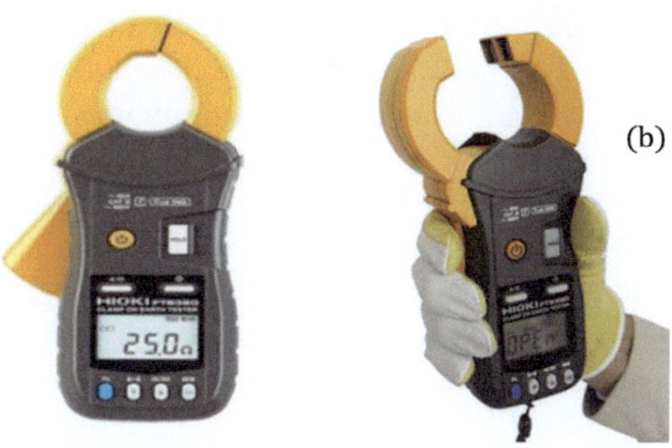

(b) The clamp opened

Fig. 10.32 Clamp-on earth resistance meter from Hioki

10.9 Conductivity Measurement

Measurement of the electrical conductivity σ of materials such as copper and aluminum is quite simple and straightforward. We take the material in standard dimension as shown in Fig. 10.33, send a current I_T through the material using the current terminal marked TC_1 and TC_2. The potential drop V_T across well-defined points TP_1 and TP_2 separated by distance d is measured. Then the resistance R_T of the sample across the points TP_1 and TP_2 is computed as

$$R_T = \frac{V_T}{I_T}\Omega. \tag{10.81}$$

The conductivity is then computed as

$$\sigma_T = \frac{l_T}{R_T A_T}\, S/m. \tag{10.82}$$

In Eq. (10.67), l_T is the length between test points TP_1 and TP_2 and A_T is the area of cross section. It is important that the test sample is homogenous, and the area of cross section is uniform between test points TP_1 and TP_2.

10.9.1 Measurement of Conductivity of Liquids

While the electrical current in materials like copper and aluminum is due to the flow of electrons, current in liquids in most cases is due to the flow of ions. Measurement of the electrical conductivity of liquids is in vogue for a century as a liquid's conductivity is an important and widely used analytical parameter. The electrical conductivity of a liquid is dependent on dissolved ionic species (salts, acids, bases, and some organic substances). However, from the conductivity measurement, it is not possible to differentiate between diverse kinds of ions, as the conductivity of a liquid combines the effect of all ions in the liquid. Major use of the conductivity of the liquid is for monitoring and surveillance of a wide range of water samples (standard ISO7888) such as drinking water, mineral water, and water used in the process. The conductivity of solutions also provides information on the solvents. For example, the conductivity of seawater is related to its salinity. For the measurement of conductivity, a current is

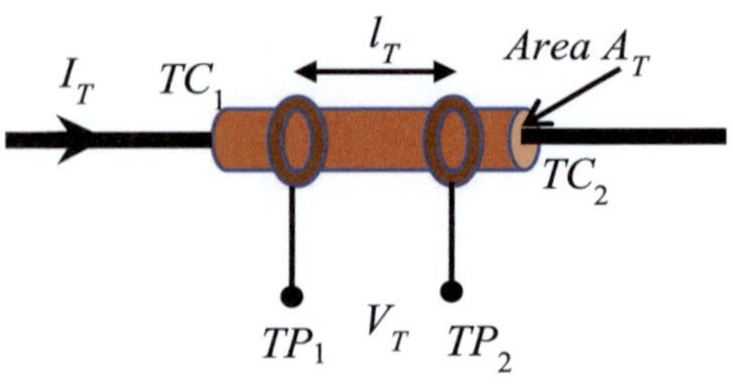

Fig. 10.33 Conductivity measurement

Fig. 10.34 Dip type conductivity cell

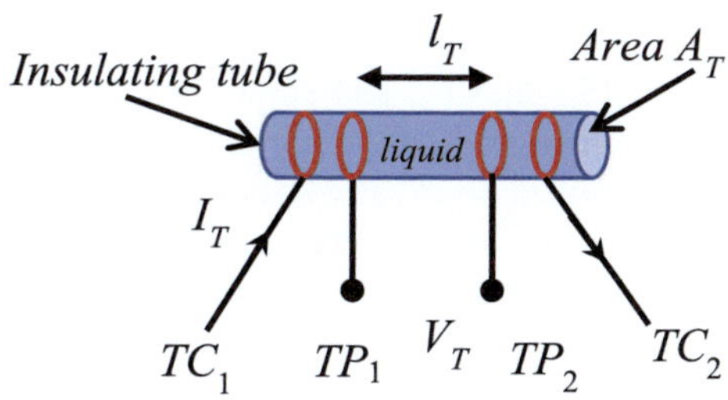

Fig. 10.35 Four-electrode conductivity cell

sent through the liquid and the voltage drop across the certain volume of the liquid is measured as indicated in Fig. 10.34.

The standard method is to use a conductivity cell with four electrodes such as the one shown in Fig. 10.35. Here the liquid whose conductivity is to be measured is filled in a standard cell and a current I_T is passed through the current electrodes TC_1 and TC_2 and the voltage V_T across potential terminals TP_1 and TP_2 is measured. Once again, the resistance R_T and the conductivity σ_T are computed using Eqs. (10.81) and (10.82). When measuring the conductivity of liquids, the term $\left(\frac{l_T}{A_T}\right)$ is known as the cell constant. While the cell constant for the structure shown in Fig. 10.35 will remain constant, the cell constants of "*dip type*" conductivity probe shown in Fig. 10.34 may vary depending on the conductivity of the liquid under test and hence require compensation to handle the variation in cell constant.

10.10 Virtual Instrumentation

In the latter half of the 1970s, the advancement of integrated circuit technology resulted in the advent of a central processing unit (CPU) in a single integrated circuit. This then resulted in the development of the personal computer (PC). Today, the personal computer is all pervasive and has become indispensable in every aspect of life, from school education to even everyday chores like shopping. The field of measurement and instrumentation is no exception. The measurement and or instrumentation accomplished with a PC is now popularly termed as "*virtual instrumentation*" (VI).

For example, if we interface an analog to digital converter (ADC) to a PC as shown in Fig. 10.36, obtain the samples of a voltage v_x connected to the ADC, compute

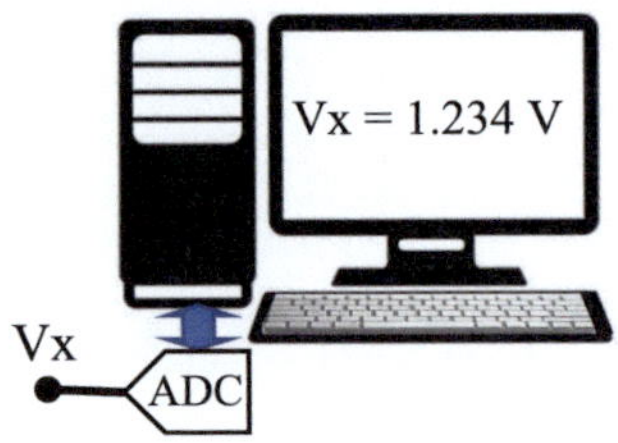

Fig. 10.36 PC with ADC

the RMS value V_x, and show its value on the display of the PC, then we will be able to measure the unknown voltage, but without an actual voltmeter. Hence, we have realized a virtual voltmeter. Thus, when measurement and or instrumentation requirements are realized using a personal computer then the resulting product is simply called the *virtual instrument*.

10.10.1 The Typical Virtual Instrument

In the early years, most digital instruments were made of hard-wired hardware (the actual circuit that performs the necessary measurement functions) and then the result of a measurement was displayed on a numeric display as we see in most digital multimeters. Later, when microprocessors and microcontrollers were developed, the digital instruments were provided with displays that included alphanumeric displays. The advent of digital storage oscilloscope brought in a paradigm shift, not only in the displays of the digital instrument, but also brought in user friendly interface and control. However, all these dedicated instruments were built with task specific hard-wired hardware and associated dedicated software. Figure 10.37 illustrates the architecture of a digital storage oscilloscope. The hardware and the dedicated control software are designed and developed for the specific task at hand, namely, the functions required for a digital storage oscilloscope.

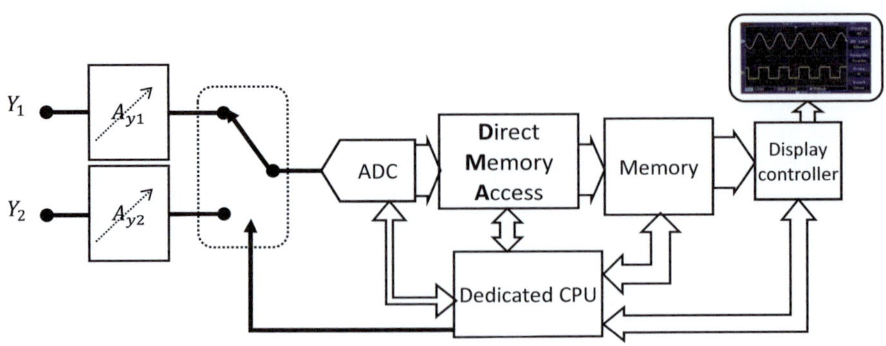

Fig. 10.37 Block diagram of a typical digital storage oscilloscope

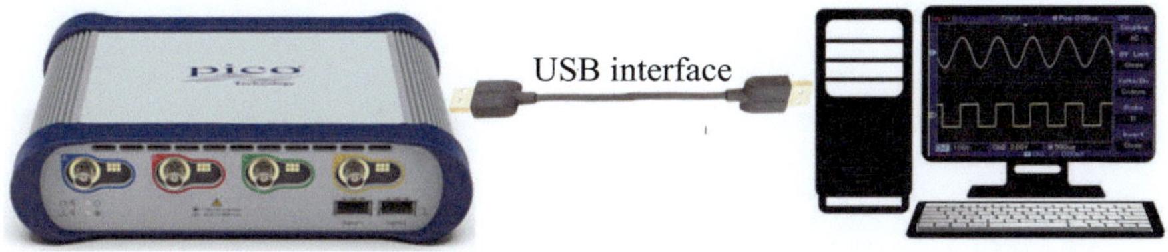

Fig. 10.38 PC based dedicated digital storage oscilloscope

Contrary to the stand-alone digital instruments, when we convert a PC to perform measurement and or instrumentation functions, we need to add and interface the required hardware to the hardware of the PC and run suitable software under the operating system of the PC. This task can be.

10.10.2 Dedicated Virtual Instrumentation

The hardware and software are tailor made for the specific task: Example is the PC-based oscilloscope shown in Fig. 10.38. Here, not only the hardware but also the software is designed and developed for the task at hand. In Fig. 10.38 the hardware that adds an external input/output unit and interfaced through the universal series bus (USB) of the PC.

10.10.3 General Purpose Virtual Instrumentation

In this type of virtual instrument, basic hardware and the software are general-purpose. One needs to add additional hardware, if required, and develop appropriate software under the general-purpose virtual instrumentation software that defines the end use of realizing the required virtual instrument.

10.10.3.1 The Hardware for Virtual Instrumentation

In both types (Dedicated and General purpose) of virtual instrument, the hardware can be interfaced to the PC as

(i) An internal device using the PCs internal input–output bus is shown in Fig. 10.39a. The PC's internal bus was initially of "Industrial Standard Architecture (ISA)" and later changed to a Peripheral Component Interface (PCI). Figure 10.39b shows a typical data acquisition card that can be inserted into one of the internal connectors (slots) on the motherboard of a PC.

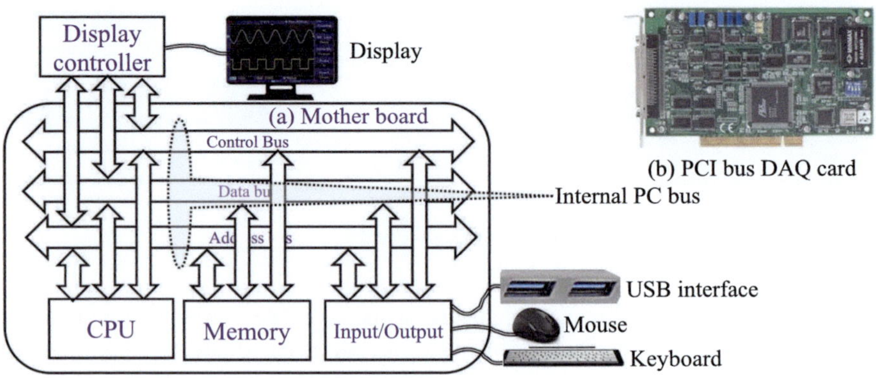

Fig. 10.39 Internal bus architecture of a PC

(ii) An external device is connected through one of the interface ports. Initially, the ports that were used for the purpose were (i) Parallel Peripheral interface and (ii) Serial port (normally conforming to the RS232 standard. These interfaces are no longer in vogue, replaced by the very popular Universal Series Bus (USB). USB-connected hardware for virtual instrumentation is gaining popularity and is the emerging norm for hardware interface. A typical USB interfaced virtual instrument hardware is shown in Fig. 10.40.

Today, both these types of virtual instrumentation hardware are popular. While PC-based dedicated instrumentation (hardware and purpose developed software) provides the optimal or best performance for the chosen virtual instrument, the general-purpose hardware, and customizable (programmable) software provides the freedom of developing a user intended virtual instrumentation that is not at all possible with the former type.

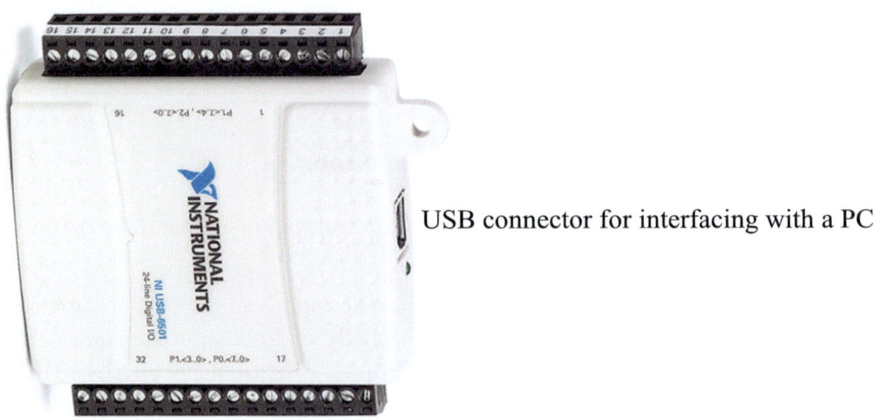

Fig. 10.40 USB interfaceable DAQ device

10.10.3.2 The Software for Virtual Instrumentation

In a virtual instrument, software plays as important a role as hardware. A PC has its own hardware containing a motherboard and peripheral devices as illustrated in Fig. 10.39a controlled by the software called the operating system such as UNIX, Windows, LINUX, ... To obtain a virtual instrument using a general-purpose PC not only additional hardware is required, but also suitable software that runs under the operating system of the PC and renders essential instrumentation functions is needed to realize a virtual instrumentation. As the PC architecture evolved, the hardware and the operating system software also were "upgraded". Thus, the software for realizing a virtual instrument must be developed such that it can operate under a chosen operating system software and its future upgrades.

Figure 10.41 illustrates the parts of the software of a virtual instrument. We see from Fig. 10.41 that the Virtual Instrumentation software has at the lowest level, the *hardware drivers.*

The hardware drivers are normally optimized to set up the hardware for the intended task and interface with the higher level VI software. The hardware drivers are normally written using the "*assembly Language*" of the CPU, compiled and run as *subroutines.*

The data collated from the instrumentation hardware (usually a data acquisition system that includes analog and digital inputs and outputs)needs to be passed on to the higher level software and that requires interfacing the data from the hardware with the system level (operating system) software and the *middleware* accomplishes this task.

Thus, *the middleware software and the system level interface* perform the functions of interfacing the virtual instrumentation tasks with the PCs internal hardware software. For example: accessing the internal clock of the PC, accessing the communication ports (parallel, serial, and Internet), and peripherals such as memory, storage media, and display.

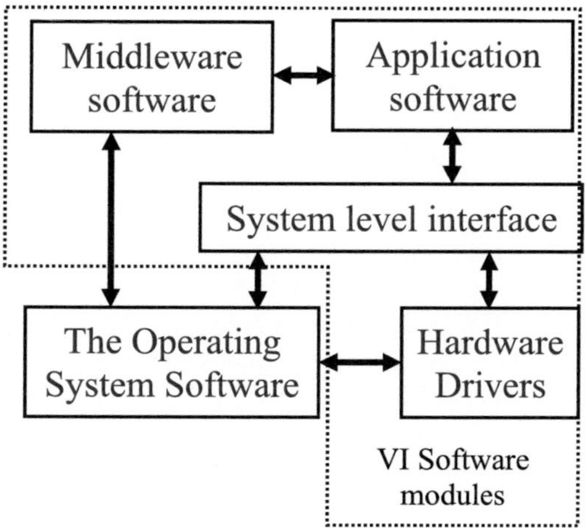

Fig. 10.41 Components of Virtual Instrument Software

At the highest level in virtual instrumentation software is the application software that accomplishes computation of data and presenting the results on the display of the PC.

The application software part of a virtual instrumentation software performs the actual tasks of measurement (and control if needed) and most importantly present the results to the end user on the display of the PC.

There are several commercial software platforms available for implementing a virtual instrument as listed below.

1. LabVIEW.
2. DASY Lab.
3. Hp VEE.

Unlike general-purpose software like C or JAWA, the tasks in virtual instrumentation software are well defined such as:

(i) Acquire samples of the signals,
(ii) Perform operations such as addition, subtraction, scaling …, and
(iii) Display the result in a particular form (may be as a digital display, analog meter type display …).

Hence almost all VI software is developed on a platform on which pre-developed software modules are interlinked to perform the required task. Each of these modules (like a subroutine on a general software platform) is identified with a graphical symbol and hence all the available VI software work as a "Graphical System Design". Thus, instead of writing codes as in a programming language like C, Jawa, or Python, in a virtual instrumentation, the software development is accomplished by interlinking different pre-defined tasks.

Each of the above virtual instrumentation platform is quite complex and requires a whole (separate) book to describe and teach the reader how to use the platform. The next section briefly explains one of the software platform called LabVIEW.

10.10.3.3 The LabVIEW

The LabVIEW software was launched in 1986 (called *Lab Windows* then) by National Instruments as a software tool to facilitate automated measurements using a data acquisition hardware interfaced to a PC. Over the years this software has evolved as a standard for learning virtual instrumentation. Many educational institutions teach LabVIEW as a part of their curriculum. LabVIEW, like all other VI software, uses a graphic interface and implement an environment that enables programming in G. G is a graphical programming language created by National Instruments that was initially developed to integrate GPIB instruments to a PC. Today, the G programming language has evolved and can integrate general-purpose data acquisition, FPGAs,

Internet of Things (IoT), and dedicated instruments that have PC interface and implement measurement, test, control, and automation applications. LabVIEW can be run on Windows, OS X (Apple), and Linux operating systems.

The LabVIEW is an acronym for **Lab**oratory **V**irtual **I**nstrument **E**nvironment **W**orkbench. As in any application, in an operating system environment, LabVIEW is a program (application or now popularly called app) that opens an operational environment consisting of:

A. ***LabVIEW environment***: The LabVIEW environment consists of LabVIEW VI manager (project explorer), the programming tools, debugging features, templates and ready-built sample examples, and an easy interface to the hardware drivers.
B. ***LabVIEW VIs***: The LabVIEW VI is a "Virtual Instrument" that enables a user interface to be built and it contains the programming code.
C. ***LabVIEW G programming***: This is the graphical programming language where the functional algorithms are built using "drag and drop" techniques.
D. ***LabVIEW data flow***: This is the core concept that determines the running order for the program.

10.10.3.4 LabVIEW Environment

The LabVIEW environment is made of two windows, namely,

1. Block Diagram and
2. The Front Panel.

1. **The block diagram** is where the virtual instrumentation program is written using the **G** language. A virtual instrument program is developed using pre-set icons interlinked to accomplish the required tasks. The strong point of the G language is a set of such interlinked program block itself be converted into a pre-set icon. The block diagram runs in the background executing the tasks of accessing the external devices, collecting the data from the hardware (external to the PC and even internal to the PC that may include data from the Front panel), processing them, and then sending the results to the front panel for displaying the same.
2. **The Front panel** is, as the name indicates, the panel that interacts with the user. Obviously, the front panel collects any input(s) to be provided by the user and displays the parameters that are programmed to be displayed.

LabVIEW is quite complex software and at the same time easy to operate if properly learnt. A complete book is required to explain LabVIEW and several such books are available. As a brief introduction, let us take a very simple example of developing a virtual instrument temperature measurement using a temperature sensor. To accomplish this task:

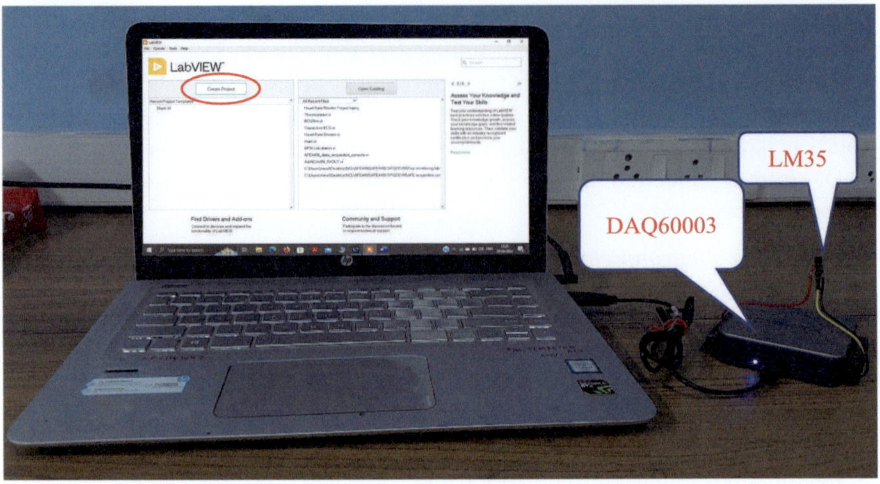

Fig. 10.42 Temperature sensor connected to the USB6003 DAQ board

(a) First, we need to select a temperature sensor. Let us choose the temperature sensor LM35 that provides a voltage output of 10 mV/°C. This voltage output must be converted to digital value and displayed. For that, we need an analog to digital converter. Let us choose the USB-based DAQ board NI—USB 6003. This board has for differential or eight differential or analog inputs AI 0 to A I7, two analog outputs AO 0 and AO 1, and 12 digital I/O pins. Figure 10.42 shows the temperature sensor connected to the DAQ board. The power supply required for the LM35 is drawn from the 5 V supply available on the DAQ 6003 module. We then connect the DAQ module to a PC (in this case a Laptop) through one of the USB ports as shown in Fig. 10.42.
(b) The next steps are programming on (i) the block diagram and (ii) the Front panel. We can do this in any order, here we start with the front panel.

First, we start the NI LabVIEW. This will show the window as seen in Fig. 10.43. To start a new virtual instrument, we need to click on "Create Project" seen in Fig. 10.43. Clicking on the "Create Project", we will get the window as shown in Fig. 10.44. Again to create a new virtual instrument (VI) we chose the option "Blank VI" by clicking on that icon to start a new virtual instrument for temperature measurement. This will result in the generation of two windows, namely "The Front Panel" and the "Block diagram" gets opened as shown in Fig. 10.45.

We now start with the front panel. On the front panel we right click to obtain a sub window opened and select the first (Top right-hand corner) icon to get the screen as seen in Fig. 10.46. Here we select the thermometer icon and place it on the front panel. This not only places the thermometer icon on the front panel but also immediately

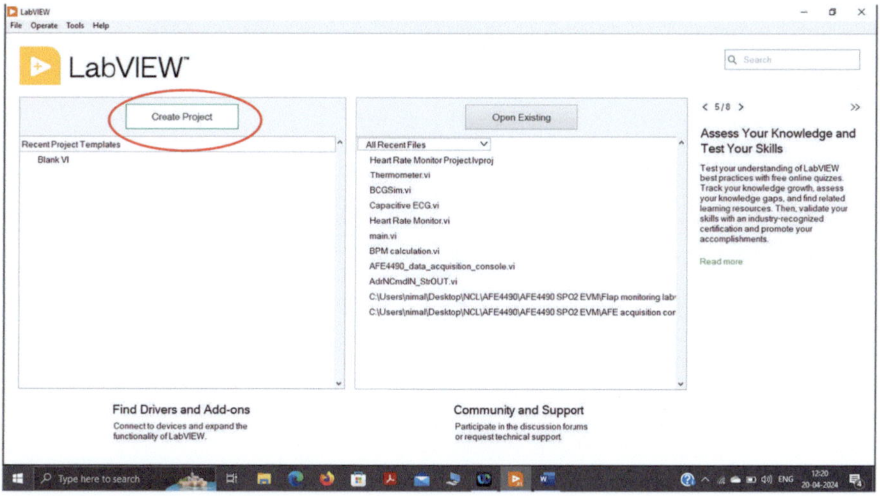

Fig. 10.43 Opening age of NI LabVIEW program

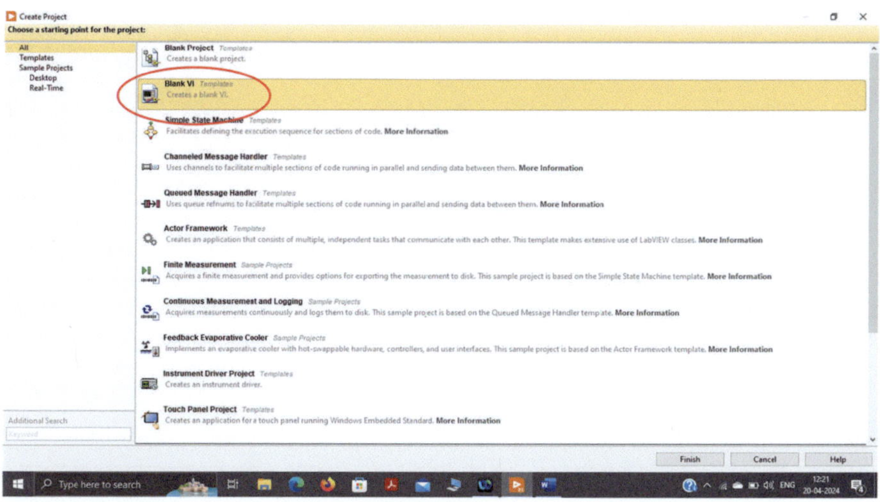

Fig. 10.44 Screen after clicking 'Create Project'

places the program graphic on the Block diagram panel as seen in Fig. 10.46. Of course, this will complete the task for the temperature measurement, but we also add a digital display as well to show the temperature value in numbers as shown in Fig. 10.47. This once again places the program graphic on the "Block diagram" panel. With these actions, we now got the displays we need for the measurement of temperature. We now move the cursor onto the Block diagram and right click on the mouse. We will get the sub window and on the sub window called the "programming"

Fig. 10.45 Screen after clicking ‘Blank VI’

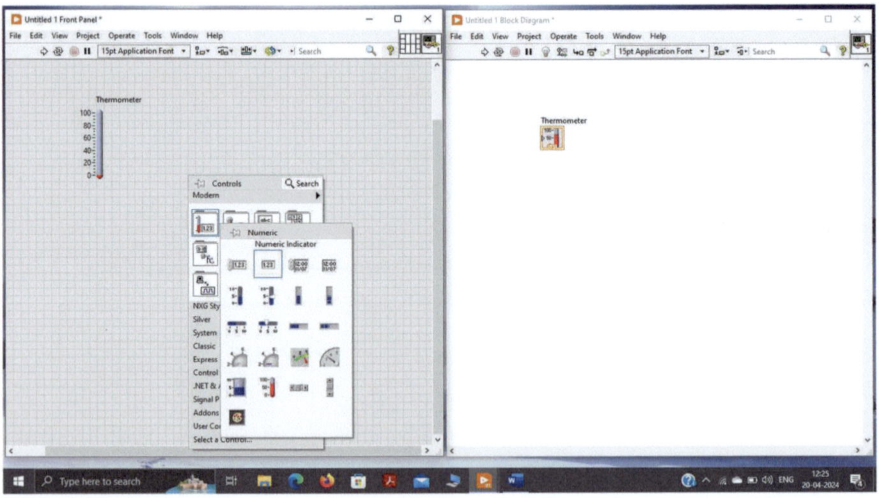

Fig. 10.46 Screen after right clicking and selecting the first icon

in which we click on the “Measurement I/O” that opens sub-sub window named “Measurement I/O” as seen in Fig. 10.47. We now click on “DAQ-Assist” that places the “DAQ-Assist” icon on the Block diagram window. Double clicking on the DAQ assist, searches for connected DAQs, recognizes that the DAQ 6003 is connected, installs the required low-level software for accessing the DAQ 6003 and then opens the sub window as in Fig. 10.48. Here we select the option “Analog input”. That would open a new sub window as shown in Fig. 10.49. We see that in Fig. 10.49

there are 8 analog inputs we can choose from and we choose a0. That would open the sub window as seen in Fig. 10.50.

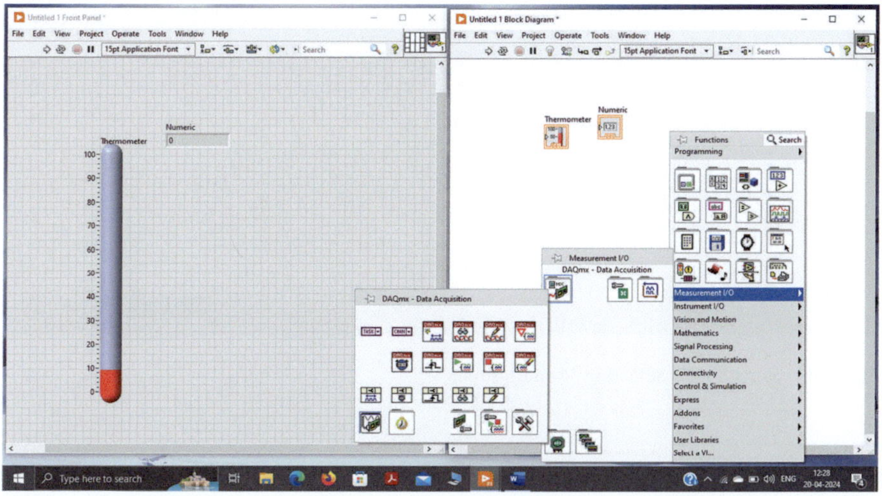

Fig. 10.47 Screen after selecting 'Numeric' and then right clicking on Block Diagram' clicking 'Blank VI'

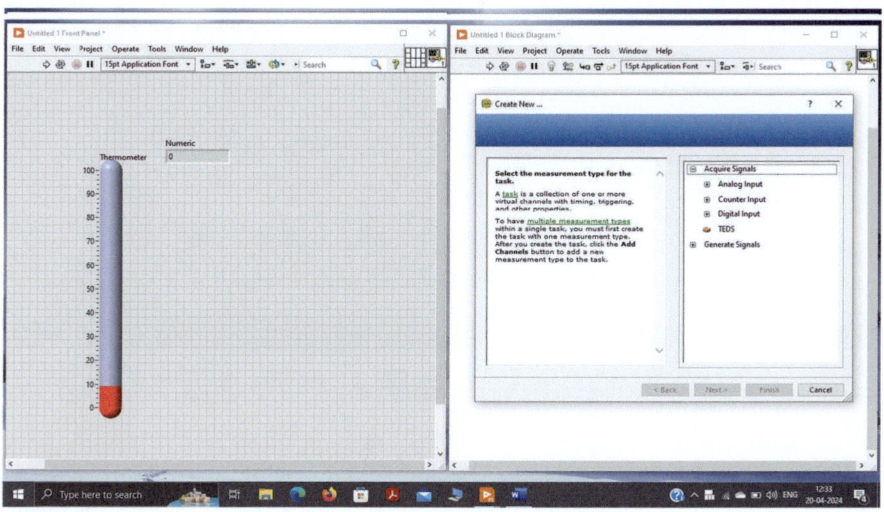

Fig. 10.48 Screen after right clicking and selecting the first icon

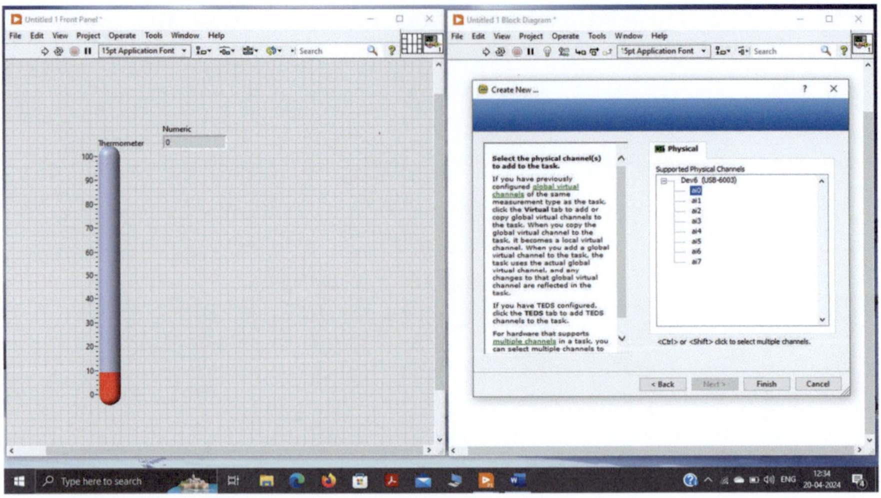

Fig. 10.49 Screen after selecting 'Analog Input'

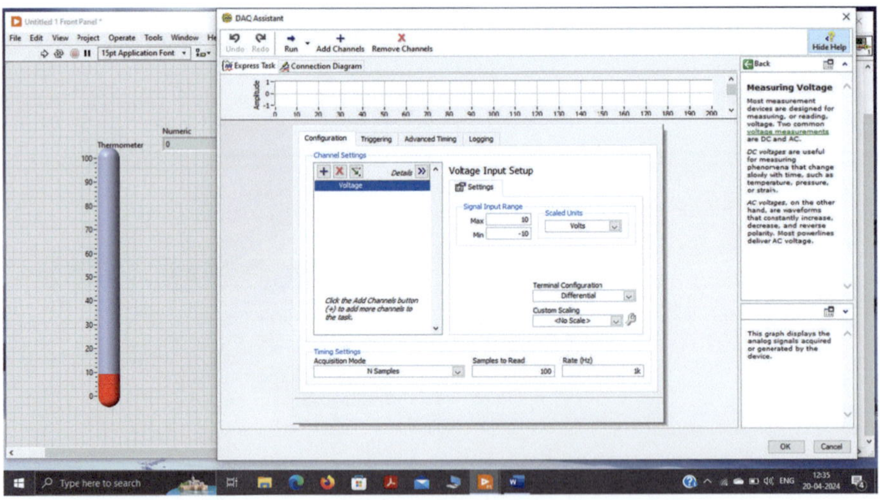

Fig. 10.50 Screen after selecting analog input (channel) a0

We now click OK to get back the Block diagram window as seen in Fig. 10.51 with the required low-level program to obtain an input from the analog (physical) input pin marked 'a0' on the DAQ 6003 which is connected to the output of the

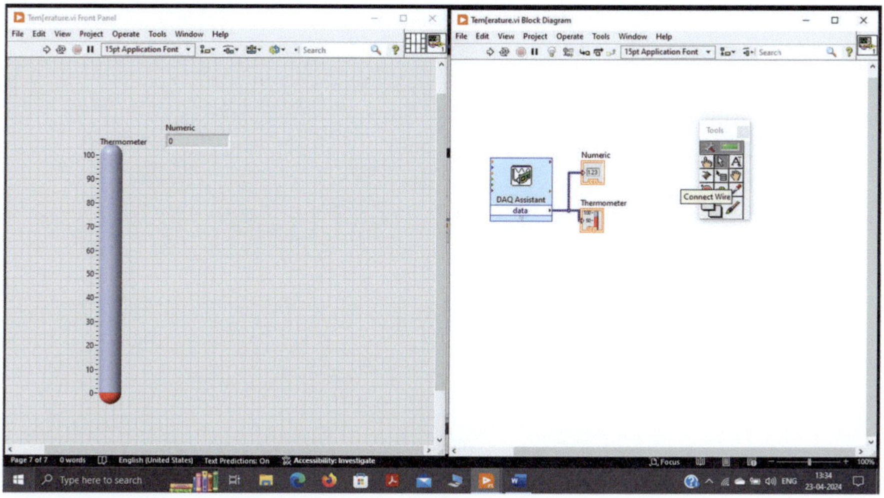

Fig. 10.51 Screen after saving the VI as temperature

temperature sensor. With this we have all the required programs (in graphic form) that are required to make "temperature measurement" with the temperature sensing IC LM35. The remaining tasks are

(1) Interconnect these sub programs.
(2) Run the overall VI to measure temperature.

  (i) To interconnect the VI icons, we open the tools palette under the "View" menu on the top ribbon, select the "Connect Wire" tool and join the data output terminal on the DAQ Assistant and the inputs of the Thermometer and Numeric blocks.

With this we save the VI by selecting the Save As option under the File menu option to get the screen as shown in Fig. 10.51. Now the VI is ready to run and to run we click on the RUN icon ⇨. With that we get the screen as shown in Fig. 10.52. We see that the Numeric display shows 0.205071. We expected that the display would show the room temperature but not the value 0.205071. This is because we have not factored in the proper multiplication constant. First we must multiply the output of the sensor by 100 to get the correct value (It may be noted that the sensor LM35 gives 0.010 V for every °C. Thus we must multiply by 100 and also we have not connected some of the inputs of the DAQ icon, namely sampling rate and number of samples to be taken. We do both and run the VI we get the screen as in Fig. 10.52. We see now that temperature measurement VI works properly.

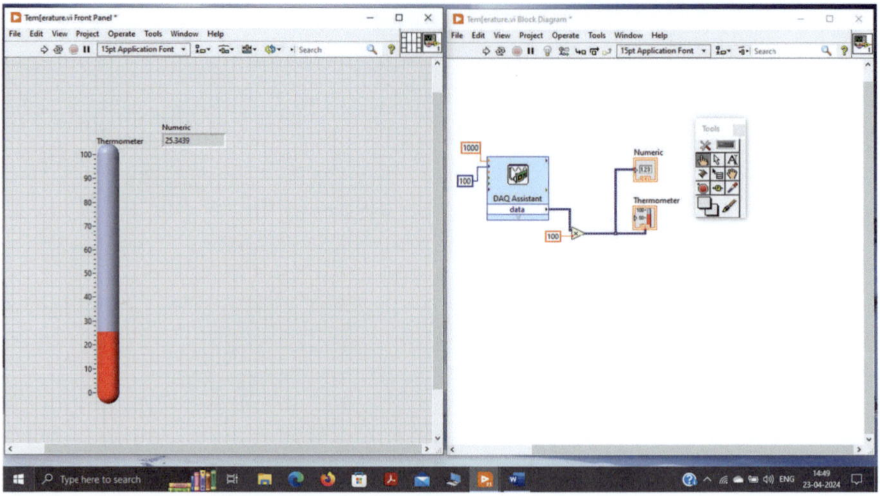

Fig. 10.52 Screen after the VI temperature is RUN once

From the foregone discussions, we see that the best way to learn "Virtual Instrumentation" is to watch a video of instructions and at the same time follow the instructions on an actual VI system. The National Instruments has a "Learning" website at https://learn.ni.com/ which is quite helpful and exhaustive.

MIX
Papier aus verantwortungsvollen Quellen
Paper from responsible sources
FSC® C105338

If you have any concerns about our products,
you can contact us on
ProductSafety@springernature.com

In case Publisher is established outside the EU,
the EU authorized representative is:
Springer Nature Customer Service Center GmbH
Europaplatz 3, 69115 Heidelberg, Germany

Printed by Libri Plureos GmbH
in Hamburg, Germany